SYSTEMATIC BOTANY

SYSTEMATIC BOTANY

FIFTH EDITION

SUBHASH CHANDRA DATTA

MSc (Cal), PhD (USA), PHI Sigma, Sigma, Xi, FBS

Department of Botany

University of Calcutta, Kolkata

Committed to Educate the Nation

NEW AGE

Publishing Globally

IN INDIA

NEW AGE INTERNATIONAL (P) LIMITED, PUBLISHERS

LONDON • NEW DELHI

Bangalore • Chennai • Cochin • Guwahati • Hyderabad • Kolkata • Mumbai

Visit us at **www.newagepublishers.com**

Published by New Age International (P) Ltd., Publishers
First Edition: 1988
Fifth Edition: 2015
Reprint: 2024

GLOBAL OFFICES

- **New Delhi** **NEW AGE INTERNATIONAL (P) LIMITED, PUBLISHERS**
7/30 A, Daryaganj, New Delhi-110002, (INDIA), **Tel.:** (011) 23253472, 23253771, **Mob.:** 9315905300
E-mail: contactus@newagepublishers.com • Visit us at www.newagepublishers.com

- **London** **NEW AGE INTERNATIONAL (UK) LTD.**
27 Old Gloucester Street, London, WC1N 3AX, UK
E-mail: info@newacademicscience.co.uk • Visit us at www.newacademicscience.co.uk

BRANCHES

- **Bangalore** 37/10, 8th Cross (Near Hanuman Temple), Azad Nagar, Chamarajpet, Bangalore- 560 018
Tel.: (080) 26756823, **Mob.:** 9315905288, **E-mail: bangalore@newagepublishers.com**

- **Chennai** 26, Damodaran Street, T. Nagar, Chennai-600 017, **Tel.:** (044) 24353401, **Mob.:** 9315905309
E-mail: chennai@newagepublishers.com

- **Cochin** CC-39/1016, Carrier Station Road, Ernakulam South, Cochin-682 016
Tel.: (0484) 4051304, **Mob.:** 9315905289, **E-mail: cochin@newagepublishers.com**

- **Guwahati** Hemsen Complex, Mohd. Shah Road, Paltan Bazar, Near Starline Hotel
Guwahati-781 008, **Tel.:** (0361) 2513881, **Mob.:** 9315905296, **E-mail: guwahati@newagepublishers.com**

- **Hyderabad** 105, 1st Floor, Madhiray Kaveri Tower, 3-2-19, Azam Jahi Road, Near Kumar Theater
Nimboliadda Kachiguda, Hyderabad-500 027, **Tel.:** (040) 24652456, **Mob.:** 9315905326
E-mail: hyderabad@newagepublishers.com

- **Kolkata** RDB Chambers (Formerly Lotus Cinema) 106A, 1st Floor, S N Banerjee Road
Kolkata-700 014, **Tel.:** (033) 22273773, **Mob.:** 9315905319, **E-mail: kolkata@newagepublishers.com**

- **Mumbai** 142C, Victor House, Ground Floor, N.M. Joshi Marg, Lower Parel, Mumbai-400 013
Tel.: (022) 24927869, 24915415, **Mob.:** 9315905282, **E-mail: mumbai@newagepublishers.com**

- **New Delhi** 22, Golden House, Daryaganj, New Delhi-110 002, **Tel.:** (011) 23262368, 23262370
Mob.: 9315905300, **E-mail: sales@newagepublishers.com**

C-23-06-13663

Printed in India at Ajit Printing Press, Delhi.
Typeset at Goswami Associates, Delhi.

NEW AGE INTERNATIONAL (P) LIMITED, PUBLISHERS
7/30 A, Daryaganj, New Delhi - 110002
Visit us at **www.newagepublishers.com**
(CIN: U74899DL1966PTC004618)

With love and respect
to
Prof. H.L. Chakravarty
M. Sc. (Cal.), M.A. (Cantab.), D. Sc. (Edin.)
F.L.S. (Lond.)
from whom I learned the
A B C of Systematic Botany

Preface to the Fifth Edition

In the present edition, the book has been thoroughly revised. Under Part One, sections on citation and authority, ICBN, systems of classification, taxonomic development in India as well as taxonomic evidence (palynology, cytology, phytochemistry, etc.) have been rewritten to bring them in line with current concept. Under Part Two, the sections on evolution of the inflorescence and origin of the angiosperms have been expanded. Under Parts Three and Four, the material on angiosperm systematics have been updated in view of the publication of Heywood's masterpiece *Flowering Plants of the World.* Moreover, bracketed key has been followed instead of the indented key; illustrative figures accompany the text and are not inserted at the end of the book.

In this major remodelling of the book, I have been helped by suggestions and kind comments from many teachers of the Indian subcontinent. It is my earnest hope that the present enlarged edition would continue to enjoy the patronage of teachers and students alike as before.

Author

Preface to the First Edition

The term systematic botany encompasses the domain not only of the higher plants, but also of the lower plants. Since it is not possible to treat adequately all the plant groups together in a single volume at present, this text is limited to a discussion of the angiosperms. The subject-matter has been divided into a number of parts—an arrangement which will find favour with the teachers and the taught alike. It is felt that the book will be of benefit to the students of the Degree (both Pass and Honours) in all the Universities of India and the neighbouring countries and may also be helpful to Postgraduate classes in Botany as well as to the study of Agriculture and allied subjects.

The author has taken into account seventy families of flowering plants prominent in India. He has not attempted to present any new system of plant classification, but has adhered closely to the outline of Bentham and Hooker. Although the term "natural order" is not used in modern botany books and is substituted by the term "family", yet the former term has been retained and used as synonymous with the latter. The diagnostic or distinguishing characters of each family are few in number and are designated in italics in the section "General Features of the Family". The information on the variety of floral structures for each particular family can be had from the sections marked "Range of Floral Structures in the Family" and "Divisions of the Family", though it is mostly emphasised in the former section. In these two sections, the name changes in genera are given as far as possible. But such alterations are not to be generally seen in standard keys for genera, such as that of Prain whose voluminous work needs thorough revision in the light of the ICBN (International Code of Botanical Nomenclature). The systematic position and relationship of each family have been mainly dealt with from the view-points of the systems of Bentham and Hooker, Enlger as well as Hutchinson in the section entitled "Position and Affinity of the Family". Where the position of a family is in dispute, all facts and figures from different systems have been put forward. In the section "Number in and Distribution of the Family", one can note an approximate estimate of the genera and species as calculated by Core. As to the number of Indian species on dicot families, the number has been taken from Chatterjee's paper on endemism. The evidence for the number of monocot species from India is Hooker's *Flora of British India*; this evidence was reluctantly utilised, as up-to-date data are not available. A list of most common plants has been given in alphabetical order under the section "Commonly Occurring Plants of the Family"; a brief description of each plant is included wherever possible. The section "Economic Aspects of the Family" bears description on economic side of each family with plants from India as well as other countries. Where the plant is not generally noticed in India, the native region of that plant is indicated in parenthesis after the scientific name. The Parts Two and Three of his treatise are concerned with the dicotyledons and monocotyledons respectively from the aforementioned angles. The other Parts will speak for themselves. In short, all efforts have been made to give a complete picture of the angiosperm systematics which appear dull and uninteresting to many students of botany these days.

Author

Contents

PART FOUR

PART FOUR

PART ONE

Principles and Practices of Taxonomy

Principles and Practices of Taxonomy

INTRODUCTION

When man first learnt to speak, he felt a desire to name and classify objects around him. As time went on, he changed into a more civilised creature and the need for communication grew acute. At present, there are actually very few things and events that we know and yet have not named them. Though the identification of objects has been an easy task, their classification have presented manifold problems. As far as the plant world is concerned, about 3,00,000 species of green plants are currently recognised, plus over 1,00,000 fungi, a few thousand bacteria and other microscopic organisms which some biologists would classify as plants" (Stace, 1980).

There is a particular branch of botany which deals with the study and description of the variation in plants "the investigation of the causes and consequences of this variation, and the manipulation of the data obtained, to produce a system of classification" (Stace, 1980). This is called *plant taxonomy.* "It is", according to Jeffrey (1982), "neither difficult nor complicated. Yet many people seem to think that it is such a learned and academic activity that they cannot possibly understand it. Others regard it as old-fashioned and out-moded, a relic of a bygone age of science that no longer demands their serious attention."

The term *systematics* is commonly synonymous with taxonomy, although some biologists prefer to draw a line of demarcation between them. In the latter case, taxonomy is limited to a classificatory investigation and systematics conveys the wider definition already given.

Heywood (1977) stated "that one of the major roles of taxonomy is to produce a classification (or system of classification) of organisms that best reflects the totality of their similarities and differences." On the other hand, Radford *et al.* (1974) mentioned the following principal aims of plant systematics:

(*i*) A convenient method of identifying, naming and describing plants;

(*ii*) A classification scheme to express phylogenetic relationships;

(*iii*) An inventory of plant resources—local, regional and continental;

(*iv*) An understanding of evolutionary process and pathway;

(*v*) An integrating and unifying role in the training of biology students, particularly the relationship between the many biological fields or types of evidence and the diversity of organisms; to be a synthesising as well as a filling and retrieval device.

Thus, plant taxonomy can be studied with both empirical and interpretative approach (Davis and Heywood, 1963). Mason (1950) considered plant taxonomy as a synthesis of the following inter-related fields:

1. **Systematic botany,** the fact finding focal point, which comprises cytological and genetical studies as well as any other techniques applied to the problem.
2. **Taxonomic system**, based on the information that are available and include the (*a*) taxonomic concepts of plant groups or taxa; (*b*) concepts of the evolutionary sequence of characters; (*c*) classification and arrangement of taxa and (*d*) phytography or description of taxa.
3. **Nomenclature**, a process of naming plants based on international rules agreed upon by botanists so as to ensure an uniform and reasonably stable system.
4. **Documentation**, which embraces the preservation of fossil or living floras in a museum or herbarium in the form of illustrations and type specimens.

By employing the above four fields, we are building up a great fund of information. These have helped us to throw light on the living, dynamic and fluctuating plant populations. There is still an immense scope for further work in this most exciting and rewarding field of biology.

HIERARCHY IN PLANT KINGDOM

Just as a continent is divided into nations, nations into countries or States, States into provinces, provinces into divisions, divisions into subdivisions, subdivisions into districts, etc., so the Plant Kingdom is divided into a number of categories which differ in rank and size.

A basic unit* in plant classification is the *species*. A species is defined as a single type of living organisms. For example, *Spider-lily* is a species and *Tiger-lily* is a species. These organisms are structurally similar and have common ancestors.

A *genus* (pl. genera) is a group of closely related species. Thus, the genus *Rosa* is made up of many species of Rose. They vary among themselves due to vegetative features, but they all possess a common quality of "roseness" with respect to reproductive characters.

A group of closely related genera is known as a *family*. Again, the Rose family is composed of several genera, among which mention may be made of the Apple genus (*Pyrus*), Peach genus (*Prunus*), Quince genus (*Cydonia*), Rose genus (*Rosa*), Strawberry genus (*Fragaria*) and others. All these have certain resemblances, yet they differ in some well defined characters. The scientific name of a family ends in "aceae"; hence, Rosaceae is the technical name of the Rose family.

A group of closely related families form an *order*, the scientific name of which ends in "ales". By way of example, it is worth mentioning that the Rose order (*Rosales*) contains the Legume family (*Leguminosae* or *Fabaceae*), Orpine family (*Crassulaceae*), Rose family (*Rosaceae*), etc. These families have certain fundamental features of floral structure, indicating a close relationship. Orders are arranged into proper units till the level of Plant Kingdom is reached. Such schemes are called *hierarchies* or *hierarchical classifications* which may be illustrated as a plan-view (*box-in-box presentation*) or as a side-view or an elevation-view (*dendrogram*).

When groups are large, they are often divided for the sake of convenience. So there may be a suborder, ending in "ineae" (*Rosineae*); a subfamily, ending in "oideae" (*Rosoideae*); a tribe, ending in "eae" (*Rosae*); a subtribe, ending in "inae" (*Rosinae*) and a subgenus as *Eurosa* or section as *Gallicanae*. Units below the rank of a species are sometimes provided as subspecies (written as ssp.), variety (var.), subvariety (subvar.), form (f.) and clone (cl.).

---

*"A *taxon* (pl. taxa) is any taxonomic grouping, such as a pkylum, a family or a species. It is a useful general term, and can be used to indicate the rank of a group as well as the organisms contained within that group" (Stace, 1980).

While there is no limit to the number of ranks contained in a hierarchy, the ICBN* has recognised twelve main ranks. The latter include Kingdom, Division, Class, Order, Family, Tribe, Genus, Section, Series, Species, Variety and Form (Stace, 1980). A complete set of modern taxonomic categories in their mutual affinity is presented below:

Kingdom
Subkingdom
Phylum
Subphylum
Class
Subclass**
Order
Suborder
Family
Subfamily
Tribe
Subtribe
Genus
Subgenus
Section
Subsection
Series
Subseries
Species
Subspecies
Variety
Subvariety
Form
Subform

NAMING OF PLANTS

Botanists refer to plants by proper scientific names and not by common or colloquial names as done by laymen. Common names are of limited value for scientific purposes. They are made up of words from the native language of the country in which the languages are spoken. Thus, in Brazil and Portugal, they are Portuguese words; in Australia, New Zealand, England and the United States, they are English words; in Egypt, Iraq, Lebanon, Jordan and Syria, they are of Arabic origin; in India, common names consist of Indian words derived from at least 14 different languages. A Dutch botanist, who is not familiar with Hindi, would be awestruck by such Hindi names as Boot, Katahar, Lashun and Papita.

* International Code of Botanical Nomenclature, the internationally agreed rule book of green plant and fungal nomenclature.

** Superorder.

Such English names as Butter-and-eggs, Forget-me-not, Goat's-beard, Hen-and-chicken, Love-in-a-mist, Milkweed, Skunk Cabbage, Snapdragon, Widow's Tears and Wood-sorrel would be meaningless to a Chinese or Japanese botanist. Another disadvantage is that a single species of plant may be designated by several common names which vary from one country to another and even within the borders of a single country. Instance can be cited for the European Waterlily which goes by 14 other English names, 44 French names, 81 Dutch names and 105 German names—a grand total of 244 names for one and the same plant! Furthermore, the name Mahogany is assigned to some 300 different types of plant. Not all plants have been given common names; out of approximately 15,000 species of orchids, only a few hundred bear common names.

Although common names may kindle one's imagination or appeal to one's sense of humour, they are often misleading or unsatisfactory. The student of plants, like any other scientist, does not restrict his investigations solely to the confines of one state. His work is read internationally and he must keep in touch with what is going on along with his line elsewhere and in other parts of the world. If he studies a certain kind of plant and publishes a report upon it in a scientific journal, his material must be identical so that other workers around the world will know what he is talking or writing about.

The scientific name of a plant consists of two separate words. The first word (*generic name*) designates the genus of the organism and begins with a capital letter, the second (*specific name* or *specific epithet*) being the species with a small letter. Scientific names should always be underlined when written by hand or typed or given in italics when printed.

The scientific names of plants are based on Latin—a language known by scholars throughout the world. In writing out the description of a plant, use is made of the *lingua franca* in vogue during the Middle Ages and not of the classical Latin of Cicero or Horace. As Latin is concise and precise, it is particularly suited to the needs of a descriptive phase of the natural sciences. Applying the Roman alphabet in Latin words "obviates the confusion which would result if the names were written indiscriminately in the characters of other alphabets" like Chinese, Greek, Hebrew or Sanskrit.

The usual objection offered to scientific names is that they are long and cumbersome. As a matter of fact, the objection lies in their unfamiliar sounds and combinations rather than in their length. The scientific names are often unusable to the laymen, since they require an exact delineation of species. While some species are clear-cut with easily recognised features, the others look so much alike that technical skill is often needed to mark the distinction.

GENERIC NAMES

The generic name is spelled or written with a capital letter. It is always a noun. It is also singular and denotes the nominative case. There are four types of origin of generic names.

1. Many genera have been named in honour of a patron of botany or horticulture or a botanist. Thus, there is a *Candollea* for Augustin de Candolle, *Einsteinia* for Albert Einstein, *Grayia* for Asa Gray, *Hookera* for Hooker, *Jeffersonia* for Thomas Jefferson, *Linnaea* for Linnaeus and *Theophrasta* for Theophrastus.
2. In many cases, generic names express some features of the plant. Examples are afforded by *Anigozanthos* (unequal flower) from the zygomorphic flowers of the Kangaroo-paw; *Callicarpa* (beautiful fruit) from the fruits of the Beauty-berry; *Gymnocladus* (naked club) from the sparse-branched stems of the Kentucky Coffee Tree; *Liriodendron* (lily tree) from the shape of the flowers of the Tulip Tree; *Oxydendrum* (sour tree) from the acid leaves of the Sourwood; *Phytelephas* (plant elephant) from the Vegetable Ivory Palm.

3. Some generic names are of fanciful, mythological or poetic origin. For examples, *Circaea* (enchanter's nightshade) refers to Circe, the enchantress; *Dodecatheon* (twelve gods) applies to a plant believed to be under the influence of the principal gods; *Nymphaea* (water-lilies) denotes the lovely water-nymphs; *Theobroma* (god's-food) alludes to the Chocolate Plant.
4. Another type of generic name includes that of aboriginal origin. The names existed in the lands where the plants were discovered and afterwards they were converted into Latin names. Here mention may be made of *Ailanthus,* from the Moluccan; *Catalpa,* from the Red Indian; *Ginkgo,* from the Chinese; *Nelumbo,* from the Ceylonese; *Ravenala,* from the Madagascarian; *Tsuga,* from the Japanese.

SPECIFIC EPITHET

The second part of the scientific name, specific epithet, has come from many sources.

1. The most common form of specific epithet is a descriptive adjective. It may indicate the (*a*) relative abundance of the plant, *e.g. fara* (rare), *vulgaris* (common); (*b*) colour of the plant or plant-part, e.g. *alba* (white), *flava* (yellow), *nigra* (black), *viridis* (green); (*c*) habitat of the plant, *e.g. aquatica* (in water), *arvensis* (in fields), *muralis* (on walls), *palustris* (in swamps), *sylvatica* (in woods); (*d*) region where the plant was found, e.g. *anglicus* (English), *canadensis* (Canadian), *gallica* (French), *mexicana* (Mexican), *nepalensis* (Nepalese), *zeylanica* (Ceylonese); (*e*) size, shape or habit of the plant, *e.g. alta* (tall), *crassa* (thick), *gigantea* (giant), *natans* (swimming), *pendulus* (hanging), *repens* (creeping), *scandens* (climbing), *tenuis* (thin); (*f*) utility of the plant, *e.g. edulis* or *esculenta* (edible), *hortensis* (grown in gardens), *officinalis* (sold in apothecaries' shops), *sativus* (sown for crops); or (*g*) other traits, *e.g. autumnalis* (of autumn), *biennis* (biennial), *foetida* (ill-scented), *hiberna* (of winter), *religiosa* (religious), *spinosa* (spiny), *toxicaria* (poisonous), etc.
2. Another type of specific epithet is the one made in honour of some renowned botanist or a person intimately linked with that species. It may be written in two ways: as a possessive adjective, *e.g. gesneriana* (*Tulipa gesneriana*), *muelleriana* (*Eucalyptus muelleriana*); or as a possessive noun, *e.g. agharkarii* (*Musa agharkarii*), *roxburghii* (*Vanda roxburghii*), etc.
3. Sometimes the specific epithet is constructed from a noun with a suffix, showing affinity or resemblance, *e.g. amaranthoides* (akin to *Amaranthus*), *bignonioides* (similar to *Bignonia*), *quercifolia* (with leaves like *Quercus*), etc.
4. The specific epithet may be a noun instead of an adjective and often a Latin word for some plant, *e.g. Alisma plantago-aquatica* (*plantago-aquatica* is an old word for Water-plantain), *Capsella bursa-pastoris* (*bursa-pastoris* was the classical name for Shepherd's-purse), *Pyrus malus* (*malus* was the ancient word for Apple), etc.
5. A type of specific epithet is a descriptive adjective constituted by combining two or more words and referring to some distinctive feature of the plant, e.g. *angustifolia* (narrow-leaved), *cordifolia* (heart-shaped leaves), *grandflora* (large-flowered), *latifolia* (broad-leaved), etc.

The specific epithet must concur grammatically with the generic name. The latter, however, differs in gender and can be usually ascertained by perusing the form of specific epithet. Generally, the masculine ending is "us", feminine "a" and neuter "on" or "um". Given below are examples showing the forms assumed by some common adjectives to denote gender:

Masculine	*Feminine*	*Neuter*
acer	acris	acre
bromoides	bromoides	bromoides
chinensis	chinensis	chinensis
japonicus	japonica	japonicum
niger	nigra	nigrum
tener	tenera	tenerum
viridis	viridis	viridis

It may not be out of place here to give some idea about possessive names. If the person honoured in a commemorative name is a man, the proper ending in gender is "*i*" or "*ii*", e.g. *baileyi, glazioui, nuttallii.* If the person so honoured is a woman, it becomes "ae", *e.g. margaretae, piersonae, olgae.* If two or more persons are honoured (such as husband and wife or two brothers), the ending will be "orum", *e.g. davisiorum.*

CITATION AND AUTHORITY

Following the name of a species, genus or other taxonomic group, there is a standardised abbreviation or initial which indicates the name of the person who first identified and described the particular taxon. Thus, the name *Solanum tuberosum* was first published by Linnaeus. He becomes the authority for that name and the name is written as *Solanum tuberosum* L. In all serious work, it is a formality to cite the authority for a particular scientific name. This is due to the fact that it avoids the confusion that might result if two or more persons, not known to each other, become the authority for the same name.

In *Asystasia gangetica* T. And., the author's initial stands for Thomas Anderson. In *Duabanga sonneratioides* Buch.-Ham., the author's initial has been abbreviated from Buchanan Hamilton. In *Clematis gouriana* Roxb., *Erioglossum edule* Bl., *Jasminum sambac* Ait. and *Xyris pauciflora* Willd., the authors' initials are derived from Roxburgh, Blume, Aiton and Willdenow respectively.

Different persons with the same name need other designations for their identification. Thus, Robert Brown is usually given as R. Br. and William Brown as Wm. Br. William Hooker and his illustrious son Joseph Dalton Hooker may be recognised by shortening the former as just Hook, and representing the latter by Hook.f. (*filius.* son). When father, son and grandson are all authors of names, it is also necessary to distinguish them. A case in point is the de Candolle family; here Augustin is written as DC., Alphonse as A. DC. and Casimir as C. DC.

It is sometimes noticed that the name has been proposed by one author, but not published by him or her. It is left to another author to publish the name. In such a case, the name of the latter is added with the letter *ex* (from). Nathaniel Wallich once named a specimen *Acer oblongum,* but prepared no description for it. It was Augustin de Candolle who studied Wallich's specimen and supplied the necessary description. Hence, the complete plant name is *Acer oblongum* Wall ex DC.

When there is a change in taxonomic rank or taxonomic position, the name of the first author must be cited in parenthesis and followed by the name of the person who effected the change. For example, *Medicago polymorpha* L. variety *orbicularis L.,* first done by Linnaeus, is elevated to the rank of a species by Allioni and thus becomes *Medicago orbicularis* (L.) All. Another example is afforded by *Althea rosea* (L.) Cav., originally called *Malva rosea* L. by Linnaeus and transferred to *Althea* by Cavanilles. The first case is an example of a change in taxonomic rank and the second one of a change in taxonomic position. Both cases are examples of the double citation.

The significance of the double citation lies in the fact that the newer name is based on the older name which is often termed as basionym. For example, *Manglietia insignis* (Wall.) Bl. was originally described as a *Magnolia* by Wallich, who validly published for it the name *Magnolia insignis.* In this case, *Magnolia insignis* Wall. is the basionym of *Manglietia insignis* (Wall.) Bl. Wherever there is a change in taxonomic position without a rank change, the double citation shows that the particular epithet is timed from the publication date of the basionym and not from that of the new combination. *Raphidiocystis chrysocoma* (Schumach) C. Jeffrey (1962) has priority over *R. welwitschii* Hoof. k. (1871) as the correct name, as its basionym *Cucumis chrysocomus* Schumach. (1827) is the oldest name; here the specific epithet *Chrysocomus* is dated from 1827. In the case of a rank change, a name does not have priority outside the rank of a taxon to which it is applied. Thus, *Cyclamen creticum* (Doerfl.) dates from 1906 and so does the specific epithet *creticum* despite the fact that *Cyclamen repandum* var. *creticum* Doerfl. is dated from 1905.

ORGANISED NOMENCLATURE

Before the middle of the 18th century, plant names were usually polynomials, *i.e.* made up of several words in a series which supplied a somewhat terse description of the organism in question. This was a cumbersome affair. This was superseded by the binomial system as adopted by Rivinus and later used by Linnaeus. The binomial system was comparatively simple in operation as long as most plantsmen followed the Linnaean principles. The discovery of new plants from recently explored areas of the world caused concern over procedures for naming these species. Therefore, elementary rules were framed to serve as a guide to botanists.

To Linnaeus goes the credit for the genuine start of a sound nomenclature for plants. The principles on which he based his system are clearly stated in his *Fundamenta Botanica*(1736) and *Critica Botanica* (1737). For instance, here are some of the rules or aphorisms proposed by him:

"213. All those plants which belong to one genus must be designated by the same generic name."

"214. All those plants which belong to different genera must be designated by different generic names."

"215. There shall be only one generic name where one and the same genus is concerned."

"216. In one and the same genus, the same generic name must be used."

"217. If one and the same generic name has been adopted to designate two different genera, it will have to be banished from one of the positions it occupies."

"218. He who establishes a new genus should give it a name."

"221. Generic names formed from two complete and distinct words are to be banished from the commonwealth of Botany."

"228. Generic names with a similar sound give a handle to confusion."

"240. Those generic names are best which show the plant's essential character or its appearance."

"249. Generic names one and a half foot long, or difficult to pronounce or, unpleasant, are to be avoided."

"250. It is unwise to use technical terms in place of genuine generic names."

However, there were no widely accepted rules governing the naming of plants. Generally speaking, the priority of publication was recognised. Most authors tried to avoid giving either the same name to the different species or different names to the same species. The rapid additions of hundreds of plants new to science gave rise to confusion. There was no index from which botanists of one country could know the names that had been proposed elsewhere by other botanists.

The publication of *Theorie elementaire de la botanique* (1813) by Augustin de Candolle was a significant event. It gave explicit instructions on matters of organography and nomenclature. It also pointed out the fallacies of practices and proposals of other contemporary botanists. This was probably the first complete and detailed set of rules on plant nomenclature.

In 1821, Steudel's *Nomenclator Botanicus* appeared. This work contained a list of the Latin names of all flowering plants known at that time, along with their synonyms.

In various countries, nomenclature patterns were set by renowned botanists. Names chosen by them gained prominence even though their application may have been influenced by personal or national sentiments. With the passage of time, the need for an international accord became increasingly apparent. It was then that Alphonse de Candolle convened an assembly of botanists to present a new set of rules.

Paris Code: The First International Botanical Congress, held at Paris in August, 1867, was aimed at the standardisation and legislation of proper nomenclature practices. About 150 American and European botanists were invited to attend and a draft of the *Lois de la nomenclature botanique* (*Laws of botanical nomenclature*) was distributed before the Congress to each botanist for preliminary study. When the botanists met, a few days were spent in the discussion of the Laws and they were adopted with slight revision. It was resolved that "these Laws, as adopted by this Assembly, shall be recommended as the best Guide for Nomenclature in the Vegetable Kingdom". These Laws are called either the Paris Code, since they were adopted at the French capital, or de Candolle rules, as they were prepared by Alphonse de Candolle.

According to the Paris Code, the starting point for all nomenclature was fixed with Linnaeus; here no date or any work was specified. The rule of priority was considered as basic with no provisions for exceptions. Attention was given to the requirements for valid publication, author citation and terms applied to categories of plants.

The Paris Code guided taxonomic activity in most countries to a considerable degree, but their application showed many inherent defects. As time went on, American and British botanists deviated from the rules and practised the unwritten law named *Kew Rule*. By this rule, if a species was transferred to another genus, the specific epithet need not be transferred to the new genus but the author was free to use a new epithet in the new combination.

Rochester Code: In the United States, a batch of botanists headed by N.L. Britton met at Rochester, New York, in 1892 and developed a set of rules to govern nomenclature. These rules, based on modifications of the Paris Code, are commonly known as the Rochester Code. Among the new recommendations were the (*a*) establishment of the type concept to ascertain the correct application of names; (*b*) strict adherence to the principles of priority; (*c*) acceptance of all binomials resulting from employment of the principles of priority, even if the specific epithet repeats the generic name (referred to as a *tautonym*) and (*d*) interpretation of priority to apply to the precedence of a name in a publication in addition to the date of publication.

Vienna Code: In the Third International Botanical Congress held at Vienna in 1905, advocates of the Rochester Code attempted to have their tenets incorporated in the revised rules. The new changes included the (*a*) establishment of Linnaeus' *Species Plantarum* (1753) as the starting point for naming vascular plants; (*b*) *nomina generica conservanda* by which generic names having a wide use would be conserved over earlier but less well known names; (*c*) banning of tautonyms and (*d*) requirement that names of new taxa be accompanied by a Latin diagnosis.

American Code: Most proponents of the Rochester Code were not satisfied with the results of the Vienna Congress and refused to accept the new rules. In 1907, they put forth a slight modification of the Rochester Code under the heading of the American Code. They did not subscribe to the principle of *nomina generica conservanda* or of the requirement of Latin diagnosis. They admitted the acceptance of the type concept. Another provision of the revised code was that "a binomial may not be used again for a plant in any way if it has been employed previously for another plant, even though the previous use may have been illegitimate."

The Rochester Code and American Code created two opposing schools of thought in the United States. One was under the leadership of Britton and his students, insisting on the provisions of the American Code. The other was under the direction of Asa Gray's pupils who followed the international rules.

International Code: The Fifth International Botanical Congress met in 1930 at Cambridge, England. It brought harmony among the major botanical factions and reconciled the basic differences between the Vienna Code and American Code. As a result of these efforts, the new rules legislated at Cambridge constituted the code of nomenclature that was truly international in name and function. Among the provisions which received the approval of both parties were the following: (*a*) the type concept should be pursued; (*b*) there should be a list of *nomina generica conservanda;* (*c*) tautonyms are not admissible and (*d*) Latin diagnosis are required after January 1, 1932.

At the Sixth International Botanical Congress, Amsterdam (1935), a few major changes in the rules were made. It was resolved that "from January 1, 1935, names of new groups of recent plants, the Bacteria excepted, are considered, as validly published only when they are accompanied by a Latin diagnosis". An attempt to select a list of *nomina specifica conservanda* was thwarted by a overwhelming vote.

The Seventh International Botanical Congress met at Stockholm in 1950. It introduced a certain number of definitions on types. At the Eighth Congress, Paris (1954), great emphasis was laid on types, but the rule of Latin came under fire. The Ninth Congress held at Montreal (1959) appointed a special committee to study the question of conservation of family names. The report of this committee was submitted at the Tenth Congress of Edinburgh (1964), from which a few points may be extracted. A.L. de Jussieu's *Genera Plantarum* (1789) is the starting point for family names. Many of the names included in the list are not a matter of dispute, but their inclusion in the list forms a ready reference providing correct names of families together with a type genus for each family. In the list of *nomina familiarum conservanda,* a few names are found with a spelling somewhat different from the long-established names, e.g. Cannabaceae for Cannabinaceae, Capparaceae for Capparidaceae, Haloragaceae for Haloragidaceae, Melastomataceae for Melastomaceae, etc. At the Edinburgh Congress, the rule of Latin for new diagnoses was settled once and for all and no amendments were proposed. The Eleventh Congress was held at Seattle in August, 1969 proposed the *Seattle Code* which was edited by F. A. Stafleu and published in 1972. The Twelfth Congress was held in July, 1975 at Leningrad, USSR and its recommendations came out in 1978. Most of the proposals submitted to the Eleventh Congress were concerned with refinements and increased precision rather than with sweeping changes, the chief issues in Seattle Code include the tautonymous designations of taxa between genus and species and below species, the perennial question of superfluous names and a reorganisation of the rules for hybrids (Henry and Chandrabose, 1980). The Thirteenth Congress was held in August, 1981 at Sydney, Australia and its recommendations are eagerly awaited.

The Leningrad Code indicates only small differences from the Seattle Code and includes the following changes: (*a*) the concept of organ genera is eliminated for fossil plants; (*b*) the Code does not apply to names of organisms treated as bacteria and does apply to all other organisms treated as plants;

(*c*) the principle of automatic typification is extended to those names of taxa above the family rank that are ultimately based on generic names and the application of the priority principle is recommended while selecting among names thus typified; (*d*) a name or combination published before 1953 without indicating the rank is considered validly published but imperative in questions of priority except for homonymy and certain names to be accepted at the varietal rank; (*e*) Art. 69 of the previous Code is modified on the basis of type method and Art. 70 and 71 dealing with discordant elements and monstrosities were deleted, but the Art. numbers are retained to facilitate the use of the Code; (*f*) the section on orthography is thoroughly rewritten; (*g*) individual paragraphs on all Articles and Recommendations are numbered in a decimal-like system, some being rearranged.

It is beyond the scope of the book to enter into a full discussion of ICBN (Stafleu *et al.*, 1978). But a few items are included here to give some idea about the Code.

Principle I

"Botanical nomenclature is independent of zoological nomenclature."

Principle II

"The application of names of taxonomic groups is determined by means of nomenclatural types.'"

Principle III

"The nomenclature of a taxonomic groups is based upon priority of publication."

Principle IV

"Each taxonomic group with a particular circumscription, position and rank can bear only one correct name, the earliest that is in accordance with the Rules, except in specified cases."

Principle V

"Scientific names of taxonomic groups are treated as Latin regardless of their derivation."

Principle VI

"The Rules of Nomenclature are retroactive when expressly limited."

Ranks of Taxa

"The principal ranks of taxa is ascending sequence are: species, genus, family, order, class, division and kingdom." (Art. 3)

"If a greater number of ranks of taxa is required, the terms for these are made either by adding the prefix sub (*sub-*) to the terms denoting the ranks or by the introduction of supplementary terms." (Art. 4)

Typification

The nomenclatural type is that element to which the name of a taxon is permanently attached, whether as a correct name or as a synonym. The nomenclatural type is not necessarily the most typical or representative element of a taxon." (Art. 7)

["A *holotype* is the one specimen or other element used by the author or designated by him as the nomenclatural type." (Art. 7)

The type of a name of a species or infraspecific taxon is a specimen or other element except in the following instance: for most non-vascular plants and small herbaceous plants, the type may consist of more than one individual which should be conserved in one preparation or a single herbarium sheet. The holotype is selected by the original author from a single collection made by the original author from a single collection made by a collector at one time and cited properly during the original publication.

e.g. Phyllanthus mukerjeeanus Mitra et Bennet in Bull. Botan. soc. Bengal 19(2): 145–146. 1966. Holotype *Bennet* 621A (CAL).]

["An *isotype* is any duplicate (part of a single gathering made by a collector at one time) of the holotype; it is always a specimen." (Art. 7)]

[For example, if a new species is based on a single gathering (consisting of three specimens and three herbarium sheets made out of them) having the same field number, the original author will designate one among them as a holotype and the remaining two will be isotypes.]

e.g. Phyllanthus mukerjeeanus Mitra et Bennet in Bull. Botan. Soc. Bengal 19(2): 145–146. 1966. Isotype *Bennet* 621 B-C (CAL).]

"A *paratype* is a specimen cited with the original description other than the holotype or isotype."

[If a new taxon is described based on more than one collection made at various times, the author will designate one collection as holotype and isotype. The rest of the collection, having various field numbers, will constitute the paratypes. For example, if the new taxon is gathered without flowers and fruits in one season and with flowers and fruits in another season, these two collections will bear two different field numbers. The author will select one collection for holotype and isotype while describing the new taxon, the next gathering cited in the protologue will form the paratype.

e.g. Polygala jacobii Chandr. in Bull. bot. Surv. India 9(1–4): 288–289. 1967.

Paratype *Rangachari s.n.* (Acc. no. 2444, MH).]

"A *syntype* is any or two or more specimens cited by the author when no holotype was designated, or any one of two or more specimens simultaneously designated as types. (Art. 7)

"A *lectotype* is a specimen or other element selected from the original material to serve as a nomenclatural type when no holotype was designated at the time of publication or as long as it is missing." (Art. 7)

[While describing the new species *Cynodon barberi,* Rangachari and Tadulingam (*J. Bombay nat. Hist. Soc.* 24:846. 1916), cited several specimens in the protologue. Since they failed to indicate one as the holotype, all specimens thus cited became syntypes. After a critical study, Jain (*Bull. bot. Surv. India* 8:204. 1967) correctly chose one of the specimens—Madras Herbarium, South India Flora No. 13715, deposited in CAL—as the lectotype.]

"A *neotype* is a specimen or other element selected to serve as a nomenclatural type as long as all of the material on which the name of the taxon based was missing." (Art. 7)

[In the new species, *Farsetia macrantha,* described by Blatter and Hallberg (*J. Bombay nat. Hist. Soc.* 26:220. 1918), three numbers— *Blatter* 5785, 7300 and 7305—were cited in the protologue. Presumably, these specimens were deposited in the Blatter Herbarium, Bombay (BLAT), where Blatter's whole collection including type specimens have been kept. While working on this species, Bhandari (J. *Bombay nat. Hist. Soc.* 72:606. 1975) could not trace the type specimens in the Blatter Herbarium. He concluded that the original material was lost or destroyed. He designated *Bhandari* 507 as neotype, "examplifying the true application of the name *Farsetia macrantha* Blatt. & Hallb. and to ascertain with certainty the sense in which the same must be used. The specimen comes from the type locality and is in complete agreement with the original diagnosis".]

"In addition to the above, it is accepted generally that a cotype is a second specimen from the same plant from which the holotype was collected. The term type, used alone and unqualified, generally refers to the holotype. A topotype is a specimen collected at the same station as was the type (i.e., from the type locality). There are other terms ending with the suffixtype, but for the most part they are concerned with ecological or genetical situations."

Priority

"Each family or taxon of lower rank with a particular circumscription, position and rank can bear only one correct name..." (Art. 11)

"For any taxon family to genus inclusive, the correct name is the earliest legitimate one with the same rank..." (Art. 11)

"A name of a taxon has no status under this Code unless it is validly published." (Art. 12)

"The application of both conserved and rejected names is determined by nomenclatural types." (Art. 14)

"When a name proposed for conservation has been provisionally approved by the General Committee after study by the Committee for the taxonomic group concerned, botanists are authorized to retain it pending the decision of a later International Botanical Congress." (Art. 15)

Effective and Valid Publication

"Publication is effected, under this Code, only by distribution of printed matter (through sale, exchange or gift) to the general public or at least to Botanical institutions with libraries accessible to botanists generally, it is not effected by communication of new names at a public meeting, by the placing of names in collections or gardens open to the public, or by the issue of microfilm made from manuscripts, type-scripts, or other unpublished material. Offer for sale of non-existed printed matter does not constitute effective publication." (Art. 29)

"A new name or combination published on or after 1 Jan. 1953 without a clear indication of the rank of the taxon concerned is not validly published." (Art. 35)

Rejection of Names

"A legitimate name or epithet must not be rejected merely because it is inappropriate or disagreeable, or because another is preferable or better known, or because it has lost its original meaning." (Art. 62)

"A name must be rejected if it has been widely and persistently used for a taxon not including its type." (Art. 69)

"A name or epithet rejected...is replaced by the oldest legitimate name or in a combination by the oldest available epithet in the rank concerned." (Art. 72)

The following types of name are unusable and are considered as illegitimate:

Synonyms: No taxa can have the same and a plant or group can have only valid name and all others are classified as synonyms. *Nomenclatural synonyms* or *homotypic synonyms* are names based on the same type. Thus, *Miliusa tomentosa* (Roxb.) Sinclair, *Saccopetalum tomentosum* (Roxb.) Hook.f. & Thoms. and *Uvaria tomentosa* Roxb. are names based on the same type and are regarded as nomenclatural synonyms.

Taxonomic synonyms or *heterotypic synonyms* are names based on various types considered to belong to the same taxon. Thus, *Hibiscus vitifolius* L. (1953) and *Hibiscus obtusifolius* Willd. (1801) are names based on two different types. A taxonomic judgement led to the proof that both types are entirely identical. Here *H. obtusifolius* becomes a taxonomic synonym of *H. vitifolius.*

Tautonyms: These are names where the specific epithet exactly repeats the generic name with or without transcribed symbol, *e.g. Malus malus.* However, *Bambusa bambos* and *Cajanus cajan* are not tautonyms; these specific epithets do not exactly repeat the generic names.

Autonyms: As a result of the proposal of another taxon, these names are automatically established for a single taxon. In 1965, the publication of the name *Crotalaria willdenowiana* ssp. *glabrifolius* Ellis automatically established the name of another *Crotalaria willdenowiana* DC. ssp. *willdenowiana* and both are based on *Crotalaria willdenowiana* DC.

Typonyms: "A name is rejected if there is an older valid name based on the same type (typonym)." The specific epithet *Asplenium vincentis* given by Herman Christ in 1897 is a typonym of *Asplenium guildangii* given by G.S. Jenman to the same species in 1894.

Metanyms: "A name is rejected when there is an older valid name based on another member of the same group (metanym)." Had the two species mentioned earlier been named from different type specimens, then *A. vincentis* might be termed a metanym.

Homonyms: "A name is rejected when preoccupied (homonym)." Identical names cannot be applied to two different taxa. If this is applied, it is a homonym in every case except the first, *e.g. Astragalus rhizanthus* Royle (1835) and *Astragalus rhizanthus* Boiss (1843). The second species is called *Astragalus cariensis* Boiss (1849).

Hyponyms: "A name is rejected when the natural group to which it applies is undetermined (hyponym)." *Adodendrum* Necker and *Calesian* Adams are examples of generic hyponyms and *Gentiana hybrida* Raf. and *Lechea furfurca* Raf. of specific hyponyms.

The aim of the Code is to secure uniform and stable system of plant nomenclature. It is no perfect and no one claims that it is, it has its short comings, but taken as a whole, it is an excellent code. It is the product of intense study by the best brains in the botanical fraternity, who for nearly a century has been studying the problems connected with the naming of plants in an effort to make the system satisfactory in all respects. (Santapau, 1965).

SYSTEMS OF CLASSIFICATION

The start of taxonomy is "undoubtedly antedate recorded history." To state that taxonomy began with Aristotle, Dioscorides or Pliny, or with anybody we know, is erroneous. The ancient civilisations of China, Egypt and India were based, to some extent, on cultivated plants. Great interest was evinced in plants of medicinal importance and these plants were studied by scholars of those days. Even among uncivilised races of the modern era, we find that they practice folktaxonomies; such people possess the equivalents of our concepts of taxa.

The endeavours to classify all known plants can be divided into four main categories (Jeffrey, 1982): *artificial systems* (upto about 1830), *pre-evolutionary natural systems* (1763-1920), *phylogenetic systems* (1883 to date) and *phenetic systems.* The earliest system of classifications was artificial systems based on certain principles of logic and pre-selected characters adopted by their authors as being significant. This was the only type of classification that could be chosen when very little was known about plants. Examples of artificial systems include those of the early herbalists (based on habit) and the sexual system of Linnaeus. In the light of increased knowledge of plants, it seemed that more useful or satisfactory groupings could be made on the basis of overall similarity and derived from the correlated variation of several characters. Though the resulting classifications were called natural systems, they are not entirely due to the continuing effect of logical principles and incomplete knowledge of the plant kingdom, they are sometimes referred to as *formal systems.* In this context, Jeffrey (*l.c.*) stated: "If any goal at all was to be assigned to the attainment of a completely natural system, it was to reveal the divine plan of creation by grouping together plants most alike in the nature of their creation." While post-Darwinian in time and pre-evolutionary in concept, the most important formal system was that of Bentham-Hooker (published 1862–1883) arising from that of A.P. de Candolle (outlined 1813) which, in turn, was based on that of A.L. de Jussieu (1789). The publication of Darwin's *Origin of Species* (1859) changed the outlook of taxonomy and provided an explanation of relationships already established. Many taxonomists argued that classifications should be based on phylogeny, since the course of evolutionary descent of its members is called. The most widely utilised system of the phylogenetic period is that of Engler (published 1886–1892) and originating from that of Eichler (outlined 1883). Following largely the procedures first expounded by Hennig (1966), a school of investigators termed *cladists* have avoided the absurdities or excesses of the earlier phylogenetic systems and evolved methods by means of which the most probable phylogeny of the constituent taxa has been constructed. Other workers called *pheneticists* emphasise the need for a classification permitting maximal generalisations to be made and based on consideration of the totality of the features of all living phenotypes. Moreover, modern systems in practice (1967 to date) are an admixture of the artificial, phylogenetic and phenetic despite the tall claims of their authors. The chief reason is perhaps the gross inequality of our knowledge of the various types of plants. Some, generally important crop plants including rice, or remarkable north temperate genera have been subjected to detailed studies by experimental and descriptive methods. Others, like lower plants and many tropical species, may be represented by an inadequately preserved solitary specimen. It is likely that a perfect phylogenetic system would coincide with a genuine phenetic one. Strangely enough, there can still be room for different systems and such differences result from differences of prior emphasis granted to various preselected features being based on different interpretations of the available evidence and deficiencies in such evidence. The differences between modern systems are possibly small and would be diminished in the future, since they would undoubtedly be modified as knowledge of plants increased. In the field of flowering plant classification, the best known modern systems are those of Takhtajan (1967) and Cronquist (1968).

Systems of classification were in existence before the birth of Christ and five periods have been recognised.

First Period: In the first period, classifications were proposed by the ancient Greeks and Romans as well as botanists for over ten centuries. They were mainly based on the habit of plants.

Ancient Greeks and Romans: The outstanding Greek naturalist was Theophrastus (370–285 B.C.), a pupil of Plato and Aristotle, who became head of the Lyceum in Athens. He is said to have written more than 2,000 treatises, of which two botanical writings (*The History of Plants* and *The Causes of Plants*) are the most important. He recognised the differences in ovary position and in polypetalous and gamopetalous corolla; distinguished trees, shrubs, undershrubs and herbs: differentiated between wild and cultivated forms, fruitful and sterile, evergreen and deciduous, terrestrial and aquatic: between monocotyledons and dicotyledons. He mentioned some 5,000 different plants in his work. But his writings had no literary charm and often suffered from the lack of scientific terms. Most of his sentences were compressed and highly elliptical, sometimes to the point of obscurity. To describe a plant or plant part in the common language would frequently take pages and tax the powers of a first-grade writer. Yet, he managed to pack into his books an amazing amount of information. Several of the names used by him were later taken up by Linnaeus and are still used in the same sense.

After Theophrastus, the name of Pliny the Elder (23–79 A.D.) is worthy of mention. He was a great Roman scholar and travelled far and wide. The name of his encyclopaedic work is *Historia Naturalis,* a series of 37 books, 16 of which deal with plants and discuss such topics as plant anatomy, timber trees, practice of horticulture and medicinal uses. Unfortunately, his studies were cut short by tragic death during the eruption of Mount Vesuvius. Although he included much factual information and his contribution to early botanical knowledge is most significant, he perpetuated several mistakes of his predecessors.

Pedanios Dioscorides (1st century A.D.), a contemporary of Pliny, was a military physician attached to the Court of Roman Emperor Nero and was interested in the medicinal properties of plants. His main contribution, *Materia Medica,* was a description of about 600 plants, ''largely from first-hand observation, and detailed their useful applications."

From the material compiled by Dioscorides, a manuscript entitled *Anicia Juliana Codex* was prepared for the daughter of a Byzantine king. It was illustrated with coloured drawings of plants.

Another botanist of the early era was Albertus Magnus (1193–1280). He was in basic agreement with Theophrastus for the placement of plants in comparison with each other. His botanical writing, *De Vegetabilis,* was popular for more than two centuries. Here for the first time, the difference between the stem structure of corticate (monocotyledons) and tunicate (dicotyledons) was recognised.

Herbalists: The decline of the Greek and Roman civilisations saw the advent of the Middle Ages when little or no significant progress was made in the field of botany, not to speak of taxonomy. The early *herbals i.e.* books containing descriptions of plants used as foods and medicines, such as the *Codex* of Dioscorides, were followed in toto for centuries without any marked improvement or alteration. The first signs of awakening came in the first half of the 16th century with the consequence that a botanical renaissance developed in Europe. This was greatly aided by the invention of printing. "Woodcuts were employed in printing illustrations of plants and these often show a remarkable degree of fidelity, originality and skill in execution" (Porter, 1967). Representatives of this period are the herbals of Otto Brunfels (1489–1534), Jerome Bock (1498–1554), Leonart Fuchs (1501–1566), Valerias Cordus (1515–1544) and others, men sometimes referred to as the German Fathers of Botany. At this time, a valuable contribution was made by Gaspard Bauhin (1560–1624). Though not used exclusively by him, he first attempted to employ a binomial system of nomenclature as evident in his *Prodromus Theatri Botanici* (1620) and *Pinax Theatri Botanici* (1623).

During the age of herbals which lasted from 1470 to 1670, an interesting development was the *doctrine of signatures.* Being handed down from the ancients, this was elaborated upon by the herbalists. This was based on the belief that plants or plant parts resemble portions of a human body and that they must have been created with the object of providing cures to the diseases of those portions. Common names given to plants at that time indicated the supposed remedial properties. The scientific names of plant still owe their origin to this doctrine. A case in point is *Hepatica* where the shape of the leaves is believed to resemble the shape of the liver. Hence, the leaves of that genus is a specific for ailments of that organ.

Transition: The transition from the botanical renaissance to the modern period produced a crop of literature. These botanists, who contributed to this literature, switched from the so called utilitarian approach to that of a scientific one. There was a shift in the development of taxonomy where we can sense a grouping towards a natural relationship of plants. At the same time, there was a growing tendency to develop a new terminology of description, a scheme of nomenclature and a system of classification. To crown all, a great influx of new plants from areas of recent geographical exploration aggravated the problems of the struggling taxonomist.

Andrea Caesalpino (1519–1603) was somewhat different from the other herbalists of his time. He studied medicine and botany at Pisa. He became Director of the Botanical Garden at Bologna and later Professor of Botany. In 1592, he received a new assignment at Rome as Physician to Pope Clement VIII. He assembled a herbarium of 768 dried and mounted plants with Italian or Latin names; this is still preserved in the Museum of Natural History at Florence. His famous contribution, *De Plantis,* was a compendium of 16 books, in which the first gave a general exposition of theoretical botany and others contained descriptions of about 1,500 plants. Although his arrangement is similar to that of the crude groupings of Theophrastus, Caesalpino used a deductive or priori reasoning. He was of the opinion that the organs of fructification are more important in classification than the habit of plants—a conclusion not appreciated by his contemporaries. He made an intensive study of seeds and fruits as well as other parts of plants and even plant activities. Thus, in his work, taxonomy is intimately linked with anatomy, morphology and physiology. He is "commemorated by the name *Caesalpinia,* a genus of handsome legumes including the Poinciana tree".

Joachim Jung (1587–1657), a brilliant German teacher, had his training in medicine and mathematics. He suggested that the classification of plants into herbs, shrubs and trees is not basic. He cleared up some of the confusion that existed between analogous and homologous structures. As a terminologist, he defined the following for the first time: perianth for the calyx instead of for the combined calyx and corolla, as now known; stamens and styles as understood today (being not aware of their functions); capitulum of the Asteraceae as composed of disc and ray florets; nodes and internodes; leaf consisting of a petiole and blade, the blade being digitate or pinnate. While doing all these, he left no publications of his own writing. But two of the students he produced revealed a record of his teaching as well as his concepts.

John Ray (1627–1705), the son of an English blacksmith, came out of Trinity College with flying colours and was ordained as a clergyman. He travelled widely in Europe and investigated plant-novelties. In his *Historia Plantarum,* he put forward a plan of classification taking about 18,000 species of plants into consideration. Under the primary divisions of woody and herbaceous plants, he set up the taxa of monocots and dicots, the classes based on fruit types and subdivided these on the basis of leaf and floral characters. "It was a system based on form and gross morphology of plant structures, and in many respects was superior to the artificial Linnaean system that came later" (Lawrence, 1951).

Joseph Pitton de Tournefort (1656–1708), a French botanist, wrote *Institutiones Rei Herbariae.* His system was inferior to that of Ray, since no distinction between phanerogams and cryptogams or monocotyledons and dicotyledons were made. But it enjoyed popularity in his time, as his arrangement facilitated easy identification of plants. Tournefort's system remained in practice till superseded by that of Linnaeus. He was also credited with the foundation of modern concepts of genera and many of his generic names (such as *Betula, Castanea, Fagus, Quercus, Ulmus*) are well known even today.

Second Period: This period was characterised by systems based on numerical classifications. In this period, the names of Carolus Linnaeus (1707–1778), Pehr Kalm (1716–1779) and Frederich Hasselquist (1722–1752) appear as prominent. The system of Linnaeus only will be discussed here.

Linnaean System: The name of Linnaeus is invariably associated with the so called sexual system of classification. The title is a misnomer, since the system is not at all sexual in the proper sense of the term. He has largely stressed the numerical relationships of the sex organs. His "was a system based on differences rather than on similarities of the sex organs". However, his system is a sort of glorified city directory. He brought into current use the binomial nomenclature, introduced terse formulae for description and fixed attention on species. His name is commemorated by "a dainty and charming little creeping plant"—*Linnaea borealis.*

In his system, Linnaeus instituted 24 classes as follows:

Class 1. *Monandria.* Stamen 1. Examples: *Lemna, Scirpus.*
Class 2. *Diandria.* Stamens 2. Examples: *Salvia, Veronica.*
Class 3. *Triandria.* Stamens 3. Example: *Iris.*
Class 4. *Tetrandria.* Stamens 4. Examples: *Mentha, Ulmus.*
Class 5. *Pentandria.* Stamens 5. Examples: *Myosotis, Primula.*
Class 6. *Hexandria.* Stamens 6. Examples: *Alisma, Berberis, Rumex.*
Class 7. *Heptandria.* Stamens 7. Example: *Aesculus.*
Class 8. *Octandria.* Stamens 8. Example: *Fagopyrum.*
Class 9. *Enneandria.* Stamens 9. Example: *Ranunculus.*
Class 10. *Decandria.* Stamens 10. Example: *Acer.*
Class 11. *Dodecandria.* Stamens 11–19. Example: *Euphorbia.*
Class 12. *Icosandria.* Stamens 20 or more, episepalous. Examples: *Rosa, Spiraea.*
Class 13. *Polyandria.* Stamens 20 or more, attached to the receptacle. Examples: *Nymphaea, Papaver.*
Class 14. *Didynamia.* Stamens didynamous. Example: *Linaria.*
Class 15. *Tetradynamia.* Stamens tetradynamous. Example: Cruciferae.
Class 16. *Monadelphia.* Stamens in 1 bundle. Examples: Geraniaceae, Malvaceae.
Class 17. *Diadelphia.* Stamens in 2 bundles. Examples: *Pisum, Lathyrus.*
Class 18. *Polydelphia.* Stamens in several bundles. Example: *Hypericum.*
Class 19. *Syngenesia.* Stamens united with the anthers. Examples: *Lobelia, Viola.*
Class 20. *Gynandria.* Stamens adnate to the carpels. Examples: *Aristolochia,* Orchidaceae.
Class 21. *Monoecia.* Plants monoecious. Examples: *Quercus, Typha.*
Class 22. *Dioecia.* Plants dioecious. Examples: *Salix, Urtica.*
Class 23. *Polygamia.* Plants polygamous. Examples: Many Compositae.
Class 24. *Cryptogamia.* Flowers concealed. Examples: Algae. Fungi, Mosses, Ferns.

Third Period: The systems of this period were based on form relationships. They were natural, as they served the practical need of classification by functioning adequately as a help to identification.

System of de Jussieu: Bernard de Jussieu (1699–1777), a French contemporary of Linnaeus, tried to lay out the Royal Gardens at Versailles according to a natural system. Working with the Linnaean system, he modified it so much that the classification almost became his own creation. However, he was not completely satisfied with his groupings and did not publish his results. His nephew, Antoine Laurent de Jussieu (1748–1836), added to it some of his own ideas and published it in *Genera Plantarum secundum Ordines naturales disposita.*

In the system of younger de Jussieu, the Plant Kingdom has been divided into 15 classes:

Acotyledones.	Plants without cotyledons	Class 1.
Monocotyledones.	Plants with 1 cotyledon	
	Stamens hypogynous	Class 2.

	Stamens perigynous	Class 3.
	Stamens epigynous	Class 4.
Dicotyledones.	Plants with 2 cotyledons.	
Apetalae	Hypogynous	Class 5.
	Perigynous	Class 6.
	Epigynous	Class 7.
Monopetalae	Hypogynous	Class 8.
	Perigynous	Class 9.
	Epigynous	Class 10.
Polypetalae	Hypogynous	Class 11.
	Perigynous	Class 12.
	Epigynous, anthers free	Class 13.
	Epigynous, anthers connate	Class 14.
Diclines irregulares.	Male and female flowers on different plants. Corolla generally absent	Class 15.

In the above system, the Acotyledones included chiefly the cryptogams as well as some aquatic flowering plants whose reproduction was unknown at that time. As given here, the Monocotyledones and Dicotyledones essentially represented the modern concepts of those taxa. The Diclines irregulars was an odd assortment of Amentiferae, euphorbias, gymnosperms and nettles.

Younger de Jussieu gave major emphasis on the cotyledons and made use of the characters of petals and stamens. But his system fell short of the modern concept with respect to the features of vascular bundles, floral axes and carpels. His system served as a basis upon which by addition and alteration the best systems have been constructed.

De Candollean System: Augustin Pyrame de Candolle (1778–1841), the senior member of a remarkable French-Swiss family, developed a system of classification which was an improvement over that of de Jussieu. His *Theorie Elementaire de la Botanique* contained 161 orders (or families now known), the number being raised to 213 in a subsequent revision by Alphonse (1806–1893). The de Candolles, father and son, and later the grandson, Casimir (1836–1918), collaborated on the preparation of a monumental work, *Prodromus systematis regni vegetabilis.* The book included all the species of vascular plants known at that time.

The outline of de Candollean system is as follows:

I. *Vasculares.* Vascular plants with cotyledons.

Class 1. *Exogenae.* Vascular bundles in a ring: Dicotyledons.

A. *Diplochlamydeae.* Both calyx and corolla present.

(*a*) *Thalamiflorae.* Polypetalous, hypogynous.

Orders 1–46: Ranunculaceae, Cruciferae, etc.

(*b*) *Calyciflorae.* Polypetalous or sympetalous, perigynous or epigynous.

Orders 47–84: Rosaceae, etc.

(*c*) *Corolliflorae.* Gamopetalous, hypogynous.

Orders 85–108: Primulaceae, Scrophulariaceae, etc.

B. *Monochlamydeae.* Calyx only present. Orders 109–128: Chenopodiaceae, Coniferae, etc.

Class 2. *Endogenae.* Vascular bundles scattered: Monocotyledons.

A. Phanerogamae. Flowers present. Orders 129–150: Liliaceae, Orchidaceae, etc.

B. Cryptogamae. Flowers absent, hidden or unknown. Orders 151–155: Ferns, etc.

II. *Cellulares.* Plants without vascular bundles or cotyledons.

Class 1. *Foliaceae.* Leafy, sexuality known.

Orders 156–157: Liverworts, Mosses.

Class 2. *Aphyllae.* Not leafy, sexuality unknown.

Orders 158–161: Algae, Fungi, Lichens.

Brown et al. Robert Brown (1773–1858), a Scottish botanist, followed the plan of de Candolle and completely ignored the Linnaean system in the *Botany of Terra Australis.* Through his studies, he showed that the female flowers in cycads and conifers are really naked ovules. He also published many papers on the geography, morphology and taxonomy of different plant groups.

Stephen Endlicher (1805–1849) of Vienna proposed a scheme of classification in his *Genera plantarum secundum ordines naturales disposita* and treated 6,835 genera. He divided the Plant Kingdom into thallophytes (algae, lichens, fungi) and cormophytes (mosses, ferns, seed-plants).

Other noteworthy systems propounded during the early part of the 19th century include those of Adolphe Brongniart (1801–1876) and John Lindley (1799–1865). Brongniart, who was the founder of palaeobotany, applied his fossil studies to plant classification. Lindley published a system based on physiological characters in *The Vegetable Kingdom.*

System of Bentham-Hooker: George Bentham (1800–1884) and Joseph Dalton Hooker (1817–1911), the British botanists associated with the Royal Botanic Gardens at Kew, set up a system of classification in their *Genera Plantarum.* A conspectus of Bentham-Hooker system is given below:

I. DICOTYLEDONS

A. Polypetalae. Petals free.

Series 1. *Thalamiflorae.* Stamens hypogynous, usually many; disc absent.

Natural orders 1–33: Ranunculaceae, Cruciferae, Malvaceae, etc.

Series 2. *Disciflorae.* Flowers with a disc which may take the shape of a cup, cushion or ring.

Natural orders 34–55: Geraniaceae, Rutaceae, Ancardiaceae, etc.

Series 3. *Calyciflorae.* Perigynous or epigynous; ovary mostly inferior.

Natural orders 56–82: Rosaceae, Myrtaceae, Umbelliferae, etc.

B. Gamopetalae. Petals united.

Series 1. *Inferae.* Ovary inferior.

Natural orders 83–91: Rubiaceae, Caprifoliaceae, Compositae, etc.

Series 2. *Heteromerae.* Ovary usually superior; carpels more than 2.

Natural orders 92–103: Ericaceae, Primulaceae, Ebenaceae, etc.

Series 3. *Bicarpellatae.* Ovary usually superior; carpels 2.

Natural orders 104–127: Scrophulariaceae, Solanaceae, Labiatae, etc.

C. Monochlamydeae. Perianth simple.

Series 1. *Curvembryeae.* Embryo curved.

Natural orders 128–134: Nyctaginaceae, Amarantaceae, etc.

Series 2. *Multiovulatae Aquaticae.* Submerged herbs.

Natural order 135: Podostemaceae.

Series 3. *Multiovulatae Terrestres.* Terrestrial plants.
Natural orders 136–138: Nepenthaceae, etc.

Series 4. *Microembryeae.* Embryo very small.
Natural orders 139–142: Piperaceae, Myristicaceae, etc.

Series 5. *Daphnales.* Ovary of 1 carpel; ovule 1.
Natural orders 143–147: Proteaceae, Lauraceae, etc.

Series 6. *Achlamydosporeae.* Pistil 1-celled; ovules 1–3.
Natural orders 148–150: Loranthaceae, etc.

Series 7. *Unisexuales.* Flowers unisexual.
Natural orders 151–159: Euphorbiaceae, Urticaceae, etc.

Series 8. *Ordines Anomali.* Relationships uncertain.
Natural orders 160–163: Salicaceae, etc.

II. GYMNOSPERMAE

Natural orders 164–166: Gnetaceae, Coniferae, Cycadaceae.

III. MONOCOTYLEDONS

Series 1. *Microspermae.* Without endosperm.
Natural orders 167–169: Orchidaceae, etc.

Series 2. *Epigynae.* With endosperm; ovary inferior.
Natural orders 170–176: Iridaceae, Amaryllidaceae, etc.

Series 3. *Coronarieae.* Ovary superior.
Natural orders 177–184: Liliaceae, etc.

Series 4. *Calycineae.* Perianth sepaloid.
Natural orders 185–187: Palmae, etc.

Series 5. *Nudiflorae.* Perianth lacking.
Natural orders 188–192: Typhaceae, Araceae, etc.

Series 6. *Apocarpae.* Carpels single or separate.
Natural orders 193–195: Alismaceae, etc.

Series 7. *Glumaceae.* Flowers in spikelets or heads.
Natural orders 196–200: Cyperaceae, Gramineae, etc.

In reality, Bentham-Hooker's system is a modification of that of de Candolle. The modifications are to be noticed in the Polypetalae, where the new series Disciflorae is interpolated between the Thalamiflorae and Calyciflorae and in the revision of those apetalous families that formed the Monochlamydeae. This system is a natural one and has been found suitable for all practical purposes. Here the Polypetalae starts with the Ranales and the monocots are derived from the dicots.

Bentham-Hooker's system is not at all phylogenetic, since it is based on a single and mostly artificial characters. Nothing has been said about the phylogeny of the dicots and monocots. The gymnosperms are treated as a third group collateral with and placed between the monocots and dicots—a point of anomaly and an item inconsistent with the current knowledge of the plant groups. The Monochlamydeae is considered apart from the Polypetalae. The Chenopodiaceae should be brought close to the Caryophyllaceae and not kept remote from each other; both of them seem to have similar floral configuration. The positions of the Orchidaceae and Scitamineae are at the beginning of the monocots; these families are not primitive, as they contain many advanced features. In the grouping of the monocots, emphasis has been laid on the relative position of the ovary and character of the perianth. The Iridaceae and Amaryllidaceae are more akin to the Liliaceae than to the Scitamineae or Bromeliaceae with which they are associated through epigynous flowers. The Juncaceae, which is connected with the Palmae, should be drawn nearer to the Liliaceae. Having an inferior ovary, the Hydrocharitaceae should be accommodated in aquatic orders with non-endospermic seeds. Moreover, a funny arrangement is the division of certain taxa on the basis of terrestrial or aquatic features.

Fourth Period: During this period, the systems of classification proposed by August Wilhelm Eichler (1839–1887), Adolph Engler (1844–1930), Charles Edwin Bessey (1845–1915), Richard von Wettstein (1862–1931), Hans Hallier (1898–1932) and John Hutchinson (1884–1972), were based on phylogeny. These botanists demonstrated a supposed series of evolutionary advancements from simple to more complex forms and endeavoured to establish the genetic as well as the ancestral relationships.

Eichler's system: A.W. Eichler, Professor of Botany at the University of Kiel, Germany, accepted the doctrine of evolution. But his system is not phylogenetic in the modern sense. Based on extensive studies on morphology, he prepared a classification of the plant kingdom in the following manner:

A. *Cryptogamae*
- Division I. *Thallophyta*
  - Class 1. *Algae*
  - Class 2. *Fungi*
- Division II. *Bryophyta*
  - Class 1. *Hepaticae*
  - Class 2. *Musci*
- Division III. *Pteridophyta*
  - Class 1. *Equisetineae*
  - Class 2. *Lycopodineae*
  - Class 3. *Filicineae*

B. *Phanerogamae*
- Division I. *Gymnospermae*
- Division II. *Angiospermae*
  - Class 1. *Monocotyleae*
  - Class 2. *Dicotyleae*
  - Subclass (*i*) *Choripetalae*
  - Subclass (*ii*) *Sympetalae*

Englerian system: The chief features of Eichler's system were adopted by A. Engler, Professor of Botany at the University of Berlin. In collaboration with Karl Prantl (1849-1893) and other associates, he wrote *Die Naturlichen Pflanzenfamilien*. A complete outline of the system appeared in Engler and Diels' *Syllabus der Pflanzenfamilien*. According to this system, the Plant Kingdom has been divided into 14 major divisions:

I. *Schizophyta*
II. *Myxomycetes*
III. *Flagellatae*
IV. *Dinoflagellatae*

V. *Heterocontae*
VI. *Bacillariophyta*
VII. *Conjugatae*
VIII. *Chlorophyceae*
IX. *Charophyta*
X. *Phaeophyceae*
XI. *Rhodophyceae*
XII. *Eumycetes*
XIII. *Archegoniatae*
XIV. *Embryophyta Siphonogama*

Subdivision A. *Gymnospermae*

Class 1. Cycadofilicales
Class 2. Cycadales
Class 3. Bennettitales
Class 4. Ginkgoales
Class 5. Cordaitales
Class 6. Coniferae
Class 7. Gnetales

Subdivision B. *Angiospermae*

Class 1. Monocotyledoneae

Order 1. Pandanales
Order 2. Helobiae
Order 3. Triuridales
Order 4. Glumiflorae
Order 5. Principess
Order 6. Synanthae
Order 7. Spathiflorae
Order 8. Farinosae
Order 9. Liliflorae
Order 10. Scitamineae

Class 2. Dicotyledonae

Subclass 1. Archichlamydeae

Order 1. Verticillatae
Order 2. Piperales
Order 3. Hydrostachyales
Order 4. Salicales
Order 5. Garryales
Order 6. Myricales
Order 7. Balanopsidales
Order 8. Leitneriales

Order 9. Juglandales
Order 10. Julianiales
Order 11. Batidales
Order 12. Fagales
Order 13. Urticales
Order 14. Podostemonales
Order 15. Proteales
Order 16. Santalales
Order 17. Aristolochiales
Order 18. Balanophorales
Order 19. Polygonales
Order 20. Centrospermae
Order 21. Ranales
Order 22. Rhoeadales
Order 23. Sarraceniales
Order 24. Rosales
Order 25. Pandales
Order 26. Geraniales
Order 27. Sapindales
Order 28. Rhamnales
Order 29. Malvales
Order 30. Parietales
Order 31. Opuntiales
Order 32. Myrtiflorae
Order 33. Umbelliflorae

Subclass 2. Metachlamydeae

Order 1. Diapensiales
Order 2. Ericales
Order 3. Primulales
Order 4. Plumbaginales
Order 5. Ebenales
Order 6. Contortae
Order 7. Tubiflorae
Order 8. Plantaginales
Order 9. Rubiales
Order 10. Cucurbitales
Order 11. Campanulatae

The main difference between Engler's system and that of Bentham-Hooker is the amalgamation of the Polypetalae and Monochlamydeae. A significant feature is the position of monocots which precede the dicots. The apetalous and catkin bearing dicots have been placed ahead of others, showing these to be primitive rather than derived taxa. The angiospermous families have been arranged in ascending pattern in accordance with the increasing complexity of the flower in general and the character of the floral envelope in particular. In this way, the Orchidaceae and Compositae are highly evolved among the monocots and dicots respectively; hence they are put at the end of each class. Moreover, the Juncaceae, Iridaceae and Amaryllidaceae are closely related to the Liliaceae.

In Engler's system, the term family is preferred to natural order and the term order to class or series of Bentham-Hooker. This system is an illustrated work with phylogenetic arrangement and modern keys, providing a means for the identification of all the known genera of plants from the primitive algae to the advanced seedplants. This system is still followed in the arrangement of families of angiosperms in nearly all floras or manuals and in most herbaria.

The system of Engler has been subjected to heavy criticism as tending to obscure the phylogeny of the angiosperms. The objectionable features include the origin of this group from the Bennettitales-like ancestors, interpretation of the simple unisexual flowers as primitive, derivation of the dichlamydeous flowers from monochlamydeous ones, derivation of the free central placentation from parietal placentation and the parietal placentation in syncarpous ovaries from axile placentation. Engler regarded the monocots to be more primitive than the dicots; it is now generally believed that the monocots arose from the dicots by way of reduction. A primitive order like the Helobiae has been placed between two more advanced orders, the Pandanales and Glumiflorae—a trait inconsistent with modern knowledge of the monocots. The Araceae is thought to be more primitive than the Liliaceae. The union of the Choripetalae and Apetalae is an undesirable quality, since it is a big jump from the Salicaceae to the Ranunculaceae. The position of the Amentiferae and Centrospermae at the commencement of dicots before the Ranales as well as the alignment of sympetalous families have brought about awkward situations. Recent investigations on fossils do not lend support to Engler's system.

Wettstein's system: R. Wettstein, Professor of Botany, University of Vienna, formulated a system that was in many respects similar to that of Engler. He believed that unisexual perianthless flowers are more primitive than bisexual flowers with perianth. According to him, the angiosperms arose from gymnosperms through *Gnetum* and monocots evolved out of the ranalian stock of the dicots. He departed from Engler in the treatment of some of the orders.

In Wettstein's system, the Cactaceae is grouped with the Aizoaceae in the Centrospermae and the Tricoccae put between the Centrospermae and Hamamelidales. The Parietales and Rhoeadales are placed side by side as closely related taxa; the former is more primitive than the latter due to the prevalence of woody habit which is earlier than the herbaceous habit. The Malvales is set after the Myrtales. Among monocots, the polycarpic Helobiae is believed as more primitive and the Pandanales as most advanced to be preceded by the Orchidaceae. The Cyperaceae and Gramineae are accommodated in separate orders.

Bessey's system: C.E. Bessey, a disciple of Asa Gray and thereafter Professor of Botany at the University of Nebraska, USA, presented a new system of classification. Though it resembles the Bentham-Hooker outline, the system is unique in several points. Contrary to the concepts of Engler and Wettstein, he did not regard the simplest naked flowers of the Amentiferae as extremely primitive. In his opinion, the angiosperms originated from gymnospermous ancestors like the Bennettitales which, in turn, arose from cycadeoid forms. He took the Ranales as most primitive angiosperm, one branch of which gave rise to the monocots and the other to the dicots. The early dicots bifurcated into two evolutionary lines: *Strobiloideae* (plants with hypogynous flowers) and *Cotyloideae* (plants with perigynous or epigynous flowers). The Strobiloideae starts with the Ranales and is followed in succession by the Malvales, Sarraceniales, Geraniales, etc.; here the last evolved order is the Lamiales which embraces the Verbenaceae and Labiatae. In the Cotyloideae, the evolutionary line is initiated by the

Rosales and after passing through the Myrtales, Loasales, etc. the line is terminated by Asterales containing the Compositae.

Hallier's system: H. Hallier of Hamburg, Germany, introduced a system which agrees with Bessey's system in many respects. This system is based on the principles that apocarpy and spiral arrangement of floral parts are more primitive than syncarpy and whorled arrangement. In support of primitiveness, he emphasised the roles of branching, simple leaf with unicostate reticulate venation and solitary large flower. He regarded the monocots as more advanced than dicots, being derived from the Lardizabalaceae.

In Hallier's system, the dicots have been divided into 29 orders under four main groups and the monocots into seven orders:

DICOTYLEDONEAE

Group I. *Proterogenes*

Order 1. Ranales
Order 2. Nepenthales
Order 3. Caryophyllinae
Order 4. Piperinae
Order 5. Aristolochiales

Group II. *Anonophylales*

Order 1. Anonales
Order 2. Hamamelinae
Order 3. Umbelliflorae
Order 4. Campanulinae
Order 5. Columniferae

Group III. *Rhodophyles*

Order 1. Cruciales
Order 2. Gruniales
Order 3. Aesculinae
Order 4. Terebinthinae
Order 5. Proteinae
Order 6. Rosales
Order 7. Rhamnales

Group IV. *Ochnigenes*

Order 1. Guttales
Order 2. Passionales
Order 3. Peponiferae
Order 4. Polygalinae
Order 5. Daphnales
Order 6. Myrtinae
Order 7. Bicornes

Order 8. Primulinae
Order 9. Santalales
Order 10. Sapotales
Order 11. Tubiflorae
Order 12. Rubiacinae

MONOCOTYLEDONEAE

Order 1. Lilliiflorae
Order 2. Artorrhizae
Order 3. Ensatae
Order 4. Enentioblastae
Order 5. Spadiciflorae
Order 6. Cyperales
Order 7. Helobiae

Certain other observations can be made about Hallier's system. Here the Papaveraceae is included in the Ranales and the Leguminosae is placed in the Aesculinae after the Sapindaceae. The Compositae is a comparatively early taxon, next to the Umbelliflorae. The Tubiflorae is derived from the Guttales and the Helobiae is considered very advanced among monocots.

Hutchinson's system: The system to grip most attention in recent years is that of J. Hutchinson, of the Royal Botanic Gardens, Kew, England. He dealt chiefly with the angiosperms, publishing his system in *The Families of Flowering Plants* and *British Flowering Plants.*

An outline of his earlier system (1926, 1934) is presented below:

Phylum I. *Gymnospermae*
Phylum II. *Angiospermae*
Subphylum I. *Dicotyledons.* Contains 76 orders and 264 families.
Division I. *Archichlamydeae.* Petals free from each other or absent, rarely united at the base, *e.g.* Magnoliales to Umbelliflorae.
Division II. *Metachlamydeae.* Petals united into a tube, *e.g.* Ericales to Lamiales.
Subphylum II. *Monocotyledons.* Contains 29 orders and 68 families.
Division I. *Calyciferae.* Calyx-bearers, *e.g.*, Butomales to Zingiberales.
Division II. *Corolliferae.* Corolla-bearers, *e.g.* Liliales to Orchidales.
Division III. *Glumiflorae.* Perianth much reduced and glumaceous, *e.g.* Juncales, Cyperales and Graminales.

In this system, Hutchinson retained the Archichlamydeae and Metachlamydeae. In the former, he recognised two phylogenetic lines, one radiating from the Magnoliales and comprising forms in which the arborescent habit is predominant and another emanating from the Ranales and embracing types in which the herbaceous habit is conspicuous. In separating the arborescent forms from the herbaceous, Hutchinson deviated from Engler who retained them in the Ranales. From these polypetalous groups of the Archichlamydeae evolutionary lines are traced through the Metachlamydeae, terminating in several climax taxa, among which are the Apocynales, Rutales, Asterales and Lamiales.

Hutchinson also proposed a theory of parallel development among the angiosperms. It has been set in Table 1.

Hutchinson's system is a phylogenetic one, but his idea of evolution of floral parts is different from that of Engler. It is based on the contention that plants with sepals and petals, associated with other floral and anatomical features, considered as also primitive, are more ancient phylogenetically than plants without sepals and petals.

Hutchinson enunciated 24 dicta for the classification of flowering plants. These dicta are mentioned here briefly:

1. Evolution is both upwards and downwards, the former tending towards preservation… and the latter to their reduction and suppression…
2. Evolution does not necessarily involve all organs at the same time...
3. Evolution has been generally consistent, and when a particular progression or retrogression has set in, it is persisted into the end of the phylum...
4. Broadly speaking, trees and shrubs are more primitive than herbs in any one family or genus.

TABLE 1. Concept of evolutionary parallelism

Dicots	*Monocots*	*Characters*
Ranunculaceae	Alismaceae	Apocarpous pistil
Cabombaceae	Butomaceae	Placentation of ovules
Ceratophyllaceae	Najadaceae	[illegible]bit
Menispermaceae	Dioscoreaceae	[illegible]al structure; climbing habit.
Aristolochiaceae	Araceae	Su[illegible] resemblance of perianth and spathe respectively
Hydnoraceae	Thismiaceae	Parasitic and saprophytic habit respectively
Hydrostachyaceae	Potamogetonaceae	Aquatic habit and spicate inflorescence
Balsaminaceae	Orchidaceae	Zygomorphic flower
Umbelliferae	Amaryllidaceae	Umbel with an inferior ovary
Asclepiadaceae	Orchidaceae	Androecium with a waxy pollen
Compositae	Eriocaulaceae	Capitate inflorescence

5. Trees and shrubs are older than climbers..
6. Perennials are older than biennials and annuals...
7. Aquatic flowering plants are derived from terrestrial ancestors and epiphytes, saprophytes and parasites are more recent than plants of normal habit...
8. Dicotyledons are more primitive than monocots.
9. Spiral arrangement is more primitive than cyclic.
10. Simple leaves are usually more primitive than compound leaves.
11. Unisexual flowers are more advanced than bisexual; dioecious plants are more recent than monoecious.
12. The solitary flower is more primitive than the inflorescence...
13. Aestivation types are evolved from contorted to imbricate to valvate.
14. Many-parted flowers (polymerous) precede and the type with few parts (oligomerous) follow...
15. Apetalous flowers are derived from petaliferous flowers.
16. Polypetaly is more primitive than gamopetaly.

17. Actinomorphy is more primitive than zygomorphy.
18. Hypogyny is usually more primitive than perigyny and epigyny is the most advanced.
19. Apocarpy is more primitive than syncarpy.
20. A gynoecium of many pistils preceded one of few pistils.
21. Seeds with endosperm and small embryo are older than seeds without endosperm and a large embryo...
22. Numerous stamens, in general, indicate greater primitiveness than does an androecium of a few stamens.
23. Separate anthers, in general, indicate greater primitiveness than does an and androecium of either fused anthers or filaments.
24. Aggregate fruits are more highly evolved than single fruits; as a rule, the capsule precedes the berry or drupe.

Recently, Hutchinson (1959) devised a new system which differs a great deal from his earlier classification (1926, 1934). His current system is based on the assumption that woody arboreal taxa are distinct and phylogenetically older than herbaceous groups. The Dicotyledoneae, according to him, arose from the Proangiosperms and was the first to make its appearance on the earth's surface. Regarding the origin of the Monocotyledoneae, he postulated that the group was derived from an early dicot like the Ranales.

In accordance with the priority of evolution, Hutchinson divided the Dicotyledoneae into the woody arboreal groups (*Lignosae*) and herbaceous groups (*Herbaceae*). As a result of such division and realignment of the orders and families, the phyletic positions of many taxa of his earlier classification have been altered considerably. He separated pairs of families regarded as closely related by all other botanists. The Nymphaeaceae and Ranunculaceae have been removed from the first order to the 55th order Ranales. The Capparidaceae has been separated from the Cruciferae, being placed in the 24th order Capparidales. The Rhoeadales constitute the 59th order after the Piperales. The Verbenaceae has been shifted to the 54th order Verbenales and the Labiatae to the 82nd order Lamiales.

A significant feature of Hutchinson's revised system is the abolition of the Archichlamydeae and Metachlamydeae. Consequently, polypetalous and sympetalous families have been intermingled. For example, the metachlamydean or sympetalous orders Loganiales and Apocynales are placed in the Lignosae after the archichlamydean or polypetalous order Sapindales. Similarly, the Solanales and Personales are placed in the Herbaceae before the Geraniales.

Hutchinson's Lignosae consists of 54 orders. It begins with the Magnoliales and is followed by the Annonales, Laurales, Dilleniales, Coriariales, Rosales, Leguminosales, etc. and ends in the Verbenales. The Herbaceae, consisting of 28 orders, starts with the Ranales and after passing through the Berberidales, Aristolochiales, Piperales, Rhoeadales, Cruciales, Resedales, etc. is terminated by the Lamiales. Hutchinson's Monocotyledoneae has not suffered any change, since the three main groups (Calyciferae, Corolliferae and Glumiflorae) are still retained except for the addition of one family (Cartonemataceae) under the Commelinales of the Calyciferae.

The system of Hutchinson gave a great fillip to phyletic thinking during the past six decades. His "... scheme has a great many useful features and ... represents a compilation by a taxonomist who probably knew a greater proportion of the world's flowering plants than any of his contemporaries" (Stace, 1980). He envisaged a monophyletic origin for the seedplants in general and the angiosperms in particular. He placed too much emphasis on habit and habitat. He insisted on the polyphyletic origin of such taxa as the Amentiferae, Asterales, Umbelliferae, Urticales, etc. He failed to supply explanatory notes as to the delimitation of the orders and/or families. He provided no recent conclusions on the phylogenies within the gymnosperms and none at all concerning plant groups below the seedplants. Recent morphological, anatomical and cytological evidence did not receive proper consideration from Hutchinson in the formulation of his dicta.

Other contemporary systems: Several botanists built systems of plant classifications other than those discussed before.

Under the authorship of Alfred Barton Rendle (1865–1938), Keeper of the Department of Botany, British Museum of Natural History, London, there appeared an excellent two-volume work, *The Classification of the Flowering Plants,* which is essentially a reflection of Engler's plan. This system has the added advantage of having the choripetalous angiospermous subdivided into two convenient categories, namely, *Monochlamydeae* (within 13 orders and 31 families) and *Dialypetalae* (with 17 orders and 95 families), while the *Sympetalae* is separated into equally convenient groups, *viz. Pentacyclicae* (flowers with five whorls) and *Tetracyclicae* (flowers with four whorls). The Pentacyclicae contains four orders and 13 families, whereas the Tetracyclicae (containing seven orders and 34 families) is again divided into two easily recognisable groups, *Superae* and *Inferae,* accordingly as the ovary is superior and inferior respectively.

Rendle postulated polyphyletic origins for the angiosperms. He laid much emphasis on wind versus insect pollination as phyletic criteria. His fundamental thesis was that the woody habit is more primitive than the herbaceous one. He exerted great care and caution in the placement of many orders and families. But the evidence for many of his views are circumstantial or even conjectural.

August A. Pulle, who was born in 1878 and worked at Utrecht, made up a system that is a modified version of Engler's classification. Like Engler, he considered the Spermatophyta as a separate division. Unlike Engler, he treated the Spermatophyta as composed of four subdivisions: *Pteridospermae, Gymnospermae, Chlamydospermae* (Gnetales), *Angiospermae.* The angiosperms are further separated into monocots and dicots. Since Pulle was of the view that the Monochlamydeae and Sympetalae are not natural taxa, the dicots are arranged into eight series as follows with the monocots taking precedence:

Class *Monocotyledonae*

- Order 1. Spadiciflorae
- Order 2. Pandanales
- Order 3. Helobiae
- Order 4. Triuridales
- Order 5. Farinosae
- Order 6. Liliiflorae
- Order 7. Glumiflorae
- Order 8. Scitaminae
- Order 9. Gynandrae

Class *Dicotyledonae*

Series I. *Casuarinales→Amentiflorae→Urticales→Centrospermales→Plumbaginales→ Primulales.*

- Order 1. Casuarinales
- Order 2. Piperales
- Order 3. Salicales
- Order 4. Garryales
- Order 5. Leitneriales

Order 6. Juglandales
Order 7. Julianiales
Order 8. Myricales
Order 9. Balanopsidales
Order 10. Hydrostachyales
Order 11. Fagales
Order 12. Urticales
Order 13. Centrospermae
Order 14. Polygonales
Order 15. Cactales
Order 16. Plumbaginales
Order 17. Primulales

Series II. *Amentiflorae→Proteales→Santalales→Balanophorales.*

Order 1. Proteales
Order 2. Santalales
Order 3. Balanophorales

Series III. *Hamamelidales→Ranales→Ebenales.*

Order 1. Hamamelidales
Order 2. Ranales
Order 3. Ebenales

Series IV. *Ranales→Aristolochiales→Rosales→Myrtales.*

Order 1. Aristolochiales
Order 2. Rosales
Order 3. Podostemales
Order 4. Myrtales

Series V. *Rhoeadales→Parietales→Guttiferales→Ericales→Campanulatae.*

Order 1. Rhoeadales
Order 2. Batidales
Order 3. Sarraceniales
Order 4. Parietales
Order 5. Cucurbitales
Order 6. Diapensiales
Order 7. Ericales
Order 8. Campanulatae

Series VI. *Rosales→Pandales→Malvales→Tricoccae→Geraniales→Terebinthales→Rhamnales.*

Order 1. Pandales
Order 2. Malvales
Order 3. Tricoccae
Order 4. Geraniales

Order 5. Terebinthales

Order 6. Rhamnales

Series VII. *Tereointhales→Celastrales→Umbelliflorae→Rubiales.*

Order 1. Celastrales

Order 2. Umbelliflorae

Order 3. Rubiales

Series VIII. *Terebinthales→Ligustrales→Contortae→Tubiflorae→Plantaginales.*

Order 1. Ligustrales

Order 2. Contortae

Order 3. Tubiflorae

Order 4. Plantaginales

Order 5. Callitrichales

Order 6. Hippuridales

Carl Skottsburg (1880–1963) of Sweden published a modification of the Engler classification. According to him, the monocots originated from an unknown primitive dicot. He interpreted apocarpy as primitive and syncarpy as advanced.

Karl Christian Mez (1866–1944), Professor of Botany, University of Koenigsberg, Germany, advanced the idea that relationships between larger groups of plants can be ascertained by analysis of their protein reactions. This physiological approach is sometimes styled as *serum diagnosis.* It consists of injecting an extracted plant protein into the body of an animal, say, a rabbit. After antibodies are formed in the rabbit, some of its serum are mixed with the protein of another plant. If a precipitate is obtained with the second protein extract, a relationship is thought to exist between the plants involved. The degree of relationship is indicated by the relative abundance of the precipitates as well as the amount of dilution of the serum and plant extract at which the precipitation will take place.

Oswald Tippo, Professor of Botany, University of Illinois, USA, published (1942) an outline of a projected classification based on "...the newest, sound developments in all the branches of plant phylogeny and in which the various groups are named in such a way as to indicate rank or degree of affinity...". He claimed no originality for his system, for it is a blend of several proposals—notably those of the non-vascular plants followed by Gilbert Morgan Smith (1885–1959) and of the vascular plants adopted by Arthur Johnson Eames (1881–1969). His system, with the exception of Engler, is applicable for the entire Plant Kingdom. But he is silent about the arrangement of the families of angiosperms. His system is presented here to give orientation to students:

Kingdom. *Plantae*

Subkingdom I. *Thallophyta.* Plant forming no embryos.

Phylum 1. *Cyanophyta* (blue-green algae)

Phylum 2. *Euglenophyta* (flagellates, euglenoids, etc.)

Phylum 3. *Chlorophyta* (green algae)

Phylum 4. *Chrysophyta* (diatoms, etc.)

Phylum 5. *Phaeophyta* (brown algae)

Phylum 6. *Rhodophyta* (red algae)

Phylum 7. *Schizomycophyta* (bacteria)

Phylum 8. *Myxomycophyta* (slime molds)

Phylum 9. *Eumycophyta* (true fungi)

Subkingdom II. *Embryophyta.* Plant forming embryos.

Phylum 1. *Bryophyta* or *Atracheata.* Plants lacking vascular tissues (mosses, liverworts).

Phylum 2. *Tracheophyta* or *Tracheata.* Plants with vascular tissues.

Subphylum 1. *Psilopsida.* Rootless leafless vascular plants.

Subphylum 2. *Lycopsida.* Plants with scale-like leaves and simple vascular tissues (club-mosses and their allies).

Subphylum 3. *Sphenopsida.* Plants with ribbed stems, scale-like leaves and simple vascular tissues (horsetails).

Subphylum 4. *Pteropsida.* Plants with large leaves and complex vascular tissues.

Class 1. *Filicineae* (ferns)

Class 2. *Gymnospermae* (cone-bearing plants, such as cycads, firs and relatives)

Class 3. *Angiospermae* (flowering plants)

Subclass 1. *Monocotyledonae*

Subclass 2. *Dicotyledonae*

With a new system in 1950, G. Gundersen based his plant classification on basic chromosome numbers along with anatomical and morphological characters. Here he did not deal with the monocots, but only with the dicots. The latter are divided into ten large groups according to the scale of evolution and consequent primitiveness: Magnoliflorae, Cistiflorae, Thea Group, Rosiflorae, Ulmus Group, Malva Group, Geranium Group, Dianthiflorae, Jasminiflorae, Rubiflorae. The value of polypetalous and sympetalous characters of flowers was ignored by Gundersen.

Fifth Period: Since 1967, a number of modern systems have been proposed. J.N. Mitra (1901–1978), a gifted Indian botanist who never occupied a high position, suggested a classification based on the idea that the evolution of angiosperms take place in both upward and downward lines. He did not claim his scheme as purely phylogenetic "but the orders and included families are arranged on probable phylogeny and natural affinities; it is rather a convenience than true phylogenetic one". The characters of unisexuality and perianthless flowers are regarded as advanced, being derived from bisexual diplochlamydeous ancestors. In this respect, he shared the views of Bessey, Hallier and Hutchinson and opposed those if Engler, Rendle and others. Mitra's arrangement is as follows:

DICOTYLEDONEAE

Group A. *Archichlamydeae*

Order 1. Ranales

Order 2. Cactales

Order 3. Aristolochiales

Order 4. Piperales

Order 5. Caryophyllales (Centrospermae)

Order 6. Polygonales

Order 7. Rhoeadales

Order 8. Sarraceniales

Order 9. Rosales

Order 10. Hamamelidales
Order 11. Hydrostachyales
Order 12. Salicales
Order 13. Leitneriales
Order 14. Juglandales
Order 15. Balanopsidales
Order 16. Casuarinales (Verticillatae)
Order 17. Urticales
Order 18. Guttiferales
Order 19. Parietales
Order 20. Malvales
Order 21. Euphorbiales
Order 22. Geraniales
Order 23. Rutales
Order 24. Celastrales
Order 25. Rhamnales
Order 26. Sapindales
Order 27. Myrtales (Myrtiflorae)
Order 28. Proteales
Order 29. Santalales
Order 30. Umbellales

Group B. *Sympetalae*

Subgroup (*a*) *Superae*

Order 1. Primulales
Order 2. Plumbaginales
Order 3. Plantaginales
Order 4. Ericales
Order 5. Ebenales
Order 6. Loganiales
Order 7. Solanales
Order 8. Boraginales (Polemoniales)
Order 9. Personales

Subgroup (*a*) *Inferae*

Order 1. Rubiales
Order 2. Cucurbitales
Order 3. Campanulales

MONOCOTYLEDONEAE

Order 1. Helobiae
Order 2. Triuridales

Order 3. Commelinales
Order 4. Liliflorae
Order 5. Spathiflorae
Order 6. Pandanales
Order 7. Juncales
Order 8. Glumiflorae
Order 9. Scitamineae
Order 10. Orchidales (Microspermae)

Cronquist, Takhtajan and Zimmermann (1966) put forward a brand new idea on plant classification involving higher taxa of the Embryobionta. They remarked that "the general system of plants and the nomenclature of higher taxa at the level of divisions and classes are now unstable and in state of confusion. Even though the principal groups of higher plants and the evolutionary relationships among them are to a large extent agreed on, botanists differ as to how to express best these concepts in a formal taxonomic system. It seems obvious that it would be useful, especially to students, if most botanists could agree on a single overall scheme". Hence, the following plan:

Subkingdom Embryobionta

Division 1. *Rhyniophyta*
Class 1. *Rhyniatae*

Division 2. *Bryophyta*
Class 1. *Anthocerotae*
Class 2. *Marchantiatae*
Class 3. *Bryatae*

Division 3. *Psilotophyta*
Class 1. *Psilotatae*

Division 4. *Lycopodiophyta*
Class 1. *Lycopodiatae*
Class 2. *Isoetatae*

Division 5. *Equisetophyta*
Class 1. *Hyeniatae*
Class 2. *Sphenophyllatae*
Class 3. *Equisetatae*

Division 6. *Polypodiophyta*
Class 1. *Polypodiatae*
Subclass 1. *Protopteridiidae*
Subclass 2. *Archaeopteridiidae*
Subclass 3. *Ophioglossidae*
Subclass 4. *Noeggerathiidae*
Subclass 5. *Marattiidae*

Subclass 6. *Marsileidae*
Subclass 7. *Salviniidae*

Division 7. *Pinophyta*

Subdivision A. *Cycadicae*

Class 1. *Lyginopteridatae*
Class 2. *Cycadatae*
Class 3. *Bennettitatae*

Subdivision B. *Pinicae*

Class 1. *Ginkgoatae*
Class 2. *Pinatae*
Subclass 1. *Cordaitidae*
Subclass 2. *Pinidae*

Subdivision C. *Gneticae*

Class 1. *Gnetatae*
Subclass 1. *Ephedridae*
Subclass 2. *Welwitschiidae*
Subclass 3. *Gnetidae*

Division 8. *Magnoliophyta*

Class 1. *Magnoliatae*
Class 2. *Liliatae*

Collecting or observing in living condition representatives of all but 44 of the 312 families accepted by him and collating evidence from vegetative anatomy, floral morphology, palynology, embryology, phytochemistry, cytology, phytogeography and palaeobotany, Thorne (1968) presented a new system of classification of flowering plants. He carried his synopsis down to the subfamily category and in the Asteraceae down to the tribe. "In this age of taxonomic inflation", his "treatment of orders, families and perhaps other taxa will probably seem rather inconclusive and conservative". There are 19 superorders and 39 orders in dicots and five superorders and 11 orders in monocots.

Subclass: DICOTYLEDONEAE

Superorder: Annoniflorae

Order: Annonales
Order: Berberidales

Superorder: Nymphaeiflorae

Order: Nymphaeales

Superorder: Sarraceniiflorae

Order: Sarraceniales

Superorder: Theiflorae

Order: Theales
Order: Nepenthales
Order: Ericales

Order: Ebenales
Order: Primulales
Order: Plumbaginales

Superorder: Rafflesiiflorae
Order: Rafflesiales

Superorder: Cistiflorae
Order: Cistales
Order: Salicales
Order: Tamaricales
Order: Capparidales

Superorder: Malviiflorae
Order: Malvales
Order: Urticales
Order: Rhamnales
Order: Euphorbiales
Order: Solanales
Order: Campanulales

Superorder: Santaliflorae
Order: Santalales
Order: Oleales

Superorder: Geraniiflorae
Order: Geraniales

Superorder: Rutiflorae
Order: Rutales
Order: Myricales

Superorder: Chenopodiiflorae
Order: Chenopodiales
Order: Batidales

Superorder: Hamamelidiflorae
Order: Hamamelidales

Superorder: Rosiflorae
Order: Rosales
Order: Pittosporales

Superorder: Proteiflorae
Order: Proteales

Superorder: Myrtiflorae
Order: Myrtales

Superorder: Gentianiflorae

Order: Gentianales

Order: Bignoniales

Superorder: Corniflorae

Order: Cornales

Order: Dipsacales

Superorder: Lamiiflorae

Order: Lamiales

Superorder: Asteriflorae

Order: Asterales

Subclass: MONOCOTYLEDONEAE

Superorder: Alismiflorae

Order: Alismales

Order: Zosterales

Order: Najadales

Superorder: Triuridiflorae

Order: Triuridales

Superorder: Liliiflorae

Order: Liliales

Superorder: Ariflorae

Order: Arales

Order: Arecales

Order: Cyclanthales

Order: Pandanales

Superorder: Commeliniflorae

Order: Commelinales

Order: Zingiberales

The above scheme was greatly modified by Thorne in 1976. While Airy Shaw (1973) recognised 539 families of seedplants including Foetidiaceae and Uapaceae "which have hitherto been included in scarcely any or no schemes of classification", Stebbins (1974) produced a classification of angiosperms with modifications of the systems of Takhtajan (1943), Thorne (1968) and Hutchinson (1973). A comparison of Cronquist (latest version 1968), Takhtajan (latest version 1969) and Hutchinson (latest version 1973) with three others was given by Becker (1973).

Dahlgren (1975) presented a highly interesting classification of angiosperms, but its construction differs from most previous attempts. It is accepted that many modern living angiospermic taxa are unlikely ancestors of other living taxa. The Magnoliales has been treated not as an ancestral taxon but as a taxon retaining a high proportion of ancestral characters. Moreover, anatomical and developmental characters along with chemical characters have been widely used in the classification. Dahlgren's scheme closely resemble the Hallier-Bessey scheme rather than the Eichler-Engler scheme. According to the new classification, the dicotyledons are divided into 27 superorders and the monocotyledons has seven superorders. According to Stace (1980), "... the application of the concepts underlying Dahlgren's scheme is likely to produce the best system of angiosperm classification...will be subjected to rigorous examination and testing by a wide range of botanists over the coming years".

DEVELOPMENT OF TAXONOMY IN INDIA

With the development of agriculture in ancient India, plant science took firm roots in our soil. In the Vedic period (2000–800 B.C.), the people of this land cultivated Barley, Cotton, Date, Melon, Millet, Wheat, etc; a knowledge of descriptive botany and elementary plant physiology was believed to be necessary at that time. If the Vedic literature is studied, one may encounter many terms used in connection with the description of external features and internal structures of plants or plant parts. Moreover, there is a mine of information concerning medicinal plants and their application against diseases. In this very remote age, crop rotation and fertiliser dressing were practised for soil improvement.

Ancient Period: From the scientific viewpoint, one of the earliest treatises dealing with plant life is Parasara's *Vrikshayurveda*. In ancient India, this book fulfilled the requirements of botany preparatory to the study of medicine. The first chapter of this book dealt with morphology (in skeletal form) of plant organs, such as root, stem, leaf, flower, fruit, etc; the second chapter was concerned with the nature and properties of soil; the third chapter took up the distribution and description of 14 forest types in India; the fourth, fifth, sixth, seventh and eighth chapters provided details of the morphology of plant organs. A system of classification was propounded, being based on the study of the comparative morphology of plants. Many families (*ganas*) were recognised and distinguished by the characters of sepals, petals, stamens, carpels and fruits.

Before 19th Century: Attracted by the vast potentialities of the country's flora, the Portuguese were the first to arrive in India and thus initiated the modern study of Indian plants. The first important contribution came from Garcia d'Orta (1565) who gave detailed accounts of some of the more striking medicinal plants from firsthand knowledge in his *Coloquios dos Simplos e Drogas da India.* The next book published was *Tractado de las Drogas* in 1578 by C. Acosta.

The Dutch succeeded the Portuguese in India and developed an interest in the plants of this country. An amateur botanist by the name of Hendrik van Rheede tot Draakenstein (1660–1699) became the Governor of the Dutch possessions in India. He made large collections of plants and published descriptions of them in *Hortus Malabaricus.* "Linnaeus saw this book when he was a student in Holland and it is mainly on Rheede's *Hortus* that Linnaeus based the nomenclature of Indian plants" (Santapau, 1958). Other interesting books appeared after this publication; mention may be made of John Burman's *Thesaurus Zeylanicus,* Nicholaus Burman's *Flora Indica* and Paul Hermann's *Paradiscus Batavus.*

John Gerard Koenig (1728–1785), a pupil of Linnaeus, came to India in 1768. He joined the Tranquebar Mission in South India as a Surgeon and naturalist. He formed a society called "The United Brothers" in order to promote the study of Indian plants. The members of the brotherhood made splendid collections in the vicinity of Madras and sent specimens to eminent botanists in Europe.

While botanical activities gained momentum in Madras, things were not lying idle in Bengal. The Royal Botanic Garden came into existence in 1787 due to the munificence of East India Company in general and to the interest of Col. Robert Kyd (1746–1793) in particular. The Garden was founded "not for the purpose of collecting rare plants as things of mere curiosity or furnishing articles for the gratification of luxury, but for establishing a stock for disseminating such articles as may prove beneficial to the inhabitants". It soon became a centre of taxonomic and phytogeographic research and an agency of plant material of economic importance between India and farthest corners of the world. Kyd became its first superintendent.

Upto 19th Century: Kyd was succeeded by William Roxburgh (1751–1815), the Linnaeus of India. By his immense zeal and scientific interest in Indian plants, he made the Botanic Garden known to the scientific world at large. Two of his monumental work include *Flora Indica and Plantae Coromandelianus*. He also made 2,382 large coloured drawings of Indian plants, but these were not published.

After Roxburgh's death. Francis Buchanan-Hamilton (1762–1829) succeeded him. He was a naturalist of great repute, making extensive tours in Nepal and elsewhere. His collections were described in *Prodromus Florae Nepalensis* by D. Don in 1825.

In 1817, Hamilton was followed by Nathaniel Wallich (1786–1854). A man of tremendous energy, he botanised in various and difficult parts of India, Malaya and Nepal. His collections were made into several sets and distributed to a number of herbaria in Europe. He himself prepared a catalogue of these collections, listing 9,148 species; this catalogue is referred to as *Wall. Cat.* (Wallich, *Catalogue of Dried Specimens*). His another great contribution was *Plantae Asiaticae Rariores*'.

William Griffith (1810–1845) officiated as Superintendent of the Garden during Wallich's absence on leave. Unfortunately, he died in 1845 while on a botanical expedition to Malaya. He spent 13 years in India, collected 9,000 species and described them. During this period, Robert Wight (1796–1872) was active in the study of Peninsular flora. He published a part of his work as *Icones Plantarum Indiae Orientalis.* He spent 35 years in India and described nearly 3,000 species of plants.

Wallich's successor was Hugh Falconer, M.D., F.R.S., a palaeontologist who held office until 1855. Early in 1858 during G. McClelland's officiating period, Sir Joseph Dalton Hooker visited the Garden on his famous journey to Sikkim and again in 1860 on his return to Calcutta. Falconer was succeeded by Thomas Thomson, a traveller and botanist of proven ability. He was the President of the Agri-Horticultural Society and co-author of the first volume of *Flora Indica.* Thomson retired in 1861 and was followed by Thomas Anderson. The latter was not only the Superintendent of the Garden and Professor of Botany, but also the first Conservator of Forests for Bengal. He was largely instrumental for the introduction and cultivation of Cinchona in India. Subsequent to Anderson's departure from India, C. B. Clarke, an Education Officer, Government of Bengal, acted as a Superintendent and during his time he began a series of botanical publications on his vast collection of plants. He was also one of the important collaborators of Hooker's *Flora of British India.* In 1871, George King took charge of the Garden and started the journal, *The Annals of the Royal Botanic Garden,* Calcutta in 1887. He was responsible for the establishment of the Botanical Survey of India in 1890. This organisation was necessary to co-ordinate the botanical work which was being done in different parts of India at that time. The inception of *The Records of the Botanical Survey of India* also owes much to the inspiration of King. Incidentally, till then, five of the seven volumes of the Flora of British India by J.D. Hooker *et al.* had been published.

Regional offices were established at Pune with T. Cooke as Director for western India, at Saharanpur with J.F. Duthie as Director for northern India and at Madras with M.A. Lawson as Director for southern India. In 1911, the office of the Reporter of Economic Products and the Industrial Section of the Indian Museum at Calcutta were also brought under the Survey.

From 1900 to Present: The first half of this period saw the fulfilment of the plan to publish local or provincial floras for the various regions which were without them until then. In course of time, due to lack of funds and manpower, the activities of the regional centres gradually shrunk and almost came to a standstill throughout the country. After the retirement of C.C. Calder in 1939, even the post of Director at Calcutta remained in abeyance and the organisation survived almost in name. The scientific personnel of the Botanical Survey comprised only the Curator, Industrial Section and a Systematic Assistant in the herbarium at Sibpur (Howrah).

Bengal continued to be at the top of taxonomy in India. At the beginning of the century, David Prain published his *Bengal Plants* and made several other important contributions. *The Records of the Botanical Survey of India*, which had put out one volume upto the end of the previous century, brought out the second volume. *The Annals of the Royal Botanic Garden* appeared in six volumes between 1900 and 1950, the latest being on *Dioscorca* by I.H. Burkill and D. Prain.

At the turn of the century, Poona was the centre of the Botanical Survey of India for the western part of the country. In Bombay itself, E. Blatter formed a school of taxonomy: in the company of his students, he collected extensively in different parts of the province. Saxton and Sedgwick made thorough explorations in Gujarat: the latter also concentrated his attention to the plants of Dharvvar. W.A Talbot studied the forest flora of the North Kanara districts. Other contributions include those from R.K. Bhide, G.A. Gammie, G.M. Woodrow, J.I. Thaker, etc.

G.S. Gamble published the first seven parts of his *Flora of Madras:* after his death in 1925, C.E.C. Fischer completed the work in 1936. The Nilgiris and other high hills of South India were explored by P.E. Fyson. P.V. Mayuranathan in 1929 put out *Flowering Plants of Madras City and* E. Barnes in 1938 added a supplement to this book.

The botanists attached to the Forest Research Institute at Dehradun were mainly concerned with the flora of our forests. Their work also extended to the Upper Gangetic Plain and other parts of North India. Here the notable contributors comprise N.L. Bor, J.N. Parker, C.E. Parkinson, M.B. Raizada and others.

H.H. Haines explored much of Chotanagpur as well as Bihar and Orissa. His work was continued by H.E. Mooney.

The flora of Assam was written by a large team of botanists under the energetic supervision of U. Kanjilal. Independently, Bor did good work on the grasses of Assam and Fischer published a number of papers on plant-novelties of the province.

Recent Position: Recent developments in the study of taxonomy are very encouraging. Our Universities are paying attention to this much neglected branch of Botany. Papers have appeared from Allahabad, Bangalore, Bombay, Calcutta, Delhi, Gauhati, Jodhpur, Meerut, Mysore, etc. Teachers and students take great interest in botanical excursions these days.

Several journals have contributed much to the knowledge of the country's flora. Among foreign journals, *Journal of the Arnold Arboretum* and *Kew Bulletin* have published some very interesting papers on Indian plants. Of the Indian journals, *Journal of the Asiatic Society of Bengal, Journal of the* Bombay *Natural History Society, Indian Forester, Indian Forest Records* and *Bulletin of the Botanical Society of Bengal* deserves special mention.

Realising the great dependence of many fields of science on raw material derived from the Plant Kingdom, the Government of India decided to revive the Botanical Survey of India. In 1952, E.K. Janaki Ammal was appointed as Officer on Special Duty to reorganise it. The Central Botanical Laboratory was the first Unit to be set up, with Janaki Ammal as the Director. Four Regional Circles at Shillong (Eastern), Dehradun (Northern), Pune (Western) and Coimbatore (Southern) were also simultaneously started for the study of plant resources in the respective regions. While the Dehradun office of the reorganised Survey had to start a new herbarium, the old herbarium of the Survey at Pune and Coimbatore and the forest herbarium of Assam at Shillong were transferred to the Botanical Survey. Consequently, on shifting of the Central Botanical Laboratory from Allahabad to Calcutta in 1962, a new regional station for Central India was initiated at Allahabad. In 1972, the Andaman and Nicobar Circle at Port Blair and the Arid Zone Circle were created. The Arunachal Pradesh Circle at Itanagar and the Sikkim Himalaya Circle at Gangtok were established in 1977 and 1979 respectively. In 1984, two more circles were established—Deccan Circle at Hyderabad, Andhra Pradesh and High Altitude Circle at Solan, Himachal Pradesh. The National Orchidarium at Shillong maintains a documented live collection of orchids.

A Scientific Programme Implementation and Evaluation Committee advises the Government on the research programmes of the Survey and periodically evaluates its functioning. The main objectives of the Survey can be summarised as follows: (*a*) exploration of unexplored and underexplored regions for new plant resources; (*b*) preparation of floras at the local, district, region and national levels; (*c*) floristic and taxonomic researches; (*d*) preservation of types and other authentic specimens by establishing herbaria and musea and exchange of plant specimens in India and abroad; (*e*) maintaining germplasm of economic, rare and interesting plants for conservation and education; (*f*) study of phytogeography and ecological changes in flora and vegetation; (*g*) investigation of threats to floras and habitats and conservation of ecosystems and (*h*) advising the Government on all matter relating to the utilisation and conservation of natural plant resources of the country.

The headquarters office of the Survey, at Calcutta co-ordinates the activities of all units according to the scientific policies of the Central Government and in consultation with senior scientists of the Survey. It maintains liaison with other research institutions of the country, *e.g.*, Council of Scientific and Industrial Research (CSIR), Indian Council of Agricultural Research (ICAR), Indian Council of Medical Research (ICMR) and Universities. It awards about 50 senior and junior research fellowships for floristic and taxonomic work. One senior scientist of the Survey is posted at the Royal Botanic Garden, Kew, acting as a Liaison Officer for reference work, loans and technical enquiries on identification and nomenclature relating to the proposed flora of Free India.

The Central National Herbarium acts as a repository of all type specimens gathered from any area in India. The Central Botanical Laboratory is engaged in some biosystematic studies involving anatomy, cytology, ethnobotany, palynology and plant physiology. In the Cryptogamic Unit, studies have been undertaken on the fern flora of South India, mosses of the Himalayas, lichens of West Bengal and taxonomic revisions of some genera. A dictionary of fern genera and. a census of threatened ferns and fernallies of India are under preparation. In the Ecology Unit, studies have been undertaken in Kanara, Coorg and coastal regions; some work has been done on leaf-sclereids and eco-anatomy. The Pharmacognosy Unit has studied bark drugs and conducted a detailed survey of *Coptis teeta* in Eastern Himalaya; the latter study helped in regulating the export trade in this medicinal herb. Limited facilities of chemical screening of plants have been organised. The high solasodin content in *Solanum viarum* was found in these laboratories.

Industrial Section at the Indian Museum, Calcutta, undertakes the study of economic plants. About 15,000 exhibits of commercial vegetable products attract researchers, students and tourists. They also constitute reference material for thousands of technical enquiries from drug dealers, exporters, Excise and Customs Department, etc.

The Documentation and Library Units provide bibliographic support in research activities. The Publication Section projects the work of the Survey through the *Bulletin of the Botanical Survey of India, Records of the Botanical Survey of India,* Annual Reports and News Letters. Many special publications are brought out from time to time. Besides over 2,200 research papers published by scientists of the Survey during the last 30 years of its reorganisation, major activities in publication include reprinting of the regional floras of Bombay (T. Cooke), Madras (J.S. Gamble), Bihar and Orissa

(H.H. Haines), Bengal (D. Prain) as well as publishing a Check List of Indian Plants, a dozen fasicles of the Flora of India, Roxburgh's Icones, Aquatic Angiosperms (Subramanyam, 1962), Illustrations of West Himalayan Flowering Plants (Rau, 1963), Flora of Rajasthan (Puri *et al.*, 1964), Phycologia Indica (Srinivasan, 1969), Dictionary of Flowering Plants in India (Santapau and Henry, 1973), High Altitude Flowering Plants (Rau, 1975), Flora of Bashahar Himalayas (Nair, 1977), Flora of Punjab Plains (Nair, 1978), Orchids of India (Rao, 1979), Flora of Jowai (Balakrishnan, 1981), Medicinal Plants (Jain, 1981), etc.

The botanical collections and publications of the Survey have been used not only by scientists of India but also monograph writers all over the world cited these research findings. More than 150 new taxa has been described and several new taxa recorded from India by scientists of the Survey. During the last 30 years, the Survey has organised the collection of more than 300,000 field numbers and the herbaria of the Survey hold over 2 million plant specimens.

The Botanical Survey is intimately connected with the assessment of impact of several developmental or other projects on ecosystems. The Survey provided material to the Government on projects related to Silent Valley, Idukki Valley, Lalpur Dam, Tehri Dam, Sutlej-Beas Link Canal, etc. and on all prospective biosphere reserves, National Parks and game sanctuaries and thus assisted the State and Union Government in many policy decisions.

Reviewing the progress of plant taxonomy in India for the last 50 years, Maheshwari and Kapil (1963) made the observations: "In the past Indian work in plant taxonomy has been purely observational and descriptive based mainly on the examination of external characters. However, Indian taxonomists now realise that substantial help can accure to taxonomy from other branches of botany ... Inspite of the feelings in the minds of younger botanists that taxonomy is a finished area of knowledge and a closed chapter, there is still a great scope for such work in India". The outlook appears promising and the future may unfold infinite possibilities for plant taxonomy in India.

TAXONOMIC EVIDENCE

According to Valentine and Love (1958), there are three stages in taxonomic study: (*a*) *exploratory phase* (involving collection and subsequent classification from a limited range of herbarium specimens); (*b*) *systematic phase* (involving extensive herbarium and field studies of a wide range of material from each taxon) and (*c*) *biosystematic phase* (involving detailed cytological and genetical studies). A fourth stage, *encyclopaedic phase* (where taxonomic data from various disciplines are pooled to construct a good phenetic classification), was appropriately added by Davis and Heywood (1963). Although Turrill (1935) stressed the continuity of these phases, he expressed the same notion differently. He described *alpha-taxonomy* (which is equivalent to the first two of the above four phases and where gross morphology is considered as the framework of plant taxonomy) and *omega-taxonomy* (which is equivalent to a perfect and ultimate phenetic system and based upon all available characters). While alpha taxonomy is the most basic field of biology, the accumulation of all evidence is finally incorporated into omega taxonomy. Though omega taxonomy is not attainable, it is the distant goal that taxonomists should aim at.

1. Embryology: "Although taxonomic characters are traditionally and conveniently morphological, in those plants where developmental processes vary discontinuously, the processes themselves may be regarded as legitimate taxonomic characters, even though the results of the different processes are apparently morphologically identical. For this reason, every detail of embryology is of potential taxonomic value, but its actual value in this respect can only be established after routine examination of the same process in all apparently related species" (Davis, 1966).

According to Maheshwari (1964) and Bhojwani and Bhatnagar (1978), the characters of taxonomic value in delimiting plant groups include the (*a*) anther; (*b*) quadripartition of the microspore mother cell; (*c*) development and organisation of the pollen grain; (*d*) development and structure of the ovule; (*e*) origin and extent of the sporogenous tissue in the ovule: (*f*) megasporogenesis and development of the embryosac; (*g*) form and organisation of the mature embryosac; (*h*) fertilisation; (*i*) endosperm; (*j*) embryo and (*k*) seed coat.

Certain families are specially marked out by their embryology. The very striking features of the Podostemonaceae is the presence of a pseudoembryosac formed by the disintegration of cell of the nucellus below the embryosac. The pollen grains occur in pairs. The ovule is tenuninucellar with two integuments, out of which the outer develops the micropyle and the inner stays greatly behind; the large basal cell of the proembryo forms a prominent haustoria. The embryosac is bisporic. There is no endosperm, the pseudoembryosac serving for embryo nutrition.

In the Onagraceae, the embryosac presents some unique diagnostic features. The megaspore mother cell gives rise to a tetrad of megaspores of which generally the micropylar cell but frequently the chalazal and sometimes both cells grow for a while. In the two nucleate embryosac, the nuclei do not part to the poles but remain together at the micropylar end of the sac. The corresponding chalazal quartet is never formed and there is neither a lower polar nucleus or nor there are antipodal cells.

In most of the angiosperms, the pollen mother cells divide to form tetrads of microspores all of which are functional and produce two or three pollen grains. After all the four microspore nuclei are formed, three of these are cut off on one side and only the fourth develops to form the tube and generative cell. All genera and species of the Cyperaceae, so far studied in Europe, Japan and India, show the presence of a single functional microspore. It is thus possible to identify a member of this family by a microscopic study of its anthers as by other floral characters.

One most important feature of the embryology of the Orchidaceae is the virtual absence of an endosperm. In some members, the polar nuclei do not even unite. If triple fusion takes place, the primary endosperm nucleus degenerates. Rarely one or two divisions occur, the maximum number on record being ten free nuclei in the embryosac of *Vanilla planifolia.* But these nuclei are consumed at a very early stage and the embryosac is entirely occupied by the embryo.

The ovule structure can be taken as an important criterion in grass systematics. In the Panicoideae, the outer integument reaches only up to half the ovule length and a well developed parietal tissue is formed from the epidermis; here the antipodals occupy the chalazal position. On the other hand, in the Pooideae, the outer integument is well developed and the nucellar epidermis does not divide periclinally; here the antipodals always take a lateral position in the embryosac.

Some spectacular examples of embryology proving useful in taxonomy may be cited. That the Empetraceae is a reduced apetalous derivative of the Ericaceae has been established on the basis of the work of Samuelsson (1913) on the embryology of the Geraniales, Sapindales, Bicornes and other groups.

Jorgensen (1923, 1925) observed the existence of a tenuninucellate ovule with a single massive integument and endosperm with micropylar and chalazal haustoria. All these are traits of the Tubiflorae. Soueges (1952) called attention to the similarities in the embryo development in *Callitriche* and *Verbena* so that the assignment of the Callitrichaceae to the Tubiflorales is quite reasonable.

While *Butomus* has an embryosac of the Polygonum type, the other genera included in the Butomaceae and all genera of the Alismaceae are characterised by an Allium type of embryosac (Johri,

1936, 1938a, 1938b). In pollen morphology too, *Butomus* does not fit in with the rest of the Butomaceae. Other features like the absence of laticiferous ducts, the occurrence of sessile linear leaves, persistent petals, anatropous ovules and straight embryo are not encountered in the Butomaceae. This is good enough to justify the transfer of *Butomopsis, Hydrocleis* and *Limnocharis,* retaining only *Butomus* in the Butomaceae.

Despite semi-inferior and bilocular ovary, monosporic tetranaceleate embryosac, lack of endosperm, Solanad type of embryo, well developed haustorial suspensor and extremely unequal cotyledons, *Trapa* was once placed in the Onagraceae; it has since been removed to a new family (Trapaceae) by Pulle (1938) and this view was fully confirmed by Ram (1956).

Airy Shaw's (1948) rash attempt to remove *Sphenoclea* from the Campanulales and place it in a separate family Sphenocleaceae lying between the Phytolaccaceae and Primulaceae has been challenged embryologically. Kausik and Subramanyam (1946) and Subramanyam (1950) indicated that whatever the superficial resemblances in *Sphenoclea* the ovules are unitegmic with an ephemeral nucellus, the endothecium is well developed, the endosperm is cellular with conspicuous haustoria and the embryo is of the Onagrad type with a straight outline. On the contrary, in the Phytolaccaceae, the ovules are bitegmic with a massive and persistent nucellus, the endosperm is free nuclear, an endothecium and endosperm haustoria are absent and the embryo is of the Caryophyllad type with a curved outline. These characters prove beyond doubt that *Sphenoclea* has no link with either the Phytolaccaceae or the Primulaceae. Furthermore, this genus exhibits a few peculiarities of its own (Scutellaria type of endosperm and four-celled micropylar and chalazal haustoria). It would be appropriate to treat it in a separate family adjacent to the Campanulaceae.

In *Cercidiphyllum,* the nucellar epidermis divides periclinally, the fertilised embryosac elongates prominently and the endosperm is cellular (Swamy and Bailey, 1949). Since none of these characters are known in the Hamamelidaceae, the genus is, therefore, assigned to a separate family (Cercidiphyllaceae). However, it is not very clear whether the new family should be included in the Ranales or a new order.

Bremekamp (1953) doubted the validity of the tribe Nelsonioideae, suggesting the transfer of *Elytaria* to the Scrophulariaceae under the Rhinanthese. Comparing the embryology of the genus with the Acanthaceae and Scrophulariaceae, Johri and Singh (1959) pointed out the presence of tenuninucellate ovules, Polygonum type of embryosac and cellular endosperm with chalazal and micropylar haustoria. Hence, embryological information does not warrant the removal of the Nelsonioideae from the Acanthaceae to the Rhinanthoideae of the Scrophulariaceae.

Narang's (1953) investigation of *Slackhousia linariaefolia* has shown that the ovule is bitegmic and tenuninucellate, the embryosac is of the Polygonum type with three multinucleate antipodal cells and the endosperm is free nuclear. Since none of these characters are recapitulated in the Rhamnaceae, Scrophulariaceae or Lobeliaceae, there is no serious objection to the inclusion of the Stackhousiaceae in the Celastrales.

Although *Pentaphragma* was removed from the Campanulaceae by Airy Shaw (1954) and placed under an unigeneric family (Pentaphragmataceae), Kapil and Vijayaraghavan (1965) felt that it correctly belongs to the first family on embryological grounds, *i.e.,* unitegmic tenuninucellate ovules, extramicropylar embryosac, cellular endosperm and Solanad type of embryo.

Studies on the embryology of *Lemna, Wolffia* and *Arisaema* confirms older views of an intimate relationship of the duckweeds with the aroids (Maheshwari, 1954, 1956a, 1956b, 1958; Maheshwari and Khanna, 1957). The presence of an integumentary operculum, cellular endosperm and the form and development of the proembryo are particularly significant in reaching this conclusion. There is greater resemblance between the embryos of the Lemnaceae and Araceae, viz. a cylindrical terminal cotyledon, vertically directed plumular apex which is completely enclosed by the hypocotyl and cotyledonary sheath except for a small pore near the hypocotyl.

The studies of Johri and associates (1957) on the Loranthaceae show that the Loranthoideae is embryologically different from the Viscoideae as regards mode of development of embryosac, endosperm, embryo and in the location of the viscid zone of the fruit and that the subfamilies should be raised to the status of families.

Due to the presence of a peculiar coenocytic phase in embryogenesis, large and elongated generative cell, long and narrow embryosac and large ovules borne on placental projections, *Paeonia* offers a contrast with the Ranunculaceae and should be accommodated in a distinct family (Paeoniaceae) as done by Yakovlev and Yoffe (1957).

The similarities in the development of the gametophytes, endosperm, embryo and seed speak for a close relationship between the Orobanchaceae and Scrophulariaceae; the former can easily be derived from the latter through forms like *Striga orobanchoides* (Tiagi, 1958).

There are some distinctive characters in *Cassytha* not shared by other members of the Lauraceae, viz. secretory type of anther tapeturn, formation of multiple embryosacs in an ovule, extension of some embryosacs beyond the nucellus and cellular endosperm (Sastri, 1962). These characters do not appear to be so remarkable as to demand the removal of the genus from the Lauraceae as suggested by Lindley (1853).

The Cactaceae agrees with the rest of the Centrospermales in having the following embryological characters: (*a*) a glandular anther tapetum whose cells become two- to four-nucleate; (*b*) microspore mother cells in which two meiotic divisions are succeeded by a simultaneous quadripartition into the microspores; (*c*) trinucleate pollen-grains; (*d*) campylotropous ovules with strongly curved funiculi and massive nucelli; (*e*) a hypodermal archesporial cell which cuts off a wall cell; (*f*) a micropyle formed by the swollen tips of the inner integument which protrude out iind approach the funiculus; (*g*) formation of a nucellar cap originating from periclinal division of cells of the nucellar epidermis; (*h*) functioning of the chalazal megaspore of the tetrad; (*i*) formation of a monosporic eight-nucleate embryosac; (*j*) functioning of the perisperm as the main storage region; (*k*) disappearance of most of the endosperm in the mature seed generally leaving merely a single-layered cap over the radicle (Maheshwari, 1964).

In *Turnera,* the ovules are anatropous, bitegmic and crassinucellate (Kaur, 1970). Here the pollen tube is persistent, endosperm is nuclear, embryo is of the Onagrad type and seeds arillate. All these features are lacking in the Loasaceae to which the genus is linked.

The Crassulaceae shows close affinity with Saxifragaceae (Subramanyam, 1970). *Crassula aquatica,* which contains the most reduced endosperm, has a mode of habit similar to the Podostemonaceae. This taxon forms a connecting link between the Crassulaceae and Podostemonaceae.

The tetrasporic eight-nucleate embryosac of *Adoxa* offers a marked contrast with the Ranunculaceae, Saxifragaceae and Araliaceae and supports an alliance between the Adoxaceae and Caprifoliaceae.

Maheshwari (1958) emphasised "that while the embryologist lays no claim to erect a phylogenetic system of his own, embryological data need to be considered along with information from other sources in order to approach a natural system of classification".

2. Anatomy: It is believed that internal structure of plants can provide more information than external morphology. However, anatomical features cannot by themselves constitute the basis of classification but can be used with advantage to supplement those morphological attributes on which classification has been built. The authors of *Die Pflanzenfamilien* were perhaps the first to make use of anatomical data for overall classification on a grand scale.

The study of woody anatomy is of value in drawing taxonomic conclusions. This prompted the allocation of *Amborella, Tetracentron* and *Trochodendron,* to their respective independent families and the institution of a new taxon, the Degeneriaceae (Bailey and Smith, 1942) and of *Brassiantha,* a new genus of the Hippocrateaceae (Smith and Bailey, 1941); the retention of *Sarcandra* in the Chloranthaceae (Swamy, 1953a); the transfer of *Idenburgia* from the Monimiaceae and its amalgamation with *Nouhuysia* (Bailey and Swamy, 1953) and of *Ascarina* with *Paracryphia* (Swamy, 1953b).

Metcalfe (1968) cited several interesting points from current developments in systematic plant anatomy. The Rapateaceac resembles the Xyridaceae in some respects, but differs in having silicabodies, tannin cells, different kinds of chlorenchyma cell and root structure. The Xyridaceae is a very heterogeneous family, having genera which are individually distinctive in their anatomy and having at the specific level characters of diagnostic value. Though the morphology of pollen grains is described to be similar, the anatomy of *Aphyllanthes* (generally ascribed to the Liliaceae) is very unlike that of the Eriocaulaceae. The Cannaceae and Marantaceae occupy rather isolated positions within the Scitaminae, yet the anatomical structure of the order is not so diverse as other monocot orders. The Cyperaceae is much more structurally diverse than is generally conceived and is not very closely related to the Gramineae. The Juncaceae is often like the Cyperaceae without silica bodies. Anatomically, there is ample justification for excluding *Cartonema* from the Commelinaceae and including the genus in a separate family, the Cartonemaceae; for erecting the Hanguanaceae for *Hanguana,* the Mayacaceae for *Mayaca* and the Heliconiaceae for *Heliconia;* for removing *Costus* from the Zingiberaceae and of raising *Orchidantha* to the rank of a family, the Lowiaceae.

Although anatomical evidence has been utilised for ascertaining the relationship of genera and families, very few attempts have been made to use them at the generic or subgeneric level. Govindarajalu and Swamy (1956) studied the petiolar anatomy of *Alangium* and noticed ranges of variability in vascular structure which are characteristic of each section, thereby showing the advantage of such investigations. In the Combretaceae, trichome anatomy is of value in the circumscription of the family to the separation of species and even varieties (Stace, 1973). In the Gramineae, the leaf epidermis and leaf cross-sectional anatomy provide significant taxonomic information (Ellis, 1976). Such features as the types of papillae and trichomes, form and distribution of silica bodies, differentiation of short cells and long cells, disposition of sclerenchyma and form and arrangement of vascular bundles are immensely helpful in the modern reclassification of grasses at the levels.

There are 31 known types of arrangement of subsidiary cells in the mature stomatal complex of vascular plants (Dilcher, 1974). At higher taxonomic levels, these distinct patterns play a big part in classification. Within the Combretaceae, the stomata arc anomocytic in the Combretoideae and paracytic in the Strephonematoideae. While the stomata are anomocytic in the Scrophulariaceae, they are diacytic in the closely related Acanthaceae. However, stomatal traits are not always reliable, *e.g.,* anisocytic, anomocytic, diacytic and paracytic exist together in the same leaf of *Lippia nodiflora* (Verbenaceae) and cotyledons of *Saintpaulia* and *Streptocarpus,* both belonging to the Gesneriaceae, contain anomocytic stomata and the mature organs anisocytic stomata.

The role of floral anatomy has been emphasised in solving taxonomic problems. According to Puri (1958), the gynoecium structure in the Capparidaceae. Cruciferae, Moringaceae and Papaveraceae is essentially alike. Though the placentation is parietal, the placental strands are inverted and occur on the inner side of the secondary marginal bundles which have normal orientation. This supports the view that the Moringaceae should be included in the Rhoeadales and that the parietal placentaion has been derived from the axile condition. As a similar phenomenon takes place in the Cucurbitaceae and Passifloraceae, it lends support to the suggested alliance between the Moringaceae and Passifloraceae.

From a study of vascular anatomy of four species of the Onagraceae, Roy (1949) observed significant variations in the vascular supply of the ovary of *Trapa,* justifying its removal to a separate family (Trapaceae). The separation of *Hydrocotyle asiatica* as *Centella asiatica* has been confirmed on anatomical grounds by Mittal who found that in species of *Hydrocotyle* ovular traces are derived from placental stands which are the fusion product of ventral traces and in *Centella* the two ovules receive their vascular supply from alternating bundles in each carpel. Gupta (1959) indicated some differences in the vasculature of calyx, corolla and placental region of *Campanula, Lobelia* and *Sphenoclea* as being of minor importance and not supporting the transfer of the last mentioned genus from the Campanulaceae.

The distribution of crystals in a given taxon is very specific and thus taxonomically very useful. According io Nair *et al.* (1977), the size and shape of crystals in the abaxial epidermis of the leaves of *Myristica fragrans* can enable us to differentiate between male snd female trees even at the sapling phase. In general, simple crystals dominate in male trees and compound ones (druses) in female trees.

In some cases, fruit wall and seed-coat anatomy have come to the rescue of taxonomists. The separation of the two genera *Ammania* and *Rotala* has puzzled taxonomists for a long time. While *Ammania* possess irregularly ruptured fruits and *Rotala* valvular capsules, these features are not evident in herbarium specimens. Whereas the cells of two main layers of pericarp are almost alike in *Ammania,* the inner layer of cells are linear, strongly lignified and transverse to the outer in *Rotala.* Chowdhury and Buth (1970) produced a key based on seed-coat anatomy to identify some of the important pulses of India.

Nodal anatomy is an important aspect of taxonomy. According to Sinnot (1914), the trilacunar node is the primitive type in angiosperms and the unilacunar node developed from it "by the loss of the two lateral gaps together with their respective traces, or by the approximation of the lateral traces to the median bundle to form a bundle composed of three traces, which is associated, therefore, with a single gap". He also stated that the multilacunar node is formed by the addition of new gaps and traces. His assumption of the primitive nature of the trilacunar node has been refuted by recent researches. It has been shown in many dicotyledons that there are unilacunar nodes with double leaf traces and that many of them belong to the primitive groups of the Ranales and the Chenopodiaceae. Bailey (1956) noted in many dicots that the vascular supply to the cotyledons consists of two leaf traces which arise from an unilacunar node. He further asserted that (*a*) the unilacunar node of certain genera of the Ranales is primitive and has not altered in course of evolution; (*b*) in certain genera of the Anacardiaceae, Leguminosae and other families, the unilacunar node has apparently been derived, by way of reduction, from a trilacunar node and (*c*) in the Epacridaceae and Chloranthaceae the trilacunar and multilacunar nodes have arisen from unilacunar node. That the evolutionary development from an unilacunar node having two leaf traces to other kinds of unilacunar nodes having one, three or more traces may occur in a single family has been demonstrated in the Chenopodiaceae. by Bisalputra (1962) and Fahn and Broido (1963). The mode of nodal vasculature is not uniform even in the same individual plant (Sehgal and Paliwal, 1974).

The foregoing makes it clear that anatomical studies of organs of flowering plants can serve as an integral part of taxonomy. Swamy and Bailey (1949) remarked that "before attempting to arrange surviving Angiosperms in phylogenetic series, it is essential to obtain reliable evidence regarding salient trends of evolutionary specialisation in the various organs and internal structures of these plants. Such evidence can be acquired only by comprehensive and time consuming investigations of the dicotyledons and monocotyledons as a *whole*".

3. Palaeobotany: Fossil record frequently furnish us with information concerning the evolutionary history of present day plants. Lam (1959) commented that no eveloutionary doctrine would be perfect without palaeobotanic evidence. As fossils of land plants are scattered and poor, difficulties often arise in the interpretation of these data.

The appearance of angiosperms changed the entire basis of fossil plant classification. A majority of fossil angiosperms, particularly leaves, seem to belong to extant families and to be related to recent genera. In some instances, where such affinity is unknown, it may be either due to a lack of knowledge of living plant structure or to poor state of preservation resulting in the absence of diagnostic features. Angiosperm families appeared and disappeared during the known course of their history and plant parts have been found in detached condition. But it is not possible to accumulate sufficient characters for recognising new families and none is known for a fossil one.

While not as numerous as leaves, fossil fruits and seeds have been found. Some of these retain anatomical structure in detail and others are preserved merely as frustrating casts or impressions. Chandler (1923) contributed to our botanical knowledge by a detailed study of the fossil and living members of *Stratiotes.* Elias (1942) detected morphological details, previously not reported for the Supeae, in fossil *Stipidium,* later seen in living *Stipa.* As the classification of grasses is based to a great extent on seed characters, Elias (1946) was fortunate enough to classify his well preserved Miocene prairie grass seeds from the high plains of northern mid America in a similar way.

It is well-known that floral structure constitutes the chief basis for finding out relationships and evolutionary trends in the classification of angiosperms. Due to their short life span and delicate nature, flowers are hardly observed as fossils. When they are fossilised, they are often crushed and important taxonomic details obscured. If they are well preserved, the chances of correlation with parent plants are few and conclusions regarding affinity remain uncertain. Conwentz (1886) placed perfectly preserved flowers in Oligocene amber from the Baltic coast near Danzig to recent genera. Shukla (1944) referred a number of well preserved petrified flowers with associated fruits from the Eocene Deccan Intertrappean Series of India to the Lythraceae, having relationship with *Decodon.*

Petrified wood is still the best known plant fossil. But this sort of investigation has been hampered for a long time by the absence of knowledge about living plant structure. Spurred on by Metcalfe and Chalk's key to living wood anatomy, the proper study of fossil angiosperm wood has started and progressed considerably.

Chesters (1964) remarked that "the palaeobotanist has been unable to study his angiosperms in the same manner and on the same basis as the botanist he is unable either to contribute to the living classification or, as yet, to create one of his own. The resemblance is so close that he is mostly able to refer his organs to living families and even genera and indeed frequently to include them in these groups under recent names. No solution can be offered to any of the many problems of angiosperm origin and evolution which worry botanists and have engendered such a large body of diverse views and opinions. He does not know, though many have attempted to guess, when, where, or how the angiosperms originated; he cannot tell whether woody or herbaceous, tropical or temperate forms arose first (although the former views are supported by a growing body of opinion); whether the monocotyledons or the dicotyledons are older; or which families are the most primitive; or even whether the angiosperms had a monophyletic or polyphyletic origin".

4. Palynology: The significance of pollen attributes in taxonomy has been realised during the last three decades (Erdtman, 1952; Wagenitz, 1955; Stix, 1960; Raj, 1961; Chanda, 1972; Nair, 1974). The outer wall of pollen grains is endowed with unique structural traits which are broadly categorised in order of their importance in phylogeny into the apertures, exine ornamentation, exine strata, shape and size. The pollen classification is based on number position character analysis, called NPC system. As a rule, the number of apertures is only one in the cryptogams and gymnosperms whenever the grains are aperturate, but the position is distal in the latter and proximal in the former and the apertures are non-trichotomous (*e.g.*, monolete) or trichotomous (*e.g.* trilete) in character. The combination of characters has the widest variation in the angiosperms. Accordingly, the pollen types are designated as sulcate (distaliaperturate), sinuate (proximaliaperturate), foraminate (globaliaperturate), etc. Twentyfive such pollen types were established by Erdtman and Straka (1961). The number of the chief aperture types is probably somewhat larger and presumably attains about 40. The aperture types are purely morphological which should be distinguished from the taxonomical types whose number is indefinitely large.

A general analysis of the apertural conditions in the plant kingdom reveals that apertures are ill developed (primorphous) in thallophytes. In the archegoniates (comprising bryophytes, pteridophytes and gymnosperms), pollen grains are trimorphous and in the angiosperms polymorphous. The above morphological situation provided a logical base for classifying the plant kingdom into the *Primorphosporatae* (Syn. Thallophyta), *Trimorphosporatae* (Syn. Archegoniatae) and *Polymorphosporatae* (Syn. Angiospermae) by Nair (1974). This analysis has thrown light on the phylogeny and evolution of the primitive angiosperms. It has been proposed that the primitive angiosperms consists of three stocks, viz. the monocot stock, magnolian dicot stock (Magnoliidae of Cronquist except for the Nelumbonaceae) and ranalian dicot stock (Ranunculiidae and other subclasses of Cronquist). The monocot and magnolian stocks are trimorphous/dominantly trimorphous and the ranalian stock polymorphous. By further analysis, it has been observed that the primitive taxa of the stock evolved at two levels—one represented by the ranalian taxa (Ranunculiidae of Cronquist) and the other by the rosalean taxa (Hamamelidae and Rosidae).

The apertural conditions of pollen grains have been looked upon as strong characters in solving taxonomic problems. It has been recommended that taxa with the same general NPC formula be grouped together and those showing aberrant NPC separately. For example, the Parietales of Engler-Diels has been probed by means of "palynological compass needle'. The NPC formula for the order is 345, *i.e.* pollen grains are 3-tremes (N_3), zonotreme (P_4) and colporate (C_5). A significant variant is the Canellaceae where the pollen has one distal colpus; here NPC is 133. Lesser divergences occur in the Frankeniaceae and Tamaricaceae (NPC, 343). Those families which deviate from the NPC formula include the Dilleniaceae, Dipterocarpaceae, Elatinaceae, Eucryphiaceae, Frankeniaceae and Tamaricaceae apart from the Canellaceae; these families have been assigned different status by Hutchinson (1959).

Plants may be unipalynous (basically similar pollen in any one taxonomic alliance at the level of family, genera or species) or multipalynous (basically dissimilar pollen). While the Gramineae is unipalynous with monoporate aperture and somewhat smooth exine surface, the Euphorbiaceae is multipalynous with various aperture forms. Though light microscopy provide the requisite data in multi-palynous taxa, its application is limited in unipalynous taxa. In the latter case, SEM (scanning electron microscopy) pictures have been immensely helpful in the study of cultivated grasses.

When the problem of Acanthaceae—Thunbergioideae is evaluated on the basis of palynological treatment, the Thunbergioideae should be given a family status. This idea arose in view of the anomotreme pollen grains studded with apertures. Such apertural characters and sporoderm nature are not found in other subfamilies of the Acanthaceae. Thus, palynological data vigorously support the elevation of the Thunbergioideae to a family rank.

It is gratifying to note that Zimmermann and Cronquist have made exclusive use of pollen features in constructing the taxonomic fabric of angiosperm classification. Erdtman (1963) listed 105 families of particular interest to support the correlation of palynology with taxonomy. Following palynological analysis, genera and section could be segregated. In fact, keys to genera and species have been constructed for use in general taxonomy.

According to Bailey and Nast (1943), "there are families of dicotyledons in which the pollen is of very considerable taxonomic significance not only in the differentiation of subfamilies and tribes but also of genera and species". A few examples will illustrate the statement. The Caryophyllales is recognised by centrospermous type of pollen with a spinulose and punctate-perforate tectum. The Malvaceae and the Compositae contain typically spinulose exine, the Plumbaginaceae verrucate pollen and the Gramineae smooth sulcate ones. Light microscopy and scanning electron microscopy have provided palynological differences correlated with features of the inflorescence, flowers and fruits between various sections of the genus *Cordia* which is the "largest perplexing group" in the Boraginaceae (Nowicke and Rideway, 1973). The sect. *Cordia* has paniculate inflorescence with infundibuliform corolla and tricolpate pollen grains with striatoreticulate exine. The sect. *varronia* has spicate globose glomerate inflorescence and small flowers with shallowly lobed corolla and triporate grains with a reticulate exine, whereas sect, *myxa* and sect. *Gerascanthus* possess paniculate inflorescence with salverform corolla and tricolpate or tricolporate grains with spinulosc exine. In view of these findings, distinct generic status has been assigned to the three groups.

Exine patterns have been known to be worthwhile in recognising the different species of a genus, e.g. *Bauhinia.* The pollen grains are pilate in *B. acuminata,* striate in *B. krugii,* spinulate in *B. malabarica,* reticulate tuberculate in *B. purpurea,* reticulate in *B. racemosa* and verrucate in *B. retusa.*

Even at the level of varieties and cultivars, pollen morphology has been found to be useful. For example, the variety *alba* of *Caltha palustris* has pantoporate grains and the variety *normalis* tricolpate grains. In segregating the cultivars of vegetable crops, it is not only the aperture which has a bearing, but the totality of characters including exine sculpturing and size is needed to be analysed statistically in order to arrive at reasonable conclusions.

The Cyperaceae and Juncaceae are closely related in view of the peculiar mode of development, of the pollen grains. In the Cyperaceae, three of the four nuclear products of meiosis abort and only one pollen grain is formed from each pollen mother cell. In the Juncaceae, pollen formation is delayed till the pollen mother has eight nuclei and the pollen shed in tight tetrads.

The number of nuclei in the pollen at the time of dispersal has been used by taxonomists (Brewbaker, 1967). The angiosperm pollen is either binucleate or trinucleate according to the precocity of division of the generative nucleus. The binucleate condition is considered as more primitive than the trinucleate. In the Centrospermae, the pollen is uniformly trinucleate. The monocot (Liliaceae) is binucleate, the apetalous and polypetalous dicots are binucleate and gamopetalous members trinucleate. Some families are heterogeneous despite the presence of two or three nuclei in pollen grains as distinctive of a particular family. Both two- and three-nucleate types occur in the same genus, *e.g., Burmannia, Ipomoea* and *Lobelia.*

The nucleate condition of the pollen grains is correlated with incompatibility systems. It has been observed that dicots with two-nucleate pollen have gametophytic type of incompatibility and those with trinucleate pollen have sporophytic type. In the Lamiaceae, two-nucleate pollen grains are usually tricolpate and three-nucleate ones are hexacolpate (Kooiman, 1972). There is also a correlation between nucleate character and stigmatic categories: wet stigmas are restricted to two-nucleate types and dry ones to three-nucleate ones.

Despite great strides, one should exercise extreme care in evaluating palynological criteria particularly when discussing relationships (Erdtman, 1943). Kuprianova (1974) aptly stated "should the definition of parallelism and convergence be based on some organs say, a morphological characteristics of pollen grains and spores, then no great decisive significance should be attached to the degree of affinity".

5. Cytology: "Any modern treatise on evolution and classification cannot exclude the use of the cytological tool. This does not mean that cytology is always an infallible criterion in determining the correct status of any taxon. Though certain cytologists seem to overestimate the value of cytology in taxonomic considerations, it has got its own limitations. "The cytological method when followed in all its details is a more trustworthy guide than any other single character in assessing relationships...it is seen that the various attributes of chromosomes like number, morphology, size, behaviour in crosses and aberrations in reproduction are all important" (Abraham, 1958).

Chromosome numbers recorded for angiosperms vary from four in the somatic cells of *Haplopappus gracilis* to 265 in those of *Poa littorosa.* A polyploid series is exemplified by the different species of *Piper* which show chromosome numbers in multiples of 26, like $2n = 52$ in *P. nigrum,* $2n = 78$ in *P. betle* and $2n = 104$ in wild species of *Piper* (Mathew, 1958*a*). While *Morus nigra* has the highest chromosome number ($2n = 308$), in *M. cathyana* there are forms with $2n = 56$, 84, 112 (Janaki Ammal, 1960). *Solanum nigrum* is a good example of the existence of a species complex, comprising diploid ($n = 12$), tetraploid ($n = 24$) and hexaploid ($n = 36$) forms (Tandon and Rao, 1966).

The classification of many families has been verified or refuted by information on chromosome number and morphology. Perhaps its best known use is in the monocotyledonous genera *Agave, Yucca* and relatives. While *Agave* and some others possess flowers with an inferior ovary, *Yucca* and others have a superior ovary. It is for this reason that the former was previously put in the Amaryllidaceae and the latter in the Liliaceae. Due to the overall similarity in these genera, they are now generally separated from their respective families and placed in the Agavaceae. This classification is supported by cytological evidence, *i.e.*, the Agavaceae is characterised by a very distinct bimodal karyotype comprising 25 small chromosomes and 5 large ones (McKelvey and Sax, 1933).

The traditional way of segregating the Ranunculaceae has been to place into the tribe Helleboreae those genera having few carpels which contain more than one seed per carpel and the Anemoneae and Clematideae those genera possessing many of several carpels which are one-seeded and indehiscent. However, the family can be conveniently divided into two tribes by means of karyotype morphology. One group of genera—*Aquilegia, Coptis, Hydrastis, Isopyrum* and *Zanthorhiza* of the Helleboreae alongside *Anemonella, Thalictrum* and *Trautvetteria* of the Anemoneae—has small chromosomes with conspicuous heterochromatic chromocentres (Gregory, 1941). The rest of the genera, including nine assigned to the Helleboreae and five to the Anemoneae and Clematideae, have large chromosomes and irregularly distributed heterochromatic regions while the first group is dominated by the basic number $x = 7$, the second group by $x = 8$.

Under the Proteaceae (Johnson and Briggs, 1963), the comparatively homogeneous subfamily Grevilleoideae has $n = 14$ small chromosomes (with $n = 10, 11, 15$ as derived karyotypes) and the heterogeneous Proteoideae $n = 7$ large chromosomes or $n = 14$ small chromosomes (with $n = 14, 13, 12, 10, 5$). However, the cytologically primitive genus *Placospernum* contain large chromosomes and $n = 7$, quite remarkable from both.

The bambosoid grass with small and many chromosomes, chloridoid grasses with small and few chromosomes, festucoid grasses with large chromosomes and panicoid grasses with medium chromosomes can be distinguished from each other (Stebbins, 1971).

In palms, two basic series exist with correspond to leaf types (Mahabale and Cheenaveeraiah, 1953). While species with $n = 16$ chromosomes possess pinnate or pinnatisect leaves, those with $n = 18$ chromosomes have palmate or palmatisect leaves (with only a few exceptions in each series). At that time, the chromosome number were known in about 60 species and 32 genera of palms. Today, the chromosome counts are available for 253 species distributed among 96 genera. However, chromosomal data have been derived from about 1/10 of the palm species. Even then, the earlier view expressed still holds good for the majority of species with pinnate leaves having $n = 16$ chromosomes and palmate leaves having $n = 18$ chromosomes. The variation in chromosome number other than $n = 16$ or 18 is expected to have taken place in the course of evolution.

There is a taxonomic confusion between the genera *Nervilia* and *Pogonia* of the Orchidaceae. Some of the species described by Santapau and Kapadia (1961) under *Nervilia* were clubbed under *Pogonia* by J.D. Hooker (1890) and Cooke (1908). Hooker distinguished between the old world and new world species of *Pogonia,* stating that the former flowered before leafing and the latter bearing leaves on flowering stems. From the work of Cheenaveeraiah and Jorapur (1966), it is now clear that the old world species are distinct and to be treated as belonging to *Nervilia.* The high chromosome in species of *Nervilia,* viz. *N. aragoana* ($2n = 72$), *N. discolor* ($2n = 72$), *N. infundibulifolia* ($2n = 54$), *N. monantha* ($2n = 144$) and *N. plicata* ($2n = 108$) and their smaller chromosomes are in striking contrast with few ($2n = 18$) and large chromosomes of the species of *Pogonia.* The chromosomal information thus indicate that *Nervilia* is distinct from *Pogonia* and the species cannot be mixed up between these genera.

Studies on the Menispermaceae indicate that cytologically *Cyclea* and *Cissampelos* are based on 12 and *Stephania* on 13 (Mathew, 1958b). It is further observed that the number $n = 13$ is distinctive of the Cocculeae which, in turn, shows smaller chromosomes. *Stephania* also has smaller chromosomes, but *Cyclea* and *Cissampelos* have chromosomes of large size. Whereas the inclusion of *Cyclea* and *Cissampelos* in the same tribe is supported, *Stephania* appears to be closer to the Cocculeae than to the Cissampelideae and may be transferred to the former.

Large chromosomes, low chromosome number and symmetrical karyotype represent a primitive status, while small chromosomes, high number and extreme asymmetry indicate advancement (Sharma, 1964). The application of these principles has yielded interesting results in the Pandanales, Alismataceae, Hydrocharitaceae, Liliaceae, Amaryllidaceae and Dioscoreaceae. Within the Geraniales, chromosome study does not justify the exclusion of the Balsaminaceae as proposed by Engler. The transfer of the-genus *Averrhoa* to a different family has been justified, but its inclusion under the Rutales can be ruled out. Likewise, the creation of the Basellaceae including *Basella* taken out by certain taxonomists from the Chenopodiceae has been supported. Similar reorientation has been suggested in the Ranales, Leguminosae, Rubiaceae and Compositae.

Although indicative of evolutionary processes, chromosomal diversity can be unhelpful taxonomically. *Colchicum,* for example, is a genus which is ecologically and morphologically homogeneous where a great variety of karyotypes is encountered (Feinbrun, 1958). In the Meliaceae, the Swietenoideae is morphologically uniform and karyologically heterogeneous (Styles and Vosa, 1971). While most genera in the subfamily appear to be palaeopolyploids, this fact plus the palaeotetraploid origin of the Genisteae and Rosoideae point out that it is not worthwhile to dismiss polyploidy as sheer 'evolutionary noise' despite progressive evolution of the angiosperms being largely at the diploid level and polyploidy occurring in 43 per cent of flowering plants.

The behaviour of chromosomes in crosses is a reliable factor in assessing relationships. Pathak (1940) made a careful karyotypic analysis of various species of *Aegilops, Secale* and *Triticum,* suggesting that the hexaploid *T. spelta* and *T. vulgare* were probably derived through hybridisation between a tetraploid wheat and *A. squarrosa.* Roy (1959) carried out detailed genome analysis of *Aegilops longissima* and *A. sharonensis.* On the basis of chromosome pairing and fertility of F_1 hybrids and the derived amphidiploidy, he thought that the two species are closely related. From a study of the karyotypes of species of *Aegilops,* Cheenaveeraiah (1962) postulated that the section *Sitopsis* should be shifted from *Aegilops* to *Triticum* or given the rank of a new genus. In interspecific crosses in *Epilobium,* chromosomal arrangements differing by one or two translocations, identify species primarily from Africa, Australasia and Eurasia (AA arrangement), North and South America (BB) and the circumboreal Alpinae (CC), thus offering a glimpse into the phytogeography and affinity of this seemingly homogeneous large group of species (Seavey and Raven, 1977). Though the species of *Anacyclus* possess similar karyotypes, these can be distinguished on the basis of Giemsa banding patterns (Ehrendorfer *et at.,* 1977).

Cheenaveeraiah and Habib (1966) studied the karyotypes in several collections involving different species of *Capsicum,* all with $2n = 24$ chromosomes. For convenience, they are categorised as A, B, C and D, representing chromosomes with satellites, median centromeres, submedian centromeres and subterminal centromeres respectively. Four features are common to most of the species: (*a*) chromosomes are medium sized; (*b*) distinctly very large or very small chromosomes not noticeable; (c) two pairs of chromosomes with median centromeres and (*d*) chromosomes with subterminal centromere, when present, are never more than one pair. Hence, the karyotypes of *Capsicum* show both uniformity and diversity and suggest a need to lump all the species into a single species.

Khoshoo (1960) engaged himself in the biosystematics of *Sisymbrium irio,* a polytypic species, where plants of various sizes are available, differing in the size and shape of leaves, flowers and fruits as well as branching pattern. According to him, variation and evolution within this species is primarily due to hybridisation and polyploidy and secondarily to gene mutation and structural changes in chromosomes.

The breeding behaviour in different species of non-tuberiferous *Solanum* has revealed that though certain species are crossable, genetic barriers exist between them resulting in hybrid invariability and considered as good species. Contrary to this, *Solanum indicum* is a highly variable species associated with a wide range of adaptation to different ecological conditions from high altitudes to low lands and evergreen forests to scrubby jungles and dry open areas (Cheenaveeraiah and Krishnappa, 1966). Populations from different localities are not without karyotypic and morphological differences in them. The chromosome number has been found to be constant with $2n = 24$ except for the occurrence of accessory chromosomes in some populations. Morphological variations mainly involve height, hair, leaf, stem, spine, inflorescence and flower. These collections can be grouped into four types which, when grown in uniform conditions, maintained their distinctiveness. At present, Type III (found in Nilgiri Hills and never on low lands) is distinct and resembles the taxonomically recognisable *S. indicum* var. *multiflora.* Type IV (adapted to low altitudes and having less rainfall) has been treated as belonging to var. *multiflora* which is not sound. Type I (found in the Western Ghats having heavy rainfall) and Type II (found in the Tirupati Hills where the forest is dry deciduous) are also distinct but showing certain intermediate characters). Therefore, *S. indicum* is a complex species with variants taxonomically recognisable and deserving to be treated as varieties.

Any chromosomal change, not causing sterility of the gametes, may result in new forms. Sharma (1956) made the interesting observation that in some vegetatively propagated members of the Araceae, Amaryllidaceae and Dioscoreaceae the same root tip show the presence of cells with chromosomes of varying size and number. He put forward the idea that changes in karyotypes of somatic tissue play a distinct role in evolution. By resorting to such a method, many species have evolved and specialised in divergent directions.

"Despite a healthy scepticism of the value of the chromosomes in easing taxonomic decision making, the annual outpouring of cytological data continues to illuminate the taxonomic study of many groups. With increasing knowledge the chromosomes will further reflect the manifold ways in which evolution has proceeded and will thus become an increasing refined tool, among the many available to the discriminating taxonomist" (Moore, 1978).

Ecology: Valentine (1978) commented that "ecological criteria are of comparatively little direct importance in taxonomy. The primary criterion, in classical taxonomy at least, has been and will doubtless remain morphological, though ecological criteria at the infraspecific level cannot be neglected. On the other hand, from an evolutionary point of view, habitat and environmental factors must play a vital part; and evolution is the fundamental process which shapes taxonomic differentiation". In flowering plants, tolerance and plasticity are widespread. The tolerance of a plant population is determined by its ability to survive and reproduce upon exposure to a range of environmental factors. The tolerance is greater when the range is wider. On the other hand, plasticity is ascertained by the degree to which the appearance of plants vary in moving from one set of factors to the other.

Ecological studies can help in determining the taxonomic status of a species. On the basis of developmental morphology, culture experiments and other analytical data, the two species of *Lindenbergia*, *L. polyantha* and *L. urticaefolia*, have been demonstrated to be ecotypes of the same species, the former being a calcicolous or a miniature form of the latter (Misra and Siva Rao, 1948). The fodder grasses, *Dichanthium annulatum* and *D. caricosum* (Pandeya, 1953), though similar in appearance, are variable in shape and size of branches, leaf and spike, depending on the environment. However, *D. caricosum* with one spike per raceme and thriving in a dry sandy place may grow into a form resembling *D. caricosum* var. *mollicomus* if the water supply is more favourable. *Euphorbia thymifolia* exhibits ecotypic differentiation in response to the calcium content of the soil (Ramakrishnan, 1960): the red ecotype is a facultative calcicole and the green one an obligate calcifuge. Pursuing the matter further, the red plant shows three physiologically distinct ecotypes. Depending on the soil exchangeable calcium, *Boerhaavia diffusa, Gomphrena celosioides* and *Mecardonia dianthera* will give rise to appropriate calcicolous and calcifugous ecotypes. Red and green ecotypes have been distinguished in *Trianthema decandra* and *T. portulacastrum. Portulaca oleracea* bears two ecotypes, *Amaranthus spinosus* three and *Cenchrus ciliaris* at least 11. *Euphorbia hirta* bears three forms: one upright form, growing in protected habitats, is a distinct ecotype from the other prostrate one; the latter also appears as a compact form of the footpath and a diffuse one of the grazed lands which are interconvertible in reciprocal transplants (Misra, 1967). Moreover, *Echinochloa colonum,* growing in and around ponds, is shown to possess a gradient of ecotypes corresponding to the moisture potential from dry to subaqueous. Similarly, *Eclipta alba, Erigeron linifolius, Evolvulus alsinoides* and *E. nummularius* contain large habitat variations. Edaphic ecotypes are noted in *Heteropogon contortus* and *Setaria glauca,* photoperiodic ecotypes in *Ageratum conyzoides* and *Xanthium strumarium,* latitudinal ecotypes in *Anagallis arvensis* and *Cassia tora* and moisture and shade ecotypes in *Leucas aspera* (Pandeya, 1967).

It is not suggested "that taxonomists should start classifying plants on the basis of their phenotypic variability because of its importance in evolution, but rather than they should be a little more critical in their description of plasticity and that where correlations in variation exist they should be pointed out.

Plastic responses are specific in relation to particular environmental influences and it is rather shocking that so little information on phenotype modifications is presented in formal taxonomic system" (Cook, 1968).

Stebbins (1970) "was attracted to the existence in certain evolutionary lines of repeated cycles based upon transfers of the function of a particular structure from protecting the ovules to aiding the dispersal of mature seeds. Each cycle is completed when protection of the dispersal unit during its early stages of development is taken over by an external structure, that previously was either poorly developed or served a different function". In the first cycle, the mature seed is the dispersal unit and the carpel or ovary wall serves only a protective function. In some genera belonging to relatively primitive families (Magnoliaceae, Ranunculaceae) and in most genera of more advanced families (Boraginaceae, Labiatae), the carpels or ovary segments have become modified to perform the act of dispersal. In many of these genera, the function of protecting developing ovaries is transferred to a synsepalous calyx. To begin the second cycle of transfer, in some Labiatae (*Marrubium*), Leguminosae (*Trifolium*), Rosaceae (*Agrimonia*) and various Rubiaceae, Valerianaceae, Dipsacaceae and Compositae, there has been the modification of the calyx into a dispersal mechanism. In certain advanced genera of the Dipsacaceae and Compositae, the involucral bracts or phyllaries do the job of dispersal. Within the Compositae, certain evolutionary lines have passed through a third cycle of transfer: the floret plus the protective bract become the unit of dispersal in *Media, Hemizonia* and various genera of the Inuleae, the outer involucral phyllaries take over the protective role in *Evax* and *Psilocarphus* and the involucral phyllaries acquire the dual function of protection and dispersal in *Ambrosia, Arctium* and *Xanthium.* Among the monocots, the first cycle of transfer begins in some representatives of the Liliales and Commelinales where the carpel has become modified into an indehiscent nutlet or berry. In both the Cyperaceae and Gramineae, the protective function is assumed by the bract which subtends the flower. In the Cyperaceae, a second cycle of transfer occurs in *Carex,* where the number of florets in the original spikelets has been reduced to one and the bract subtending this floret, through fusion of its margins, has been modified into the perigynium. In the Gramineae, the start of a second cycle is accompanied by the suppression of floret formation in its lower bracts, resulting in sterile lemmas or glumes. While the fertile lemma serves both the protective and dispersive functions in the Bambuseae, Chlorideae, Eragrosteae, Festuceae and Hordeae, the sterile bracts has taken over the protective function in the Agrostideae, Arundineae, Aveneae, Phalarideae, Stipeae and Panicoideae. Similar cycles of transfer from the protective to the dispersive function, accompanied by corresponding structural modification, have taken place in the Chenopodiaceae (*Atriplex, Chenopodium*), Nyctaginaceae (*Mirabilis*), Leguminosae (*Trifolium*) and Labiatae (*Monardella*). A suggestion has been offered that the ancestors of angiosperms passed through a stage in which unitegumentary ovule, homologous to the anatropous cupules of the Caytoniales, was the dispersal unit and the cupule wall, homologous to the outer integument, had a protective function. There is also a hypothesis that cycles of transference progress in groups that are primarily adapted to montane habitats with loose sandy or gravelly soils and generally in arid or semi-arid climates. These cycles are "probably produced by alternating selective pressures under conditions which promote a succession of altered compromises between the conflicting selective pressures favoring large seed size, rapid seed development and efficient seed transport. The distribution of these cycles among the various orders and families of angiosperms is in accord with this hypothesis".

Phytochemistry: Chemical characters of plants can be used in classification readily as such familiar features as those of leaf arrangement, indumentum and chromosome number. The application of chemistry to taxonomy has brought to light during the past twenty-five years or so an exciting field, *chemical plant taxonomy* (*plant chemotaxonomy, chemosystematics* or *plant chemistry*), for which one may consult Hegnauer (1962–69), Alston and Turner (1963), Mirov (1963), Swain (1963, 1966), Leone

(1964), McNair (1965), Harborne (1970), Gibbs (1974), Bendz and Santesson (1974), Smith (1976) and Cagnin *et. al.* (1977). The recent growth of phytochemistry as a popular field of research is mostly due to the (*a*) development of rapid analytical techniques which allow screening samples or many species within a short time; (*b*) the realisation that behind the ubiquitous presence in plants of several vital biochemical routes, there is a tremendous variation between taxa in many other less important pathways and (*c*) the present belief that data from as many sources as possible should be employed in plant classification (Stace, 1980).

In a phytochemical study, Hegnauer (1967) recognised six chief stages reconciling the need for informed chemical procedure and a project design satisfying formal taxonomic requirements: (*a*) choice of a group, taxonomic survey and sound sampling; (*b*) choice, mastery and modification of suitable chemical techniques in a pilot survey; (*c*) full analysis of all material; (*d*) interpretation and comparison with data from all other sources; (*e*) adjustment of classification as necessary and (*f*) treatment of any evolutionary relationship as indicated by the new data.

Mentzer (1966) suggested a biogenetic classification for phytochemical research based on the natural relationships between the various plant constituents. The three broad categories of constituents include the (*a*) *primary metabolites, e.g.,* organic acids, amino acids, chlorophylls and sugars; (*b*) *secondary metabolites, i.e.,* those lacking nitrogen and not involved in basic metabolism of the cell and (*c*) *semantides, i.e.,* information carrying molecules (*e.g.,* DNA is a primary semantide, RNA a secondary semantide and proteins are tertiary semantides). In order to meet the various requirements, these categories can be further split into groups, families, subfamilies, sections, etc. However, the alkaloids will not fit into this classification. Sometimes the primary and secondary metabolites are called *micromolecules* and the semantides along with the larger polysaccharide as *macromolecules.*

Evidence from Amino acid Distribution: At one time, investigators largely ignored the amino acids. Before World War II, some non-protein amino acids were found. Of those typical of certain leguminous species, abrine, canavanine, 3, 4-dihydroxyphenylalanine, djenkolic acid, mimosine and willardine deserve special mention. After World War II, many new amino acids and peptides have been isolated from the species of *Lathyrus* and *Vicia.* Infact, Bell (1971) constructed intriguing infrageneric classifications using biochemical data. Seven infrageneric groups were recognised by him, each with a different amino acid or a host of associated amino acids.

Another kind of amino acid evidence is based on the restricted distribution of single amino acids. Despite supreme efforts, canavanine has not been detected outside the Lotoideae under the Leguminosae (Turner and Harborne, 1967). Although only 60 percent of the species so far examined prove to contain it, genes for the formation of this amino acid seem to be limited to this taxon. Birdsong *et al.* (1960) considered the prevalence of canavanine as an advanced character, as it was unknown from the primitive tribes Podalyrieae and Sophoreae. Subsequent discovery of this amino acid in both of these tribes may even justify the primitiveness of the two tribes. Still another instance of limited distribution is that of azetidine-2-carboxylic acid, being present in the Agavaceae, Amaryllidaceae and Liliaceae: this amino acid is also reported from the legumes *Delonix* and *Peltophorum.* The abundance of this amino acid in the lilialean families may denote some taxonomic significance.

Tschierech and Hanelt (1967) created four groups in *Vicia* on the basis of the occurrence of canavanine, β-cyanoalanine, unidentified amino acids A, B and C and the replacement of non-protein amino acids by high concentrations of arginine. These groups tally with classifications based on cytological and morphological data.

Simola (1968) compared the amino acid pools of different species of *Lathyrus* by studying different organs at four developmental stages. The quantity of proline is greater in *L. maritimus* than in *L. niger* and *L. sylvestris*. They all contain pipecolic acid, but in *L. maritimus* it is present in high concentrations and this species can also synthesize 5-hydroxypipecolic acid. *L. maritimus* also contains abundant γ-methylglutamic acid, which is not found in the other two species. Homoserine and α, γ-diaminobutyric acid are distinctive of *L. sylvestris*. *L. maritimus* does not contain these two compounds, but *L. niger* has homoserine in leaves and stems. Homoarginine occurs only in traces in the shoot of *L. maritimus,* despite the fact that it is the precursor of lathyrine, a compound enriching all other organs of this species. Lathyrine accumulates in the wintering organs.

There are non-protein amino acids which are widely distributed and which occur in plants which are only distantly related. For example, acetyl ornithine has been identified as the main amino acid in Fumarioideae under the Fumariaceae. It has subsequently been found in ferns and grasses. Though the predictive value of the occurrence of the amino acid is extremely low, the occurrence may explain that different plants can probably produce the same amino acid by different biochemical routes.

Evidence from Polysaccharides: In view of their complexity and diversity, perhaps polysaccharides offer the greatest hope for taxonomic evidence. So far they have been examined from a systematic viewpoint, although, in principle at least, they might carry some coded information if we know how to read it. This is partly due to the fractionation and isolation procedures for polysaccharides frequently consist of many laborious phases, any of which can bring up non-taxonomic variations.

Studying water soluble polysaccharides in seeds of 22 grass species, MacLeod and McCorquodale (1958) observed considerable taxonomic variations. All the Festuceae contain lesser or greater amounts of β-glucosan, but *Festuca* and *Lolium* are distinctive in having an unusual trisaccharide. On the other hand, the Bromeae forms a very natural tribe containing only fructosans and no raffinose.

From a taxonomic standpoint, the sugar alcohols have been studied by Plouvier (1963). Sorbitol distribution is in accordance with the taxonomy of the Rosaceae, supporting the transfer of *Ulmaria* the Rosoideae (usually without sorbitol) from the Spiraeoideae which contains genera with sorbitol. Quercitol is exceptionally common in the Menispermaceae, although it is found elsewhere. While it is present in all 35 species of *Quercus* so far tested, it is absent from *Castanea* and *Fagus* of the Fagaceae. Pinitol is likewise widespread but is particularly common in the Caryophyllaceae (existing in 83 out of 85 species tested). It is found in *Magnolia* alone of genera in the Magnoliaceae and not in closely related families.

Comparing linear and branched hemicellulose fractions of species of the Leguminosae and Gramineae, Gaillard (1965) demonstrated prominent differences between the two families. While branched polymers of the grasses contain a high proportion of xylose, those of the legumes have comparatively more arabinose, galactose and uronic acid. Uronic acid is attached to xylose in the grass polymer and arabinose in the legume polymer.

Some taxonomic claims have been made for the distribution of seed 'amyloids' among storage polysaccharides. When amyloid is treated with iodine solution and sodium sulphate, a characteristic blue colouration is imparted. Amyloids consist of a principal chain of glucose with side chains embracing galactose and xylose. Structures of these compounds have been fully worked for *Annona muricata* and *Tamarindus indicus,* being basically similar. Kooiman 1960 detected amyloid in 16 families of the dicotyledons and none of the monocotyledons in a test involving over 2,500 species. Though the amyloid families Leguminosae and Acanthaceae are not closely related, amyloid distribution follows taxonomic lines within these two families. Despite amyloid being found in many genera of the Caesalpinioideae, all compounds are restricted to the Amherstieae, Cynometreae and Sclerobiae. In the Lotoideae, a few species bear amyloid. Whereas some taxonomists would transfer *Pazeoina* to a monotypic family, it is the solitary amyloid producing genus among 30 tested from the Ranunculaceae.

Evidence from Secondary Metabolites: Plants accumulate in their cells a variety of constituents, sometimes considered as waste products of plant metabolism. They range from the odoriferous essential oils to the cathartic anthraquinones and the coloured anthocyanins and flavones of flower petals. Since they vary immensely in their distribution, they offer an excellent set of characters for utility in a chemical approach to plant systematics.

The secondary constituents arise from carbon dioxide, employing very specific biosynthetic routes. It is very probable that at a very closely related phylogenetic level some related biosynthetic pathways are common. The fatty acid pathway seems to be very general and leads directly to fatty acids which cannot be characterised as secondary constituents, though they are often stored or accumulated. The mevalonic acid pathway leads from acetic acid via mevalonic acid to the isopentenyl units, the building blocks of rubber, steroids, terpenoids, etc. One general route, shikimic-prephenic acid pathway, leads via shikimic acid to a variety of aromatic compounds, such as lignins and phenylalanine. Another route, the polyketide pathway, leads from acetic and malonic acids to a vast array of natural products.

(*i*) *Alkaloids:* Among the most attractive chemical compounds for taxonomic purposes are alkaloids which are not universally accumulated by plants. Alkaloids are somewhat characteristic of particular families and one can refer to Amaryllidaceae alkaloids, Berberidaceae alkaloids, Leguminosae alkaloids, Ranunculaceae alkaloids, Solanaceae alkaloids and so on. These taxa are so distinct that the "mere accumulation of alkaloids does not imply a taxonomic relationship at the family level" (Smith, 1976).

"Alkaloids characterising species of a particular taxon are frequently of the same chemical or biogenetic group, suggesting that related plants share the same pathways of alkaloid synthesis. If this is so, there is some reason to consider alkaloid content as a source of taxonomic evidence." Thus, both the Papaveraceae and Fumariaceae accumulate isoquinoline alkaloids and always bear protopine. The members of the Berberidaceae—*Berberis, Bongardia, Caulophyllum, Epimedium, Gymnospermium, Leontice, Mahonia, Nandina* and *Podophyllum*—contain berberine, palmatine and related alkaloids. Alkaloids of the berberine type also occur in the Annonaceae (palmatine in *Enantia),* Hydrastidaceae (berberine and hydrastine), Menispermaceae, Ranunculaceae (*Aquilegia, Coptis, Thalictrum, Zanthorhiza*) as well as Fumariaceae and Papaveraceae. All these families are closely related and the presence of alkaloids of the berberine type does not pose a problem. The Leguminosae contain lupin (quinalizidine) alkaloids and the Solanaceae the tropane derivatives. The Compositae and Gramineae form different types of alkaloid. But the Rutaceae is probably the most versatile of all the families of higher plants, containing as many as nine distinct types.

In the Leguminosae, there are instances of alkaloid restriction and possible taxonomic value. Dipiperidyl lupin alkaloids such as hystrine occur only in *Genista* (Genisteae) which has long been considered as related to *Adenocarpus* producing adenocarpine, isoorensine and santiaguine. But *Ammodendron,* normally placed in the Sophoreae, lacks the matrine alkaloids which are so typical of the tribe. When one type of alkaloid is replaced by another, this is indicative of the fact that *Ammodendron* deserves to be transferred to the Genisteae.

There is a very narrow distribution of certain alkaloid molecules. Coniine is restricted to a few umbellifers and morphine only to *Papaver somniferum,* but alkaloids of the strychnine group are usually associated with a few species of *Strychnos* (Loganiaceae).

Sometimes identical or chemically closely related secondary products are found in biologically unrelated organisms and should not be considered to have taxonomic significance, *e.g.,* anabasine in *Anabasis* and *Nicotiana,* bufotenine in toad (*Bufo*) and *Piptadenia,* dendrolasin in ant (*Lasius*) and *Torreya africana,* nicotine in *Nicotiana* and *Equisetum:* Nicotine also occurs in Solanaceae (*Duboisia,*

Salpiglossis), Crassulaceae (*Sedum*), Leguminosae (*Mucuna*), Asclepiadaceae (*Asclepias*), Scrophulariaceae (*Herpestis*) and Compositae (*Eclipta, Zinnia*).

Completely new chemical compounds have sometimes been noticed in hybrids. While examining the *Solanum* hybrid, produced by crossing two South American tuberiferous species, *S. demisum* and *S. stoloniferum,* a new "hybrid" alkaloid was found by Harborne (1968).

The heartwood of several species of *Sorbus* contains a lignan xyloside lyoniside and hydroxydiphenyl derivatives, the aucuparins. While aucuparin occurs in the wood of a Guttiferae, *Kielmeyera coriacea,* there is no lyoniside in it. It has a series of xanthones common to many other species of the Guttiferae. Lyoniside has been noted in the leaves of an Ericaceae, *Lyonia ovalifolia.* It is clear that the co-occurrence of a compound with quite different types of compounds (a flavone and an alkaloid, a tropolone or an acetylene) is taxonomically more interesting than the co-occurrence of a compound and a simpler derivative of it (a flavone and its methyl ether glucoside).

(*ii*) *Phenolics:* From a taxonomic viewpoint, no other type of secondary metabolites has proved so popular as the phenolic compounds (Smith, 1976). These compounds are simply extracted from plant material, easily separated by paper chromatography and readily identified by location reagents. Though phenolics has an aromatic ring with hydroxyl substituent, most of them are more complicated than phenol itself and can be derived by combination, polymerisation and substitution. While the majority of them have no known functions in plants, some are involved in the inhibition of pathogenic fungi and others cover the most important flower pigments.

Bate-Smith (1962) concentrated on the phenolics of vegetative tissues and seeds, obtaining an interesting correlation between the presence of leucoanthocyanins and the woodiness of plants. Leucoanthocyanins are not found at all in the arboreal members of the Bignoniaceae, Oleaceae, Scrophulariaceae and Verbenaceae and present only in the Apocynaceae, Asclepiadaceae, Caprifoliaceae and Rubiaceae, whereas the Primulaceae and Plumbaginaceae are rich in leucoanthocyanins. Moreover, methoxycinnamic acids are correlated with the herbaceous habit. The difference in physiology between woody and herbaceous plants is reflected in these correlations. The presence or absence of woodiness is not likely to be an acceptable basis for a natural classification, despite Hutchinson's recognition of the woody Lignosae and herbaceous Herbaceae as two plant groups among dicots.

The most important taxonomic evidence stems from the mutually exclusive nature of anthocyanins on the one hand and the betacyanins and betaxanthins (collected called betalains) on the other. The metabolic pathway of anthocyanins is separate from that of anthocyanins. But betacyanins and betaxanthins are interconvertible and structurally related. They occur in only ten families of angiosperms, most of which have been treated under the Centrospermae on the basis of morphological and anatomical characters. The newer definition of the Centrospermae differs from the old, *i.e.,* the Cactaceae, usually and unsatisfactorily placed in a monotypic order of its own (Cactales or Opuntiales) contains betalains as well as the Caryophyllaceae and Molluginaceae lack betalains and possess anthocyanins. Though the two families of the latter group are closely related to betalain containing families, some botanists are still convinced the anthocyanin- and betalain-containing Centrospermae do not represent distinct taxa at any level.

(*iii*) *Flavonoids:* Flavonoids can be used as valid taxonomic markers due to their widespread distribution, chemical stability, structural variability and ease of detection. Harborne (1966) listed flavonoid characters as primitive, advanced and isolated (Table 2).

TABLE 2. Primitive, advanced and isolated flavonoid characters

Primitive	Advanced gain mutations	Isolated
(1) 3-Deoxyanthocyanidins	(1) Complex *O*-glycosylation (including acylation of sugars)	(1) Replacement of anthocyanin by betacyanin
(2) Flavonols	(2) 6- or 8-Hydroxylation	(2) Shift of flavonoid B ring to 3-position: isoflavone formation (Leguminosae)
(3) Leucoanthocyanins	(3) O-methylation	(3) 2′-Hydroxylation
(4) Chalcones, flavones and dihydrochalcones	(4) Oxidation of chalcones aurones	(4) Elimination of 5-hydroxyl group
(5) *C*-substitution (methylation, phenylation and glycosylation as well as biflavonyl formation)		
	Loss mutations	
	(1) Replacement of flavonols by flavones	
	(2) Elimination of leucoanthocyanidins	
	(3) Elimination of trihydroxylation	

The flavonoid chemistry of the monocotyledons suggests four categories of families in an evolutionary sequence with the first category embracing the Typhaceae and possessing flavonols alone as the most primitive and the fourth category containing the Gramineae and producing tricin as the most advanced. The most startling discovery is the indication that the Palmae is related to the Gramineae, despite the fact that palms and grasses differ greatly in gross morphology and palms are generally regarded as a more primitive group. The presence of five leaf flavonoids—glycosyl flavones, tricin, luteolin, bisulphate complex and 5-glucoside—link the two families. The Juncaceae exhibit many resemblances in leaf flavonoids, though tricin appears to be lacking. Another major deviation from the grasses is that inflorescence flavonoids (desoxyanthocyanins) have been detected in five species of *Juncus* and *Luzula.* While the Gramineae and Cyperaceae are usually placed very near each other in most classificatory systems, they have the same basic flavonoid pattern. However, there is a minor difference: bisulphates, prevailing in a few grasses, have not been detected in sedges. In the two groups, inflorescence pigments vary: grasses contain cyanidin based pigments and sedges the aurone (aureusid in).

While simple flavones are widely distributed, isoflavones are restricted to a few plant families. The different species of *Spirodela* are sometimes unidentifiable on morphological grounds, they are extremely distinctive in their flavonoid chemistry. In sharp contrast, *Psoralea,* containing a number of species recognisable morphologically, has a recurrent pattern of flavonoid chemistry specific for the genus. Both *Spirodela* and *Psoralea* possess glycoflavones.

The Parietales is taxonomically difficult, since it represents a heterogenous group of families. Flavonoid bisulphates are uniformly present in the Bixaceae, Frankeniaceae and Tamaricaceae; occasionally in the Cistaceae and Guttiferae and otherwise lacking. These compounds are correlated with the distribution of ellagic acid in those families. The only exception is the Cochlospermaceae which bears ellagic acid in three species of *Cochlospermum, viz. C. orinocense, C. religiosum* and *C. vetifolium.* The other chemical data too fit in with a grouping of the families into three: those with bisulphates uniformly present; those occasionally present and those lacking them.

Gossypetin in the yellow flower pigment of *Gossypium.* It is known as a flower pigment in the Malvaceae and a number of other families such as Ranunculaceae (*Ranunculus*), Ericaceae (*Rhododendron*) and Primulaceae (*Primula*). It is consistently present in all known taxa of the Empretaceae—a family related to the Ericaceae on morphological basis. Gossypetin is a taxonomic marker of significance in the Ericaceae, connecting the two tribes Phyllodoceae and Rhodoreae under the subfamily Rhododendroideae. This piece of chemical evidence, along with the presence of flavonol 5-methyl ethers, point out the need to put these tribes more closer than earlier taxonomic treatments.

Chiranthodendron pentadactylon, a member of the Sterculiaceae, revealed the presence of cyanidin 3-glucoside in sepal pigment and gossypetin as 3-glucuronide in leaves of the plant. This is not a matter of surprise, as the Sterculiaceae is closely related to the Malvaceae which is the original source of gossypetin. Other chemical links between the two families include the presence of cyclopropene fatty acid (sterculic acid) in the seed fat of both groups. Gossypetin has also been detected in another member of the Sterculiaceae, *e.g.,* flowers and leaves of *Fremontia californica.*

Another class of unusual structures, 6-hydroxyflavonoids, is of systematic interest. The fact that 6-hydroxyflavones are much more common than 6-hydroxyflavonols and 6-hydroxyflavonoids occur predominantly in the families of theTubiflorae and the Compositae indicate a switchover in the specificity of the enzyme oxidising the A-ring from the 8- to the 6-position. The switchover may be due to 6-hydroxy compounds being more effective than 8-hydroxy compounds as toxins to predating animals. This switchover is interesting in the sense that those expecting to have 6-hydroxyflavonoids are devoid of them and those not expecting have them. Their presence in two families of the Centrospermae and the Polygonaceae is amazing, but these are not components of a generalised distribution pattern and are isolated occurrences. If the subclass Magnoliidae is considered evolutionarily advanced, the presence of 6-hydroxyflavonoids is rather surprising in it. The same flavonoid structure is produced in the Betulaceae, a comparatively primitive family, where the occurrence can be linked to function. They are produced in woody buds of *Alnus* and *Betula*— presumably offering protection from microbial infection and not in flowers or leaves. In the subclass Rosidae, 6-hydroxyflavonoids are much less striking and almost expected. Many of these families in general and the Leguminosae in particular display signs of evolutionary advance in terms of morphology.

Even at lower levels of classification, the 6-hydroxyflavonoid character can be utilised. It has been noticed that 6-hydroxyluteolin occurs in the genus *Valerianella,* particularly in the more advanced Mediterranean and North American species and not in the Asiatic species. Since such chemical evidence correlate well with morphology and cytology, a new treatment of the genus is suggested. The same compound has been found useful at the species level while comparing flavonoid patterns in northern and southern populations of *Plantago maritima.* However, populations of northern hemisphere plants are characterised by 7-glucoside of 6-hydroxyluteolin.

The division of the Gesneriaceae into the Old World species being anisocotylous and the New World species isocotylous has been supported by the chemistry of leaf and petal pigmentation in some 80 species. A class of pigment, 3-deoxyanthocyanin, is almost present in all the isocotylous species and absent in every anisocotylous species. Similarly, the two South American species of *Eucryphia* (Eucryphiaceae)—*E. cordifolia* and *E. glutinosa*—can be differentiated from the three Australian species—*E. lucida, E. milliganii, E. moorei*—by their leaf flavonoids. Other examples of correlation between geography and flavonoids are provided by *Pyrus* (Rosaceae) and *Crocus* (Iridaceae).

Hybrid plants usually contain the chemical constituents of both parents. Chromatographic studies of a hybrid extract will show a pattern of spots somewhat identical with a synthetic mixture of extracts of the two parents. Such studies of leaf flavonoids have been widely used for the identification of hybrids among natural populations of *Baptisia* (Alston and Turner, 1963). In the legume genus, three subgeneric patterns are recognised: the first group of species (*B. leucophaea)* is dominated by flavone monoglycerides like apigenin and luteolin 7-monoglycosides; in the second group (*B. sphaerocarpa*), flavone diglucosides are most important and in the third group (*B. leucantha*) flavonols apparently replace the flavones. It is thus possible to group the species of *Baptisia* on the basis of their chemical affinities and indicate their phylogenetic relationships.

(*iv*) *Terpenoids*: Volatile constituents, especially terpenes, have been utilised in the study of systematic problems. Whereas flavonoids are generally more numerous and widespread than terpenoids, they are more difficult to work with at the technical level, at least as contrasted with the comparatively few monoterpenes which can be rapidly identified by combined gas-liquid chromatography (GLC) and mass spectrometry. Moreover, terpenoids are more numerous as to types within a given species. For example, *Passiflora edulis* is known to contain 250 discrete peaks by using GLC. Such a mass of data, if fully quantified and identified, will impress a systematic botanist. Though most terpenoids have been examined for their taxonomic potential, distributional data are not yet complete for most plant groups. However, in certain cases, terpenoid evidence has aided taxonomy.

The classification of *Eucalyptus* has presented great difficulty on the basis of gross morphology. Baker and Smith (1920) divided the genus into larger groups containing different oil combination which are correlated with leaf venation and bark structure: (*a*) butterfly venation shows a very high yield with phellandrene and piperitone as main components; (*b*) obtuse feather venation shows a low yield with pinene as the dominant compound and (*c*) more acute venation, having marginal veins, shows a slightly higher oil yield with cineole and pinene as chief constituents. That terpene composition is a valid morphological trait in the analysis of introgression within a group has been indicated by a survey of monoterpenes in 19 species of *Salvia* (Emboden and Lewis, 1967).

Sesquiterpene lactones constitute a group of bitter tasting principles, being exemplified by lactucin and lactucopikrin of the Compositae. In this family, the oxidation level of sesquiterpenes is sometimes specific to a tribe, subtribe or even a genus. The absence of sesquiterpenes in the Astereae may be rare or genuine. When *Ambrosia* and its allies—*Iva, Franseria* and *Xanthium*—are taken away from the Heliantheae, the residual elements are similar to the Helenieae in terms of lactone distribution. Diversity in lactone production resides in the allied forms of *Ambrosia.* There is a need to construct a separate tribe or a distinct family in order to accommodate all these genera.

Sesquiterpene lactone characters have contributed helpfully to both population and hybridisation studies in the Compositae. Eight major components were identified by McMillan *et al.* (1976) in leaf extracts of *Xanthium strumarium* by thin layer chromatography (TLC) or nuclear magnetic resonance (NMR) spectroscopy. In F_1 hybrids, produced between plants belonging to Old and New World morphological complexes, the inheritance of these components were followed. Old World populations contain xanthinin or xanthinosin with related structures or both, while New World populations contain xanthinin or its stereoisomer xanthumin. Plants of the *Chinense* complex (from South Louisiana), thought to be the probable source of *Chinense* populations in Australia and India, produce xanthumin. Some of the Indian plants are known to contain xanthumin derivative in addition to xanthumin itself. A likewise mixed sesquiterpenoid pattern is revealed by F_1 hybrids between Old World plants and pure *Chinense* plants, suggesting a natural introgression of indigenous Old World and introduced *Chinense* populations. Xanthonin, produced by Old World *chinense* populations, is lacking in the Louisiana *Chinense* populations; this is common in Old World populations of *Xanthium.* However, xanthosin formation is almost or completely suppressed in hybrids involving xanthinosin producing plants of Asia and xanthinin or xanthumin producing plants of America.

The significance of the monoterpenoid cyclopentanoid lactones called iridoids has been reviewed by Bate-Smith and Swain (1966). Among these compounds, asperuloside is particularly common in the Rubiaceae. Aucubin is frequently noted in the Cornaceae, Scrophulariaceae, Orobanchaceae and some allies. On the basis of iridoid distribution, some taxonomic changes have been suggested. The aucubin containing genus *Buddleia* must be removed from the Loganiaceae to the Buddleiaceae, with a position near the Scrophulariaceae. While morphological evidence is adduced to support the transfer of the Garryaceae from the apetalous group to the neighbourhood of the Cornaceae, interestingly both produce aucubin. A minor modification in Takhtajan's classification can lump all iridoid producing orders into a single group, indicating a possible common origin.

(*v*) *Oils, fats and waxes:* McNair (1929) related the nature of oils and fats to taxonomy, suggesting that the iodine number of lipids is higher in more advanced taxa. The highly evolved Compositae do not contain lipid that are unsaturated, while the lowly algae possess very unsaturated lipids. In view of the fact that lipids are inconstant in absolute composition, characters relating to their intact structure would be of doubtful value in taxonomy.

As a possible source of taxonomic importance, the fatty acid composition of lipids has been tried by Shorland (1963). On the basis of major fatty acids released by hydrolysis of their lipids, he recognised groups of families containing (*a*) linolenic-rich seed fats; (*b*) linoleic-rich/oleic-rich seed fats; (*c*) seed fats rich in linoleic and oleic acid, with linolenic acid or a conjugated polyethanoic acid as principal components; (*d*) seed fats with palmitic, oleic and linoleic acids as principal components and (*e*) seed fats with characteristic acids other than or additional to oleic, linoleic and palmitic acids. However, almost all fatty acids are noticed in all groups at low concentrations and oleic is the main fatty acid in most groups. Distantly related families sometimes occur under the same group and a family may appear in more than one group. But the Palmae is a rare example of a family where the proportions of fatty acids in lipids are constant in both species and genera (Hilditch, 1952).

Fatty acids specific to particular plant groups may be helpful in taxonomy. The Olacaceae and Santalaceae are characterised by ximenyric acid. Erucic acid is almost completely restricted to the Cruciferae and petroselinic acid to the Umbelliferae. Chaulmoogric and some related acids seem to characterise certain species of the Flacourtiaceae.

Taxonomic significance may be attached to the occurrence of unusual fatty acids. One case involves the presence of linolenic acid and octadecatetraenoic acid in both seed and leaf lipids in eight members of the Boraginaceae. But the same unusual fatty acids are present in ten species of the unrelated Caryophyllaceae. Malvalic acid exists in members of the Malvaceae. Capric acid, once believed to be the chief component in *Ulmus,* is richly represented in seed lipids of *Zelkova* (Ulmaceae) as well as members of the Lauraceae and Lythraceae.

In the Gramineae, alkane variation in cuticular waxes serves as a useful source of taxonomic information. Within the genera of the Crassulaceae, there is a general uniformity in proportions of alkane hydrocarbon constituents. On the basis of wax characters, *Eucalyptus* species of the Australian arid zone form a homogeneous group. In 22 tuber bearing species of *Solanum,* wax alkane variation can be related to morphology, cytogenetics and protein analysis. While wax alkanes appear attractive as taxonomic characters, Martin-Smith *et al.* (1967) admitted as follows: "A systematic investigation into the possible influence of season, climate, geographical distribution and the kind and age of organs on the composition of plant surface waxes is essential before the method can be accepted without qualifications."

Evidence from Proteins: There are serious and interpretative difficulties in protein analysis, but recently methods have been elaborated for the separation and isolation of individual proteins. It is very likely that comparative protein studies will provide a ready taxonomic evidence, as within a decade or so the number of structurally known proteins of similar biological function from species of different taxa is bound to increase. Comparative studies of enzymes may yield data about differences in primary structure which can be traced to evolutionary adaptations.

By the use of disc electrophoresis, Desborough and Peloquin (1969) separated the soluble tuber proteins of the species of *Solanum* where all 14 major bands did not occur consistently. If consistently present (100% bands) are treated alone, species specific patterns are depicted. Only single bands are always present in the protein fraction in *S. acule* and the group *Stenotomum* of *S. tuberosum.*

Proteins of the wheat group has revealed a close correlation between genome constitution and storage protein characters (Johnson and Hall, 1965). As evident in diploid *Triticum monococcum* (AA), amphiploid wheat types *T. aestivum* (AABBDD) and *T. dicoccum* (AABB) has almost all the proteins of the A genome. They also share proteins of the B genome, which, however, is of uncertain origin. The high probability of the ancestry of *T. aestivum* from *Aegilops squarrosa* (DD) has been demonstrated by mixing their proteins artificially which possess the same electrophoretic properties as *T. aestivum* extracts.

Comparing electrophoretic mobilities of glutamic and formic dehydrogenase from 103 leguminous species, Thurman *et al.* (1967) detected activity by the reduction of a tetrazolium salt to insoluble formazans. While no taxonomic changes were proposed, a heterogenous tribe Genisteae produced greater enzyme variation than more homogenous Trifolieae and Viciae. That the position of *Cicer* is somewhat apart from other genera of the Vicieae—*Lathyrus, Lens, Pisum* and *Vicia*—was confirmed by a different isozyme pattern in the first genus in contrast to characteristic isozymes of glutamic dehydrogenases occurring in species of the other four genera.

Evidence from Semantides. Semantides constitute primary sources of taxonomic information, since they include the primary genetic information itself (DNA), or the secondary (RNA) or tertiary (protein) derivatives of it. Since it has not been possible to sequence DNA or RNA molecules in a routine manner, nucleic acid traits are concluded indirectly by DNA hybridisation. In short, DNA extracted from one organism is treated to turn it into a single stand polynucleotide chain; the amount of annealing (reassociation) with likewise DNA from another taxon which takes place upon mixing the two extracts is an estimate of relationship of the nucleotide sequences.

The relative homology of DNA and RNA of various organisms is useful in a taxonomic study (McCarthy and Bolton, 1963; Hoyer *et al.,* 1964) and as a possible screening method for interfertility of species (Bendich and Bolton, 1967). All sixteen possible hybrids are made from repetitious DNA of four different grass species fitting well with the usually accepted classification of the four genera. Barley, Rye and Wheat DNAs (Triticeae) indicate more base sequence complementary to each other than they do to oats (Aveneae), suggesting that the position of *Avena* is somewhat isolated from the other three genera. The interactions of DNAs of cereal fruits show that Wheat DNA (75%) is more similar to Rye DNA (100%) than to Barley DNA (57%).

Comparisons based on DNA hybridisation have been reported for taxa belonging to the Leguminosae. The degree of repetition in the DNA is different, the percentage of rapidly reassociating DNA varying from 15 to 38 (Chooi, 1971). Four of *Vicia* species reveal a high level of homology with each other and *V. faba* in competition experiments. *V. sativa* is an exception, as it competes very poorly with *V. faba* DNA.

RNA has been studied by the technique of DNA-RNA hybridisation in which the quantity of association occurring between the DNA of one organism and a fraction of the RNA (usually ribosomal RNA) of another organism is considered as a basis of similarity. By this technique, Mabry (1976) studied plants of the Centrospermae, inferring that the Caryophyllaceae (which contains anthocyanins only) is related to the betalain producing families but the relationship is not so close as the latter are to each other.

Pollard (1964) noted the specificity of ribosomal RNA in Cabbage, Cauliflower, Celery, Corn and Parsnip. He showed that 28 S and 18 S ribosomal RNA from these taxa have distinct base compositions which are characteristic of the species.

All thirteen species of *Cucurbita* possess mainband and satellite DNA whose densities are constant throughout the genus (Goldberg *et al.*, 1972). When ribosomal RNAs of *Cucurbita* are hybridised with DNAs, there is a considerable variation (1.4–3.1) occurring among these species. Whereas dissimilar species-*C. andreana* and *C. sororia*-have the same proportion (1.7) of ribosomal DNA. Similar species-*C. andreana* and *C. maxima*-have different proportions with 1.7 and 3.1 respectively.

Percentage homologies, produced as a result of DNA hybridisations, have been utilised to construct the relationship chart. The results point out that *Cucurbita palmata,* a xerophyte, is most distantly related to the other four species used as homologous taxa. This finding is in accord with data from fattyacid and cucurbitacin distribution as well as the concept that xerophytic cucurbits are genetically isolated from the remainder of the genus. Interestingly, DNAs of *C. iundelliana, C. maxima* and *C. pepo* are alike. According to Whitaker and Bennis (1965), *C. lundelliana* is a progenitor of the other two species and each is more similar to *C. lundelliana* in nucleotide sequence than either is to another.

Other Evidences: Saville (1954) opined that parasite relationship is a valuable biochemical tool when properly used. With a reasonable degree of probability, mycologists have been able to work out the evolutionary trends of some genera in the Uredinales. Such evolutionary trends suggest affinity or comparative age in host plants. According to Leppik (1953), a new rust form becomes regularly adjusted "to some new group of host plants of its origin". The evolution of fungi is irreversible in the sense that "advanced" rusts would net return to "primitive" hosts. Applying this approach to *Carex* and related genera, through a perusal of some of the smuts, indicates the presence of a previously unrecognised natural subgenus of the group and calls for its taxonomic improvement. It has also been established that within each major host group a smut tends to restrict itself to a few closely related species. By employing evolutionary trends in *Puccinia* teliospores, it has been possible to arrange host plants in the Saxifragaceae in a roughly chronological sequence. The value of such a procedure lies in the natural classification of parasites and accurate identification of hosts.

Occasionally, parasitism and symbiosis can give rise to interesting metabolites. Certain orchids like *Laroglossum* and *Orchis,* live symbolically with a fungi (*Rhizoctonia*), the symbiosis being controlled by the formation of specific phenols of phenanthrene type (hircinol, orchinol). When grown on *Ipomoea batatas,* the ascomycete *Cerastostomella fimbriata,* produces a series of unusual isoprenoid compounds including the sesquiterpene ipomoeamarone. If the wood of *Prunus jamasakura,* containing flavonoids, is infested by *Polystictus versicolor,* it bears large quantities of the lignan iso-olivil. While this compound is foreign to both the host and parasite, it is a normal constituent of *Olea cunninghamia.*

Conclusions: Of late, plant chemistry has made its inroads to taxonomy (Chennaveeraiah, 1983). There is a tendency to consider chemical characters as more fundamental than other characters, but it has only a limited use. Perhaps it is incoherent to think that smaller and more cryptic the character, the more important it is for indicating relationship. Chemical characters may exhibit a chemical relationship in the same way as morphological characters do for morphological relationship. The value of individual chemical constituents varies from one group to the other and from one compound to another. Much labour is involved in the study which has been branded as luxury diversifications or expressions of biochemical virtuosity. The attractiveness of chemical characters lies in the fact that sometimes the chemical evolution of a character has taken place in a particular way and this kind of evidence may be helpful in understanding the course of evolution rather than phylogeny.

Biostatistics: Some investigators have looked into the botanical landscape through mathematical glasses. A biometrical correlation between vegetative and floral features has been indicated (Sporne, 1948). Among dicotyledons, certain characters are associated, *i.e.*, they occur together more frequently than they would if they were only randomly distributed within the group. As some of these characters were more abundant in plants from Cretaceous and Tertiary deposits than among present day dicots, it has been concluded that the associated characters are primitive ones. Further development of these ideas led to the introduction of simple statistical tests of significance by Chalk (1937). By means of 2×2 contingency tests, values of X^2 were deduced from which one could examine the probability of the observed frequencies of association occurring by chance. It was possible to find which characters are correlated, *i.e.*, statistically associated. Beginning in 1939, Sporne has been applying essentially the same techniques as those, of Chalk to the widest possible range of dicot characters and the work is still continuing. Data have accumulated about 40 characters, out of which 26 are significant indicators of evolutionary status.

The 259 families recognised by Diels (1936) were listed on recording sheets in alphabetical order. While one sheet would record the presence/absence of character X, another the presence/absence

TABLE 3. Correlations among 26 characters in dicotyledons (After Sporne, 1977)

Character correlations in dicotyledons (1976)	1 Woody habit	2 Scalariform end-plates	3 Scalariform side-walls	4 Apotracheal parenchyma	5 Unstoreyed wood	6 Leaves alternate	7 Stipules present	8 Secretory cells	9 Leucoanthocyanins	10 Flowers unisexual	11 Flowers actinomorphic	12 Petals free	13 Stamens pleiomerous	14 Carpels pleiomerous	15 Carpels free	16 Axile placentation	17 Pollen binucleate	18 Seeds arillate	19 Two integuments	20 Integument bundles	21 Ovules crassinucellate	22 Endosperm nuclear	23 Pollen pauciaperturate	24 Tapetum glandular	25 Ellagitannins present	26 Aluminium accumulation
1 Woody habit		✚	•	×	×	✚	+	✚	✚	✚	✚	✚	✚	✚	•	✚	✚	+	✚	+	✚	✚	✚	+	✚	✚
2 Scalariform end plates	✚		✚	✚	✚	•	•	•	✚	+	✚	•	•	•	•	•	✚	•	•	+	•	•	✚	•	•	✚
3 Scalariform side-walls	•	✚		•	✚	•	•	•	•	✚	•	✚	•	•	•	✚	•	+	•	•	•	•	•	•	+	✚
4 Apotracheal parenchyma	×	✚	•		✚	✚	•	•	✚	+	✚	•	✚	•	•	•	✚	•	•	✚	•	−	✚	•	•	✚
5 Unstoreyed wood	×	✚	✚	✚		•	•	•	•	✚	•	•	•	•	•	•	•	•	•	•	−	•	✚	•	•	✚
6 Leaves alternate	✚	•	•	✚	•		•	+	•	+	+	+	✚	+	•	•	•	✚	+	+	•	✚	+	•	•	•
7 Stipules present	+	•	•	•	•	•		✚	✚	+	+	✚	✚	✚	+	✚	+	✚	✚	+	✚	✚	•	•	•	•
8 Secretory cells	✚	•	•	•	•	+	✚		✚	+	✚	✚	✚	✚	+	•	+	+	✚	✚	✚	✚	•	•	•	•
9 Leucoanthocyanin	✚	✚	•	✚	•	•	✚	✚		✚	✚	✚	✚	+	•	•	✚	•	✚	✚	✚	✚	•	•	✚	✚
10 Flowers unisexual	✚	+	✚	+	✚	+	+	+	✚		✚	+	+	+	+	•	+	+	+	+	✚	+	+	•	•	•
11 Flowers actinomorphic	✚	✚	•	✚	•	+	+	✚	✚	✚		✚	✚	✚	•	+	•	•	✚	•	✚	+	✚	✚	•	•
12 Petals free	✚	•	✚	•	•	+	✚	✚	✚	+	✚		✚	✚	•	✚	•	✚	✚	+	✚	✚	✚	✚	✚	•
13 Stamens pleiomerous	✚	•	•	✚	•	✚	✚	✚	✚	+	✚	✚		✚	✚	+	•	✚	✚	✚	✚	✚	•	•	•	•
14 Carpels pleiomerous	✚	•	•	•	•	+	✚	✚	+	+	✚	✚	✚		✚	✚	•	✚	✚	+	✚	•	✚	•	✚	•
15 Carpels free	•	•	•	•	•	•	+	+	•	+	•	•	✚	✚		×	•	+	✚	+	•	•	✚	•	•	•
16 Axile placentation	✚	•	✚	•	•	•	✚	•	•	•	+	✚	•	✚	×		•	+	•	+	•	•	+	•	•	•
17 Pollen binucleate	✚	✚	•	✚	•	•	+	+	✚	+	•	•	•	•	•	•		•	•	+	•	•	•	✚	•	•
18 Seeds arillate	+	•	+	•	•	✚	✚	+	•	+	•	✚	✚	✚	+	+	•		✚	✚	✚	✚	•	•	•	•
19 Two integuments	✚	•	•	•	•	+	✚	✚	✚	+	✚	✚	✚	✚	✚	•	•	✚		•	✚	✚	•	•	✚	•
20 Integument bundles	+	+	•	✚	•	+	+	✚	✚	+	•	+	✚	+	+	+	+	✚	•		✚	✚	•	•	•	✚
21 Ovules crassinucellate	✚	•	•	•	−	•	✚	✚	✚	✚	✚	✚	✚	✚	•	•	•	✚	✚	✚		✚	•	•	•	•
22 Endosperm nuclear	✚	•	•	−	•	✚	✚	✚	✚	+	+	✚	✚	•	•	•	•	✚	✚	✚	✚		•	•	•	•
23 Pollen pauciaperturate	✚	✚	•	✚	✚	+	•	•	•	+	✚	✚	•	✚	✚	+	•	•	•	•	•	•		•	•	•
24 Tapetum glandular	+	•	•	•	•	•	•	•	•	•	✚	✚	•	•	•	•	✚	•	•	•	•	•	•		•	•
25 Ellagitannins present	✚	•	+	•	•	•	•	•	✚	•	•	✚	•	✚	•	•	•	•	✚	•	•	•	•	•		•
26 Aluminium accumulation	✚	✚	✚	✚	✚	•	•	•	✚	•	•	•	•	•	•	•	•	•	•	✚	•	•	•	•	•	
27 Pre-Oligocene (Muller)	✚	✚	+	✚	•	•	✚	✚	+	•	•	•	✚	•	•	•	•	•	✚	✚	•	✚	−	•	•	✚
28 Pre-Tertiary (C.G.H.)	✚	+	•	✚	−	✚	✚	+	✚	✚	•	•	•	•	•	•	✚	•	+	✚	✚	✚	✚	•	•	✚
29 Rain-forest	✚	•	•	+	−	•	✚	✚	✚	✚	•	•	+	✚	•	✚	•	✚	•	✚	•	✚	•	•	•	✚

of character Y. By superimposing the two sheets and viewing them under transmitted light, those families with characters X and Y would be obvious and would be recorded on a third sheet. Say, the number of families with character X is x and the number with character Y is *y*, then the expected number (m') with both would be XY/259 on the basis of a null hypothesis. If the real occurrence (m) is greater than m', the characters X and Y are stated to be positively correlated; if less, then negatively correlated. Greater the difference between m' and m, the more significant the correlation. By calculating the value of X^2, the value of p can be ascertained from tables where p represents the probability that the difference between m' and m has occurred purely by chance.

In Table 3, characters are numbered from 1 to 26, where the degree of correlation between them is shown by various symbols. Positive correlations are indicated by plus signs and negative by minus signs. Bold symbols indicate correlations of significance greater than 50:1 and faint symbols those of significance between 50 : 1 and 20 : 1. A dot implies absence of correlation and a cross meaningless correlation.

These characters are rearranged in seven groups, the number in each being shown in the circles (Fig. 1), as follows: growth habit (1), vessels (2), leaves (3), flower (5), pollen (3), seeds (5), and biochemistry (3). Lines indicate intergroup correlations whose number in each case is depicted as a fraction of the total possible number. In each group, the number of intragroup correlations is shown in

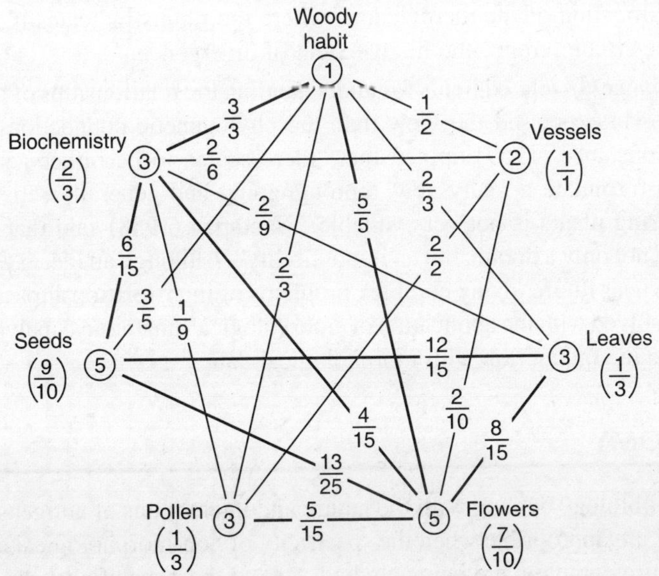

Fig. 1 Correlations among 22 characters summarised from Table 3. The characters are rearranged into seven groups, the number in each being shown in the circles. Lines indicate intergroup correlations, whose number in each case is shown as a fraction of the total possible number. Alongside each group, the number of intragroup correlations is shown in brackets (After Sporne, 1977).

brackets. Among the five flower characters, there exist seven correlations out of a possible ten; such intragroup correlations can be explained on the basis of functional association. Between the three leaf characters and five seed characters, there are twelve correlations out of a possible fifteen. Since these cannot be explained on the grounds of functional association, it is very likely that they result from evolutionary trends. The same statement holds good for most of the intergroup correlations, as they are between characters which occur in various regions of the plant, at different times in its development and different phases of its life cycle. If these correlations arise from evolutionary trends, some knowledge of the fossil record is valuable in establishing the direction in which they have proceeded. By examining the fossil record, it is possible to discover which characters are presently more common than the earlier record and which less common. The commoner of two opposing character states is thought to be the more primitive.

By use of these data, an *advancement index* has been constructed for each family. The score of 0 (most primitive) to 100 (most advanced) is based on the proportion of advanced characters that a family has. The average advancement index for the 15 families with exclusively multiaperturate pollen grains is found to be 77.2 and the families with restricted pauciaperturate pollen is 54.4. Similarly, the 50 families with ellagitannins in their tissues have a low average advancement index 47.3 in comparison to an average for the world flora of 56.6 and the 37 families where aluminium accumulators occur possess an average of 49.0. Furthermore, the Magnoliaceae, as defined by Cronquist, has the lowest advancement index (20) and four families—Callitrichaceae, Hippuridaceae, Hydrostachyacae and Phrymaceae—the highest (100). Among the most advanced families, there are at least two dissimilar types of flower: the extremely reduced type as in the Callitrichaceae and Hippuridaceae and gamopetalous zygomorphic type as in the Martyniaceae, Pedaliaceae and Phrymaceae. Several families, occurring in different orders, are almost equally primitive. However, no single family is alive today that can be treated as the ancestor of all other dicotyledons. Sporne (1974) put forward a circular representation of the phylogenetic classification of the dicotyledons where the radius indicates the advancement index and the position on the circumference the likely extent of divergence.

Despite the advances in related fields for understanding the relationships of plant groups, Sprague (1940) and Turrill (1951) expressed the view that, the phylogenetic connections of angiosperms are totally speculative. Constance (1957) opined that "there seems to be rather general agreement that sufficient evidence to formulate a really new, thoroughgoing and generally satisfactory phylogenetic arrangement of flowering plants is not yet available." Santapau (1958) said that "to adopt these new methods is to my mind not only a dream, but an impossibility". Although an ideal system of classification may not emerge in the near future, many complex problems of inter-relationships among different plant groups are likely to be solved with the application of embryological, anatomical, palynological, cytological and chemical evidence in conjunction with morphological data.

SEROTAXONOMY

Serology is a branch of biology dealing with the nature and interactions of antigens and antibodies. The serological procedures are appropriate when the specificity of non-morphological attributes, whether in an individual or an entire complex, are being studied. Due to their specificity, the antigen components of plants represent characteristic of taxonomic importance. Both structural and reserve proteins can be used in the field of systematics, as long as they belong to the same group and the same organs are always compared. Although a large mass of data has been based on proteins extracted from seeds and stem tubers, algal cells, fern spores, pollen grains, fruits and leaves can be employed as satisfactory antigenic material for systematic investigations.

In order to understand some of the principles and concepts of the serological phenomenon, it is necessary to provide annotated terminology (Fairbrothers, 1968).

1. **Antigen:** "antigenic material, antigenic system, or a single antigen molecule (complete or incomplete hapten)."
2. **Antigenic Material:** "substance capable of inducing the formation of antibodies (serum globulin) under appropriate conditions and which has the ability to react with the antibodies."

3. **Determinant Groups:** "the portion of the antibody molecule that reacts (combines) with a portion of the antigenic material."
4. **Precipitins:** "antibodies capable of combining with and reacting upon antigenic material (or haptens) because of certain determinant groups."
5. **Precipitin Reactions:** "serological reactions between antibodies and soluble antigenic material resulting in the formation of a precipitate."
6. **Immunoprecipitating Systems:** "precipitin reactions observable as band lines, or arcs in diffusion gel techniques."
7. **Reference Antigenic Material (Antigen) (R-Ag):** "the antigenic material used to immunize the antibody producer."
8. **Reference Reaction (R-Re):** "the reaction between an antiserum and the antigenic material used to stimulate its formation."
9. **Cross Reacting Antigenic Material (Antigen) (X-Ag):** "the antigenic material, other than the reference antigenic material which will react serologically with the antibodies."
10. **Cross Reaction (X-Re):** "the reaction between an antiserum and any antigenic material (or hapten) other than the antigenic material used in its formation."
11. **Specificity of Antiserum:** "the ability to react (antiserum reactivity) only with determinants contained in the reference antigenic system."
12. **Fidelity of Antiserum:** "the degree to which an antiserum reacts with all the different determinants of the reference antigenic material."

Several serological methods are available. The most common method is the use of serial dilutions of the antigenic material and a constant concentration of the antibody, the rates and amounts of precipitation being a function of time and temperature.

1. **Quantitative Precipitin Methods:**
   (*a*) *Heidelberger and Kendall Method.* The amount of antigen and antibody is measured by determining the difference between nitrogen content of the antigenic material and the precipitated complex.
   (*b*) *Boyden Procedure* (*BP*). A normal titration curve or the complete titration curve is a measure for estimating serological correspondence.
   (*c*) *Moritz Procedures* (*MP*). Here less of the antigenic and antibody material are necessary to obtain a value which is roughly equivalent to the whole curve required in the BP.
   (*d*) *Quantitative Ring* (*Layering*) *Precipitation Reaction* (*QRP*). A mixture of the Heidelberger and Boyden methods, this method measures turbidity at an antigen-antibody interface.
2. **Qualitative Precipitin Methods:** These methods result from the diffusion of antigens and antibodies in semi-solid media, being a migration from a zone of higher concentration to that of a lower concentration. In semi-solid medium, gelatin was once used; now-a-days, this has been replaced by agarose, various types of agar, polyacrylamide and alginate gels. Some techniques involve the use of an electric field to separate the extracts of antigenic material before antisera are placed (Immunoelectrophoresis or I.E.A.). Moreover, the techniques of diffusion precipitation testing may be the single diffusion method or the double diffusion method (which may again be performed as a one dimension or a two dimension method). Upon incubation, the antibodies react with their appropriate antigens forming precipitation zones (arcs, bands, lines).

3. **Electrophoresis:** In general, electrophoresis is a migration of proteins across a voltage gradient.

Disc electrophoresis (D.E.) is a case of zone electrophoresis which detect similarities and differences of protein molecules based on two specific properties. The use of acrylamide gel columns ensures the separation of protein mixture into narrow bands. The test sample is placed on the top of the column containing two gels, the upper or stacking gel having larger pore size allows the initial sorting of proteins into a closely layered stack of discs and the lower or separating gel facilitates these protein fractions to spread out into bands. The separated protein bands can be located by applying selected stains.

Disc immunodiffusion technique is a combination of D.E. and serological double diffusion method on the same slide. Here the location of precipitin bands is compared with an identical stained and non-stained D.E. gel column.

By the use of starch gel electrophoresis, the fractionated protein components can be detected as stained bands.

The serological and electrophoretic methods have been used to detect similarities and dissimilarities between organs of the same individual, population, races, varieties, subspecies, species, genera, tribes, families and orders.

Using precipitation reaction as an indicator of relationship in the Festucoideae, Fairbrothers and Johnson (1959) found antisera to proteins of a species giving little or no reaction with proteins from species estimated to be remotely related and intense reactions with proteins from species judged to be closely related. They also observed *Dactylis glomerata, Festuca rubra* and *Poa pratensis* showing strong serological correspondence with *Bromus inermis* and could separate six species of *Bromus* serodiagnostically. By applying immunological techniques to the taxonomy of *Solanum,* it has been possible to arrange 38 species into groups which is in accord with orthodox taxonomy and cytological affinity (Gell *et al.,* 1960). Hall and Johnson (1962) reported the value of an electrophoretic analysis of the amphidiploid of *Stipa viridula* × *Oryzopsis hymenoidea* and the parental species. Using starch gel electrophoresis, Coulson and Sim (1964) compared wheat endosperm patterns for eight species of *Triticum* with *Aegilops squarrosa* and obtained patterns for 34 varieties of *Triticum vulgare.* This technique has also been used in comparative studies of the seed proteins of amphidiploid species of *Brassica* (Vaughan and Waite, 1967*b*) and seed proteins of selected species of *Brassica* and *Sinapis* (Vaughan and Waite, 1967*a*). Vaughan *et al.,* (1966) compared the albumin and globulin fractions of seeds of three *Brassica* species, supporting the separation of the taxa into distinct species and pointing to a closer correspondence between *B. campestris* and *B. oleracea* than between either of these and *B. nigra.* The distribution of serological characteristics in the Ranunculaceae indicates close resemblance between *Aconitum-Delphinium, Actaea-Cimicifuga, Anemone-Clematis* and *Ranunculus-Myosurus* and suggests a common ancestry for *Aquilegia, Leptopyrum* and *Thalictrum* (Jensen, 1967). Using pollen collected from four taxa of *Typha,* Lee and Fairbrothers (1967) noticed essentially the same correspondence or similarities and dissimilarities with the serological data and D.E. data. These two types of data support the view that the cultivated *Magnolia denudata* and *M. liliflora* are mostly self sterile clones and often produce only hybrid seeds as a result of intercrossing (Pickering and Fairbrothers, 1967). Thurman *et al.* (1967) compared the electrophoretic mobilities of formic and glutamic dehydrogenoses in the Leguminosae at the levels of tribes and species. Studying the albumin and globulin fractions of the seeds of a wide range of *Brassica* and *Synapis* species, Vaughan and Denford (1968) presented their results in the form of a three dimensional model. They discovered that *B. nigra* is closer to *Sinapis* species than all other *Brassica* species and that *B. napus, B. juncea* and *B. carinata* are amphidiploids of the putative parents, *B. nigra,*

B. campestris and *B. oleracea*. Working on the serological assessment of relationships in the Solanaceae, Hawkes and Tucker (1968) recorded strong serological relationship between *Datura, Hyoscyamus, Nicotiana, Salpiglossis* and *Solanum,* demonstrated in *Datura* large differences at species level being accompanied by a clear change in the seed protein content and produced in *Schizanthus, Salpiglossis* and genera of the Scrophulariaceae faint cross-reactions. Kloz and Klozova (1968) reported on the occurrence of two protein characters in *Phaseolus* which can help in the selection of proteins in plant breeding research. Immunoelectrophoretic and serological double diffusion methods reveal patterns of protein similarities inconsistent with presumed connections among 35 taxa of the Triticeae and Agrosteae (Johansson and Hillebrand, 1969). A serological investigation of the intrageneric relationship in *Viburnum* was made by Hillebrand and Fairbrothers (1969) who indicated serological correspondence with the sections of the genus derived from analyses of gross morphological characters. In this work, representatives of the most primitive section show least serological correspondence with those of the most advanced section. Phytoserological systematic survey of the Caprifoliaceae by Hillebrand and Fairbrothers (1970*a*) points to a monophyletic or polyphyletic origin of the family as presently composed. According to Hillebrand and Fairbrothers (1970*b*), the serological correspondence between the Caprifoliaceae and Cornaceae suggests a closer relationship between these two families than between the Caprifoliaceae and the tested genera of the Rubiaceae. In making an electrophoretic comparison of six species of *Hesperaloe* and *Yucca,* Smith and Smith (1970) obtained individuality of protein patterns for each species. From serological evidence, it appears that *Bromus arvensis* and *B. scoparius* may be the diploid progenitors of *B. hordeaceus* or perhaps close allies of its progenitors (Smith, 1972).

Despite significant contributions made in recent years, some people are still sceptical about the value of serological evidence in modern taxonomy. "What exactly the varying degrees of serological correspondence tell us about relationship is very difficult to judge. If serological relationship could be shown to be more or less accurate reflection of overall genetic relationship, then clearly these methods would be of the greatest importance. That this is so, however, is very far from being established (Davis and Heywood, 1963).

NUMERICAL TAXONOMY

Adanson, a French botanist, was the first to hit upon a plan for assigning numerical values to the similarity between organisms. Until now, no useful procedure has been developed for such work. However, *taximetrics* or *numerical taxonomy* has come to the limelight in recent years (Sneath and Sokal, 1973).

Numerical taxonomy is an attempt to assign numerical values to the degree of similarity between taxonomic units and to build a classification on this basis. It can be, according to Sokal and Sneath (1963), defined as "the numerical evaluation of the affinity or similarity between taxonomic units and the ordering of these units into taxa on the basis of their affinities".

Numerical taxonomy has been the object of heated controversy during the last twenty years or so. The controversy lies not so much in the method as'in the merits or demerits of the concept. The idea is to obtain an index which is of universal application and valid. The degree of overall similarity of two or more organisms, represented by numerical values, will considerably help the traditional taxonomist who aims at creating a classification and the modern biosystematist who is trying to gain an understanding for the reasons leading to the similarities and dissimilarities between the taxa. "When combined with genetic, cytological, ecological and chemical studies, these methods promise a realistic estimation of the course of evolution" (Solbrig, 1970). Talking in a different plane, "offer a method of dealing with material that is fragmentary… Numerical techniques the linking of fossil stems and leaves, inflorescences and stems, etc., and even the linking of fossils with present day plants" (Cullen, 1968).

Stace (1980) enumerated seven main benefits of numerical taxonomy over conventional taxonomy; these resemble the concepts and methods of Adanson and hence are called *neo-Adansonian principles:*

1. a given classification will be better if based on more characters and the information content in the taxa is greater;
2. every character has equal weightage in creating natural taxa;
3. overall similarity between any two entities is a function of their individual similarities in each of the many characters upon which they are compared;
4. since character correlations differ in the groups of organisms under study, distinct taxa can be recognised;
5. assuming certain evolutionary mechanisms and pathways, phylogenetic conclusions can be drawn from the taxonomic structure of a group and character correlations;
6. taxonomy is practised and viewed as an empirical science;
7. classifications are based on phenetic similarity.

Any numerical taxonomic method contains the following five basic stages (Dale, 1968):

1. Selection of operational taxonomic units (OTU) or entities to be classified.
2. Selection of properties whose values provide a description of OTU. The property values may be subjected to transformations to establish some desirable feature.
3. Some measures of likeness between the OTU and/or previously erected groups of OTU.
4. Definition of rules applied to organise the OTU into groups.
5. Derivation of rules of allocation, permitting new OTU to be allocated to suitable group.

Three major types of indices have been developed for determining similarity between the taxonomic entities:

1. **Coefficients of Association:** These chiefly refer to characters of two state units expressed as + or – (or 1 or 0). A simple form is a numerical index S for the similarity of each pair of organisms examined which is as follows:

$$S = \frac{N_s}{N_s + N_d},$$

where N_s = number of positive features shared by any two OTUs; N_d = number of features positive in one OTU and negative in the other.

2. **Coefficients of Correlation:** These do not have to be scored in a plus or minus way, but their true values can be used. These are obtained by calculating the product moment correlation in the following manner:

$$r_{jk} = \frac{\sum_{i=s}^{n} (X_{ij} - \bar{X}_j)(X_{ik} - \bar{X}_k)}{\sum_{i=1}^{n} \sqrt{(X_{ij} - \bar{X}_j)^2 (X_{ij} - \bar{X}_k)^2}}$$

where j and k stand for the two units under comparison and X_{ij} stands for the value of the character i in unit j and X_{ik} for the value of character i in unit k; $\bar{X}_j$ and $\bar{X}_k$ stand for the mean for all characters in units j and k respectively; n is the number of characters.

3. **Coefficients of Distance:** A species is supposed to occupy a point in a multidimensional space, the number of dimensions being equal to the number of characters. The coefficient of distance between any two taxa is a measure of the distance between the points in this multidimensional space.

In order to describe each OTU, it is imperative to use a list of properties with their associated values. The sequence of properties can be fixed and the results given by a vector. By grouping vectors for each OTU, the familiar data *matrix* is secured. Vector representation leads to sorting rules like remote neighbour and close neighbour, but may be complicated by weighting of properties.

It is necessary to find out how lists can be constructed. The user will furnish the list structure. But two problems will arise: some means of combining the list structure descriptions of OTU and some measure of likeness are required. While the combination should be additive, a valuable statistic for *multistate data* is an important coefficient which is defined in the following way. If n OTU has the first value and m the second, the information can be arrived at by

$$(n + m) \log (n + m) = n \,.\, \log (n) - m. \log (m).$$

A measure of disorder, this can be easily extended to properties having more than two states which will include "not known" for missing values.

Properties may have values which are not only *multistate* (nominal scale), but also *quantitative* (interval or ratio scales), *ranked* (ordinal scale) or some other esoteric scale (ordered interval scale). A good measure of likeness can allow as many scales as possible and scaling can avoid some troubles.

Weighting is a delicate topic, causing a great deal of controversy among taxonomists. Though an objective weighting has been proposed, it has been of little use. The major difficulty is inherent in circularity of weighting, "the results supporting the weights and the weights producing the results".

Most of numerical taxonomy methods have been dominated by hierarchical sorting routes, there being three basic types. In *polythetic similarity,* the pair of OTU most alike are found, fused and simlarities with other OTU calculated; the next most similar pair is then located and the process repeated. *Monothetic dissimilarity* shows the use of a single character division, as in a dichotomous key. *Dissimilarity polythetic* methods attempt to divide the OTU into the two most different groups; here weights can be changed after each division. Apart from the choice of route, the other major sorting problem is the provision of a "level of importance" for any fusion or division and methods have been devised to estimate the route as a level indicator.

The primary object of numerical taxonomy is the organisation of data and the taxonomist has only contact with the procedures through the results obtained. To produce output in the most acceptable form, it will mean storing information not used in the analysis, such information, which include previous names, authorities, dates and bibliographic references, will be presented to the user on request. More use of graphic output appears desirable and it will be possible to exhibit diagrams and dichotomous keys to help identification.

With the use of two or more methods, it is possible to test correspondence. But these methods do not show the utility of any results. Random data have been used and can be classified into groups. It is difficult to assess the importance of this effect, as how the random data were prepared are not specified. However, usefulness can be measured by predicting random variables and collecting a frequency distribution of prediction probabilities.

While providing conclusions relating to the botanical problems of numerical taxonomy, Cullen (1968) asserted that the subject "is in a rapid state of development, and today's dogma may well be tomorrows heresy; new classifications produced by numerical means are still few of us to be able to decide precisely what the problems are ··· We should realize that Numerical classifications are not likely to supplant orthodox ones—they will either confirm them, or, if very different, exist side by side with them... The search for large number of unit characters will lead to the discovery of many facts about plants which, whether taxonomically useful or not, will be important biologically."

BIOLOGICAL SPECIES CONCEPT

The concept was advanced by Mayr (1957) and gained general acclaim through the advocacy of Stebbins, Clausen, Grant and others in botanical taxonomy (Heywood, 1974). This concept underscores reproductive isolation which is, in turn, a by-product of genetic divergence.

The biological species, according to Grant (1957), is defined as ' a community of cross fertilising individuals linked together by bonds of mating and isolated reproductively from other species." In short, a biological species is formed by "groups of interbreeding natural populations that are reproductively isolated from such other groups." In this concept, two aspects are stressed: (*i*) interbreeding between members of the same species and (*ii*) reproductive isolation between the members of different species.

Since the bulk species of the angiosperms are outbreeding, this aspect allows a good quantity of gene exchange between populations and accounts for the variability within species; this aspect contributes little by way of isolation. Inbreeding, on the other hand, reduces variability by curtailing gene exchange and aids in spreading homozygous population under the condition that there are no harmful influences for the homozygosity (Bradshaw, 1972).

Continued self fertilisation will create deleterious effects on the genetic make up of populations. Under fluctuating or unreliable conditions, selfing has proved helpful over interbreeding by ensuring pollination which leads to rapid growth in the population (Levin, 1971). This has been demonstrated in populations of *Agrostis tenuis* and *Anthoxanthum odoratum* which can tolerate heavy metals (Antonovics, 1968). In this case, autogamy is a contrivance for metal tolerance by insulating these populations against gene exchange from non-tolerant forms. However, selfing can evolve for reasons other than isolation. In *Armeria maritima,* the high degree of self fertility in metal tolerant populations compared to the normal ones is not subject to gene exchange from non-tolerant populations and has probably evolved in response to low density in colonising state. *Portulaca* is another example where autogamy has led to the production of several pure line populations which are isolated to such an extent from one another that each one of them has been assigned a separate binomial.

Interspecific incompatibility, one of the corner stones of the biological species concept, is an isolating mechanism between species that has been facilitated by evading fertilisation in the prezygotic phase. This is evident in *Linum, Lycopersicon, Nicotiana, Petunia* and *Solanum.* One of the main causes of this incompatibility of the related species in the lack of correlation between the length of the pollen tube and length of the style. Kostoff (1943) listed 68 interspecific crosses in *Nicotiana* which failed in view of the pollen tubes not being able to reach the ovules. In such crosses, generally the maternal styles are very long for the pollen tube to travel the path to the ovules. However, such crosses with those species possessing shorter pollen tube as females are reportedly fruitful.

Still another type of infertility has been recorded in the genus *Datura* where the pollen tube grows faster in the style of the same species and frequently bursts in the styles of other species. In comparison with the rate of domestic pollen tube growth, that of the alien pollen differs from one recipient to another.

Discordant evolution of genotype and phenotype poses an entirely different problem. Genotype does not undergo divergence and phenotype does in polymorphic species, whereas genotype diverges and phenotype remains the same in cryptic species. Such cases frequently confuse many practising plant taxonomists and correct decisions are difficult to arrive at. In evolution, the basic unit is a breeding population. A fundamental tenet of the biological species concept is the criterion of reproductive isolation. In asexually reproducing organisms, where there are neither breeding populations nor reproductive isolation, interbreeding is not relevant. Among angiosperms, several genera exhibit the phenomenon of apomixis. As a consequence, the biological species concept cannot be applied to these cases. The problem is so acute that "the most satisfactory solution in taxonomic practice seems to be a dualistic one .... It consists in defining the term species biologically in sexual organisms and morphologically in asexual ones". While agreeing with this statement, Maheshwari (1967) proposed the terms 'biospecies' and 'phenospecies' respectively for them.

Although the value of the concept lies in population biology and involves fertility tests; such tests have been limited to a few selected individuals and the results then applied to the entire population. In allopatric populations, these tests can be performed only under experimental conditions and are of mere academic interest because gene exchange does not take place in natural populations. Furthermore, local breeding populations differ considerably in their genetic properties and may have various kinds of kinship with such other populations. Among higher plants, there are many natural interspecific and intergeneric crosses which present formidable problems to the evolutionary taxonomists.

During the last three decades, the concept has stirred the imagination of biologists. In the domain of botany, the concept has remained largely conceptual rather than operational. Even Mayr (1957) admitted the hazards in applying the concept to animal organisms. Heywood (1974) observed that the "general acceptance of a concept is, of course, a different matter from acceptance in practice". Even the so called experimental studies which the concept itself encouraged lead to the interference that most species are not evolutionary units held together by gene flow and that efforts to accommodate all variations into units so defined are misguided. Simultaneously, the validity of the concept has been challenged on operational grounds by numerical taxonomists. Sokal and Crovollo (1970) jumped to the conclusion that the biological species concept is "neither operational nor heuristic are of practical value". They went to the extent of suggesting that this concept be abandoned in favour of the phenetic species concept which is the most suitable one to be associated with the taxonomic category 'species' (Raven, 1976).

HERBARIUM

The herbarium is a store house of plant specimens arranged in the sequence of an accepted classification. It is, in the words of Fosberg (1946), "a great filing system for information about plants, both primary in the form of actual specimens of the plants and secondary in the form of published information, pictures and recorded notes".

Herbaria are associated with research institutes, scientific societies, botanic gardens, universities and colleges. There are also many private herbaria. They may represent a local collection, the flora of a country or of a continent or more. They "... may be variously restricted in scope; their specimens vary from a few hundred to several millions" (Stace, 1980).

In *Index Herbariorum* Part I (*Regnum Vegetabile*, Vol. 92, 1974), over 1600 of the world's most important herbaria are listed. Part II of the same publication contains a list of the most important plant collectors and present locations of their specimens. A list of the nine large herbaria of the world, with approximate number of specimens in each, is given below:

Royal Botanic Gardens, Kew (K)	6,500,000
V.L. Komarov Botanical Institute, Leningrad, (LE)	6,500,000
Museum National d'Histoire Naturelle, Paris (P)	5,000,000

British Museum (Natural History), London (BM)	5,000,000
Conservatoire et Jardin Botaniques, Geneva (G)	5,000,000
Universite de Lyon, Lyon (LY)	3,800,000
Harvard University, Cambridge, USA (A + FH + GH)	3,800,000
Instituto Botanico della Universita, Florence (FI)	3,300,000
Universite de Montpellier, Montpellier (MPU)	3,300,000

A few words on the herbaria of this country. The present herbarium of the Indian Botanic Garden, known as the Central National Herbarium, dates from 1832. Previous to this date, collections were made by the herbarium. But they were taken to Europe by Wallich and distributed by him to various institutions. The herbarium houses the valuable collections of Clarke, Hooker, Thomson and others. From the quantity and quality of its specimens, it is the first herbarium in India and one of the most important in Asia. It includes about 10,000 typematerials and over 12,000 Wallichian specimens. Approximately, 12,000 photographs of Indian and other authentic specimens from Kew Herbarium, British Museum and Linnean Herbarium, about 4,000 drawings/illustrations and microfilms from 26 European herbaria have been procured.

Besides the Central National Herbarium, plant specimens arc also preserved in the different herbaria of India. The Herbarium of the Industrial Section at the Indian Museum, Calcutta, has over 50,000 specimens of useful plants; the National Botanic Gardens at Lucknow contain nearly 50,000 specimens and the Blatter Herbarium at Bombay possesses over 100,000. The Eastern Circle Herbarium located at Shillong has well about 115,000 specimens. The number of specimens in the Western Circle Herbarium at Poona has shot up to 125,000. The Southern Circle Herbarium at Coimbatore has piled up 190,000 specimens. Dehradun is the seat of the famous herbarium of the Forest Research Institute, containing about 300,000 specimens. The Northern Circle Herbarium at Dehradun and the Central Circle Herbarium at Allahabad contains about 60,000 and 45,000 specimens respectively. During the last 25 years, the Botanical Survey of India has organised the collection of more than 300,000 field numbers and the herbaria of the Survey hold over 2 million specimens.

Nearly 250 new taxa have been described and new reports of distribution covering more than 500 species have been from India.

The aim of " ... herbaria is ... to present a picture, by means of representative specimens, of the composition of the modern plant world ... herbaria must be the chief basis for future monographic and phytogeographical studies. But they also serve many others beside the taxonomist; the economic botanist, the ethnobotanist, the morphologist, the geneticist, and students of many other disciplines must seek much of their basic data in the collections and publications of systematic botanist."

FIELD AND HERBARIUM TECHNIQUES

With a view to handling fresh plants for pressing into herbarium specimens, one can (*a*) press each plant in a portable press composed of a pair of collecting frames, a few blotters and folded sheets of old newspaper or (*b*) accumulate the material in a special metal box called *vasculum* and press them as soon as possible or (*c*) carry the gathered specimens in light pack basket or back pack or paper sack and press them after returning to camp headquarters. A polyethene bag makes an excellent container for fresh plants and so does a square of oil cloth.

A *field notebook* is a must in a collecting trip. The purpose is to record information of taxonomic value not apparent from the specimen. Dates of collection, altitude, habitat, locality, plant size, branching, root system, flower colour, associated species and other items of interest should be incorporated into the field notes.

Collecting: The specimens should be collected to represent the species, the most satisfactory one being in the flowering stage and from a normal habitat. In herbaceous plants, it is essential to collect the subterranean parts. In some cases, bulb, corm, tuber or other underground portions are necessary for identification. For this reason, the collector should carry cutting and digging implements. "A geology pick, a heavy screwdriver, a sharpened tire iron, an unusually strong trowel, or almost any other sharp instrument may suffice" (Benson, 1957). A pocket knife is recommended for collecting shrub and tree specimens and a pair of pole type pruning shears for securing cones from a coniferous tree.

Pressing: The placing of specimens in the press is of vital concern, since the look of a permanent specimen will depend on how it has been pressed and dried. Due to the bulky nature of plants, it is customary to press them flat or trim them suitably. Every specimen should be arranged so as to appear as natural as possible and to avoid the overlapping of organs. The leaves must not be crumpled or piled on the top of one another. A few leaves may be turned over to show the lower and upper sides in the final specimen. Bulky organs may be sliced open or hollowed out to facilitate drying, but cotton batting or paper padding are often helpful around them. Showy flowers are "best taken to pieces and their parts pressed separately". Those flowers which have inflated corollas can be kept from being crushed by introducing a wad of cotton within the corolla. Some flowers need the use of *flimsies, i.e.,* folders of absorptive tissue paper or even flimsy newspaper above and beneath them.

Temporary labels should be placed in each specimen, after which the folder is closed and kept in the press, with a blotter underneath and above it.

Drying: In very humid climates, it is desirable to dry the plants before the tissues undergo decay. Under such conditions, the plants are dried with *artificial heat* (from incandescent bulbs, hot steam pipes, kerosene lanterns or petrol pressure cookstoves). Moderate heat is maintained in order to prevent decolourisation or side effects on the specimen. A sheet of corrugated material (which is a sheet of cardboard or thin metal like aluminium with fluted ducts), acting as a ventilator, is inserted between each pressing paper and its specimen. Safe and quick drying makes the specimen natural.

Mounting: After drying, the specimens are attached to sheets of herbarium paper of standard size (25.4 cm × 40.6 cm or 29.2 cm × 41.9 cm). Sometimes 100 percent rag paper is utilised, but it is not very cheap and a paper of lower rag content is usually preferred. Specimens are mounted in various ways, but the most satisfactory method involves the use of white gummed cloth strips (*Holland cloth*), "supplemented by a non-staining glue, which one applies to the flat surfaces by means of a spatula or by laying the specimen on a gluecoated glass plate and then transferring it to mounting sheet". These strips are to be used in such a way as not to obscure critical parts on the specimen. Since cellulose tape is not permanent, it should be avoided. Wire staples can also be employed for fastening specimens to sheets, but they tend to rust. Loose fragments, dissections, etc. are placed in paper packets or envelopes which are in turn, affixed to one corner mounting paper with pin.

Recently, many herbaria have adopted a new plastic aid in mounting specimens (Archer, 1950; Rollins, 1955). To begin with, ethyl cellulose and a resin are dissolved in a solvent made up of methyl alcohol and toluene. The result is a thick, syrupy and adhesive liquid.

This is applied by means of a slim nosed squeeze bottle or pressure oil gun. The method is rapid and gives a pleasing appearance to the mounted specimens.

Herbarium label is an indispensable part of the finished specimen. It varies in size and format, but is generally about 6.3 cm × 10.1 cm. It should be neat and legible. It is usually attached to the lower left hand corner of the sheet. It should bear the following data: (*a*) a banner headline to show the region of collection, accompanied by the name of the institution or person to designate ownership; (*b*) the scientific name of the plant with author citation; (*c*) the locality of collection, preferably with a small map with the spot encircled; (*d*) the habitat; (*e*) the name of the collector; (*f*) the collector's field observations; (*g*) the date on which the collection was made and (*h*) the name of the person who identified the plant (Porter, 1967).

Frequently, corrections are made in the name changes or systematic position on a separate sheet along with the name of persons suggesting them. It is then pinned or pasted near the previous label; these are called *annotation labels.*

Filing: Herbarium specimens are filed according to a particular system of plant classification. Those of the genus are placed within a *genus cover.* When there are several sheets of a given species, they are placed in a genus cover aside from the other species of that genus. In general, the genus cover has the name of the genus written on the lower left edge. Where the cover is restricted to specimens of a single species, the genus initial and species names are written on the lower right hand margin. It is customary to arrange the genera and species in alphabetical order under each family.

In major herbaria, the specimens are kept in specially built steel *herbarium cases* compartmented into pigeonholes and supplied with tight fitting doors. An index card is displayed outside each door.

Preserving: As herbarium specimens are damaged by insect attacks, they ought to be preserved by the use of *repellants* or *insecticides.*

Repellants are chemicals which prevent infestation by insect, but are not necessarily lethal to the insect. The two chief repellants used in herbarium management include napthalene compounds (naptha flakes) and paradichlorobenzene. To be effective, a small muslin bag is filled with a mixture of two parts napthalene and one part paradichlorobenzene and placed in a herbarium case. These bags are refilled with the mixture year after year.

Insecticides kill the insect by being digested or by contact. The *digestive poisons* are salts of mercury or arsenic and are applied by dipping the specimens (before mounting) in suitable concentration of the chemical. The *contact poisons* comprise cyanide, paradichlorobenzene, carbon disulphide, mixture of ethylene dichloride and carbon tetrachloride or dichlorodiphenyl trichloroethane (DDT). The first four are applied in the form of a gas by keeping the herbarium specimens in an enclosed chamber. The fifth chemical is dusted or sprinkled on the plants as soon as they are dried and removed from the plant press.

Handling: As herbarium sheets are flexible, judgement should be exercised in handling and examining the specimens. Carelessness can spoil valuable collections in no time. They should not be bent or turned over like the pages in a book. They should be picked up with both hands. Once the proper step has been attained through experience, single hand is generally sufficient. Sometimes, type specimens are segregated from the general collection in order to avoid undue wear and tear.

Herbarium specimens are often sent out on loans by major herbaria to enable workers to make corrections and revisions in the nomenclature or identification. The borrower should carefully adhere to the loan policy and supply adequately prepared annotations.

One way to build up a good collection at a little cost is through exchange of specimens among collectors and institutions. Those earmarked for exchange are unmounted, but should be provided with suitable labels.

Every precaution must be taken to prevent loss or ruin owing to fire. This means that the herbarium specimens should be stored in a fire proof building.

BOTANIC GARDENS

There arc 525 botanic gardens in the world. They "are of value not only to botanists, horticulturists, nurserymen, seedsmen, landscape gardeners, home gardeners, and foresters but also to thousands of tourists. Most of them are open to the public, and many offer instruction with special lectures and demonstrations to meet the needs and interests of individuals in all walks of life".

Among the old and well-known botanic gardens are the following, the year of establishment being given in brackets:

Botanic Garden of Padua, Padua, Italy (1545);
Botanic Garden of Heidelberg, University of Heidelberg, Germany (1593);
National Museum of Natural Histroy, Paris, France (1635);
Botanical Garden and Museum, Berlin, Germany (1646);
Royal Botanic Garden, Edinburgh, Scotland (1670);
Botanic Garden of the Academy of Sciences, Leningrad, Russia (1712);
Botanical Garden and Botanical Institute of the University of Vienna, Vienna, Austria (1754);
University Botanic Gardens, Cambridge, England (1762);
Indian Botanic Garden, Howrah, India (1787);
Botanic Garden of Innsbruck, University of Innsbruck, Austria (1793);
Botanic Garden of Harvard University, Cambridge, Mass., USA (1807);
Botanic Garden of Rio de Janeiro, Rio de Janeiro, Brazil (1808);
Botanic Gardens, Munich, Germany (1809);
Royal Botanic Gardens, Paradeniya, Ceylon (1810);
Botanic Gardens of New South Wales, Sydney, Australia (1816);
Government Botanic Gardens, Buitenzorg, Java (1817);
Botanic Garden, Washington, D.C., USA (1820);
Botanic Garden, Hamburg, Germany (1821);
Royal Botanic Garden, Kew, England (1841);
Missouri Botanical Garden, St. Louis, Mo., USA (1859);
Arnold Arboretum of Harvard University, Jamaica Plains, Mass., USA (1872);
The New York Botanical Garden, New York, N.Y., USA (1895);
Brooklyn Botanic Garden, Brooklyn, N.Y., USA (1910);
Montreal Botanic Garden, Montreal, Canada (1936).

In these botanic gardens are planted species from every corner of the globe, being sorted out according to habitats or by families. They have green house and outdoor plantings. Their libraries may contain reference books, pamphlets, monographs, manuals, floras, photographs and the herbaria may be extensive. For example, the Royal Botanic Garden, Kew, Surrey, England, of 288 acres, has a library of over 44,000 volumes, a herbarium of 6,500,000 specimens, an arboretum and fruticetum together of 7,000 species and varieties, 8,000 species growing out of doors and 13,000 species growing under glass. In 1965, the building of Jodrell Laboratory was completed; this provided working conditions for cytologists, geneticists and pliysiologists. There is only one seedbank at Wakehurst Place, the satellite garden of the Kew Gardens, which devotes itself entirely to the conservation of genetic resources of wild species.

This account will not be complete unless something is mentioned about the Indian Botanic Garden, Howrah, near Calcutta. It is situated on the west bank of the River Hooghly and covers an area of 110 hectares. Its present plant population is over 12,000. These plants are grouped on a geographical basis, *i.e.,* the typical plants of a particular country are planted in a particular area and are growing in harmony with the climatic conditions congenial to them. The library of the Garden contains over 30,000 books.

The Indian Botanic Garden is probably the largest in Asia. It has been an active centre of botanical and horticultural research in India. It is famous for its plant introduction work, about 80 species being successfully introduced during the last 70 years. Some of the achievements of the Garden authorities include the cultivation of Aloe, Cocoa, Coffee, Cardamon, India rubber, Henbane, Japanese Mulberry, Sarsaparilla, Tapioca, improvement of Indian Cotton, introduction of Cinchona, Mahogany, Tea, utilisation of Jute, etc. Sir J.D. Hooker aptly remarked that the Garden has "contributed more useful and ornamental tropical plants to the public and private gardens of the world than any other establishment before or since".

The main centre of attraction in the Garden is the great Banyan tree which is one of the largest trees in size in the world. Today, it is an object of wonder and admiration. It looks like a forest in miniature, rather than an individual tree and this appearance is caused by the hundreds of aerial roots which hang down from the branches to the ground, attaining striking development as subsidiary trunks. There are 1,825 such aerial roots actually rooted in the ground. The circumference of its canopy is more than 245 m and the maximum height attained by one of its branches is 29 m. The age of the tree is considered to be over 250 years.

Besides the Banyan tree, the interesting show spots in the Garden are the small Palm house, large Palm house, Orchid house, Fernary, Branching Palms, Screw pine collection, Pinetum and Bamboo collections, Cacti and succulent collection, giant Waterlily of the Amazon, green meadows, array of winter annuals, delightful walks, pleasant lakes and picturesque vistas. Germplasm collections of *Bougainvillea, Citrus, Jasminum, Nymphaea* and *Vitis* and a section on medicinal plants have been developed in recent years.

TAXONOMIC LITERATURE

Since taxonomy is basically a descriptive science, its literature is vast, varied and versatile. According to Porter (1967), "it runs the gamut from ponderous volumes to obscure notes in periodicals and even letters of correspondence between workers". The steady flow of documents presents the taxonomist with a vexing problem as to ascertain what has been published in his field of interest. In this connection, he may consult *Bulletin of the Torrey Botanical Club* and *Taxonomic Index* (a regular part of *Brittonia)* which lists titles of taxonomic publications. Another fertile source of desirable literature is noticed at the end of research papers in the form of references or bibliographies. Abstracts of published paper and texts are available in *Biological Abstracts, Bulletin Signaletique, Current Advances in Plant Science, Excerpta Botanica* (Section A), *Referativnyi Zhurnal* (*seriya Botanika*) and *Plant Breeding Abstracts.* All these information can be accumulated in card files, each card being keyed to author or subject. From an up to date file of this sort, one can obtain the pertinent data with a minimum of time and effort.

Publications dealing with taxonomy have always occupied a prominent position in botanical literature, the classification of which is a difficult task. An attempt will be made here to supply representative examples of the taxonomic literature.

1. **Classics:** The work of Theophrastus, Dioscorides, Brunfels, Cordus, Ray, etc. profoundly influenced the development of plant taxonomy and are regarded as landmarks in the history of botany.

2. **Manuals or Floras:** The plants of an area, in the aggregate, are referred to as the *flora.* We may speak about the flora of Japan or flora of Rajasthan or flora of the Eastern Himalaya. The term is also synonym of the botanical manual. A flora or manual of Assam is a description of all the plants known to occur in that State. The species in a flora or manual are arranged according to one or another of the available systems (Engler, Bentham-Hooker, etc.), giving for each plant the scientific name, author citation, reference to the source of the original publication, synonymy and geographical distribution within the area in question.

An usual part of the flora or manual is the *determinator* or *diagnostic key,* a device to identify an unknown species easily and quickly. Confronted with series of *couplets* or mutually exclusive pairs of *leads* or statements, the user arrives at the name of the unknown by following the right track. Two kinds of construction are in practice. The plants used in the following example are common genera of the Capparidaceae, using exactly the same data.

Indented key (all possibilities derived from the first lead are tackled before the second lead is stated).

Leaves simple	
Plants armed, climbing	1. *Capparis*
Leaves compound	
Plants unarmed, not climbing	
Flowers white; gynophore present	2. *Gynandropsis*
Flowers yellow; gynophore absent	3. *Polanisia*

Bracketed key (two leads of each couplet exist together)

(1) Leaves simple	(2)
(1) Leaves compound	(3)
(2) Plants armed, climbing	*Capparis*
(3) Plants unarmed, not climbing	(4)
(4) Flowers white; gynophore present	*Gynandropsis*
(4) Flowers yellow; gynophore absent	*Polanisia*

Keys may be *artificial* (based on clear cut and conspicuous characters) or *synoptic* (based on a bare minimum or diagnostic characters which may be cryptic or difficult to determine). Artificial keys are of two chief types: *single access* or *sequential keys* (which possess a single commencing point) and *multi-access* keys (which can be started at any point). Multi-access keys are usually produced on separate *punched cards* and not on pages in a book.

3. **Notes and Observations:** A survey of the taxonomic literature will reveal the existence of notes and observations concerning the range of plant forms, deviations from the normal structure and the new species or genus. This kind of information is significant, although the publication is often unsuitably titled and of varied nature.

4. **Monographs and Revisions:** A monograph is a treatise, covering a family or genus with valuable information of morphological or taxonomical nature. Though a monograph should include the group as it is found all over the world, the term is often loosely used and is applied to treatments limited to a continent, a country or even a small area. For this type of work, the term revision should be employed. Moreover, a revision is based on the studies of herbarium specimens. A monograph should embrace morphology, embryology, anatomy, palynology, cytology and ecology of the group, together with its taxonomy.

5. **Popular Treatments:** Those who would like to consult nontechnical work might find pleasure in flower books. A merit of such books is that they enable any person to identify the common plants or ornamentals by means of pictures alone. A demerit is that they do not cover all the plants of a given area and are generally incomplete. *The Macmillan Wild Flower Book* by C.J. Hylander and E.F. Johnston as well as *Wild Flowers of America* are good samples of highly illustrated flower books.
6. **Botanical Serials:** A huge amount of taxonomic literature comes out in the serial publications of educational institutions, learned societies and other organisations. Those out at regular intervals are termed *periodicals* and those out irregularly *bulletins.* It has been estimated that there are about 1,000 serials publishing articles on plant taxonomy, the most important among them being *Baileya, Brittonia, Kew Bulletin, Madrono, Rhodora, Taxon* and *Torreya.*
7. **Index:** They aim to index all plant names. They serve to pinpoint the source of the original publication of a name; to state to which order, family or tribe a given species may belong and to learn if a particular name has been applied before. The following list is a part and parcel of a taxonomist's library:

   Index Kewensis. Oxford, 1893–1895 and supplements to date;

   Index Holmensis. Sweden, since 1969;

   Gray Herbarium Card Index. Cambridge, Mass., USA, since 1873;

   Genera Siphonogamarum. Berlin, 1900–1907;

   Index Londinensis to Illustrations of Flowering Plants, Ferns and Fern Allies. Oxford, 1920–1941:

   Guide to the Literature of Botany. London, 1881 (reprinted in 1964);

   A Bibliography of Eastern Asiatic Botany. Jamaica Plain, Mass., USA, 1938.
8. **Dictionaries:** Dictionaries and glossaries form the nucleus of any taxonomic library. A botanical glossary is a list of botanical terms arranged in an alphabetical order and provided with their interpretations. To taxonomists, Lindley's *Glossary of Botanical Terms,* Asa Gray's *Structural Botany,* Le Maout and Decaisne's *A General System of Botany* and Stearn's *Botanical Latin* are helpful for the definition of the technical terms. On the other hand, a botanical dictionary is an alphabetical list of all known genera of certain taxa with their descriptions, e.g. *A Dictionary of Flowering Plants, and Ferns* by J.C. Willis.
9. **Rules:** Since 1867, botanists have met somewhat regularly at the international level and agreed upon legislation in plant names. These rules and regulations are contained in the ICBN = *International Code of Botanical Nomenclature.* They are subject to clarification or revision at each Botanical Congress and a new edition is then prepared. "The ICBN does not adjudicate upon taxonomic matters" and "merely lays down the criteria for naming a taxon whose circumscription, position and rank have been taxonomically decided" (Stace, 1980).
10. **Maps and Cartography:** Though this item may seem out of place here, taxonomists should have a knowledge of maps and their availability for the area under investigation or should be able to prepare maps based on their studies involving particular taxa.

SELECTED BIBLIOGRAPHY

ABRAHAM, A. 1958. Cytology and taxonomy. *Mem. Indian botan. Soc.,* No. 1, pp. 24–29.

ALSTON, R.E. and TURNER, B.L. 1963. *Biochemical systematics.* Prentice-Hall, Englewood Cliffs, New Jersey.

ARBER, A. 1938. *Herbals: Their Origin and Evolution.* 2nd edition. Cambridge University Press, Cambridge.

ARCHER, W.A. 1950. New plastic aid in mounting herbarium specimens. *Rhodora,* **52:** 298–299.

BAILEY, I.W. 1953. The anatomical approach to the study of genera. *Chron. Bot.,* **14:** 121–125.

BAILEY. I.W. 1956. Nodal anatomy in retrospect. *J. Arnold Arb.,* **37:** 269–287.

BATE-SMITH, E.C. 1962. The phenolic constituents of plants and their taxonomic significance. *J. Linn. Soc. (Bot.),* **58:** 95–173.

BEAMAN, J.M., ROLLINS, R.C. and SMITH, A.H. 1965. The herbarium in the modern university. *Taxon,* **14:** 113–133.

BECKER, K.M. 1973. A comparison of angiosperm classification systems. *Taxon,* **22:** 19–50.

BEHNKE, H.D. 1975. The bases of angiosperm phylogeny: Ultrastructure. *Ann. Mo. Bot. Gdn.,* **62:** 647–663.

BENDICH, A.J. and BOLTON, E.T. 1967. Relatedness among plants as measured by the DNA-agar technique. *Pl. Physiol.,* **42:** 959–967.

BENSON, L. 1957. *Plant Classification.* D.C. Heath, Boston.

BENSON, L. 1962. *Plant Taxonomy: Methods and Principles.* Ronald Press, New York.

BESSEY, C.E. 1915. The phylogenetic taxonomy of flowering plants. *Ann. Mo. Bot. Gdn.,* **2:** 109–164.

BISALPUTRA, T. 1962. Anatomical and morphological studies in the Chenopodiaceae. III. The primary vascular system and nodal anatomy. *Aust. J. Bot.,* **10:** 13–24.

BREWBAKER, J. 1967. The distribution and phylogenetic significance of binucleate and trinucleate pollen grains in angiosperms. *Am. J. Bot.,* **54:** 1069–1083.

BROWN, W.H. 1938. The bearing of nectaries on the phylogeny of flowering plants. *Proc. Am. Phil. Soc.,* **79:** 549–595.

CARLQUIST, S. 1961. *Comparative Plant Anatomy.* Holt, Rinehart and Winston, New York.

CHESTERS, K.I.H. 1964. Fossil plant taxonomy. *In*: W.B. Turrill (ed.). *Vistas in Botany,* Vol. 4, pp. 239–297. Pergamon Press, London.

CLAUSEN, J., KECK, D.D. and HIESEY, W.W. 1939. The concept of species based on experiment. *Am. J. Bot.,* **26:** 103–106.

COLE, A.J. (ed.). 1969. *Numerical Taxonomy.* Academic Press, London and New York.

CONSTANCE, L. 1951. The versatile taxonomist. *Brittonia,* **7:** 225–231.

CONSTANCE, L. 1957. Plant taxonomy in an age of experiment. *Am. J. Bot.,* **44:** 88–92.

CONSTANCE, L. 1964. Systematic botany—an unending synthesis. *Taxon,* **13:** 257–273.

CORE, E.L. 1955. *Plant Taxonomy.* Prentice-Hall, Englewood Cliffs, New Jersey.

CRONQUIST, A. 1968. *Evolution and Classification of Flowering Plants.* Thomas Nelson, London.

CULLEN, J. 1978. A preliminary survey of ptyxis (vernation) in the angiosperms. *Notes R. bot. Gdn. Edinb.,* **37:** 161–214.

DAHLGREN, R. 1975. A system of classification of the angiosperms to be used to demonstrate the distribution of characters. *Bot. Notiser,* **128:** 119–147.

DALE, M.B. 1968. On property structure, numerical taxonomy and data handling. *In:* V.H. Heywood (ed.). *Modern Methods in Plant Taxonomy,* pp. 185–197. Academic Press, London and New York.

DAVIS, G.L. 1966. *Systematic Embryology of the Angiosperms.* John Wiley, New York.

DAVIS, P.H. and HEYWOOD, V.H. 1961. *Principles of Angiosperm Taxonomy.* Oliver and Boyd, Edinburgh and London.

DIETERLE, J.V.A. 1960. Sandbags as a technical aid in mounting plants. *Rhodora,* **62:** 322–324.

DRESS, W.J. 1955. On the genera of scientific plant names. *Baileya,* **3:** 50mes. *Rev. Gen. Bot.,* **68:** 83–110.

DUFOU, O. 1961. Structure cytologique du pollen de quelques Angiospermes. *Rev. Gen. Bot.,* **68:** 83–110.

EAMES, A.J. 1963. Floral anatomy as an aid in generic limitation. *Chron. Bot.,* **14:** 126–132.

ELIAS, M.K. 1942. Tertiary prairie grasses and other herbs from the high plains. *Geol. Soc. Am. Spec. Pap.,* **41:** 1–176.

ELIAS, M.K. 1946. Taxonomy of Tertiary flowers and herbaceous seeds. *Am. Midl. Nat.,* **36:** 373–379.

EMBODEN, W.A., Jr. and LEWIS, H. 1967. Terpenes as taxonomic characters in *Salvia* section *Audibertia. Brittonia,* **19:** 152–160.

ERDTMAN, G. 1952. *Pollen Morphology and Plant Taxonomy: Angiosperms.* Almqvist and Wiksell, Stockholm.

EYDE, R.H. 1975. The bases of angiosperm phylogeny: Floral anatomy. *Ann. Mo. Bot. Gdn.,* **62:** 521–537.

FAEGRJ, K. 1937. Some fundamental problems of taxonomy and phylogenetics. *Botan. Rev.,* **3:** 400–423.

FAHN, A. and BROIDO, S. 1963. The primary vascularization of stems and leaves of the genera *Salsola* and *Suaeda* (Chenopodiaceae). *Phytomorphology,* **13:** 156–165.

FAIRBROTHERS, D.E. 1977. Perspectives in plant serotaxonomy. *Ann. Mo. Bot. Gdn.,* **64:** 147–160.

FERNALD, M.L. 1942. Some historical aspects of plant taxonomy. *Rhodora,* **44:** 21–43.

FOSBERG, F.R. 1946. The herbarium. *Sci. Mo.,* **63:** 429–434.

GIBBS, R.D. 1965. A classical taxonomist's view of chemistry in the taxonomy of higher plants. *Lloydia,* **28:** 279–299.

GIBBS, R.D. 1974. *Chemotaxonomy of Flowering Plants.* McGill Queen's Univ. Press, Montreal.

GILMOUR, J.S.L. and TURRILL, W.B. 1941. The aim and scope of taxonomy. *Chron. Bot.,* **6:** 217–219.

GOTTLIEB, L.D. 1977. Electrophoretic evidence and plant systematics. *Ann. Mo. bot. Gdn.,* **64:** 161–180.

HALLIER, H. 1905. Provisional scheme of the natural (phylogenetic) system of flowering plants. *New Phytol.,* **4:** 151–162.

HARBORNE, J.B. 1970. *Phytochemical Phylogeny.* Academic Press, London.

HAWKES, J.G. (ed.). 1969. *Chemotaxonomy and Serotaxonomy.* Academic Press, London and New York.

HEGNAUER, R. 1962–69. *Chemotaxonomie der Pflanzen.* Vols. 1–5. Birkhauser Verlag, Basel.

HEGNAUER, R. 1967. Chemotaxonomy—past and present. *Lloydia,* **28·** 267–278.

HEGNAUER, R. 1967. Chemical characters in plant taxonomy: some possibilities and limitations. *Pure appl. Chem.,* **14:** 173–187.

HENRY, A.N. and CHANDRABOSE, M. 1980. *An Aid to the International Code of Botanical Nomenclature.* Today and Tomorrow's, New Delhi.

HENNIG, W. 1966. *Phylogenetic Systematics.* Tr. by D.D. Davies and R. Zangerl. University Illinois Press, Urbana.

HESLOP-HARRISON, J. 1956. *New Concepts in Flowering Plant Taxonomy.* Harvard University Press, Cambridge, Mass.

HEYWOOD, V.H. (ed.). 1968. *Modern Methods in Plant Taxonomy.* Academic Press, London and New York.

HICKEY, L.J, and WOLFF, J.A. 1975. The bases of angiosperm phylogeny: Vegetative morphology. *Ann. Mo. Bot. Gdn.,* **62:** 538–589.

HOYER, B.H., McCARTHY, B.J. and BOLTON, E.T. 1964. A molecular approach in the systematics of higher organisms. *Science,* **144:** 959–967.

HUTCHINSON, J. 1926. *The Families of Flowering Plants.* I, *Dicotyledons.* Macmillan, London.

HUTCHINSON, J. 1934. *The Families of Flowering Plants.* II, *Monocotyledons.* Macmillan, London.

HUTCHINSON, J. 1948. *British Flowering Plants.* Macmillan, London.

HUTCHINSON, J. 1969. *Evolution and Phytogeny of Flowering Plants.* Academic Press, London and New York.

HUTCHINSON, J. 1973. *The Families of Flowering Plants,* 3rd edn. Oxford University Press, London.

JANAKI AMMAL, E.K. 1960. The effect of the Himalayan uplift on the genetic composition of the flora of Asia. *J. Indian botan. Soc.,* **39:** 327–334.

JEFFREY, C. 1982. *An Introduction to Plant Taxonomy.* 2nd edn. Cambridge University Press, Cambridge.

JENSEN, U. 1967. Die Verwandtschaftsverhaltnisse innerhalb der Ranunculaceae aus serologischer Sicht. *Ber. dt. Bot. Ges.,* **79:** 407–412.

JOHANSSON, H. and HILLEBRAND, G.H. 1969. A serological and electrophoretic comparison of proteins in crude saline extracts of the Triticeae and Agrosteae. *Hereditas,* **63:** 429–443.

JOHNSON, A.M. 1931. *Taxonomy of the Flowering Plants.* Century, New York.

JOHRI, B.M. and SINGH, H. 1959. The morphology, embryology and systematic position of *Elytaria acaulis* (Linn. f.) Lindau. *Bot. Notiser.,* **112:** 227–251.

JONES, K. 1978. Aspects of chromosome evolution in higher plants. *Adv. Bot. Res.,* **6:** 119–194.

JORGENSEN, C.A. 1925. Zur Fragc der systematischen Stellung der Callitrichaceen. *Jb. wiss. Bot.,* **64:** 440–442.

KAPIL, R.N. and VIJAYARAGHAVAN, M.R. 1962. Embryology and systematic position of *Pentaphragma horsefieldii* (Miq.) Airy Shaw. *Curr. Sci.,* **31:** 270–272.

KAUSIK, S.B. and SUBRAMANYAM, K. 1946. A contribution to the life history of *Sphenoclea zeylanica* Gaertn. *Proc. Indian Acad. Sci.,* B. **23:** 274–278.

KHOSHOO, T.N. 1960. Biosystematics of *Sisymbrium irio* complex. *J. Indian Botan. Soc.,* **39:** 217–226.

KOUPRIANOVA, L.A. 1954. Sur la phylogenie des monocotyledones (d' apres les donnees palynologiques). *Essais de botanique.* Editions Acad. Sci. U.R. SS. Tome **1:** 111–128.

KUBITZKI, K. (ed.). 1977. *Flowering Plants. Evolution and Classification of Higher Categories.* Springer-Verlag, Vienna.

KUNDU, B.C. 1966. Intensification of studies on plant taxonomy. *Sci. Cult.,* **32:** 114–119.

LAKHANPAL. R.N. 1958. Palynology and taxonomy. *Mem. Indian botan. Soc.,* No. 1, pp. 19–23.

LAM, H.J. 1948. Classification and the new morphology. *Acta Biotheor.,* **8:** 107–154.

LAM, H.J. 1959. Taxonomy: general principles and angiosperms. *In:* W.B. Turrill (ed.). *Vistas in Botany,* Vol. 1, pp. 3–75. Pergamon Press, London.

LAWRENCE, G.H.M. 1951. *Taxonomy of Vascular Plants.* Macmillan, New York.

LEPPIK, E.E. 1953. Some viewpoints on the phylogeny of rust fungi.I. Coniferous rusts. *Mycologia,* **45:** 46–74.

LINDLEY, J. 1853. *The Vegetable Kingdom.* B. Evans, London.

MABRY, T.J. 1977. The order Centrospermae. *Ann. Mo. Bot. Gdn.,* **64:** 210–220.

MACLEOD, A.M. and McCORQUODALE, H. 1958. Water soluble carbohydrates of seeds of the Gramineae. *New Phytol.,* **57:** 168–182.

MAHESHWARI, P. 1964. Embryology in relation to taxonomy. *In:* W. B. Turrill (ed.). *Vistas in Botany.* Vol. 4, pp. 55–97. Pergamon Press, London.

MASON, H.L. 1950. Taxonomy, systematic botany and biosystematics. *Madrono,* **10:** 193–208.

MATHEW, P.M. 1958a. Studies on Piperaceae. *J. Indian botan. Soc.,* **37:** 155–171.

MATHEW, P.M. 1958b. Studies on the Menispermaceae. *Proc. Indian Acad. Sci.,* B. **47:** 274–286.

McCARTHY, B.J. and BOLTON, E.T. 1963. An approach to the measurement of genetic relatedness among organisms. *Proc. nat. Acad. Sci. (USA),* **50:** 156–164.

MERRILL, E.D. 1943. Some economic aspects of taxonomy. *Torreya,* **43:** 50–64.

METCALFE, C.R. 1946. The systematic anatomy of the vegetative organs of the angiosperms. *Biol. Rev.,* **21:** 159–172.

METCALFE, C.R. 1954. An anatomist's views on angiosperm classification. *Kew. Bull.,* **9:** 427-440.

MEZ, C. 1926. Die Bedeutung der Serodiagnostik fur die stammesges-chichtliche Forschung. *Bot. Arch.,* **16:** 1–23.

MIROV, N.T. 1963. Chemistry and plant taxonomy. *Lloydia,* **26:** 117–124.

MISRA, R. 1967. Form, function and factors in ecology. *J. Indian botan. Soc.,* **46:** 114–153.

PALSER, B.F. 1975. The bases of angiosperm phylogeny: Embryology. *Ann. Mo. botan. Gdn.,* **62:** 621–646.

PANDEYA, S.C. 1967. Contributions to botany by Prof. R. Misra. *J. Indian botan. Soc.,* **46:** 289–313.

PATHAK, G.N. 1940. Studies in the cytology of cereals. *J. Genet.,* **39:** 437–467.

PENNELL, F.W. 1948. The taxonomic significance of an understanding of floral evolution. *Brittonia,* **6:** 301–308.

PERCIVAL, M.S. 1961. Types of nectar in the angiosperms. *New Phytol.,* **60:** 235–281.

PORTER, C.L. 1967. *Taxonomy of Flowering Plants.* W.H. Freeman, San Francisco and London.

PULLE, A. 1938. *Compendium van de Terminologie, Nomenclatuur en Systematick der Zaadplanten.* Utrecht.

PURI, V. 1958. Floral anatomy and taxonomy. *Mem. Indian botan. Soc.,* No. 1, pp. 15–18.

RADFORD, A.E., DICKISON, W.C., MASSEY, J.R. and BELL, C.R. 1974. *Vascular Plant Systematics.* Harper and Row, New York.

RAM, M. 1956. Floral morphology and embryology of *Trapa bispinosa* Roxb. with a discussion on the systematic position of the genus. *Phytomorphology,* **6:** 312–323.

RAMAKRISHNAN, P.S. 1961. Calcicole and calcifuge problems in *Euphorbia thymifolia* Linn. *J. Indian Botan. Soc.,* **40:** 66–81.

RENDLE, A.B. 1904. *The Classification of Flowering Plants,* Vol. 1, *Gymnosperms and Monocotyledons.* Cambridge University Press, Cambridge.

RENDLE, A.B. 1925. *The Classification of Flowering Plants.* Vol. 2, *Dicotyledons.* Cambridge University Press, Cambridge.

RIBEREAU-GAYON, P. 1972. *Plant Phenolics.* Oliver and Boyd, Edinburgh and London.

ROBINSON, T. 1963. *The Organic Constituents of Higher Plants—Their Chemistry and Interrelationships.* Burgess, Minneapolis.

ROGERS, D.J. 1963. Taximetrics—New name, old concept. *Brittonia,* **15:** 258–290.

ROLLINS, R.C. 1953. Cytogenetical approaches to the study of genera. *Chron. Bot.,* **14:** 133–139.

ROLLINS, R.C. 1955. The Archer method of mounting herbarium specimens. *Rhodora,* **57:** 294–299.

ROLLINS, R.C. 1957. Taxonomy of higher plants. *Am. J. Bot.,* **44:** 188–196.

ROLLINS, R.C. 1965. On the bases of biological classification. *Taxon,* **14:** 1–6.

ROMEIKE, A. 1978. Tropane alkaloids—occurrence and systematic importance in angiosperms. *Bot. Notiser,* **131:** 85–96.

ROY, R.P. 1959. Genome analysis of *Aegilops sharonensis. Genetica,* **29:** 331–357.

SAMUELSSON, G. 1913. Studien uber Entwicklungsgeschichte der Blutten einiger Bicornes-Typen. *Svensk. Bot. Tidskr.,* **7:** 97–188.

SANTAPAU, H. 1962. The present state of taxonomy and floristics in India after independence. *Bull, Botan. Surv. India,* **4:** 209–216.

SAVILLE, D.B.O. 1954. The fungi as aids in the taxonomy of flowering plants. *Science,* **120:** 583–585.

SHARMA, A.K. 1956. A new concept of a means of speciation in plants. *Caryologia,* **9:** 93–103.

SHARMA, A.K. 1954. Cytology as an aid in taxonomy. *Bull. Botan. Soc., Bengal,* **18:** 1–4.

SHUKLA, P. and MISRA, S.P. 1979. *An Introduction to Taxonomy of Gymnosperms.* Vikas Publishing House, New Delhi.

SIMOLA, L.K. 1968. Comparative studies on the amino acid pools of three *Lathyrus* species. *Acta Bot. Fennica,* **81:** 1–62.

SINNOTT, E.W. 1914. Investigations on the phylogeny of angiosperms. I. The anatomy of the node as an aid in the classification of angiosperms *Am. J. Bot.,* **1:** 303–322.

SIVARAJAN, V.V. 1984. *Introduction to Principles of Plant Taxonomy,* Oxford & IBH Publishing Co., New Delhi.

SMITH, A.C. 1957. Fifty years of botanical nomenclature. *Brittonia,* **9:** 2–8.

SMITH, P.M. 1976. *The Chemotaxonomy of Plants.* Edward Arnold, London.

SMITH, P.M. and BAILEY, I.W. 1941. *Brassiantha,* a new genus of Hippocrateaceae from New Guinea. *J. Arnold Arb.,* **22:** 389–394.

SNEATH, P.H.A. and SOKAL, R.R. 1973. *Numerical Taxonomy.* W.H. Freeman, San Francisco.

SOLBRIG, O.T. 1970. *Principles and Methods of Plant Biosystematics.* Macmillan, London.

SOSA-Bourdouil, C. 1954. Composition chimique des pollens et Systematique. *Rap. et. Comm., Huitieme Congress International de Botanique, Paris. Soc.*, **6:** 251–253.

Soueges, R. 1952. L'albumin et l'embryon chez le *Callitriche vernalis* Kuetz. (*C. vernalis* L.). *C.R. Acad Sci., Paris,* **248:** 453–456.

SPORNE, K.R. 1974. *The Morphology of Angiosperms.* Hutchinson University Library, London.

STACE, C.A. 1980. *Plant Taxonomy and Biosystematics.* Edward Arnold, London.

STAFLEU, F.A. *et al.* 1978. *International Code of Botanical Nomenclature* Bohn, Scheltema and Holkema, Utrecht.

STEBBINS, G.L. 1970. Transference of function as a factor in the evolution of seed and their accessory structures. *Israel J. Bot.*, **19:** 59–70.

STEBBINS, G.L. 1974. *Flowering Plants. Evolution above the Species Level.* Edward Arnold, London and Harvard University Press, Cambridge, Mass.

STREET, H.E. (ed.). 1978. *Essays in Plant Taxonomy.* Academic Press, London.

SUBRAMANYAM, K. 1950. A contribution to our knowledge of the systematic position of the Sphenocleaceae. *Proc. Indian Acad. Sci.,* B. **31:** 60–65.

SWAIN, T. (ed.). 1963. *Chemical Plant Taxonomy.* Academic Press, London and New York.

SWAIN, T. (ed.). 1966. *Comparative Phytochemistry.* Academic Press, London and New York.

TAKHTAJAN, A.L. 1959. *Die Evolution der Angiospermen.* Gustav Fischer, Jena.

TAKHTAJAN, A.L. 1966. *Systema et Phylogenia Magnoliophytorum.* Nauka, Moscow and Leningrad.

TAKHTAJAN, A.L. 1969. *Flowering Plants. Origin and Dispersal.* Tr. by C. Jeffrey, Oliver and Boyd, Edinburgh.

TANDON, S.L. and RAO, G.R. 1966. Interrelationship within the *Solanum nigrum* complex. *Indian J. Genet. Pl. Breed.*, **26:** 130–141.

THOMAS, H.H. 1936. Palaeobotany and the origin of angiosperms. *Botan. Rev.*, **2:** 397–418.

THORNE, R.F. 1976. A phylogenetic classification of the Angiospermae. *Evol. Biol,* **9:** 35–106.

TIPPO, O. 1942. A modern classification of the plant kingdom. *Chron. Bot.*, **7:** 203–206.

TIPPO, O. 1946. The role of wood anatomy in phylogeny. *Am. Midl. Nat.*, **36:** 367–372.

TURESSON, G. 1922. The species and variety as ecological units. *Hereditasy,* **3:** 100–113.

TURRILL, W.B. 1942. Taxonomy and phylogeny. *Botan. Rev.*, **8:** 247–270.

TURRILL, W.B. 1951. Modern trends in the classification of plants. *Taxon,* **1:** 17–19.

WALKER, J.W. and DOYLE, J.A. 1975. The bases of angiosperm phylogeny: Palynology. *Ann. Mo. bot. Gdn.*, **62:** 664–723.

WALTERS, S.M. 1961. The shaping of angiosperm taxonomy. *New Phytol.*, **60:** 74–84.

WATSON, L., WILLIAMS, W.T. and LANCE, G.N. 1966. Angiosperm taxonomy: a comparative study of some numerical techniques. *J. Linn. Soc. (Bot.)*, **59:** 491–501.

WETTSTEIN, R. 1930–35. *Handbuch der Systematischen Botanik.* Ed. 4. Deuticke, Leipzig und Wien.

YOUNG, D.J. and WATSON, L. 1970. The classification of the dicotyledons: A study of the upper levels of the hierarchy. *Aust. J. Bot.*, **18:** 387–433.

PART TWO

Special Topics on Angiosperms

Special Topics on Angiosperms

THIS PART deals with certain topics which have their bearing on the morphology, embryology and phylogeny of the angiosperms.

PHYLLODE THEORY

A.P. de Candolle (1827) posed the problem whether the linear outline and parallel venation of the monocotyledon leaf might not be ascribed to their phyllode nature. The theory, strongly supported by Arbei (1930), tries to ascertain the real nature of monocotyledonous leaves. The phylloid nature is looked upon as an adaptation in the development of the aquatic habit.

The sheathing leaves of some dicotyledon leaves (Umbelliferae) closely resemble the monocot condition. The petioles of such a genus as *Triglochin* is reminiscent of the terete leaves of *Allium, Sagittaria* and *Sisyrinchium.* The sheathing leaf bases and petioles possess parallel venation. The species of *Sagittaria* exhibit stages in the flattening of terete leaves by two series of vascular bundles oriented with phloem towards the dorsal and ventral surfaces. The vascular bundles are in two opposite rows and in opposite directions, resulting in bifacial leaves. In some flattened leaves, the two series come to one plane but alternate irregularly in orientation. This type of vascular anatomy is found in the leaves of the Hydrocharitaceae, Pontederiaceae, etc. Apparently, dorsiventral leaves are also found by its anatomical structure to be a modified cylindrical leaf.

Not only the blade, but also the petiole may be reduced in the course of evolution. In *Hyacinthus, Doryanthes* and *Dracaena,* the petiole is reduced to a vestigial structure and the leaf base forms a photosynthetic structure. In *Hemerocallis,* the petiole is completely lost and leaf base forms the whole leaf.

Along with the reduction, there is an expansion of the tip of the phyllode to constitute a secondary blade. This is evident in the Alismataceae, Scitaminales, *Eichhornia, Smilax,* etc. In such instances, the blade is petiolar in nature with the venation commonly intermediate between the parallel and reticulate types.

Exception is to be noted in the case of a palm leaf where the early stages of ontogeny show that it is a complete leaf with sheath, petiole and blade. Its development follows the pattern of a typical dicot leaf.

It has been questioned whether the phyllode theory can be applied to *Dioscorea pentaphylla,* a monocotyledonous plant bearing palmately compound leaves. Moreover, the theory pre-supposes lamina, leaf base and petiole as distinct entities. Many a morphologist at present devote their attention to meristematic processes occurring during leaf ontogeny (Sporne, 1974).

LEAFSKIN THEORY

It has been suggested that the outer tissues (skin) of the shoot axis are made up of decurrent leaf bases (Saunders, 1922). The skin is formed by the downward continuation and development of the leaf base as the internode elongates. The skin has been demonstrated for some groups, both dicots and monocots.

It is often stated that the epidermis and hypodermis are foliar in nature and not axial. This condition is reported to exist in *Globba bulbifera.* Here the axis is not a separate unit. It is formed of the leafskin along with the true stem. In a transverse section, the axis reveals a pseudo cortex which belongs to the leafskin.

The stem of *Fritillaria imperialis* shows colour and texture variations, the lower portion being green and thick and the upper violet and thin. The lower region of the axis is interpreted as consisting of the prolonged leaf sheath and the axis proper.

In the rhizomes of *Corallorhiza* and other monocots, the leafskin enters into the surface organisation and formation of the axis texture.

In *Danae racemosa,* the leaves are reduced to ephemeral scales. Here the leaf bases can be traced down the internodes as ridges.

The so called *squamulae intravaginales* present in some members of the Helobiales have been explained as axillary in nature. But they are really appendages of basal regions of skin of the next leaf above.

EVOLUTION OF THE INFLORESCENCE

There are three different views to account for the origin of the inflorescence. The first view, held by Nageli (1884), Celakovsky (1892) and Pilger (1922), postulates that the primitive type is a panicle. Goebel (1931) explained how the various types of inflorescence might evolve from this type by reduction. The second view, supported by Parkin (1914), proposes the construction of large compound inflorescence from the solitary flower by the addition of flowers by axillary growth. The third view, accepted by Rickett (1944), states that the dichasium represents the primitive type giving rise to other types of inflorescence by slight structural modifications.

According to Rickett, the dichasium gives rise to such inflorescence types as bostryx, cincinnus, scorpioid cyme, helicoid cyme, etc. by the elimination of lateral buds. By the apparent elimination of internodes, the dichasium may form umbel, capitulum and corymb. The origin of spike or raceme involves "the reduction of individual dichasia to as little as a single flower accompanied or not by similar reduction of leaves to subtending bracts".

The most recent attempt to take a broader view of the problem of inflorescence morphology has been made by Troll (1969). According to him, all diverse forms of inflorescences can be put under two major types called *polytelic* and *monotelic.* The difference between the two types is based on the difference in the termination of the axes of complex inflorescences termed *synflorescences.* The distinguishing feature of a *polytelic synflorescence* is that the main axis ends in a group of flowers designated as the principal inflorescence, below which there are lateral shoots of *paracladia* repeating the morphological organisation of the terminal inflorescence and are called *coinflorescences.* Without exception, polytelic synflorescences are of the open type, since the apices of both the principal inflorescence and coinflorescences do not produce terminal flowers and all flowers terminate secondary axes. In the *monotelic synflorescence,* a terminal flower is formed at the tip of the principal and lateral axes. So monotelic synflorescences correspond to the closed type of inflorescence. Under the type of synflorescences both paniculate and various categories of thyrsoid inflorescences are included. Polytelic inflorescences are exemplified by *Capsella bursa-pastoris* where inflorescences are simple racemes and by the more complex cymoid synflorescences of *Galeopsis* and other members of the Labiatae.

The primitive order Ranales contains both solitary flower and flower cluster. If the ancestry of the angiosperms is traced to the Glossopterids, the solitary flower may not be preferred to the flower cluster. In his gonophyll hypothesis, Melville (1962) regarded the epiphyllous flower cluster to be primitive. This is in contradiction with the current idea. Evidence from morphology, anatomy and cytology support the primitiveness of the Ranales. The solitary flower, representing the primitive state, is seen in certain woody genera—*Calycanthus, Degeneria, Dillenia, Eupomatia, Himantandra* and *Magnolia.* But panicles are dominant in woody plants as well as some herbaceous families (Ranunculaceae, Rosaceae and Liliaceae). The solitary flower of *Magnolia* is determinate and the bulk of "large much branched inflorescence are at least in part determinate". In *Annona* and related genera, there are cymose flower clusters and the solitary flowers are followed by bracts along the peduncle with or without abortive flowers; this is a proof of the fact that solitary flowers are developed by reduction.

ALTERNATION OF GENERATIONS

In the life cycle of an organism which reproduces sexually, there is an occurrence of two phases or generations. One of these generations produces the gametes and is termed the *gametophyte generation.* The other, which produces the spores, is known as the *sporophyte generation.*

The components of the angiosperm gametophyte and sporophyte are shown below:

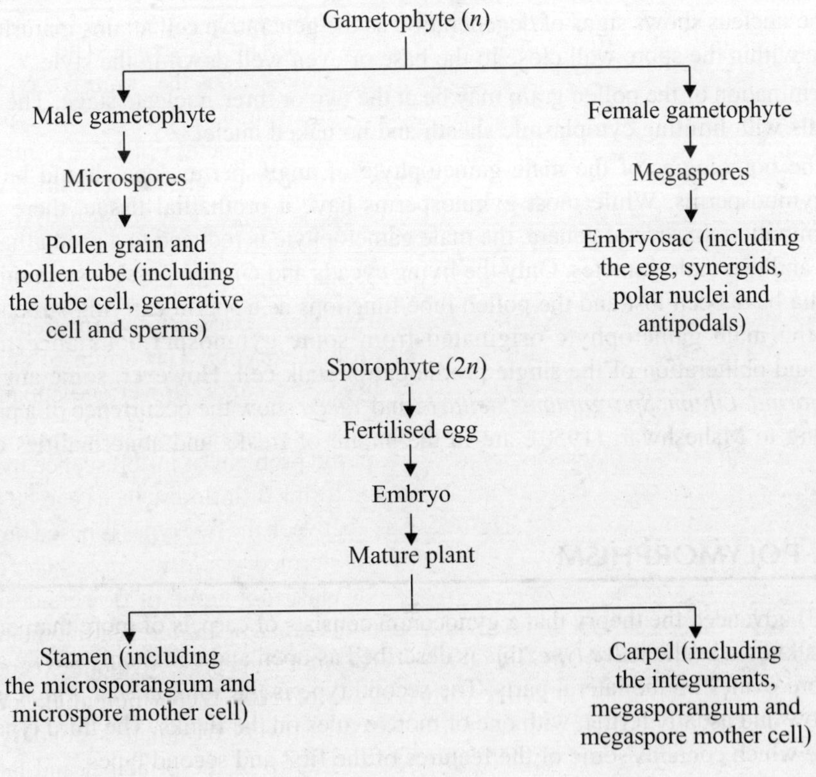

MALE GAMETOPHYTE

The development of the male gametophyte is uniform in angiosperms. The microspore is the first cell of the gametophyte generation. When newly formed, it has a centrally placed nucleus, dense cytoplasm and delicate wall. The cell increases rapidly in size and volume. The nucleus comes to a peripheral position and a central vacuole develops. Germination usually occurs some days or weeks after maturation, but in tropical plants it is immediate. In temperate plants, the sporogenous tissue may overwinter at the spore mother cell stage, mature microspore stage or at the two cell stage of the gametophyte. The first division of the microspore gives rise to a large vegetative or tube cell and a small spermatogenous or generative cell. The vegetative nucleus has a prominent nucleolus, whereas the generative nucleus has a small nucleolus or none. When freed from the spore wall, the generative cell becomes ellipsoid or fusiform.

The formation of pollen tube on the stigma takes place immediately (*Saccharcum, Sorghum*) or within 5–10 min (*Taraxacum, Zea*) or after several hours to even after two days. The grain swells and the intine protrudes outside through the germpores. One of the protrusions develops into a pollen tube. The contents of the grain pass into the tube. The tube usually pushes between the cells of the stigmatic surface and inner tissues of the stigma and style. In some taxa, the tube forms a canal from the stigma to ovules by destroying the adjacent tissues. The tube may lie deep within the stylar tissue or move through the protoxylem.

The pollen tube may penetrate through the chalaza and is called *chalazogamy.* It may enter through the integuments and is termed *mesogamy.* Sometimes, the entrance of the tube is by the micropyle and is known as *porogamy.*

The tube nucleus shows signs of degeneration as the generative cell attains maturity. Its division may take place within the spore wall close to the base or even well down in the style.

The germination of the pollen grain may be at the two or three nucleate stage. The male gametes are definite cells with limiting cytoplasmic sheath and no naked nuclei.

As to the homologies of the male gametophyte of angiosperms, one should be aware of the condition in gymnosperms. While most gymnosperms have a prothallial tissue, there is a tendency towards its elimination. In several genera, the male gametophyte is reduced to a prothallial or stalk cell, a tube nucleus and two male gametes. Only the living cycads and *Ginkgo* possess swimming sperms. In the rest, the cilia have been lost and the pollen tube functions as a sperm carrying structure. It is quite possible that the male gametophyte originated from some gymnospermous ancestor by further simplification and obliteration of the single prothallial or stalk cell. However, some angiosperms like *Atriplex, Eichhornia, Lilium, Sparganium, Stellaria* and *Yucca* show the occurrence of a prothallial cell. These, according to Maheshwari (1950), are in the nature of freaks and abnormalities of little or no significance.

CARPEL POLYMORPHISM

Saunders (1923) advanced the theory that a gynoecium consists of carpels of more than one basic type. The first type is known as the *valve type*; this is described as open and commonly sterile and, if sterile, with one or more ovules on the lateral parts. The second type is the *solid carpel* which is regarded as columnar, narrow and usually fertile, with one or more ovules on the flanks. The third type is *semisolid* or *pseudo valve* which contains some of the features of the first and second types.

On the basis of the theory, the gynoecium of the Cruciferae consists of four to eight or more carpels of two types: two valve and two solid. Similarly, the gynoecium of the Liliaceae consists of six carpels rather than three in two whorls, one valve and one solid. Pseudo valves are noticed in the Papilionoideae. Pseudo valves in conjunction with valve carpels are present in the Berberidaceae and in combination with solid carpels in the Orchidaceae. In accordance with the theory, *Arachis* and *Scorpiurus* contain 14 carpels, *Brassica* 16 and *Raphistrum* 40–50. These genera are generally considered to have one or two carpels.

There is an anatomical basis for the establishment of the three types of carpel. The presence of a carpel is determined by the vascular supply to the gynoecium or individual carpel. The pair of bundles or double bundles on the ventral side of many carpels is thought to be the result of a longitudinal split in preparation for the dehiscence along the suture. The number of vascular supply to a gynoecium indicates the original number of carpels present and the origin of these traces at two levels shows two whorls of carpels.

The value of the theory lies in the vast amount of knowledge about floral morphology contributed by its promoter. The inherent defect is that no line of demarcation between the types of carpel has been described. Eames (1961) pointed out that the theory misinterpreted the stellar and trace structure in the floral receptacle as well as the vascularisation of carpels. Again, the vasculature of flowers is not interpreted in relation to that of the plant body. Moeliono (1970) did not demand that the classical carpel should be completely abandoned but merely suggested that it is inappropriate to the families embracing the Centrospermae.

FEMALE GAMETOPHYTE

The megaspore mother cell is larger than the surrounding cells; it gives rise to a linear tetrad of megaspores through a meiotic division. The germination of the megaspore is accompanied by the enlargement of its wall and increment of cytoplasm content as well as nucleus size. The typical eightnucleate monosporic embryosac is formed by five successive divisions. Starting with one nucleus, we get eight nuclei which are arranged in three groups; three lie in the chalazal end, two near the centre and three in the micropylar end. The three at the chalazal end, known as antipodals, are walled. The two median nuclei, the polar nuclei, stay unwalled. Of the three micropylar nuclei, the middle one (egg) is larger and the two lateral (synergids) are walled by delicate membranes. The antipodals and synergids are usually short lived, but sometimes they may persist and become haustorial forming tubes that extend into the adjacent tissues.

Types: The female gametophyte or embryosac of angiosperms are classified on the basis of two features. The first is the number of megaspore nuclei which participate in the development of an embryosac and the second is the number as well as arrangements of the nuclei in the mature embryosac. Of the four nuclei, only one develops into an embryosac and the other three disintegrates: this is the *monosporic type.* Sometimes, two megaspore nuclei take part in the formation of an embryosac and this is called the *bisporic type.* In still others, where all the four nuclei may give rise to the embryosac, the type is termed *tetrasporic.* The monosporic embryosacs are grouped into two types: eight-nucleate and four-nucleate. The bisporic types are usually eight-nucleate, while the tetrasporic types show much variations.

The most common type is the *Polygonum* or *normal type.* Out of the four megaspores formed by meiosis, the upper three degenerate and the lowermost undergoes three successive divisions. Hence, from one two, from two four and from four eight nuclei produced. These nuclei are distributed as three at the micropylar end (to form the egg and two synfergids), three at the chalazal end (to form the antipodals) and two in the centre (to form the polar nuclei).

In the *Oenothera type,* the third division is omitted. Here the mature embryosac has only four nuclei which organise into two synergids, one egg and one polar nucleus. Neither the chalazal polar nuclei nor the antipodals are formed.

In the *Allium type*, the megaspore mother cell divides to form two dyad cells. The upper dyad is much smaller and soon degenerates. The lower dyad forms eight nuclei by successive divisions. These nuclei are arranged as in the *Polygonum type.*

In the three of the tetrasporic types, 16 nuclei are formed as a result of two divisions following megasporogenesis. Depending on the polarity and organisation of the nuclei, we arrive at the *Peperomia type* where the egg and one synergid lie at the upper end of the sac, eight nuclei fuse to form a central cluster and the remaining six are cut off at the periphery of the lower part; *Penaea type* where the nuclei lie in four distinct quarters arranged crosswise in the embryosac wall and *Drusa type* where three nuclei lie at the micropylar end, two at the central region and 11 at the base.

In the fourth of the tetrasporic types, the *Fritillaria type,* there is a crowding of the three of the megaspore nuclei into the chalazal end and an union of their spindles in the next division. The result is a secondary four-nucleate stage with two haploid nuclei at the micropylar end and two triploid nuclei at the chalazal end. The consequence of the next division is eight nuclei, out of which four are haploid and four triploid; these are arranged as in the *Polygonum type.*

The fifth of the tetrasporic types in which the egg is alone at the micropylar end, the *Plumbagella type,* has a four-nucleate sec. Here one of the two central polar nuclei and the solitary antipodal are triploid. There are no synergids.

The last two tetrasporic eight-nucleate types, the *Plumbago type* and *Adoxa type,* are characterised by the four megaspore nuclei dividing to form eight nuclei. In *Plumbago,* the organisation is tetrapolar, i.e. one at one end and the other two near the centre. In *Adoxa,* the arrangement is bipolar and the nuclei lie as in the *Polygonum type;* here the number of nuclei is attained by the third rather than the fifth division.

Besides the above well established types, the embryosacs of a few plants remain aberrant or unclassified. These include *Balsamita vulgaris, Chrysanthemum, cinerariaefolium* and *Limnanthes douglasii.*

Morphological Nature: Schnarf (1936) assumed the monosporic eight-nucleate embryosac as the most primitive. This idea is supported by the fact that this type is of wide occurrence in the angiosperms and that in the pteridophytes and gymnosperms the female gametophyte is monosporic. Moreover, all the other types can be easily derived from it.

Though Maheshwari (1950) suggested that the normal type of embryosac is primitive and all others derived, Hjelmqvist and Grazi (1964) were able to secure four different patterns of behaviour in the same species by temperature alternation. Hence, too much importance should not be attached to embryosac types.

Several theories have been put forward to explain the structure of the embryosac. The first theory, proposed by Chamberlain (1935), seeks the derivation of the angiosperm embryosac from that of the gymnosperm. The second theory, the Gnetalean theory, advanced by Hofmeister (1858) and Strasburger (1900), interprets that all the nuclei of the embryosac are potential eggs. The third theory, advocated by Porsch (1907), considers that the embryosac consists of two archegonia without any prothallial tissue. The fourth theory, described by Schuroff (1928), represents the micropylar quartet as remnants of the two archegonia and the chalazal quartet as entirely prothallial.

ENDOSPERM

The endosperm is formed by triple fusion or double fertilisation, i.e. union of the second male nucleus with the two polar nuclei. It is a food storage tissue, providing nourishment to the developing embryo and germinating seed. Since the nuclei entering into the triple fusion are haploid, the endosperm is usually triploid. It is suppressed in the Podostemonaceae and Orchidaceae, but it is strictly diploid in the Onagraceae. In *Peperomia, Penaea* and *Plumbago,* the endosperm is tetraploid.

Depending on the differences in the method of formation of early stages, the endosperm may be classified into three general types. In the *nuclear type,* the first and usually several succeeding nuclear divisions are without immediate wall formation. In the *cellular type,* the nuclear divisions are accompanied by wall formation. In the *helobial type,* which is intermediate between the two types, the first division of the endosperm nuclei results into two chambers, one of which is usually free nuclear and the other with walled cells.

Among 288 families of angiosperms 161 show nuclear, 72 cellular and 17 helobial endosperm (Davis, 1966). Cellular endosperm is largely confined to dicot families, but occurs only in the Araceae and Lemnaceae among monocots. Of the 17 families indicating helobial endosperm, 14 are monocots.

The morphological nature of the angiosperm endosperm is a topic for discussion. Hofmeister (1859) regarded it to be a gametophytic structure whose growth and differentiation is arrested till the pollen tube entered the embryosac. Le Monnier (1887) viewed it as a second embryo, serving as food for the zygotic embryo. Strasburger (1900) and others interpreted it as a delayed complex type of nutritive gametophytic tissue.

Affinity and Evolution: The phylogenetic sequence of the various types of endosperm has not been worked out. Which type is more primitive or more advanced is not known with certainty. Coulter and Chamberlain (1903) assumed the cellular to be more primitive on the basis that "even when the endosperm begins with free nuclear divisions, a rudimentary cell plate often appears, suggesting derivation from an endosperm in which nuclear division was followed by cell formation". To Schnarf (1931), the derivation from the nuclear to the cellular type appeared more plausbible. Difficulty arises when both types occur side by side in a primitive order like the Ranales and in an advanced order like the Campanulales. Moreover, highly evolved taxa, such as the Rubiaceae and Orchidaceae, contain nuclear endosperm.

Among dicotyledonous families, nuclear endosperm occurs exclusively in 60 percent and a mere eight percent are mixed (Sporne, 1974). Nuclear endosperm is primitive, being correlated with several undoubtedly primitive features. Moreover, nuclear endosperm is comparatively more abundant among taxa known to have been in existence by the end of the Cretaceous and rain forest families. However, cellular endosperm is more primitive than nuclear among monocotyledons (Lowe, I960). Since the two conclusions are based on similar statistical methods, it is difficult to explain such a discrepancy. According to Stebbins (1971), perhaps many evolutionary trends are reversible and a trend from nuclear to cellular endosperm may be one of them.

It has been suggested that the spatial condition in the embryosac is related to the occurrence of the nuclear or cellular endosperm. Whereas short and broad embryosacs possess nuclear endosperm, long and narrow embryosacs are characteristics of the cellular endosperm. The phragmoplast disappears in the case of a broad embryosac, since it cannot form a complete partition across the sac. This explanation may be reasonable, but it is not adequate for those instances where the nuclear and cellular types are present in the same species and even in the same ovary or where the first longitudinal wall divides the sac from one end to the other.

Rao (1938) attempted to correlate the endosperm type with the rate of embryo growth. Plants with slow embryo growth have cellular endosperm and those with rapid embryo growth nuclear endosperm. Several examples can be cited in which such correlations are absent, the notable one being *Impatiens* where the endosperm is cellular and has rapid embryo growth.

EMBRYO

After fertilisation, the zygote undergoes a period of rest. The first division of the zygote is followed by wall formation. In the earlier stages of development, there is no essential difference between the embryo of the dicotyledons and monocotyledons. For this reason, Johansen (1950) did not separate the dicot embryos from the monocot ones. Generally, six types are distinguished:

(I) Division of the zygote longitudinal	*Piperad type*
(1) Division of the zygote transverse	(2)
(2) Terminal cell divides by a longitudinal wall	(3)
(2) Terminal cell divides by a transverse wall	(4)
(3) Basal cell plays little or no part in embryo development	*Onagrad type*
(3) Both basal and terminal cells participate in embryo development	*Asterad type*
(4) Both basal and terminal cells participate in embryo development	*Chenopodiad type*
(4) Basal cell plays little or no part in embryo development	(5)
(5) Basal cell usually forms a suspensor of two or more cells	*Solanad type*
(5) Basal cell divides no further and forms a large suspensor cell	*Caryophyllad type*

Soueges (1951) developed a classificatory scheme of angiosperm embryos. According to him, the segmentation of the terminal cell of the two celled proembryo, origin of the four celled proembryo, behaviour of the basal cell and the relative contributions of each cell of the four celled proembryo all lead to the formation of mature embryo. The wisdom of applying these laws to cover embryos of the whole plant kingdom has been questioned in view of the fact that recent cell lineage studies in embryogenesis have indicated significant deviations from the accepted patterns of division and disposition of proembryo cells (Raghavan, 1976).

In the monocots, the cotyledon is terminal and the stem-tip lateral. However, the cotyledons are lateral and the stem-tip terminal in the dicots. It is often difficult to derive the monocot condition from the dicot one. One concept is that one of the cotyledons became retarded in development at an early stage. The other view is that the single cotyledon originated by a union of two originally distinct primordia. Still another idea is that the monocot state arose by a division of labour between the two primordia; one became reduced to form the first plumular leaf and the other retained the cotyledonary position and function.

Some monocotyledons have a second vestigial cotyledon, *e.g. Commelina, Dioscorea, Tamus* and *Tinantia.* But taxa with one cotyledon are frequent among dicotyledons, *e.g. Abronia, Carum bulbocastanunu, Corydalis cava, Erigenia bulbosa, Ranunculus ficaria* and several members of the Gesneiriaceae. A third cotyledon is often present in *Acer* and *Juglans.* The magnolian line in the Ranales is reported to contain a high percentage of polycotyledony having 87 percent with three cotyledons, 13 percent with four and none with two cotyledons.

APOMIXIS

Apomixis is the replacement of sexual reproduction by an asexual process which does not involve any nuclear fusion. It is of four types. In the first or non-recurrent type, the megaspore cell undergoes meiosis and a haploid embryosac is born. The new embryo may be formed from the egg *(haploid parthenogenesis)* or from some other cell of the gametophyte (*haploid apogamy*). The resulting plants are sterile with a single set of chromosomes, *e.g., Datura, Solanum,* etc.

In the second or recurrent type, the embryosac may develop from a cell of the archesporium (*generative apospory*) or from some other part of the nucellus (*somatic apospory*). All the nuclei of the embryosac are diploid, since there is no reduction in chromosome number.

The embryo may arise from the egg (*diploid parthenogenesis*) or from some other cell of the gametophyte (*diploid apogamy*). In the third type, *adventitive embryony* or *sporophyte budding,* embryos are formed only from the cells of the nucellus or from the integument, *e.g. Hiptage bengalensis, Mangifera indica,* etc., In the fourth type, as in *Globba bulbifera,* there is a formation of bulbils or other vegetative propagules in place of the flowers and this type is not important from the embryological standpoint.

Apomixis occurs in at least 40 families of angiosperms, being common in the Compositae, Gramineae and Rosaceae. According to Davis (1972), it "is merely a sophisticated form of vegetative reproduction and facultative apomicts bear the same relationship to evolution as sexually reproducing plants with ancillary reproductive by vegetative means".

AGE OF THE ANGIOSPERMS

From the late Triassic rocks in Greenland, we have the evidence of the oldest known angiosperm fossil (*Furcula*) which is a leaf. Other fossil flowering plants were found in Jurassic rocks from Scandinavia, Germany, France, Scotland, England, etc. These discoveries led to the concept that the angiosperms came very early in the Mesozoic or late in the Palaeozoic. The angiosperms were widespread, though not yet dominant during the middle and late Mesozoic. By the time of middle Cretaceous (near the end of the Mesozoic), many families like the Fagaceae, Leguminosae, Palmae, Rosaceae, Salicaceae, Tiliaceae and Typhaceae came into existence. It may be recalled that the frequency of angiosperm fossils, which are mostly leaf impressions, occasionally woods and rarely fruits and seeds, increases gradually from Middle Cretaceous onward. Palynologists failed to find recognisable angiosperm pollen in pre-Cretaceous rocks, though even upland plants are occasionally represented in the lowland pollen calendar today. But critical examination cast doubts on reports of pre-Cretaceous angiosperms and taxonomic identities of Cretaceous leaves.

During the Cenozoic, the angiosperms dominated the bulk of the land habitats. Only the extensive *taiga* (northern coniferous forest) of Asia and Canada as well as high mountains were still swayed by gymnosperms. From the tropical areas, they spread into the temperate and arctic regions. During the Permian and Triassic, in the words of Axelrod (1952); the "(1) angiosperms probably were in existence ... that (2) they probably occupied upland regions remote from lowland sites of deposition, that (3) they were distributed in the diverse environments of the tropical zone, that (4) at least some of them were ancestral to modern primitive types of monocotyledons and dicotyledons which are not represented in tropical regions, and that (5) they were then undergoing quantum evolution from proangiosperms which occupied upland regions". The present day patterns of angiosperm distribution were largely established in the early Cenozoic. The late Cenozoic was accompanied by several vegetational changes in the temperate zone, but tropical vegetation experienced essential stability. At the time of Pleistocene, mass migration of the angiosperm and (gymnosperm) types took place in front of the four major glaciers that moved across North America and Eurasia from the north.

On the basis of a detailed study, Bailey (1949) recommended Northern Australia, New Guinea, New Caledonia and Fiji as the probable ancestral home of angiosperms. It is the living floras of this region extending northward to southern China that have yielded and are continuing to yield the richest crop of missing links in the chains of angiosperm phylogeny. For example, most of the vessel-less angiosperms have been reported from this region. The celebrated anatomist went on to assert: "Furthermore, these living floras southern latitudes have already yielded more structurally primitive dicotyledons than have all the known fossil floras of northern latitudes. Therefore, I am inclined to question the assumption of a northern origin for the angiosperms at least until the Tertiary and Mesozoic rocks of these southern islands have been adequately explored."

PHYLY OF THE ANGIOSPERMS

The primitive orders of dicots and monocots are characterised by groups of families or single families that do not bear any close relationship to each other. There are about ten lines in the Ranales alone, while the monocots have three to four. This great diversity in the primitive families suggests a polyphyletic origin, *i.e.,* the dicots or monocots arose at different times from different primitive stocks and reached their present similarities through convergent or parallel evolution. Two strong points in favour of the polyphyletic origin include the variety of perianth and nature of carpel. In more primitive taxa, we find a perianth of one, two or several bud scales with or without petaloid organs. The closed carpel is not an early angiosperm trait, but carpel closure became established through a general trend in specialisation.

It also seems likely that the angiosperms represent a monophyletic group, *i.e.*, the group originated from a single ancestor at a single time in the past. A great merit of the theory is the common possession of the eight-nucleate embryosac and secondary endosperm. Other supporting evidence include the position, arrangement and type of floral appendage, stamen with four sporangia in two pairs, simple monocolpate pollen, follicular carpel and anatropous ovule. However, all evidence for monophylesis is based on modern plants and on embryological features for which there is no direct fossil evidence and the fossil record favours polyphylesis. Consensus of opinion at the present time appears to support the view that Angiospermae as a group are monoplyletic and for families or groups of families polyphyletic. Cronquist (1968) held that the distinctive syndrome of angiosperm features was acquired by parallel evolution from a set of similar gymnospermous taxa.

The differences between the monocots and dicots reflect the fact that they evolved separately for many millions of years. According to some authorities, the origin of the monocots is "one of the great unsolved problems in phylogeny". Others felt that there is sufficient evidence to warrant considering the monocots as an off shoot from the Ranales. Still others believed that both the monocots and dicots were derived from proangiosperm ancestors.

ORIGIN OF THE ANGIOSPERMS

The origin of the angiosperms is an abominable mystery which botanists have been trying to solve for many decades. Since the fossil record is equivocal and sparse, theories regarding the probable ancestors are highly speculative. Several groups have been proposed as the possible precursors of angiosperms.

1. **Isoetes-Monocotyledon Theory:** When Linnaeus (1775) first described *Isoetes,* he mistook it for a seedplant and even described its seeds (spores) and fruits (sporangia). The linear leaves and cormous habit of *Isoetes* have been correlated with the simpler aquatic monocots. Campbell (1895, 1928) went to the extent of deriving the monocotyledons from *Isoetes* via the Ophioglossaceae. But most other botanists consider his views as fantastic.

2. **Pteridosperm Theory:** The pteridosperms have been postulated as a possible ancestral stock of the angiosperms on the basis of vascular histology and stellar structure. In the more primitive ferns and angiosperms, sporangial development is eusporangiate. The laminar placentation of the angiosperms recalls the soral distribution of the ferns. A basic nodal structure of two traces to one gap and amphiphloic stele are common to both the groups. But it is difficult to explain the simple angiosperm ovule in terms of the complex pteridosperm seed. Perhaps seed bearing was not an essential criterion in the ancestral type. Again, the seed-like structures developed in several unrelated lines.
3. **Caytonialean Theory:** The Caytoniales, a Mesozoic group of fossil seedplants, has been looked upon as a possible ancestor of the angiosperms. In the Caytoniales, the ovule bearing organs are enclosed in a carpel-like sheath—a condition representing an early stage in carpel evolution. But the caytonialian organ and angiosperm carpel are entirely different structures. The caytonialian organ is strongly pinnately veined and the angiosperm carpel palmate. The single integument of the Caytoniales is unlike the two integuments of most angiosperm ovules.
4. **Bennettitalean Theory:** The angiosperm flower and cycadeoid cones are alike in gross morphology. The microsporophyll of the angiosperms is a laminar organ, while that of the Bennettitales is pinnately compound. The angiosperm megasporophyll is also laminar with many ovules on the adaxial surface and the megasporophyll of the Bennettitales is stalk-like with a terminal ovule. The megasporophylls of the Bennettitales are separated by several scales—structures unknown in the angiosperms. The angiosperm stem has a small pith, thick vascular cylinder and thin cortex with leaf traces not girdling or lateral ones only slightly girdling; the bennettitalean stem has a large pith, thin vascular cylinder and thick cortex with strongly girdling leaf traces. The xylem rays of the angiosperms consist of two kinds of cell in which marginal cells are conspicuous, whereas the xylem rays of the Bennettitales are made up of one type of cells only and lack the marginal cells.
5. **Coniferales-Amentiferae Theory:** The Amentiferae is reckoned as an example of primitive angiosperms. The naked flowers and anemophily of the Amentiferae are suggestive of the cones and pollination of the conifers. With two pollen sacs, the angiosperm stamen is regarded as a replica of the bisporangiate microsporophyll of many conifers. The flowers of the Amentiferae have been noted to be more specialised than primitive (Eames, 1961). The simplicity of flowers of the Salicaceae, Betulaceae and other families is the simplicity of reduction rather than that of primitiveness. The anatomical condition of the wood in these families is viewed to be quite advanced.
6. **Gymnosperm-Gnetales-Angiosperm Theory:** The Gentales is a transitional group from the gymnosperms to the angiosperms. *Gnetum* is in possession of an angiosperm-like leaf, angiosperm-like female gametophyte, naked seeds and vessel bearing wood. But the ovules of the Gnetales in general and the female gametophyte in particular are peculiar. Moreover, the vessels of the Gnetales are unique in structure; their perforations have been derived from round pitted tracheids, rather than the scalariform pitted ones, which are typical of the angiosperm vessels.
7. **Gonophyll Theory:** According to Melville (1963), the basic component of the ovary of the angiosperms was a leaf with an epiphyllous fertile branch called the *gonophyll*. The evidence was in favour of the ovary being formed by alternating whorls of foliar and fertile organs. There was an interesting comparison of the primitive gonophyll with the fructifications of *Glossopteris* where fertile segments were fixed to the midrib of leaves. Additional points, supporting the theory, include the existence of nectaries and coronae on petals and prevalence of epiphyllous

inflorescences in a number of angiosperm families. After examining the structure of androecium of a number of supposed primitive families of angiosperms, Melville concluded that this structure evolved on similar lines. When he compared the anther with the supposed male fructifications of *Glossopteris,* the fructifications failed him.

8. **Pentoxylalean Theory:** Meeuse (1961) proposed that "the genus *Pandanus* is a direct descendant of a group of plants practically identical with the known fossil members of the Pentoxylales". He listed a number of comparable features, especially in the stem anatomy, morphology and anatomy of leaf, male and female inflorescences, pollen grains and ecology. He believed that other angiosperms such as the Helobiae, Nymphaeales, etc. may have originated independently of the Pandanales. However, these resemblances may be the consequence of parallel evolution.
9. **Durian Theory:** The theory derives its name from the durian (*Durio zibethinus* of the Bombacaceae) which is accepted by Corner (1949) as the surviving model of the primitive angiosperms. The durian is a cauliflorous tree, bearing attractively coloured large loculicidal spiny capsules with fleshy arillate seeds. There are about 45 diverse angiospermous families containing arillate genera which are mostly of tropical distribution. Within these families, one can trace a reduction series form genera with arillate fruits to genera having fruits of other types. It is usually possible to correlate the reduction with a gradual increase in the power of dormancy, *i.e.*, arillate fruits/ seeds are not dormant and other types of fruit show an increased dormancy and offer a greater opportunity for dispersal.

Complex problems are often complicated further by imaginative theories and nowhere is this more true than is speculations about the origin of angiosperms (Pant and Kidwai, 1971). Under the circumstances, it would be best to concentrate on searching for new evidence in living plants and in the rocks of Upper Jurassic and Lower Cretaceous.

EVOLUTIONARY LINES IN THE ANGIOSPERMS

If we assume the angiosperms as representing the culmination of an extensive evolutionary development, it would seem that evolution has been active within the class itself. This means that the taxa of angiosperms are related and it is possible to arrange them in a sequence indicative of their affinity.

At the foot of the angiosperms, there lies the Casuarinales. One of the features causing us to think that it is a very primitive order is the presence of naked flowers with ovules containing many embryosacs. In the rest of the Apetalae, where the ovule contains a single embryosac, the perianth is either absent or bract-like, *e.g.,* Piperales to Julianiales or the perianth is sepaloid, *e.g.* Urticales to Centrospermales. These simpler stages precede the relatively high phase of development now represented by the ranalian type where the perianth is differentiated into calyx and corolla.

The position of the Cactales is not clear to many taxonomists, but it forms a connecting link between the Centrospermales and Ranales. The most familiar family of the Ranales is the Ranunculaceae where we notice the presence of numerous floral parts arranged in close spirals upon the receptacle.

From the ranalian plexus, there are at least three main lines of ascent. The first line of advance comprises the Rhoeadales, Sarraceniales, Parietales, Malvales, Geraniales, Sapindales, Rhamnales and some members of the Tubiflorales. The second line culminates in the Campanulales through such orders as the Rosales, Myrtales, Umbellales and Rubiales. The third line includes the Liliales and Orchidales.

In the first line, there has been no change from the hypogynous state. The occurrence of syncarpy as well as reduction in the number of floral parts are early phenomena. Other types of coalescence and the shift to zygomorphy have appeared somewhat later, but still early in the line. From the Ranales, parallel evolution of the Rhoeadales and Sarraceniales have taken place. The Parietales is related to the Rhoeadales with respect to the gynophore development in the Capparidaceae and Passifloraceae. The origin of the Malvales can be traced through the Bombacaceae. The Geraniales resembles the Malvales as regards the general floral construction. The Sapindales is allied to the Geraniales and the Rhamnales to the Sapindales. In the Solanaceae, one can find perfect and regular flowers which exhibit the coalescence of sepals, petals, and carpels; here the stamens are adnate to the corolla tube. The most advanced family in the line is shown by the Labiatae where the flowers are still hypogynous, although they are irregular. Here is a reduction in the number of floral parts to four or two; the sepals, petals and carpels are united. Not only the stamens are adnate to the corolla tube, but also especially adapted for insect pollination.

The second line of ascent involves a change from the hypogyny to the epigyny through the perigyny. The Rosales is characterised by a wide range of floral structures. Evolutionary changes within the order include the gradual coalescence, reduction in the number of floral parts and alteration from the hypogyny to slight perigyny to true epigyny. In the Myrtales, the perigyny passes into the epigyny with the hypanthium formation. A further advance from the Myrtales is the Umbellales where the flowers are epigynous and exhibit syncarpy. Still higher up in the line is the Rubiales which is closely related to the Cornaceae (Umbellales) in such features as the cymose umbel-like inflorescence, epigyny, suppression of the calyx, etc. The climax is reached in the Campanulales whose flowers have such advanced features as the reduction in the number of floral parts, modification of the floral parts (pappus), irregular flowers, some imperfect flowers, and evidence of coalescence in each whorl and adnation.

Some primitive monocotyledonous plants, particularly the members of the Alismataceae, have much in common with the Ranunculaceae. The Liliaceae is distinguished from the Ranales in all of the distinct monocot characters. Besides, there is a reduction in the number of floral parts and a connation of carpels. The Iridaceae and Amaryllidaceae show slight advance over the Liliaceae in that they have epigynous flowers. The Gramineae is believed to have originated from the Liliaceae. The primitive features of the Gramineae comprise the hypogyny, actinomorphy and mere coalescence of carpels. The loss of perianth parts as well as reduction in the number of stamens and carpels have made the Gramineae to appear more advanced than the Liliaceae. The Orchidaceae is said to represent the most advanced family among the monocots; here the flowers are irregular, but very showy and highly specialised for insect pollination.

SELECTED BIBLIOGRAPHY

ARBER, A. 1930. Root and shoot in the angiosperm: a study of morphological categories. *New Phytol.,* **29:** 297–315.

AXELROD, D.I. 1952. A theory of angiosperm evolution, *Evolution,* **6:** 29–60.

BARNARD, C. 1961. The interpretation of the Angiosperm flower. *Aust. J. Sci.,* **24:** 64–72.

CELAKOVSKY, L.F. 1892. Gedanken über eine zeitmasse Reform der Theorie der Blutenstande. *Bot. Jb.,* **16:** 33–51.

CHAMBERLAIN, C.J. 1935. *Gymnosperms, Structure and Evolution.* University of Chicago Press, Chicago.

COULTER, J.M. and CHAMBERLAIN, C.J. 1903. *Morphology of Angiosperms.* D. Appleton, New York.

CROIZAT, L. 1964. Thoughts on high systematics, phylogeny and floral morphogeny, with a note on the origin of angiosperms. *Candollea,* **19:** 17–96.

DAVIS, G.L. 1972. Apomixis. *Adv. Pl. Morph.* pp. 126–132, Savita Prakashan, Meerut.

DE CANDOLLE, A.P. 1827. *Organographie végétale.* Paris.

DOYLE, J.A. 1978. Origin of angiosperms. *Ann. Rev. Ecol. Syst.,* **9:** 365–392.

EAMES, A.J. 1961. *Morphology of the Angiosperms.* McGraw-Hill, New York.

GOEBEL, K.I.E. 1931. Blutenbildung and Sprossgestaltung. Anthokladian and infloreszenzen. (Suppl. 2 to *Organographie der Pfianzen,* ed. 3).

HOFMEISTER, W. 1858. Neue Beobachtungen über Embryobildung der Phanerogamen. *Jb. f. wiss. Bot.,* **1:** 82–186.

HOFMEISTER, W. 1859. Neue Beiträge zur Kenntnis der Embryobildung der Phanerogamen. I. Dikoytledonen mit ursprunglich einzelligem, nur durch Zelltheilung wachsendem Endosperm. *Abh. Konigl. Sachs. Ges. Wiss.,* pp. 535–672.

JOHANSEN, D.A. 1950. *Plant Embryology.* Chronica Botanica. Waltham, Mass.

JOSHI, A.C. 1947. Carpel histogenesis and carpel morphology. *J. Indian Botan. Soc,* **26:** 63–74.

KRASSILOV V.A. 1977. The origin of Angiosperms. *Botan. Rev.* **43:** 143–176.

LE MONNIER, G. 1887. Sur la valeur morphologique de lalbumen chez les angiosperms. *de Botan.,* **1:** 140–142.

LORCH, J. 1963. The carpel—case history of an idea and a term. *Centaurus,* **8:** 269–291.

MAHESHWARI, P. 1950. *An Introduction to the Embryology of Angiosperms.* McGraw-Hill, New York.

MAJUMDER, R.B. 1963. An *Introduction to Spermatophyta.* Merit Publishers, Calcutta.

MEEUSE, A.D.J. 1975. Origin of the Angiosperms—Problem of Inaptitude. *Phytomorphology,* **25:** 373–379.

MELVILLE, R. 1963. A new theory of the Angiosperm flower. *Kew Bull,* **17:** 1–50.

N̈AGELT, C.W. VON. 1884. *Mechanisch-physiologische Theorie der Abstammungslehre,* Munchen.

PARKIN, J. 1914. The evolution of the inflorescences. *J. Linn. Soc,* **42:** 511–582.

PILGER, R.K.F. 1922. Über Verzwcigung and Blutenstandsbildung beiden Holzgewaschsen. *Bibl. Botan.,* **23:** 1–38.

PORSCH, O. 1907. *Versuch einer phylogenetischen Erklarung des Embry asackesund der doppelten Befruchtung der Angiospermen.* Jena.

PANT, D.D. and KIDWAI, P.F. 1971. The origin and evolution of flowering plants. *J. Indian Botan. Soc.,* **50A:** 242–274.

PURI, V. 1967. The origin and evolution of angiosperms. *J. Indian Botan. Soc,* **46:** 1–14.

RAGHAVAN, V. 1976. *Experimental Embryogenesis in Vascular Plants.* Academic Press, London.

RAO, V.S. 1938. The correlation between embryo type and endosperm type. *Annls. Bot.* (*N.S.*), **2:** 535–536.

RICKETT, H.W. 1944. The classification of inflorescences. *Botan. Rev.* **10:** 187–231.

SAUNDERS, E.R. 1922. The leafskin theory of stem. *Annls. Bot.* **36:** 135–165.

SAUNDERS, E.R. 1923. *Carpel Polymorphism.* Cambridge University Press, Cambridge.

SCHNARF, K. 1931. *Vergleichende Embryologie der Angiospermen.* Gebriüder Borntraeger, Berlin.

SCHNARF, K. 1936. Contemporary understanding of embryo sac development among angiosperms. *Botan. Rev.* **2:** 565–585.

SCHÜROFF, P.N. 1928. Über die Emtwicklung des Eiapparates der Angiospermen. *Ber. dt. Botan. Ges.,* **46:** 560–572.

SCOTT, R.A., BARGHOORN, E.S. and LEOPOLD, E.B. 1960. How old are the angiosperms? *Am. J. Sci.,* 258A: 284–299.

SPORNE, K.R. 1974. *The Morphology of Angiosperms.* Hutchinson University Library, London.

STRASBURGER, E. 1900. Einige Bemerkungen zur Frage nach der doppelten Befruchtung bei Angiospermen. *Bot. Ztg.,* II, **58:** 293–316.

TROLL, W. 1969. *Die Infloreszenzen.* Bd. 2, Teil 1. Gustav Fischer, Stuttgart.

TUTIN, T.G. 1952. Phylogeny of flowering plants: fact or fiction? *Nature* (*Londs.*), **169:** 126–127.

PART THREE

Systematic Survey of the Dicotyledoneae

Systematic Survey of the Dicotyledoneae

The two subclasses of angiosperms can be separated by a number of characters as given in Table 4.

TABLE 4. Delimitation of two subclasses of angiosperms

Characters	*Dicotyledoneae*	*Monocotyledoneae*
Mature root system	Primary or adventitious or both	Entirely adventitious
Growth form	Either herbaceous or woody	Mostly herbaceous, a few arborescent
Intrafascicular cambium	Usually present	Usually absent
Vascular bundles	Arranged in a ring, surrounding a pith	Generally scattered or in 2 or more rings
Leaf venation	Mostly reticulate	Mostly parallel
Number of floral parts	Multiples of 4 or 5	Multiples of 3
Pollen	Typically triaperturate or triaperturate-derived type	Uniaperturate or uniaperturate-derived type
Embryo	Cotyledons 2 (rarely 1, 3 or 4); endosperm present or lacking	Cotyledon 1: endosperm often present

The Dicotyledoneae forms the bulk of angiosperms and are much more successful today than the Monocotyledoneae. There are about 200,000 species of dicotyledons and 50,000 of monocotyledons (Sporne, 1974). The dicots are considered here as more primitive than the monocots which group is a later evolutionary offshoot from a primitive proangiosperm stock.

The dicots have been variously treated. Bessey (1915) regarded the dicot families to number 255 and placed them under 22 orders. Engler and Diels (1936) accepted 258 families as distributed among 44 orders. In an earlier endeavour (1926), Hutchinson created 264 families and 76 orders; these were later (1959) raised to 342 families and 76 orders remaining constant. Benson (1957) mentioned 266 families and 56 orders. Porter (1967) took into account 71 families and 37 orders. Cronquist (1968) described 293 families and 56 orders. Mitra (1974) detailed 138 families and 31 orders, although he proposed more taxa. Heywood (1978) recognised 249 families and 58 orders. The present author deals with 160 families and 33 orders. All these evidence go to show that there is a lack of agreement as to the composition of the dicots and consequently about the interrelationship of the various groups contained therein.

In this book, the Dicotyledoneae is divided into three broad groups: Apetalae, Polypetalae and Sympetalae.

APETALAE

The Apetalae is an artificial assemblage of orders and families, numbering 13 and 33 respectively. Here the flowers apparently have no perianth or have a single perianth.

The main diagnostic characters of the various orders under the Apetalae are provided in the key below (Lawrence, 1951; Core, 1955). It should be noted that this key represents broad statements about diversified taxa and does not indicate exceptions to the general situation.

(1) Ovules with many embryosacs	*Casuarinales*
(1) Ovules with a single embryosac	(2)
(2) Perianth none or bract-like	(3)
(2) Perianth sepaloid or petaloid	(8)
(3) Flowers mostly perfect, borne in spikes	*Piperales*
(3) Flowers imperfect staminate ones borne in catkins	(4)
(4) Ovary inferior	(5)
(4) Ovary superior	(6)
(5) Plants monoecious; carpels 2	*Juglandales*
(5) Plants dioecious; carpels 3	*Julianiales*
(6) Fruits 1-seeded	*Leitneriales*
(6) Fruits 4- to many-seeded	(7)
(7) Fruits 4-seeded	*Balanopsidales*
(7) Fruits many-seeded	*Salicales*
(8) Ovary inferior	(9)
(8) Ovary superior	(10)
(9) Stamens opposite to perianth segments and united with them	*Santalales*
(9) Stagmens alternate with perianth segments and free from them	*Aristolochiales*
(10) Flowers perfect or imperfect; carpel 1	*Proteales*
(10) Flowers mostly perfect; carpels 2 or more	(11)
(11) Leaves usually ochreate	*Polygonales*
(11) Leaves not ochreate	(12)
(12) Placentation mostly at apex of loculus	*Urticales*
(12) Placentation basal or free central	*Centrospermales*

Order 1. *Casuarinales*

The order contains a solitary family, the Casuarinaceae. The characters of this order are the same as those for the family.

Casuarinaceae (Beef-wood Family)

GENERAL FEATURES OF THE FAMILY

Habit	— Evergreen shrubs or trees.
Stem	— Woody, grooved, *jointed, verticillate.*
Leaves	— *Whorled, scale-like,* linear or lanceolate, united at base to form a toothed sheath.
Inflorescence	— Male flowers in spikes and females in spherical heads.
Flowers	— Unisexual, *perianthless,* standing at axil of a bract and surrounded by 2 bracteoles.
Androecium	— *Stamen 1* (in male flowers), central; anthers 4-celled, basifixed, dehiscence by vertical slits.
Gynoecium	— Carpels 2 (in female flowers), ovary superior, originally bilocular but ultimately unilocular by suppression at anthesis; style 1, very short; stigmas 2, linear.
Fruits	— 1-seeded winged nut, *enclosed by woody bracteoles to form a dry cone-like structure;* embryo straight; endosperm absent.

POSITION AND AFFINITY OF THE FAMILY

The Casuarinaceae is an isolated taxon under the Apetalae. It was thought by Engler as the most primitive of dicots, constituting the **A**, shoot with male inflorescences and fruits; **B**, portion of male inflorescence showing two whorls; **C**, diagram of a whorl of male flowers (*p*, perianth; *a*, bracteole; *b*, bract); **D**, male flower before elongation of filament; **E**, female inflorescences with protruding stigmas; **F**, female flower (*a*, bracteole); **G**, diagram of female flower; **H**, fruit; I, lower portion of fruit with L.S. of seed (*r*, radicle; *c*, cotyledon); **J**, portion of L.S. of ovary showing functional ovule and course of pollen tube; **K**, T.S. of a branchlet (*j* and *o,* inner and outer rings of bundles; *s*, sclerenchyma; *p*, palisade tissue). (**A**, after Poisson; **C**, **D** and **G**, after Engler; **J**, after Treub; **B**, **E**, **F** & **H**–**K**, after Rendle).

Verticillatae. It also formed the eighth family of Rendle's Monochlamydeae and the first family of Core's Archichlamydeae.

At first, Bessey (1897) took the Casuarinaceae as a member of the Sapindales; later (1915), he made it look like a "leafless representative of and derived from the Hamamelidaceae within the Rosales". Hallier (1905) treated it as a part of his Amentacees. Anatomically, according to Tippo (1938), the family illustrates a highly advanced element originating from the Hamamelidaceae. Hjelmqvist (1948) held the view that the family is not a component of the Amentiferae*, though it is a primitive dicot. The primitiveness of the family is ascribed to the presence of apparent catkin-like male inflorescence, simple anemophilous flowers and large rayed wood. Studying the wood anatomy of 30 different species of *Casuarina,* Moseley (1948) concluded that the family is "moderately rather than highly specialized". This is evident from the lack of libriform fibres and predominance of simple perforation of vessels (which are thin walled and circular in outline), metatracheal xylem parenchyma and alternate vascular pitting. Cytological data indicate the heterogeneity in the number of chromosomes in the different species of *Casuarina* (Barlow, 1959). Hutchinson (1959) derived the family from Hamamelidales, considering it as a climax family. Eames (1961) favoured a relationship of the Casuarinaceae with the Betulaceae. This is based on the similarity of pollen structure, two trace stamen character and other floral aspects.

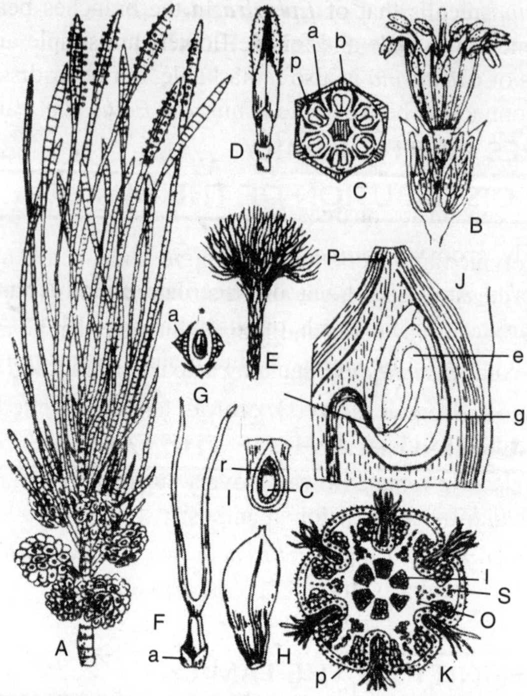

Fig. 2 CASUARINACEAE. *Casuarina equisetifolia* J. R and G Forst

Lam (1961) went to the extent of removing *Casuarina* from the angiosperms to be included in the protoangiosperms along with *Ephedra, Welwitschia* and *Gnetum.* The monotypic genus *Casuarina* provides a rather mysterious chemical correlation in producing biflavonyls—compounds atypical of angiosperms but similar to many conifers, ginkgo and cycads. It has an extremely low relative proportion of syringyl subunits, found only in certain conifers like *Podocarpus* and not in angiosperms. While taking into account the natural affinities of *Casuarina,* it is difficult for a taxonomist to ignore these interesting chemical features of the plant (Alston and Turner, 1966). Whether such chemical parallelisms form valid phylogenetic links are far from clear. Cronquist (1968) opined that there is nothing obvious in the floral or vegetative morphology that cannot justify derivation from a generalised fagalean ancestor, but entertained the possibility that the similarities reflect parallel changes from a common ancestry in the Hamamelidaceae. Chanda (1969) stated that the Casuarinaceae is a stenopalynous family with little pollen morphological variation: pollen grains betuloid, aspidote, oblate to suboblate, three porate (sometimes four to five porate). On the basis of palynotaxonomy involving 40 species and subspecies of *Casuarina,* he asserted that the genus is neither primitive nor a derivative of the hamamelidaceous ancestors. There is also a striking palynological resemblance between the Casuarinaceae and Juglandaceae.

Hutchinson (1969) conceived the Casuarinaceae "as an extreme reduction adapted mostly to dry climatic conditions, but with no very relations still living. The cone-like infructescence resembles that of the Pandanaceae in the Monocotyledons, but should not be regarded as more than a parallelism, nor has it anything to do with the cones of the Pinaceae, the resemblance being entirely superficial".

* "The orders Salicales to Fagales, inclusive, have been grouped by some botanists in the superorder Amentiferae" (Core, 1955).

The habit of *Casuarina* recalls that of *Ephedra* in the branches bearing alternating whorls of scale-like leaves. In both genera, the male and female flowers are simple and reduced. The equisetoid appearance of the branchlets of *Casuarina* is also remarkable, but other dissimilarities are so basic and many that no concept of a connection between the genus and *Equisetum* can be upheld.

NUMBER IN AND DISTRIBUTION OF THE FAMILY

The Casuarinaceae is a monotypic family, containing a single genus (*Casuarina*) with about 65 species. It is mainly developed in northeast Australia but also distributed in Fiji, New Caledonia, Malaysia and Mascarenes Islands. "Casuarinas", in the words of Humphries (1978), "are well adapted to very dry habitats in the region of high temperature and low rainfall and are thus remarkably xeromorphic."

COMMONLY OCCURRING PLANT OF THE FAMILY

Australian Pine or Beef-wood Tree (*Casuarina equisetifolia* J.R. and G. Forst.) is a big tree, usually planted in avenues.

ECONOMIC ASPECTS OF THE FAMILY

C. equisetifolia is grown on the coasts of India in order to reclaim sandy dunes. Valuable timbers are produced by *C. cunninghamia* and *C. stricta.* Various species yield fine grained wood suitable for furniture, gates, shingles, staves, turnery and wagons and the bark of some is utilised for tanning.

Order 2. *Piperales*

The order is characterised by the mostly or nearly all herbaceous plants, simple leaves, racemed or spicate inflorescence and largely bisexual flowers often with bracts and without perianth.

The Piperales show no geographic or ecologic unity beyond the fact that many of the members occur in moist tropical forests, but they do not occupy such habitat by excluding other groups.

The order, as believed by Engler, signifies one of the most primitive dicot forms and hence treated by him near the beginning of the group. Core as well as Lawrence followed Engler in the placement of the order. Both Hallier and Hutchinson considered it to be a terminal and lateral offshoot of the Ranales. Rendle thought that it is related to the Polygonales by the trimerous flower. "It is to be noted that in both the Polygonaceae and the Piperales the ovule is characteristically solitary (not always as in Piperales) and orthotropous" (Benson, 1957). An affinity has been suggested by Lotsy between the Piperales and Araceae, the item of resemblance being the inflorescence.

Cronquist (1968) claimed that the "Piperales are Magnoliidae with reduced, crowded flowers and orthotropous ovules". The order retains the primitive uniaperturate pollen of its magnolialean ancestor. It also resembles the Magnoliales in possessing ethereal oil cells, but differs in having predominantly herbaceous plants. Within the Magnoliales again, the Lactoridaceae probably represents the closest approach to the Piperales. However, *Lactoris* cannot be regarded as ancestral to the Piperales. It has well developed vessels, while *Sarcandra* of the Chloranthaceae is devoid of vessels. Both the Piperaceae and Saururaceae bear merely scanty endosperm and a copious perisperm (which is, however, not known in the Magnoliales). The Chloranthaceae stands more or less apart from the other two families and can perhaps be included in the Magnoliales were it not for the fact that this family has reduced perianth and orthotropus ovules. However, these characters also make the Chloranthaceae somewhat anomalous in the Piperales.

Murty (1959) studied the development of floral organs of some members of the Piperales. He rejected the earlier idea that the bract in *Peperomia* typifies a staminode. According to him, the bract is hypopeltate and the solitary bundle in its stalk is a much reduced midrib. The placentation in *Saururus* has been variously interpreted as axile, basal, lateral or parietal; this is shown by Murty to be marginal in character. Furthermore, he stated that two ventral bundles pass through free margins of the carpel. Another evidence in support of marginal placentation is the occasional presence of inverted bundles in the free margins. Raju (1961), while studying the morphology and anatomy of the Saururaceae, declared that the laminar placentation of *Saururus* represents the basic condition from which the parietal placentation of *Anemopsis* and *Houttuynia* is derived.

The Piperaies, according to Core (1955), is composed of three families as follows:

(1) Stamens 1–3 ..... *Chloranthaceae*
(1) Stamens 1–10 ..... (2)
(2) Carpels 2–5, united; ovary 1-celled ..... *Piperaceae*
(2) Carpels 3–4, free or united; ovary 2-or more-celled ..... *Saururaceae*

Piperaceae (Pepper family)

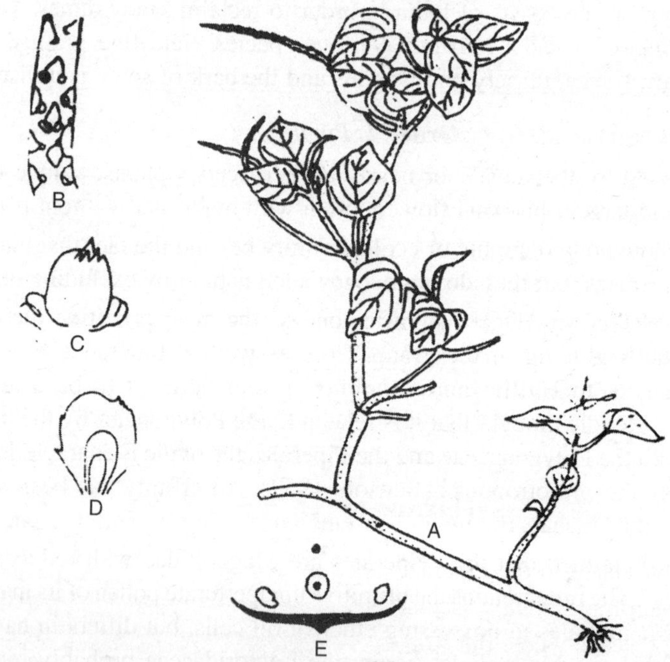

Fig. 3 PIPERACEAE. *Peperomia pellucida* (L.) HBK. **A**, portion of plant; **B**, inflorescence; **C**, flower; **D**, L.S. of ovary

GENERAL FEATURES OF THE FAMILY

Habit — *Succulent herbs,* shrubs or trees.

Stem — Vascular strands distinct, scattered and not arranged in concentric rings.

Leaves — Alternate, petiolate, simple, entire, fleshy; stipules lateral adnate when present.
Inflorescence — *Dense spikes* or umbellate spikes.
Flowers — *Minute, bracteate, naked,* bisexual or unisexual, hypogynous, regular.
Androecium — Stamens 1–10; anthers 2-celled or apparently 1-celled, dehiscence longitudinal.
Gynoecium — Carpels 2–5, united; ovary superior; *ovule basal, orthotropous, 1 in each loculus.*
Fruits — Drupe.
Seeds — Small, embryo minute; endosperm copious.

NUMBER IN AND DISTRIBUTION OF THE FAMILY

The Piperaceae is composed of 12 genera and about 1,400 species, inhabiting mostly rainforests.

COMMONLY OCCURRING PLANTS OF THE FAMILY

Peperomia pellucida (L.) HBK. and *P. tetraphylla* (Forst.) Hook. and Arn. are herbaceous weeds, found in waste places; the former is slender and the latter stout.

Piper, hamiltonii C. DC, a shrub of North Bengal, has 5- to 7-nerved leaves.

ECONOMIC ASPECTS OF THE FAMILY

Piper betle is used as a masticatory, the fresh leaves being chewed with Betel nuts. *P. caninum* yields a type of volatile oil called 'Kababchini', utilised with Betel leaves. The wood of *P. chaba* forms a pungent condiment. The dried unripe fruits of *P. cubeba* (East Indies) produce 'cubebs', valued in the treatment of catarrh and as a stimulant. Long pepper is derived from *P. longum* and *P. retrofractum* (Java). The roots of *P. methysticum* (Polynesia) constitute 'Kavakava', an intoxicating beverage. The ripe fruits of *P. nigrum* serve as a source of white pepper, whereas black pepper is made from the unripe fruits of the same species. The stems of *P. tiliaefolium* (Mexico) are used as walking sticks.

Some species of *Macropiper, Peperomia* and *Piper* are grown for their ornamental foliage.

Saururaceae
(Lizard's-tail Family)

The family is closely related to the Piperaceae, but differs from it in the presence of a multicarpellate gynoecium. In the Saururaceae, the vascular strands in the stem are united into a single ring; this is in contrast with the Piperaceae where the strands are distinct.

The family includes five genera and seven species, prevailing in wet soils: these are all natives of eastern Asia and North America.

Houttuynia cordata Thunb., an erect herb with a spike having four large bracts below it, is seen in Khasi Hills, Meghalaya.

The leaves of *H. cordata* (Vietnam) are used as salads and for treating eye diseases.

Chloranthaceae
(Chloranthus Family)

The family is characterised by the opposite stipulate leaves, connate petiole bases and inferior ovary with a single ovule.

The family comprises five genera and 65 species of the tropics and southern temperate regions.

Chloranthus glaber is raised domestically for its leaves and red fruits. *C. inconspicuus* (China) yield 'chulan', useful in perfuming tea. The leaves of *C. officinalis* (Indonesia) are used to make a beverage.

Order 3. *Salicales*

Salicaceae
(Willow Family)

GENERAL FEATURES OF THE FAMILY

Habit — Shrubs or trees, dioecious*.

Leaves — Alternate, simple, stipulate.

Inflorescence — *Catkins* with stamens in male flowers and carpels in female flowers.

Flowers — Unisexual, sessile or shortly pedicellate, hypogynous, *subtended by bracts and a disc or gland.*

Perianth — *Absent.*

Androecium — Stamens 2 to many, free or united; filaments long.

Gynoecium — Carpels 2, united; ovary superior, unilocular; *ovules many, anatropous; placentation parietel.*

Fruits — Bi-or tetra-valved capsule.

Seeds — *Tiny, plumose-hairy;* testa thin; embryo straight; endosperm absent.

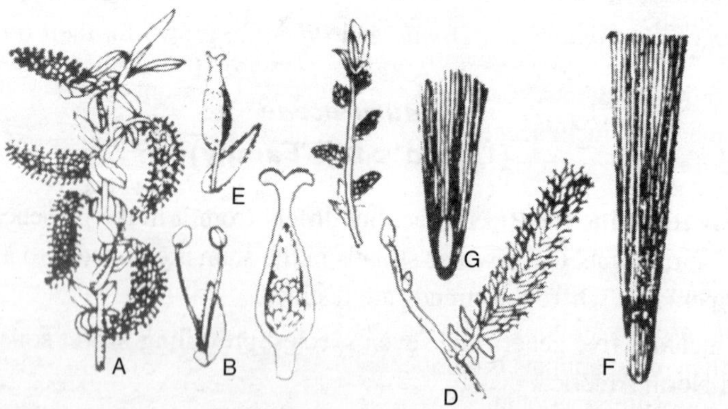

Fig. 4 SALICACEAE. *Salix alba* L. **A**, shoot with male inflorescence; **B**, male flowers; **C**, female flowers; **D.** shoot with female inflorescence; **E.** female flower; **F**, seed; **G**, V.S. of seed. (**A–G**, after Baillon)

* However, *Salix medemii* is a monoecious plant.

POSITION AND AFFINITY OF THE FAMILY

Taxonomically, the family is isolated. There is a tendency to refer the family to the Amentiferae in view of the perianthless flowers being aggregated into catkins. Engler regarded the Salicaceae as one of the most primitive among the dicotyledons due to simple perianthless unisexual flowers. He included the family in his third order of the subclass Archichlamydeae. Following his basis, Rendle treated the family as primitive and placed it in a lowly position under the Monochlamydeae. Hutchinson put the family in an order by itself between the Hamamelidales and Leitneriales. In the arrangement of Bentham-Hooker, the Salicaceae appeared under the Ordines Anomali.

From extensive studies on the Salicaceac, Fisher (1928) concluded that the simplicity of the flowers is a consequence of reduction. He suggested that the flowers possess a perianth of certainly one and probably two whorls which is or are now represented by the cup-like gland of *Populus* and the finger-like gland or glands of *Salix*. He also explained that the bract "does not subtend the flower… but is a leaf upon the pedicel of a formerly stalked ancient flower." Hjelmqvist (1948) thought that the cup- or finger-like gland in the flowers has been "formed by the reduction of an undifferentiated envelope and that it is not appropriate to designate them as perianth". Eames (1961) opined that the glands signify the lost perianth, since they contain the same vascular supply as the perianth. Hutchinson (1969) "noted that *Populus* is anemophilous and has deteriorated, or at any rate has not evolved into many species or forms, whilst *Salix* has remained entomophilous and perhaps in consequence has become much the larger and more successful genus. The species of *Salix,* like those of *Rubus, Rosa* and many other advanced genera, appear to be still in a state of flux, hybridising freely, a condition usually the reverse of primitive".

Van Tieghem (1900) sought a link of the family with the Piperaceae. Bessey (1915) established an affinity of the Salicaceae with the Tamaricaceae. However, the presumed relationship with the Tamaricaceae appears to be completely superficial from morphological and ecological view points. Fisher (1928) expressed the view that the family is an advanced phyletic group of unknown ancestral relationships. Brown (1938) stated that the Salicales is a florally reduced derivative of the Violales, pointing out some resemblances in nectary structure. The pollen morphology of the Salicaceae is akin to some of the Tamaricaceae and Flacourtiaceac, but similar pollen also occurs in some of the Rosales and Campanulales and in the Platanaceae. Lawrence (1951) remarked that the Salicaceae appears "to be more specialized than most or all other amentiferous families and the only one of an order that stands well apart from other orders of dicots." Cronquist (1968) saw "a forerunner of the salicalean gynoecium in some members of the Hamamelidaceae which" has "capsular fruits with parietal placentation".

NUMBER IN AND DISTRIBUTION OF THE FAMILY

The Salicaceae represents four genera and 350 species. They tend to occur in wet places, such as in bottom lands and along streams, although not solely confined to such habitats. "Neither wind distribution of the seeds nor growth in moist habitats is unique to the Salicales, nor is there any structural or ecological need for these two features to be associated" (Cronquist, 1968).

Though the members of the family are cosmopolitan in distribution, they abound in four centres of thc temperate zone, such as the Pacific North America, the area around Behring's Sea, Central Europe and the Himalaya. However, they are absent in Australia and Malaya Archipelago.

COMMONLY OCCURRING PLANTS OF THE FAMILY

Poplar (*Populus alba* L.) and Willow (*Salix tetrasperma* Roxb.) inhabit the hills.

ECONOMIC ASPECTS OF THE FAMILY

The Salicaceae is of great economic importance. The wood of various species of *Populus* is used for floor boards and paper pulp. The wood of *P. tremula* is used for making match boxes and match sticks. *Salix alba* and *S. nigra,* both of the Mediterranean area, yield valuable timber. *S. alba* is also the source of 'salicin', a medicinal glucoside which is an ingredient of 'aspirin'. The wood of *S. alba* var. *caerulea* is used for making cricket bats. The twigs of *S. amygdalina* (Eurasia) and *S. viminalis* (Europe to Siberia) are good in making baskets. A few plants are cultivated as ornamentals, *e.g., S. babylonica, S. caprea, S. cinerea, S. daphnoides, S. pentandra,* etc.

Order 4. *Balanopsidales*

Due to the conglomeration of advanced and primitive characters, the position of the order is controversial. It has been thought to be related to the Juglandales or Euphorbiaceae.

The order is represented by a solitary family, the Balanopsidaceae. The family contains two genera, *Balanops* and *Trilocularia*; the former is found in New Caledonia and Queensland and the latter in Polynesia.

There are no reported economic uses for the Balanopsidaceae.

Order 5. *Leitneriales*

The order, according to Lawrence (1951), is a much reduced relic and not a primitive group. The affinity of the order has been suggested with the Salicales or Juglandales, but it is more likely that some groups in the Rosales or Geraniales or the Hamamelidaceae in the Rosales are its allies (Benson, 1957). Willis (1966) remarked that the order is an interesting relic of very obscure affinities.

The order contains only the Leitneriaceae, represented only by Florida Corkwood (*Leitneria floridana*), a small tree native to muddy saline swamps of south-eastern United States. The tree yields light wood used as floats for fishing nets.

Order 6. *Juglandales*

There is little agreement among taxonomists as to the number of families contained in the Juglandales. Engler considered it a monotypic order with the Juglandaceae alone. This idea was also shared by both Lawrence (1951) and Core (1955). Rendle (1925) treated the order as comprising the Myricaceae and Juglandaceae. Benson (1957) accommodated the Rhoipteleaceae and Juglandaceae within the order. According to Hutchinson (1959), the order includes the Rhoipteleaceae, Juglandaceae and Picrodendraceae. Cronquist (1968) defined the order as consisting of three families and as attempted by Hutchinson. Mitra (1974) proposed five families under the order: Juglandaceae, Fagaceae, Betulaceae, Julianiaceae, Myricaceae. The present author supports Mitra's delimitation of the order, but feels that the Julianiaceae should be accommodated in a separate order.

As with most other orders of the Amentiferae, the affinity of the Juglandales is not certain. The order may be primitive as considered by Engler and Prantl, but it shows a high degree of specialisation.

The floral range of the Juglandales may be summarised as follows:

(1) Ovary superior	*Myricaceae*
(1) Ovary inferior	(2)
(2) Leaves pinnately compound	*Juglandaceae*
(2) Leaves simple	(3)
(3) Male and female flowers in catkins	*Betulaceae*
(3) Male flowers in catkins, females subtended by an involucre which becomes cupular in fruit	*Fagaceae*

Myricaceae
(Bayberry Family)

GENERAL FEATURES OF THE FAMILY

Habit	—	Shrubs or trees, deciduous or evergreen.
Leaves	—	Alternate, short petioled, simple, rarely pinnatifid, stiff, *contain yellow, aromatic resin glands.*
Inflorescence	—	*Dense catkins.*
Flowers	—	Bracteate, *perianthless,* mostly unisexual.
Androecium	—	*Stamens* 2–20 (in male flowers), often 4; anthers basifixed, dehiscence by vertical slits.
Gynoecium	—	Carpels 2 (in female flowers); *ovary superior, unilocular; ovule 1, orthotropous; style 2-branched.*
Fruit	—	Drupe, *covered with a whitish wax secretion.*

POSITION AND AFFINITY OF THE FAMILY

The conspicuous peltate resinous glands of the leaves of the Myricaceae are homologous to similar glands in the Juglandaceae and Fagaceae. Cronquist (1968) commented thus: "The Myricaceae appear to be allied to both the Fagales and Juglandales, but they would be peripheral and aberrant in either order. At the present state of our knowledge it seems most useful to maintain the Myricaceae in an order of their own."

NUMBER IN AND DISTRIBUTION OF THE FAMILY

This is a small family of two genera and 35 species, being distributed in the subtropical and temperate regions.

COMMONLY OCCURRING PLANT OF THE FAMILY

Myrica esculenta Buch-Ham. ex D. Don, an evergreen dioecious tree with pubescent branchlets, is found in Khasi Hills, Meghalaya.

ECONOMIC ASPECTS OF THE FAMILY

The berries of *M. cerifera* and *M. pennsylvanica* are the source of an aromatic wax, used for candle and soap. The fruits of *M. esculenta* are edible. *M. gale* yields tannic acid.

GENERAL FEATURES OF THE FAMILY

Habit	—	Shrubs or trees, resinous, aromatic.
Leaves	—	Alternate or opposite, imparpinnate, exstipulate.
Inflorescence	—	Catkins.
Flowers	—	*Unisexual.*
Perianth	—	In male flowers, 3- to 6-parted and in female 4-parted.
Androecium	—	Stamens 3–100; anthers basifixed.
Gynoecium	—	Carpel 1; ovary *inferior*, 1-*locular*; style short; *stigma plumose.*
Fruit	—	*Drupe-like nut, surrounded by a leathery or fleshy husk.*
Seeds	—	Embryo very large; cotyledons fleshy, 4-lobed, foliaceous or corrugated; endosperm none.

Juglandaceae (Walnut family)

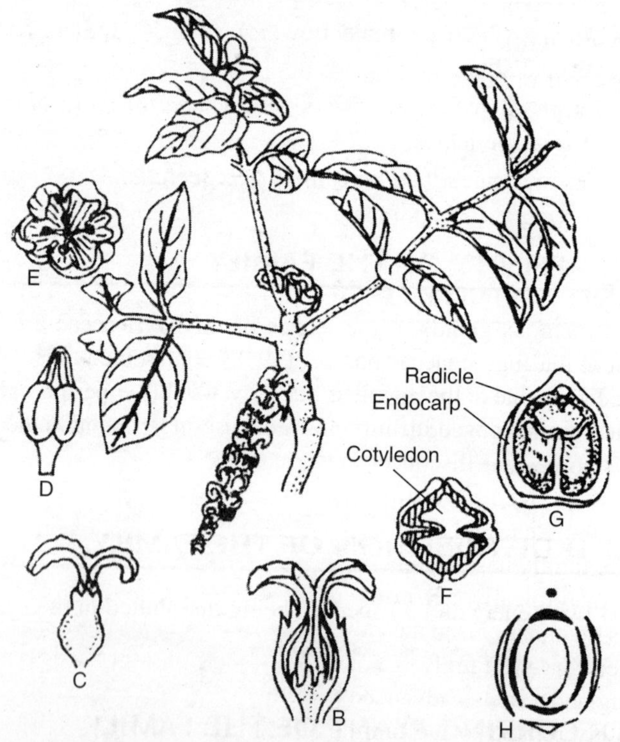

Fig. 5 JUGLANDACEAE. *Juglans regia* L. **A**, portion of shoot; **B**, L.5. of female flower; **C**, female flower; **D**, stamens; **E**, male flower; **F**, T.S. of nut; **G**, L.S. of nut; **H**, diagram of female flower. (**B**, after Nawaschin; **C–H**, after Rendle)

FLORAL RANGE IN THE FAMILY

The Juglandaceae shows variations in the type of catkin and floral organisation. In the first type, the inflorescence is a panicle of clustered erect or pendulous male catkins associated with a female central catkin, *e.g.*, some species of *Engelhardtia* and *Pterocarya.* In the second type, we find a solitary terminal pendulous female catkin which, by way of reduction, conforms to a few flowered erect spike (*Carya, Juglans*). The third type is a triclustered male catkin on old wood, as in *Alfaroa.* The fourth and last type takes the form of a simple lateral male catkin, as in some species of *Juglans.*

The male flowers show a wide range in the development of perianth as well as in the number of stamens. The perianth is absent in *Carya;* here the stamens are 4–10, but sometimes 2 or 3. In *Juglans,* the perianth is 3- to 4-parted; here the stamens may be 20 in the lower flowers of the catkin and 6 or 8 in the upper flowers. *Pterocarya* is in possession of 4-parted perianth, with stamen number varying between 8 and 16.

The structure and development of female flower as well as the relation of ovule to carpel has been a subject of dispute. According to van Tieghem, the vascular supply to each floral leaf comes out from the stem stele at the base of the flower; hence, the ovary is superior and not inferior. Nicoloff stated that the ovule is a development from the apex of floral axis and the carpels do not contribute to its formation. Benson and Welsford (1909) suggested that the ovule is parietal and anatropous. Nast (1935) interpreted the ovule as a modified axile type. Manning (1940) contended that the ovule is derived from an axile condition where the ovules are clustered near the centre of a normal or incomplete partition. If a macroscopic examination of the young ovary is made, the ovule will appear basal and orthotropous.

As to the position of the carpels, it is either median (*Engelhardtia, Juglans, Pterocarya*) or transverse (*Carya, Pterocarya*).

The bract and a pair of bracteoles are united to various degrees. In *Engelhardtia*, the bract and bracteoles are united at the base of the ovary to form a trilobed wing in the fruit. In *Juglans,* the bract is united to a little above the ovary and the bracteoles are united still higher to constitute a toothed rim. The bracts and bracteoles are almost free in *Pterocarya;* here the bracts disappear in the fruit, but the bracteoles form a broad wing in the centre.

POSITION AND AFFINITY OF THE FAMILY

The phyletic position of the Juglandaceae has been debated at length. Both Bessey and Hutchinson felt that the family evolved from one of the members of the Sapindales. However, Hutchinson (1969) changed his mind and regarded the supposed affinity of the family to be superficial. "The absence of stipules from Juglandaceae is somewhat disconcerting, for most of the immediate family stocks from which they seem to have been derived have stipulate leaves". In view of the fact that stipules are very rarely present in the Caesalpinioideae and are not even constant in the Rosaceae, they have been suppressed in the Juglandaceae. The family has been included in the Terebinthales by Hallier on the basis of its origin from the Rutaceae. Anatomically, according to Tippo (1938), the family is much more advanced than indicated by Engler's treatment. From the view points of wood anatomy and pollen morphology, Heimsch (1942, 1944) reported that the family is not allied to the Anacardiaceae and Julianiaceae. Hjelmqvist (1948) placed the Juglandaceae as advanced over the Myricaceae. Lawrence (1951) suggested that the Juglandaceae is closely related to the families under the Urticales; his idea is based on the resinous condition, pinnate leaves and floral morphology. Core (1955) believed that the family is akin neither to the Salicales nor to the Fagales. Benson (1957) opined that the family is of common origin with the Hamamelidaceae.

NUMBER IN AND DISTRIBUTION OF THE FAMILY

A family of about seven genera and 50 species, the Juglandaceae abounds in the north temperate regions and mountains of the northern tropics.

COMMONLY OCCURRING PLANTS OF THE FAMILY

Engelhardtia spicata Bl. and *Juglans regia* L. are large trees; the former is found in subtropical Himalaya and the latter in temperate Himalaya.

ECONOMIC ASPECTS OF THE FAMILY

The Juglandaceae is of considerable economic importance. The seeds arc rich in oil and several species are edible, *e.g.*, Pecan Nut (*Carya, illinoinensis*), Hickory Nut (*C. laciniosa, C. ovata*), Mockernut (*C. tomentosa*) Butternut or White Walnut (*Juglans cinerea*), Black Walnut (*J. nigra*) and English Walnut (*J. regia*). The species of *Carya, Engelhardtia* and *Juglans* produce valuable timber. The species of *Pterocarya* are raised for their ornamental value.

Betulaceae
(Birch Family)

GENERAL FEATURES OF THE FAMILY

Habit — *Shrubs or trees, with perulate buds.*
Leaves — Alternate, simple, stipulate.
Inflorescence — Male flowers in pendulous catkins and females in cylindrical cone-like spikes.
Flowers — *Unisexual.*
Androecium — Stamens 1–4; anthers basifixed.
Gynoecium — Carpels 2, united; *ovary inferior.*
Fruit — *1-seeded nut, subtended by foliaceous bracts.*

DIVISIONS OF THE FAMILY

The Betulaceae is divided into two subfamilies:

Subfamily I. *Betuloideae.* Male flowers in 3-flowered dichasia in axil of a bract and have a perianth. Female flowers lack a perianth. Examples: *Alnus* and *Betula.*

Subfamily II. *Coryloideae.* Male flowers solitary, without a perianth. Female flowers have a perianth. Examples: *Carpinus, Corylus*, etc.

NUMBER IN AND DISTRIBUTION OF THE FAMILY

There are six genera and approximately 170 species in the Betulaceae, mostly distributed in the cooler portions of northern hemisphere but also in tropical Andes and Argentina.

COMMONLY OCCURRING PLANTS OF THE FAMILY

Himalayan Black Cedar (*Alnus nitida* Endl.), a deciduous tree, is found in western temperate Himalaya.

Paper Birch (*Betula utilis* D. Don) is a tree with the bark peeling off in horizontal flakes or plates.

ECONOMIC ASPECTS OF THE FAMILY

The Betulaceae contains many useful plants. *Alnus glutinosa* is used for preparing high grade charcoal. *A. nitida, A. rubra, Betula lenta, B. lutea, B. pendula, B. pubescens, Carpinus betulus, C. caroliniana* and *Ostrya carpinifolia* produce valuable timber. Birch beer is made from the sugary sap of *Betula* sp. and 'oil of betula' from twigs of the same. Some species are popular as ornamentals, such as *B. alba and B. papyrifera.* The bark of *B. papyracea* is used for making canoes. The outer bark of *B. utilis* is employed as a material for writing upon. Edible Hazelnuts are derived from *Corylus americana* and *C. cornuta* as well as Filberts from *C. avellana, C. colurna* and *C. maxima.*

Habit — Shrubs or trees, deciduous or evergreen.
Leaves — Alternate, simple, stipulate.
Inflorescence — Male flowers usually in catkins and females in few flowered clusters.
Flowers — Unisexual.
Perianth — Present, 4- to 7-lobed in male flowers and 4- to 8-lobed in female flowers.
Androecium — Stamens 4–40; anthers basifixed.
Gynoecium — *Carpels 3, united; ovary inferior.*
Fruit — *1-seeded nut, invested by a cupule*.*

Fagaceae
(Beech family)

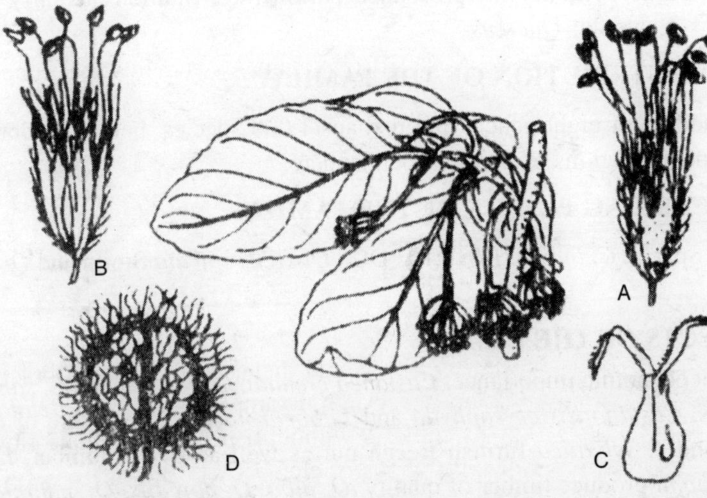

Fig. 6 FAGACEAE. *Fagus sylvatica* L. **A**, male flower; **B**, V.S. of male flowers; **C**, female flower; **D**, female inflorescence. (**A–D**, after Baillon)

* While Castanea has 3 fruits per cupule, *Fagus* has 2 fruits and *Quercus* has 1 fruit. In *Fagus*, the cupule splits into 4 lobes from the beginning. In *Castanea*, the cupule is divided into 4 valves at maturity. Brett (1964) thought that the 4-partite cupule represents a reduction from the three 3-lobed cupules condensed together. The elimination of inner lobes results in the 5-lobed cupule of *Trigonobalanus*, which after reduction, gives rise to the cupule of *Quercus*. The whole cupule of *Lithocarpus* with 1 flower is equivalent to one of the three theoretical 3-lobed cupules. When the central cupule, along with adjacent lobes, is eliminated, 2 lateral cupules remain, each represented by 2 lobes, *e.g.*, 4-valved cupule of *Castanea* and of *Fagus* (where the central flower is not present).

POSITION AND AFFINITY OF THE FAMILY

Wettstein, Rendle and Hutchinson accepted the view that the Betulaceae and Fagaceae are more closely related to each other than to other families of the Amentiferae and hence retained them in the Fagales. Bessey considered the two families along with the Juglandaceae and derived them from sapindaceous ancestors. On the basis of floral morphology and anatomy, Berridge (1914) inferred that the Fagaceae sprang from epigynous rosaceous ancestors. As a result of the studies on wood anatomy, Tippo (1938) stated that the family originated from tricarpellate stocks ancestral to the existing Hamamelidaceae. Hjelmqvist (1948) retained both Fagaceae and Betulaceae within the Amentiferae, but segregated them as two orders. Cronquist (1968) treated the two families "as derived from a common ancestor which would not fit well into either of them".

DIVISIONS OF THE FAMILY

The Fagaceae is divided into three subfamilies:

Subfamily I. *Castaneoideae.* Inflorescence catkin-like. Stamens usually 12. Stigma punctiform. Example: *Castanea.*

Subfamily II. *Fagoideae.* Flowers in axillary clusters, rarely solitary. Fruit 2- or 4-angled. Example: *Fagus.*

Subfamily III. *Quercoideae.* Inflorescence catkin-like. Stamens usually 6. Stigma broad. Example: *Quercus.*

NUMBER IN AND DISTRIBUTION OF THE FAMILY

The Fagaceae is made up of eight genera and perhaps 1,000 species, being distributed largely in the subtropical and temperate regions of northern hemisphere.

COMMONLY OCCURRING PLANTS OF THE FAMILY

Castanopsis armata Spach., *C. indica* (Roxb.) A. DC., *Quercus spicata* Smith and *Q. truncata* King are trees.

ECONOMIC ASPECTS OF THE FAMILY

The family is of great economic importance. *Castanea crenata*, *C. dentata* and *C. sativa* provide useful timber and chestnuts. *Castanopsis crysophylla* and *Lithocarpus densiflorus* are of ornamental value. *Fagus grandifolia* and *F. sylvatica* furnish Beech nut as well as Beech lumber. *L. densiflorus* and *Nothofagus cunninghami* produce timber of quality. *Q. alba, Q. borealis, Q. macrolepis, Q. montana, Q. vallonea,* etc., yield tannin. Oak lumber is derived from *Q. bicolor, Q. coccinea, Q. garryana, Q. imbriearia, Q. laevis Q. palustris, Q. phellos, Q. stellata, Q. virginiana,* etc., *Q. suber* is the source of commercial cork.

Order 7. *Julianiales*

The order contains a solitary family (Julianiaceae) with two genera, *Amphipterygium* (Mexico) and *Orthopterygium* (Peru). It "may be primitive as suggested by Engler and Prantl,but more likely it is not" (Benson, 1957). Despite the pinnately compound leaves and superficial amentiferaean traits, the family is not related either to the Juglandaceae or other taxa of the Amentiferae (Hjelmqvist, 1948). Rendle (1925) placed the group between the Juglandales and Fagales. Hutchinson (1959) included the family in his 49th order Sapindales lying between the Meliales and Loganiales. The presence of alternate imparipinnate exstipulate leaves, dissimilar male and female flowers, broad stigmatic lobes and unitegmic ovules recall strikingly what are obtained in the Juglandaceae. While vegetative characters show no resemblance to the Fagales, the floral characters speak for an alliance with this order than with the Juglandales. The staminate inflorescence and flowers as well as minute globose pollen grains are suggestive of the species of *Quercus*. The pistillate flowers are crowded in an involucre, indicating the condition of *Castanea* and *Fagus*. A relationship can be traced to the Anacardiaceae by the items of foliage, resinducts, reduced unisexual flowers, single exalbuminous seed and embryo form as well as anatomical features. The Julianiaceae is unique by the possession of the ovule ascending from the loculus-base, fruit enclosed in the enlarged involucre and seeds without endosperm.

The bark of *Amphipterygium adstringens* is used medicinally as an astringent.

Order 8. *Urticales*

Hallier (1905) placed the order in the Terebinthinae and Bessey (1915) in the Malvales. The order is equivalent to the Urticiflorae of Rendle (1925) as well as Engler and Diels (1936). Under this order, Engler treated families like the Ulmaceae, Rhoipteleaceae, Moraceae and Urticaceae. Rendle conceived the order to contain also four families; here the Rhoipteleaceae was omitted and the Cannabinaceae included. Tippo (1938) indicated that the Eucommiaceae belongs to this order and so did Benson (1957). Withner (1941) excluded the Rhoipteleaceae from this order and put the family in the Juglandales. Lawrence (1951) considered only three families (Ulmaceae, Moraceae, Urticaceae) in the order, the Cannabinaceae coming within his second family. Hutchinson (1959) created six families in the Urticales; Ulmaceae, Cannabinaceae, Moraceae, Urticaceae, Barbeyaceae, Eucommiaceae. Cronquist (1968) inserted five families by leaving out the Eucommiaceae.

According to Bechtel (1921), the members of the Urticales are a combination of primitive and advanced features. He postulated that the order is not remote from primitive entomophilous ancestors. The order is generally accepted as a natural taxon, but its position "is at the culmination of a distinct line of descent from a protoangiospermous plexus from which also ranalian descended". It is also regarded as phyletically advanced and a terminal group of an evolutionary line from which the Fagales has been derived. Porter (1967) suggested that the Urticales is "of ancient origin and dubious affinities".

The Urticales is distinguished by the presence of androecium with few to several stamens and bicarpellate unilocular superior ovary with a solitary ovule. The usual four families of the order can be identified by referring to the following synoptic key of Core (1955):

(1) Fruit an achene	(2)
(1) Fruit not an achene	(3)
(2) Plants armed with stinging hairs, not aromatic	*Urticaceae*
(2) Plants erect or climbing, aromatic	*Cannabinaceae*
(3) Fruit multiple; sap milky	*Moraceae*
(3) Fruit a drupe, nut or samara; sap not milky	*Ulmaceae*

Ulmaceae
(Elm Family)

GENERAL FEATURES OF THE FAMILY

Habit	—	Shrubs or trees; *sap not milky.*
Leaves	—	Alternate, petiolate, stipulate, *unequal sided.*
Inflorescence	—	Cymose, fasciculate or solitary.
Flowers	—	Bisexual or unisexual.
Perianth	—	*Present,* of 4–8 sepaloid segments, somewhat united.
Androecium	—	Stamens equal in number and opposite to perianth segments; *filaments erect in bud;* anthers 2-celled, opening by longitudinal slits.
Gynoecium	—	Carpels 2; ovary superior, unilocular; *ovule amphitropous or anatropous, pendulous from apex of ovarian chamber.*
Fruit	—	*Drupe or samara.*
Seeds	—	Embryo straight; cotyledons conduplicate, flat or sinuous; endosperm little or none.

DIVISIONS OF THE FAMILY

The Ulmaceae is separated into two tribes, according to the nature of fruit and seed:

Tribe (*i*) *Celtideae.* Fruit a reduced drupe. Embryo curved with folded or rolled cotyledons. Endosperm sometimes present. Examples: *Celtis, Trema,* etc.

Tribe (*ii*) *Ulmeae.* Fruit a winged nut. Embryo straight with flat cotyledons. Endosperm absent. Examples: *Haloptelea, Ulmus,* etc.

NUMBER IN AND DISTRIBUTION OF THE FAMILY

A family of 16 genera and about 2,000 species, the Ulmaceae extends its sway throughout the tropical and extra tropical regions of the world.

COMMONLY OCCURRING PLANTS OF THE FAMILY

Celtis tetrandra Roxb. and *Haloptelea integrifolia* (Roxb.) Planch are trees.

Charcoal Tree or Indian Nettle Tree [*Trema orientalis* (L.) Bl.] is a small twiggy tree, marked with scars.

ECONOMIC ASPECTS OF THE FAMILY

The Ulmaceae is important from the economic stand point. *Celtis australis, C. occidentalis, C. procera* and *Ulmus thomasi* produce timber. Some species of *Celtis* are appreciated for their edible fruits. The fragrant wood of *Planera abelicea* is used as a substitute for sandal wood. The wood of *Trema orientalis* is useful in making charcoal for gunpowder; the fruits of this plant are edible. Several species are cultivated domestically as ornamentals, *e.g.*, *U. americana, U. carpinifolia, U. glabra,* etc. *U. rubra* yields mucilaginous inner bark, which finds application in the manufacture of medicinal troches. The wood of *Zelkova serrata* (Japan) is used for making elegant carved boxes.

Moraceae
(Mulberry Family)

GENERAL FEATURES OF THE FAMILY

Habit	—	Shrubs or trees, with *milky sap.*
Leaves	—	Alternate or opposite, simple, entire, stipulate.
Inflorescence	—	Heads, racemes, spikes, etc.
Flowers	—	Unisexual, 4-merous.
Perianth	—	Present, usually of 4 segments, free or united.
Androecium	—	Stamens equal in number and opposite to perianth segments; *filaments inflexed or erect in bud;* anthers 2-celled, versatile, dehiscing lengthwise.
Gynoecium	—	Carpels 2, often 1 abortive; ovary inferior or superior; ovule solitary, pendulous from apex.
Fruit	—	*Multiple.*
Seeds	—	Embryo curved, surrounded by a fleshy endosperm.

FLORAL RANGE IN THE FAMILY

The inflorescence is fundamentally cymose, but it is often greatly modified. In *Broussonetia, Cudrania* and *Maclura,* the female inflorescence is a globose head. The union of pedicels and peduncles results in a laminate receptacle whose ventral surface is studded with minute sessile flowers, as in *Dorstenia.* The cymes may coalesce to develop into a fleshy hollow axis, bearing flowers on the interior surface, *e.g.*, *Ficus.* The inflorescence of *Morus* is condensed into a pendulous amentiferous structure.

The apparent spike inflorescence in *Artocarpus* is formed by fusion and coalescence of several fertile branches (Sharma, 1965). The cauliflorous flowering buds arise exogenously on the young and endogenously in the old stems. Due to total shifting of flowering and fruiting to old limbs and the main trunk, the cauliflory existing in *A. heterophyllus* is believed to be of advanced type. Pundir (1965) noted that the first crop of the syconia in *Ficus glomerata* is formed by axillary buds of the uppermost leaves. Some exogenously derived dormant buds become active and form the flowering shoots on old branches. Simultaneously, there starts the *de novo* production of secondary buds which are interpreted as endogenous. Of these, one becomes a leader and acts as a cauliflorous shoot.

Although the stamens are 4, the number is sometimes reduced to 1 or 2 (*Artocarpus, Dorstenia, Ficus, Morus*).

Of the two carpels, the hinder one is variously aborted. It is sometimes developed (*Artocarpus*) or sometimes not indicated (*Chlorophora*). It may be represented by an empty ovarian chamber, as in *Morus.* In many other genera, the presence of a hind carpel is shown only by a style resembling an anterior one.

DIVISIONS OF THE FAMILY

The Moraceae is divided into three subfamilies:

Subfamily I. *Artocarpoideae.* Leaves spirally folded in bud. Stamens erect in bud. Ovule apical. Example: *Artocarpus.*

Subfamily II. *Conocephaloideae.* Leaves spirally folded in bud. Stamens erect in bud. Ovule apical or basal, orthotropous. Example: *Conocephalus.*

Subfamily III. *Moroideae.* Leaves not spirally folded in bud. Stamens inflexed in bud. Ovule apical. Example: *Morus.*

NUMBER IN AND DISTRIBUTION OF THE FAMILY

The Moraceae comprises 75 genera and about 3,000 species. They largely abound in the tropical regions, but a few spread over the temperate zone.

COMMONLY OCCURRING PLANTS OF THE FAMILY

Sack Tree or Uppas Tree [*Antiaris toxicaria* (Pers.) Leschen.] is a stately gigantic glabrous tree whose trunk is buttressed.

Bread Fruit Tree [*Artocarpus altilis* (Park.) Fosberg] is a tree with pinnatifid leaves.

Chaplash (*Artocarpus chaplasha* Roxb.) is a large deciduous tree with globose or oblong inflorescence.

Jack fruit Tree (*Artocarpus heterophyllus* Lamk.) and Angeli Wood Tree (*A. hirsutus* Lamk.) are evergreen trees.

Conocephalus suaveolens Bl. is a large climber.

Cudrania cochinchinensis (Lour.) Kudo and Masam. ex. Sauer is a scandent shrub.

Banyan Tree (*Ficus bengalensis* L.), Indian Rubber Tree (*F. elastica* Roxb.) and Pipal Tree (*F. religiosa* L.) are large trees.

Ivy Fig (*Ficus stipuilata* Wall.) is a herbaceous plant which climbs on walls.

White Mulberry (*Morus australis* Poir.) is a moderate sized deciduous tree.

Phyllochlamys spinosa Burm. is a large spinous shrub.

Streblus asper Lour, is a small rigid scraggy evergreen tree.

ECONOMIC ASPECTS OF THE FAMILY

The bark of *Antiaris toxicaria* is used for making garments and sacks; the stem juice is the source of an arrow poison. Edible fruits are derived from *Artocarpus altilis, A. heterophyllus, A. lakoocha, Ficus carica, F. cunia, F. glomerata, F. hispida, F. sycomorus, Morus nigra, Treculia africana*, etc. The wood of *Artocarpus chaplasha, A. hirsuta, F. bengalensis* and *F. religiosa* is employed for construction purposes. *Brosimum aubletii* and *Piratinera guianensis* produce curiously streaked wood. The fruits of *B. alicastrum*

(West Indies) are cooked and eaten. The latex of *B. galactodendron* (Cow Tree of Venezuela) is sweet and nutritious. *Broussonetia papyrifera* forms good fibres for manufacturing paper from the inner bark and is largely cultivated throughout Polynesia for making 'tapa cloth'. The latex of *Castilla elastica* forms 'Panama rubber' and *C. ulei* 'Caucho rubber', while that of *F. elastica* makes 'Indian rubber'. *Chlorophora excelsa* yields valuable timber. 'Fustic', a natural yellow pigment, is extracted from *C. tinctoria* (tropical America). The roots of *Conocephalus suaveolens* are used for poulticing itch. *Dorstenia brasiliensis* is a drug plant, the roots being used as a stimulant and tonic. Shellac is obtained from *F. laccifera.* The bark of *F. nekbudu* (Mozambique) serves as a source of 'mutshu cloth'. The wood of *Maclura aurantiaca* is suitable for making bows. *Morus australis* is grown for the leaves which are fed to silkworms. The wood of *Streblus asper* is used for making small articles and cart wheels; the twigs are utilised as baits for catching prawns; the leaves serve as sand paper.

Urticaceae
(Nettle Family)

GENERAL FEATURES OF THE FAMILY

Habit	—	Herbs or undershrubs; sap not milky; epidermal cells contain cystoliths.
Leaves	—	Alternate or opposite, simple, stipulate, often *armed with stinging hairs.*
Inflorescence	—	Cyme, head or solitary.
Flowers	—	Minute, unisexual, regular.
Perianth	—	Absent or present, when present of 4–5 connate or free parts.
Androecium	—	Stamens as many as perianth parts and opposite to them, *inflexed in bud* and *exploding when ripe*.
Gynoecium	—	Carpel 1; ovary inferior or superior; ovule erect; *style single, with a brush-like stigma.*
Fruit	—	Achene or drupe.
Seeds	—	Embryo straight, surrounded by an oily endosperm.

DIVISIONS OF THE FAMILY

The Urticaceae is separated into five tribes:

Tribe (*i*) *Boehmerieae.* Stinging hairs absent. Bracts absent. Stamens 4–5. Example: *Boehmeria.*

Tribe (*ii*) *Forskohleae.* Flowers naked. Stamen 1. Example: *Forskohlea.*

Tribe (*iii*) *Parietarieae.* Bracts united in an involucre. Flowers sometimes bisexual. Example: *Parietaria.*

Tribe (*iv*) *Procrideae.* Stinging hairs absent. Perianth of female flowers 3-merous; Example: *Pilea.*

Tribe (*v*) *Urereae.* Stingingh airs present. Leaves alternate or opposite. Example: *Urtica.*

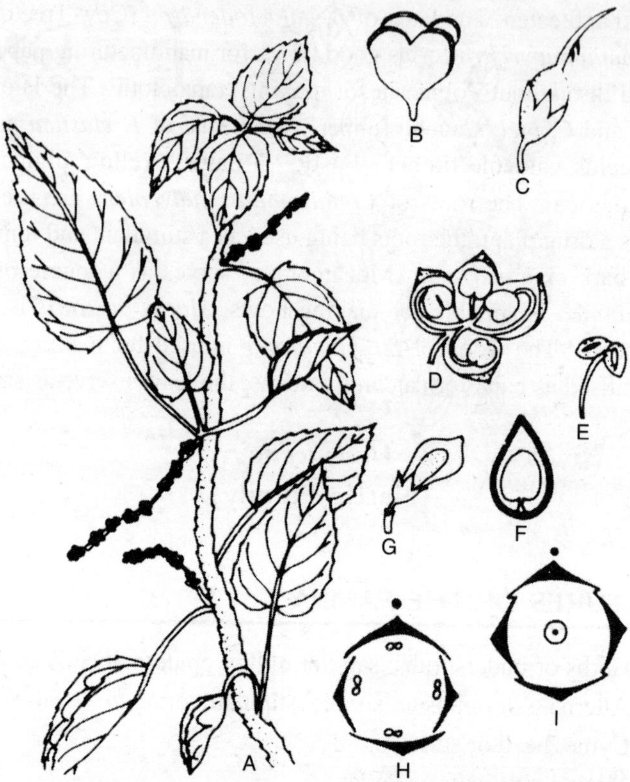

Fig. 7 URTICACEAE. *Fleurya interrupta* (L.) Gaud. **A**, portion of plant; **B**, male flower; **C**, female flower; **D**, corolla split open; **E**, stamen; **F**, L.S. of ovary; **G**, fruit; **H**. diagram of male flower; **I**, diagram of female flower

NUMBER IN AND DISTRIBUTION OF THE FAMILY

The Urticaceae embraces 45 genera and over 1,000 species. They are mainly prevalent in the tropical and subtropical areas of the globe.

COMMONLY OCCURRING PLANTS OF THE FAMILY

China Grass or Rhea Grass [*Boehmeria nivea* (L.) Gaud.] is a low hairy shrub, having a perennial rootstock and herbaceous branches.

Nettle [*Fleurya interrupta* (L.) Gaud.] is a weed with stinging hairs.

Nilgiri Nettle (*Giraradina zeylanica* Decne) is a tall stout coarse erect herb.

Parietaria debilis Forst. f., a prostrate herb of the Himalaya at an altitude of 2,134–3,658 m, thrives on rocks in shady places.

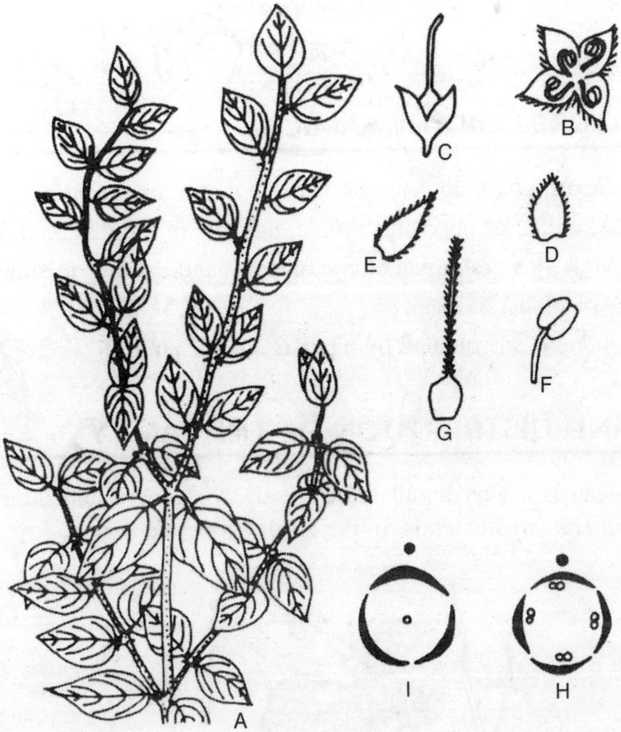

Fig. 8 URTICACEAE. *Pouzolzia zeylanica* (L.) Benn. **A**, portion of plant; **B**, male flower; **C**, female flower; **D**, perianth of male flower; **E**, perianth of female flower; **F**, stamen; **G**, L.S. of ovary; **H**, diagram of male flower; **I**, diagram of female flower

Artillery Plant or Gunpowder Plant [*Pilea microphylla* (L.) Liebm.], seen on damp walls, scatters a volley of pollen grains from the anthers when the plant is shaken.

Pouzolzia zeylanica (L.) Benn., a marshy herb, bears samaroid fruits.

ECONOMIC ASPECTS OF THE FAMILY

Boehmeria nivea yields the toughest and most silky fibre known; it is particularly suitable for making paper pulp for the manufacture of bank notes. The fruits of *Debregeasia hypoleuca* are edible and are used for flavouring. The fibres, obtained from *Girardina zeylanica* and *Villebrunea integrifolia,* are used for cordage, ropes, sack cloth and similar articles. The root of *Pouzolzia tuberosa* is eaten. *Urera baccifera* (tropical America) is used as a cattle hedge. The young tops of *Urtica dioica* and *U. urens* form the source of edible greens. A few members are raised for their decorative foliage, *e.g., Pellionia, Pilea Soleirolia* and *Urera.*

Cannabinaceae
(Hemp Family)

GENERAL FEATURES OF THE FAMILY

Habit — *Herbs, climbing or erect, with no milky sap, aromatic.*
Leaves — Alternate or opposite, simple, palmate or palminerved, stipulate.
Inflorescence — Male flowers in paniculate racemes and females in spikes.
Flowers — Apetalous, axillary.
Fruit — *Achene*, surrounded by a persistent perianth.

NUMBER IN AND DISTRIBUTION OF THE FAMILY

The Cannabinaceae is a very small family with two genera and three species, out of which *Cannabis* is tropical and *Humulus* extends to the north temperate region.

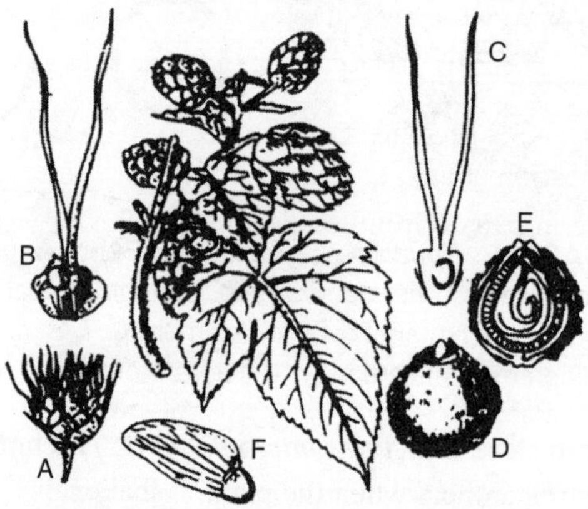

Fig. 9 CANNABINACEAE. *Humulus lupulus* L. **A**, female inflorescence; **B**, female flower; **C**, V. S. of female flower: **D**, fruit. **E**, V.S. of fruit; **F**, seed. (**A–F**, after Baillon)

ECONOMIC ASPECTS OF THE FAMILY

The leaves of *Cannabis sativa* are used for making the intoxicating beverage called 'bhang' or 'siddhi', the resinous exudation is smoked as 'charas'; the young inflorescences are smoked as 'ganja'; the seed oil serves as a luminant and in the preparation of paints, varnishes and soap; the bark yields 'hemp' which is in demand for cables, tarpaulins, etc.

Hops (*Humulus lupulus*) are widely cultivated in Europe for the spikes which are used to give flavour and sparkle to beer.

Order 9. *Proteales*

The order contains a single family, the Proteaceae. The characters of this order are the same as those for the family.

Proteaceae
(Silk-oak Family)

GENERAL FEATURES OF THE FAMILY

Habit — Shrubs or trees.
Leaves — Alternate, opposite or whorled, simple, *exstipulate*.
Inflorescence — *Showy bracteate heads,* racemes or spikes.
Flowers — Bisexual, irregular or regular, hypogynous.
Perianth — *Tetramerous, petaloid, coloured, valvate, variously split when open.*
Androecium — *Stamens* equal in number, opposite to *perianth-segments and adnate to them*; anthers 2-celled, free.
Gynoecium — Carpel 1; ovary superior; ovules 1 or more; *placentation parietal.*
Seeds — Winged; *endosperm* absent.

FLORAL RANGE IN THE FAMILY

The various types of inflorescence occurring in the family have been derived from panicle by reduction and condensation or simplification (Venkata Rao, 1965).

The flowers are borne in pairs (Grevilleoideae) or are solitary (Persoonioideae). Although they are bisexual, unisexual flowers are seen in *Leucadendron.* They are usually regular, but irregular in *Lomatia* and other genera.

Out of 4 stamens, 2 are 2-celled and the rest 1-celled: this condition is noticed in some species of *Conospermum* and *Synaphea.* In *Symphyonema,* the stamens are free from the tepals.

At the base of the ovary, one may find various forms of nectariferous scales. These structures, according to Venkata Rao (1962), consist of three free lobes alternating with perianth segments. However, Kausik (1938) and Haber (1959) suggested that these organs represent reduced inner perianth members.

DIVISIONS OF THE FAMILY

The Proteaceae is divided into two subfamilies:

Subfamily I: *Grevilleoideae.* Flowers in pairs borne on apparent racemes or spikes. Ovules many, sometimes 2. Fruit a several seeded follicle. Examples: *Banksia, Hakea, Lomatia,* etc.

Subfamily II: *Persoonioideae.* Flowers single at axil of bracts. Ovules 1, 2 or few. Fruit a 1-seeded drupe. Examples: *Leucadendron, Persoonia, Protea,* etc.

POSITION AND AFFINITY OF THE FAMILY

Basically, the Proteaceae is not a primitive taxon. But there seems to be no agreement among botanists as to its proper position. It formed the 15th series of Engler's Archichlamydeae, near the Urticales and before the Santalales. Accepting Engler's opinion of primitiveness of the order, Hallier derived it from the Proberberideae. Bessey cast doubt on the relationship of the Proteaceae which was placed in the Sapindales. Rendle also pointed out that it is difficult to associate the order with other taxa. Lawrence stated that the order is not closely related to any other group. Hutchinson regarded it to be a terminal group originating from stocks ancestral to the Thymelaeaceae. As the Proteaceae bears close affinity to the Thymelaeaceae, Mitra (1974) imparted the position of the Proteales after the Myrtales and before the Santalales. That the Proteaceae is fairly advanced has been revealed by the studies of Haber (1959); he proved that the occurrence of four types of ovule (anatropous, basal, orthotropous, pendulous), reduction of petals into nectariferous scales and occasional zygomorphy with orninthophilous flowers are signs of advancement. In its inflorescence, the family "often mimics both Asteraceae (Compositae) and Pinaceae, in *Protea* and *Leucadendron* respectively, while the superficial resemblance of its fruits with those of some Fabaceae (Papilionaceae) has caused some authors to suggest an affinity with the last mentioned family" (Hutchinson, 1969).

NUMBER IN AND DISTRIBUTION OF THE FAMILY

A large family of 62 genera and over 1,000 species, the Proteaceae is highly developed in the drier regions of southern hemisphere, mainly Australia and South Africa.

COMMONLY OCCURRING PLANTS OF THE FAMILY

Australian Silky Oak (*Grevillea robusta* A. Cunn. ex R. Br.), often grown in gardens, parks and roadsides, is a dominant tree of Mount Abu, Rajasthan.

Hakea sericea Schrad. is a conifer-like plant, bearing acicular leaves and axillary clusters of white or pinkish flowers.

ECONOMIC ASPECTS OF THE FAMILY

The Proteaceae furnishes us with many beautiful ornamentals like *Embothrium coccineum, Grevillea robusta, Hakea laurina, Leucadendron argenteum, Protea mellifera* and *Telopea speciosissima. Faurea macnaughtonii* and *Knightia excelsa* yield good timber. The fruits of *Gevuina avellana* and *Macadamia ternifolia* are edible. The bark of *Leucospermum conocarpodendron* is used for tanning leather.

Order 10. *Santalales*

The Santalales is mostly a natural group, but botanists are divided in their opinion as to its exact phyletic position. Bessey set up the families of the order in the Celastrales, deriving them from rosaceous ancestors. Hutchinson considered the order as originating from the Celastrales. Rendle treated the order as closely related to the Proteales, from which it differs chiefly in the ovary being inferior.

Due to the presence of parasitic members, Schellenberg (1932) held the view that the Santalales is not a primitive taxon. "It is not certain how much of the vegetative reduction in the Santalales has positive survival value in correlation with the parasitic habit, and how much has merely been permitted by the relaxation of selection, resulting in the accumulation of loss mutations. It may be suggested that the ovular reduction is a mere side effect of the vegetative reduction, being controlled by some of the same genes" (Cronquist, 1968).

The order is distinguished by the flowers generally unisexual, perianth segments small and opposite to adnate stamens and seeds with endosperm.

Different systematists have regarded the Santalales as composed of families ranging from three to ten.

Engler and Diels	:	Olacaceae, Opiliaceae, Octoknemataceae, Grubbiaceae, Santalaceae, Myzodendraceae, Loranthaceae.
Rendle	:	Santalaceae, Loranthaceae, Balanophoraceae, Cynomoriaceae.
Lawrence	:	Olacaceae, Santalaceae, Balanophoraceae.
Core	:	Santalaceae, Olacaceae, Loranthaceae, Balanophoraceae.
Benson	:	Olacaceae, Loranthaceae, Santalaceae, Opiliaceae, Octoknemaceae, Grubbiaceae, Myzodendraceae.
Hutchinson	:	Loranthaceae, Grubbiaceae, Santalaceae, Misodendraceae, Balanophoraceae.
Cronquist	:	Medusandraceae, Dipentodontaceae, Olacaceae, Opiliaceae, Grubbiaceae, Santalaceae, Loranthaceae, Misodendraceae, Balanophoraceae, Cynomoriaceae.

The present author, following Core (1955), takes into account four families with the under mentioned characters:

(1) Non-green total parasites	*Balanophoraceae*
(1) Green partial parasites	(2)
(2) Ovules undifferentiated from placenta	*Loranthaceae*
(2) Ovules differentiated from placenta	(3)
(3) Perianth segments dissimilar with a distinct calyx and corolla	*Olacaceae*
(3) Perianth segments similar, sepaloid or petaloid	*Santalaceae*

Olacaceae
(Malla-tree Family)

The family is characterised by the exstipulate leaves with parchment like rough texture, flowers with a distinct calyx and corolla, superior ovary and ovules differentiated from the placenta.

The Olacaceae is divided into three subfamilies:

Subfamily I: *Anacolosoideae.* Ovules with 1 or 2 integuments, anatropous with a dorsal raphe, micropyle facing upwards. Example: *Ximenia.*

Subfamily II: *Olacoideae*. Ovules without integuments, anatropous with a dorsal raphe, micropyle facing upwards. Example: *Olax.*

Subfamily III: *Schoepfioideae.* Ovules without integuments, atropous, pendulous, micropyle facing downwards. Example: *Schoepfia.*

A family of about 25 genera and 250 species, the Olacaceae is widespread in the Old World tropics.

Anacolosa densiflora Bedd., from Western Ghats to Anaimalai southwards, is a lofty tree with white fragrant flowers.

Miquelia gibba Baill., *Natsiatum herpeticum* Buch-Ham. ex Arn. and *Olax scandens* Roxb. are climbing shrubs.

The walnut-like fruits of *Coula edulis* is eaten fresh, roasted or boiled. *Harmandia coccinea* is sometimes grown as an ornamental hothouse plant. The seeds of *Olax gambecola* (West Indies) are used as a condiment. The wood of *Ximenia americana* is suitable for carving and used as a substitute for sandalwood.

GENERAL FEATURES OF THE FAMILY

Habit — Herbs, shrubs or trees, *non-parasitic or semi-parasitic on roots.*

Leaves — Alternate or opposite, simple, exstipulate, glabrous.

Inflorescence — Head, raceme or spike.

Flowers — Small, bisexual or unisexual, regular.

Perianth — Simple, sepaloid or petaloid, united below into a tube, segments 4 or 5, valvate.

Androecium — Stamens equal in number and opposite to perianth segments; a tuft of hairs present behind each stamen; anthers 2-celled, at or near base of a filament.

Grynoecium — Carpels 3–5; *ovary somewhat inferior or nearly superior, unilocular, ovules differentiated from placenta*; stigma capitate or lobed.

Fruit — Achene or drupe.

Seeds — Embryo straight; cotyledons terete; endosperm fleshy.

Santalaceae
(Sandalwood family)

Fig. 10 SANTALACEAE. *Santalum album* L. **A**, V.S. of flower; **B**, sepal and stamen; **C**, fruit; **D**, V.S. of fruit. (**A–D**, after Hutchinson)

DIVISIONS OF THE FAMILY

The Santalaceae is separated into three tribes:

Tribe (*i*) *Anthoboleae*. Ovary inferior or superior. Ovules not fully differentiated from placenta. Pedicel becomes swollen and fleshy with fruit development. Examples: *Anthobolus, Exocarpos,* etc.

Tribe (*ii*) *Santaleae*. Ovary inferior. Receptacle shallowly cup-or saucer-shaped, lined with a nectar secreting disc. Examples: *Acanthosyris*, *Santalum*, etc.

Tribe (*iii*) *Thesieae.* Ovary inferior. Receptacle deeply cup shaped without a disc. Examples: *Arjona, Thesium,* etc.

POSITION AND AFFINITY OF THE FAMILY

In Bentham-Hooker's arrangement, the Santalaceae appeared in the series Achlamydosporeae of the subclass Monochlamydeae. Engler included the family in the suborder Santalineae of the Santalales under the Archichlamydeae. Hutchinson placed it in the Santalales after the Olacales and before the Rhamnales.

The Santalaceae is closely allied to the Loranthaceae in some characters of perianth, ovary and seeds. While the plants of the Santalaceae are non-parasitic or parasitic only on roots, those of the Loranthaceae are aerial parasites. The flower of the Santalaceae has an ovary with three ovules, whereas the ovary of the Loranthaceae flower has no distinct ovule. It is very likely that the Santalaceae and Misodendraceae are the nearest related taxa.

There is strong superficial resemblance of the Santalaceae in foliage and inflorescene to the Myrtaceae (Richardson, 1978).

Data from anatomy and embryology have helped to assign systematic position of two genera of the Santalaceae—*Calyptosepalum* and *Exocarpos.* The first genus is not propitious for the family. According to Swamy (1949), it has a trilacunar node, enlarged vessel member, predominant uniseriate tall vascular rays, apotracheal xylem parenchyma and an unique floral vasculature. The second genus was includcd in a separate family (Sarcopodacees) near the Taxaceae in the gymnosperms by Gagnepain and Boreau (1946). Lam (1948) commented that the "*Exocarpos*... probably represents a transition between the protoangiospermous gymnosperms and the Monochlamydeae". Swamy's work suggested no link between *Exocarpos* and the gymnosperms, since he demonstrated the presence of three trace unilacunar nodes, true vessels and specialised type of sieve tubes in this genus. Ram (1959) showed that the flowers of *Exocarpos* possess perfect angiospermous features and is unlike *Casuarina* or *Salix* in its embryology. Here the archesporial mother cell functions as a megaspore mother cell, embryosac is of the Polygonum type, endosperm is cellular with a chalazal caecum and pericarp is derived from the ovarian wall. "Though somewhat aberrant, *Exocarpos* should definitely be assigned to the Santalaceae rather than to the Salicaceae or Casuarinaceae" (Maheshwari and Kapil, 1963).

NUMBER IN AND DISTRIBUTION OF THE FAMILY

The Santalaceae consists of about 35 genera and 400 species. The members of this family are widely distributed in the temperate as well as tropical zones of the world and concentrated in relatively dry areas.

COMMONLY OCCURRING PLANT OF THE FAMILY

Sandalwood Tree (*Santalum album* L.) is found in South India and is often grown in gardens.

ECONOMIC ASPECTS OF THE FAMILY

The fruits of *Acanthosyris falcata* as well as *Exocarpos cupressiformis* and tubers of *Arjona tuberosa* are edible. The species of *Buckleya* and *Pyrularia* are often cultivated for their showy flowers.

Colpoon compressum, Eucarya spicata, Santalum freycinetianum and *S. yasi* yield valuable wood. The inner part of the wood, the heartwood, of *S. album* is employed for boxes, cabinet work, coffins, combs, fancy articles, furniture, picture frames, etc. The sawdust of heartwood is made into scented cakes and sticks to be burnt as incense. An essential oil, distilled from the heartwood and roots, is used in perfumery. The outer part of sapwood does not contain any essential oil and is not of any value except as firewood.

GENERAL FEATURES OF THE FAMILY

Habit	—	*Herbs, aerial parasitic**.
Flowers	—	Bisexual or unisexual, regular; *receptacle cup shaped.*
Perianth	—	Simple or petaloid, free or united, large, coloured; segments in 2 or 3 whorls, sometimes with a slightly toothed or irregular rim (*calyculus***).
Androecium	—	Stamens equal in number and opposite to perianth segments, epiphyllous.
Gynoecium	—	*Carpel 1*; *ovary inferior, unilocular*; *loculus ill defined* and *distinct ovules absent.*
Fruit	—	Pseudocarp.
Seeds	—	Embryo large; cotyledons appressed face to face; endosperm fleshy.

Loranthaceae
(Mistletoe family)

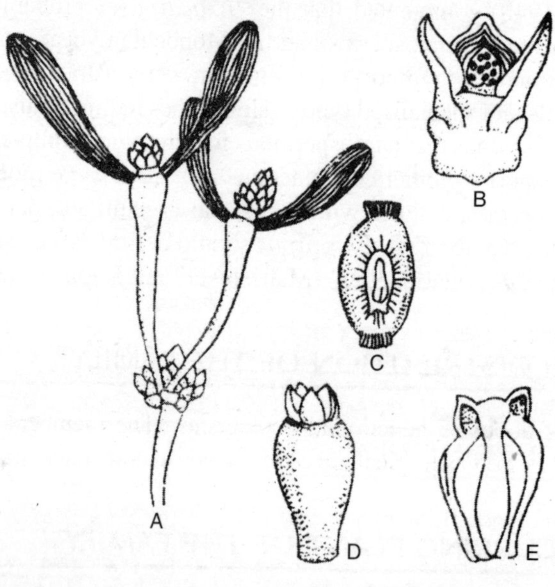

Fig. 11 LORANTHACEAE. *Viscum album* L. **A**, portion of shoot; **B,** L.S. of flower bud; **C**, L. S. of seed; **D**, fruit ;**E**, L. S. of fruit. (**A–E**, after Le Maout and Decaisne)

* A few root in the earth like *Nuytsia floribunda,* a small tree of Western Australia.

** It represents a vestigial whorl of bracts of suppressed flowers of a lateral branch inflorescence which has become adnate to the inferior ovary (Venkata Rao, 1963).

DIVISIONS OF THE FAMILY

The Loranthaceae is divided into two subfamilies:

Subfamily I: *Loranthoideae.* Flowers bisexual or unisexual. Calyculus present. Examples: *Loranthus, Struthanthus,* etc.

Subfamily II: *Viscoideae*. Flowers unisexual. Calyculus absent. Examples: *Arceuthobium, Viscum,* etc.

POSITION AND AFFINITY OF THE FAMILY

The Loranthoideae and Viscoideae show great variations in the floral structure. Moreover, the mode of development of the embryosac, endosperm and embryo and location of the viscous part of the fruit are different in the two subfamilies.

The Loranthoideae contains bisexual or unisexual flowers generally associated with a calyculus. The ovary is characterised by a collenchyma pad and contains a mamelon which may or may not be lobed. The viscid zone of the fruit is located outside the vascular bundles of the corolla. Other features peculiar to the Loranthoideae include the Polygonum type of embryosac, protrusion of tip of the embryosac into the style, formation of composite endosperm and vertical division of the zygote.

The Viscoideae contains only unisexual flowers where the calyculus is invariably absent. The viscous layer of the fruit is situated within the vascular bundles of the perigone. The bisporic embryosac with a slow but steady curvature, endosperm developing from a single embryosac and embryo without a suspensor are unique to the Viscoideae.

In view of the above differences, Johri and associates (1957) suggested that the Loranthoideae is distinct from the Viscoideae. These differences, according to them, warrant the elevation of the status of the subfamilies to that of families—Loranthaceae and Viscaceae.

NUMBER IN AND DISTRIBUTION OF THE FAMILY

The Loranthaceae comprises at least 35 genera and about 1,300 species. The numbers of this family are found in wooded regions of the tropics and extends into extratropical areas.

COMMONLY OCCURRING PLANTS OF THE FAMILY

Arceuthobium minutissimum Hook, f., parasitic on the twigs of *Pinus wallichiana* A.B. Jackson, is the smallest known dicotyledonous plant.

Dendrophthoe falcata (L. f.) Etting. and *Macrosolen cochinchinensis* (Lour). Van Tiegh. are found on Mango and other trees.

Eurasian Mistletoe (*Viscum album* L.) "was used as a ceremonial plant by the Druids and the other early Europeans, when arose the modern custom of kissing under the mistletoe at Christmas" in western countries. Another popular Christmas Mistletoe is *Phoradendron flavescens* (North America).

Balanophoraceae
(Balanophora Family)

The family represents *non-green total parasites.* The vegetative organs are reduced to branched tuberous reddish or yellowish rhizomes which are fixed by means of suckers to the roots of a host tree. Sometimes within the rhizome, the inflorescence develops and then projects above the ground. The flowers are minute and unisexual, the male being with 3–4 stamens and the female with 1–3 carpels. *The inferior ovary contains 1–3 ovules* which are naked, sometimes reduced to an embryosac and often *united with the ovarian wall.* The fruit is a nut. The seed contains abundant endosperm which encloses an undifferentiated embryo.

The Balanophoraceae is akin to the Cynomoriaceae with which it is often merged (Mabberley, 1978). Since these plants are notable for their reduced structure and curious development, it is hard to reconcile with the view that they are end products of convergent evolution or they represent natural assemblage. Although they have links with the Hydnoraceae and Rafflesiaceae, they are probably unrelated to other parasitic groups. They are related to the Olacaceae, Opiliaceae and Santalaceae. A "true relationship, albeit a distant one, may be with *Gunnera* (Haloragaceae)".

In the Balanophoraceae, there are 18 genera and 120 species which are prevalent in the savannahs and tropical woods.

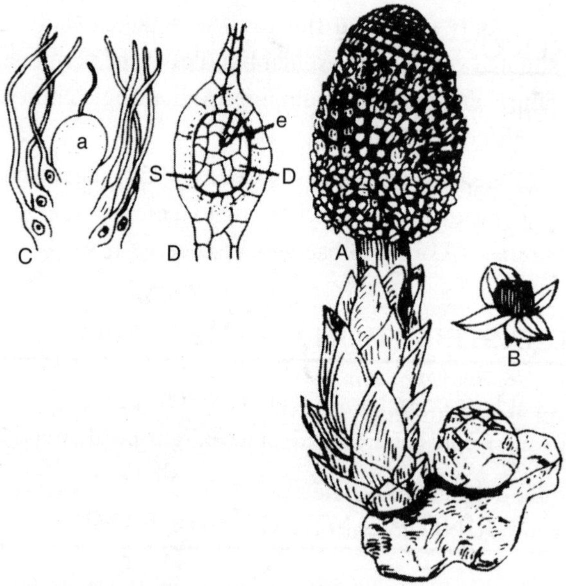

Fig. 12 BALANOPHORACEAE. *Balanophora dioica* R. Br. **A**, male plant; **B**, male flower; **C**, V. S. of portion of fertile spike showing sterile apex *(a)*; **D**, V. S. of fruit (**s**, seed; *e*, embryo; *n*, endosperm). (**A–D**, after Rendle)

Balanophora dioica R.Br. is found in Khasi Hills, Meghalaya.

In Java, waxes are extracted from the plants of the Balanophoraceae and are burnt for lighting. Sometimes the plants are reported to have aphrodisiac properties.

Order 11. *Aristolochiales*

A great diversity of views exists as to the proper position of the Aristolochiales. According to Engler, Warming and Rendle, the order is next to the Santalales. Hallier treated it as originating from the Berberidaceae via the Lardizabalaceae. Bessey considered it as the most advanced taxon under the Myrtales—an order sprung from the Rosales. Wettstein placed the Aristolochiales within the Polycarpicae on the basis of its annonaceous ancestry. Hutchinson set up the Aristolochiales as a separate order, adding to it also the Nepenthaceae. He held the opinion that the order represents "probably reduced Berberidales by way of Menispermaceae".

In this order, the perianth is uniseriate and petaloid, stamens numerous and free and ovary inferior with axile or parietal placentation.

The families of the order, according to Core (1955), may be categorised as follows:

(1) Flowers unisexual	*Rafflesiaceae*
(1) Flowers bisexual	(2)
(2) Climbing shrubs or herbs, not parasitic	*Aristolochiaceae*
(2) Parasitic herbs	*Hydnoraceae*

Aristolochiaceae
(Birthwort Family)

GENERAL FEATURES OF THE FAMILY

Habit	—	Low herbs or climbing shrubs.
Stem	—	*Softly woody, with broad medullary rays.*
Leaves	—	Alternate, petiolate, simple, entire, *exstipulate.*
Inflorescence	—	Racemose or solitary.
Flowers	—	Bisexual, irregular or regular.
Perianth	—	*Petaloid, enlarged to form a bell-like or trumpet-like structure,* usually highly coloured.
Androecium	—	*Stamens 6–36, set around apex of ovary or stylar column;* anthers free or united, dehiscing longitudinally.
Gynoecium	—	Carpels 4–6: *ovary inferior*; ovules many in each loculus.
Fruit	—	Capsule, dehiscing septicidally.
Seeds	—	Embryo small; endosperm fleshy.

DIVISIONS OF THE FAMILY

The Aristolochiaceae is divided into three tribes:

Tribe (*i*) *Aristolochieae.* Perennial herbs or shrubs. Flowers irregular, solitary or in cymes or racemes. Petals absent. Stamens in one row, united with style. Ovary inferior. Fruit a capsule. Examples: *Aristolochia*, *Euglypha* and *Helostylis.*

Tribe (*ii*) *Bragantieae.* Shrubs or semishrubs. Flowers regular in cymes or racemes. Petals absent. Stamens in one or two rows, free or united with style. Ovary inferior. Fruit a capsule. Examples: *Apana* and *Thottea.*

Tribe *(iii)* *Sarumeae.* Perennial herbs. Flowers regular, solitary. Petals present or absent. Stamens in two rows, free or united at base of filaments. Ovary inferior or semi-inferior. Fruit a capsule or follicle. Examples: *Asarum* and *Saruma.*

NUMBER IN AND DISTRIBUTION OF THE FAMILY

Seven genera and about 625 species constitute the Aristolochiaceae which is distributed throughout the hot and temperate zones of the world.

COMMONLY OCCURRING PLANTS OF THE FAMILY

Aristolochia bracteolata Lamk. is a decumbent slender perennial. *A. tagala* Cham, is a twining perennial shrub.

ECONOMIC ASPECTS OF THE FAMILY

The family is of some economic importance for ornamentals, e.g., *Aristolochia clematitis, A. gigas, A. grandiflora, A. macrophylla, A. orninthocephala* and *Asarum europaeum.* The roots of *Aristolochia indica* are used by snake charmers for catching snakes. The roots of *A. serpentaria* and *Bragantia wallichii* serve as a cure for snake bites.

Hydnoraceae (Hydnora Family)

This is a small family of herbs, *parasitic on the roots of various shrubs and trees.* The scale leaves are absent. The flowers are solitary, subsessile, bisexual and rather large. The perianth is very thick. The ovary is inferior and Unilocular; *the ovules are very numerous from apical or parietal placetae.* The fruit is baccate. The seeds are minute and countless; the endosperm is copious.

The family contains only two genera—*Hydnora* with about eight species in Africa and the Malagasy Islands and *Prosopanche* with one species in the Pampas of Argentina.

Rafflesiaceae (Rafflesia Family)

This is a family of *parasitic herbs.* The vegetative organs are reduced to mycelia which ramify through the tissues of the host. Produced above the ground as adventitious shoots from the mycelium, the *flowers are unisexual, regular and minute to very large.* The perianth is 4- to 5-parted; the male flowers have many stamens and the female 4, 6 or 8 carpels, with an inferior ovary; *the ovules are numerous on parietal placentae*. The fruit is a succulent berry. The seeds are minute with a hard testa, undifferentiated embryo and oily endosperm.

The Rafflesiaceae is divided into four tribes:

Tribe (*i*) *Apodantheae.* Flowers unisexual, solitary, small. Stamens in 2 or 4 rings on central column. Ovary inferior with 4 placentae or 1 continuous placenta. Example: *Apodanthes.*

Tribe (*ii*) *Cytineae.* Flowers unisexual, in racemes. Stamens in 1 ring. Ovary inferior with 8–14 placentae. Example: *Cytinus.*

Tribe (*iii*) *Mitrastemoneae.* Flowers bisexual, solitary. Example: *Mitrastemon.*

Tribe (*iv*) *Rafflesieae.* Flowers unisexual, solitary, large. Stamens in 1 ring. Ovary inferior with many irregular chambers. Example: *Rafflesia.*

"The closest relatives of the Rafflesiaceae are undoubtedly the Hydnoraceae, which differ in having root-like structures, no bracts below the flowers, bisexual flowers and stamens borne on a column. Some botanists prefer to isolate *Mitrastemon* as a separate family placed between the Rafflesiaceae and Hydnoraceae. The affinities of this group are very uncertain. Most authorities link the Rafflesiaceae with the Aristolochiaceae because they have a similar perianth, but the evidence for this association is not conclusive" (Kupicha, 1978).

The Rafflesiaceae is primarily a subtropical family of about nine genera and 500 species, being found in the Old World.

The flower of *Rafflesia arnoldi* (Sumatra) is the largest in the Plant Kingdom, measuring a metre in diameter and weighing upto 8 kg.

Order 12. *Polygonales*

This is a monotypic order with the characters of its single family, the Polygonaceae.

GENERAL FEATURES OF THE FAMILY

Habit	—	*Herbs with swollen nodes,* sometimes climbing.
Leaves	—	Alternate, simple, entire, *stipulate* (stipules ochreate*).
Flowers	—	Small, bisexual, regular, hypogynous.
Perianth	—	Simple, sometimes petaloid, segments 3–6, persistent.
Androeciura	—	Stamens 5–8, opposite to perianth-segments; anthers 2-celled, dehiscing longitudinally.
Gynoecium	—	Carpels 2 or 3, united; *ovary triangular, superior, unilocular; ovule solitary, erect, orthotropous.*
Fruit	—	*Biconvex or triangular nut.*
Seeds	—	Embryo s-shaped; endosperm mealy.

* Absent in *Antigonon, Brunnichia, Chorizanthe, Eriogonum* and *Koeniga.*

Polygonaceae
(Buckwheat family)

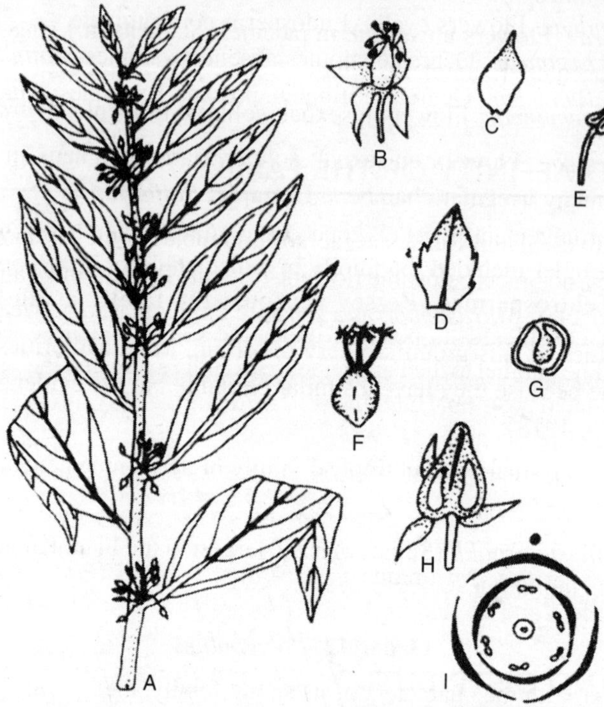

Fig. 13 POLYGONACEAE. *Rumex dentatus* L. **A**, portion of shoot; **B**, flower; **C**, perianth of outer whor; **D**, perianth of inner whor; **E**, stamen; **F**, carpel; **G**, L.S. of ovary; **H**, fruit; **I**, floral diagram

FLORAL RANGE IN THE FAMILY

The flowers arc generally anemophilous, but entomophilous in *Fagopyrum* and some species of *Polygonum.* They are usually bisexual, but unisexual in *Oxyria digyna.* The perianth is not spirally arranged, but whorled in *Polygonum bistorta* and *P. orientate.* The inner perianth segments are suppressed in *Koeniga.* The perianth segments may be persistent and form winged covering in the fruits of *Rumex.* The perianth lobes are basally winged with the fruit at the top, as in *Afrobrunnichia erecta.* The fruiting perianth is solid or closed over the nut in *Emex, Oxygonum* and *Symmeria.* It may be beset with bristles in the margin, as in *Harpagocarpus snowdenii.* Sometimes, the inner whorl of stamens are suppressed (*Polygonum diospyrifolium*). The stamens are numerous in *Calligonum* and are inserted on a central disc in the male flowers of *Ruprechtia* and *Symmeria.* The gynoecium is tetramerous in *Calligonum.*

DIVISIONS OF THE FAMILY

The Polygonaceae is divided into three subfamilies, each with two tribes:

Subfamily I: *Coccoloboideac.* Endosperm ruminate.

Tribe (*i*) *Coccolobeae.* Flowers bisexual. Examples: *Coccoloba, Muehlenbeckia,* etc.

Tribe (*ii*) *Triplaridzac.* Flowers unisexual. Example: *Triplaris.*

Subfamily II: *Polygonoideae.* Flowers somewhat cyclic. Endosperm non-ruminate.

Tribe (*i*) *Atraphaxideae.* Shrubs. Example: *Calligonum.*

Tribe (*ii*) *Polygoneae.* Herbs. Examples: *Fagopyrum, Polygonum,* etc.

Subfamily III: *Rumicoideae.* Flowers cyclic. Endosperm non-ruminate.

Tribe (*i*) *Ergoneae.* Ochreate stipules absent. Examples: *Antigonon, Eriogonum,* etc.

Tribe (*ii*) *Rumiceae.* Ochreate stipules present. Examples: *Rheum, Rumex,* etc.

POSITION AND AFFINITY OF THE FAMILY

In Bentham-Hooker's arrangement, the Polygonaceae is the last member of the Curvembryeae under the Monochlamydeae. Engler included the family in a distinct order (Polygonales), lying between the Aristolochiales and Centrospermae. Bessey accepted the family as an advanced taxon of the Caryophyllales. Hallier considered its position under the Caryophyllales, seeking its origin directly from early Ranales on a line parallel to the Papaveraceae. Rendle expressed the view that the Polygonaceae occupies a somewhat isolated position. Hutchinson opined that the Polygonales is a degraded and reduced type of the Caryophyllales, descending from the Ranales.

The Polygonaceae is related to the Amaranthaceae, Chenopodianceae and Nyctaginaceae in the plan of floral structure, but differs from them by the presence of ochreate stipules, triangular ovary, solitary erect ovule and s-shaped embryo. The family is allied to the Urticaceae by having stipulate leaves and orthotropous ovules. An unique feature of the family, as suggested by Randle, is "the multiplication of the stamens and the relation between the two-whorled trimerous and the cyclic perianth".

NUMBER IN AND DISTRIBUTION OF THE FAMILY

The Polygonaceae embraces about 30 genera and 750 species. The members of the family are natives of the north temperate zone, although a few are found in the tropics.

COMMONLY OCCURRING PLANTS OF THE FAMILY

Railway Creeper (*Antigonon leptopus* Hook. & Arn.) is a garden climber, bearing panicles of pink or white flowers.

Calligonum polygonoides L. is a slow growing nearly leafless shrub.

Coccoloba uvifera L. and *Homalocladium platycladum* (F. v. Muell.) Bailey are characterised by flattened leaf-like stems.

Polygonum barbatum L., *P. glabrum* Willd., *P. orientate* L. and *P. serrulatum* Lag. are weeds in ditches and damp places.

Polygonum plebejum R. Br. is a diffusely branched prostrate herb with pinkish flowers.

Rumex dentatus L. is a marsh weed whose perianth segments have a white tubercled midrib.

ECONOMIC ASPECTS OF THE FAMILY

Coccoloba uvifera supplies edible fruits as well as gum kinos. The starchy seeds of *Fagopyrum esculentum*, called 'buckwheat', form an article of diet. *Polygonum tinctorium.* is the source of a blue dye. The rootstock of *Rheum officinale* (China) and *R. palmatum* (Turkey) is the source of 'drug rhubarb'. The succulent acid petioles of *R. rhaponticum* (USA) are in demand in making pie and sauce. The young shoots of *Rumex conglomeratus, R. crispus, R. obtusifolius* and *R. vesicarius* are used as edible greens. *R. hymenosepalus* (Mexico and USA) yields 'canaigre', utilised in tanning. A few plants are grown ornamentally, *e.g. Antigonon leptopus, Atraphaxis frutescens, Coccoloba uvifera, Muehlenbeckia axillaris and Polygonum aubertii.*

Order 13. *Centrospermales*

The taxon is distinguished by the perianth typically biseriate, ovary superior and unilocular, placentation basal or free central (with reference to the name of the present order) and embryo coiled or curved (the order called Curvem bryeae by some authors).

The order formed the 20th series of Engler's Archichlamydeae and 13th order of Rendle's Monochlamydeae. Wettstein followed Engler concerning the position of the order, but also incorporated the Cactaceae into it. Bessey designated the order as the Caryophyllales and supported its ranalian ancestry. According to Hallier, the order arose from ranalian Podophylleae. Hutchinson created two orders, the more primitive Caryophyllales (derived from the Ranales and containing the Elatinaceae, Molluginaceae, Caryophyllaceae, Ficoideae and Portulacaceae) and the advanced Chenopodiales (derived from the caryophyllaceous stock and containing the Barbeuiaceae, Phytolaccaceae, Gyrostemonaceae, Agdestidaceae, Petiveriaceae, Amaranthaceae, Cynocrambaceae, Batidaceae and Basellaceae).

The Aizoaceae is primarily succulent and the Chenopodiaceae, Phytolaccaceae, Basellaceae and Portulacaceae possess many succulent members. Since most of the chenopods are halophytes with or without succulence, halophytism and succulence have been considered by Cronquist (1968) as "two expressions of the same physiological evolutionary potentialities".

Anomalous secondary thickening resulting in the formation of concentric vascular bundles is almost a family character in the Chenopodiaceae, Amaranthaceae and Aizoaceae. This feature is not reported in the Basellaceae and Portulacaceae. It frequently turns up in the Nyctaginaceae and is more sporadic in other families.

Common embryological features of the centrospermalian plexus include the tetrasporangiate nature of anther, monocotyledonous type of anther wall development, glandular tapetum with multinucleate cells, simultaneous cytokinesis of pollen mother cells, occurrence of an air space at the base of two integuments, formation of parietal tissue, occurrence of Polygonum type of embryosac development, ephemeral antipodals, development of nuclear endosperm and Caryophyllad or Chenopodiad type of embryo development (Maheshwari Devi and Pulliah, 1975).

A distinct group of chemical compounds, betalains—betacyanins as vascular pigments in blue-violet to red series and betaxanthins as those in the yellow to orange-red series—have been known in most of the families. The Caryophyllaceae is strikingly different in lacking betalains and having anthocyanins. However, both betacyanins and anthocyanins are noted in the flowers of a species of *Mesembryanthemum.* But "no individual plant, and indeed no genus or even family of plants, has yet been found to contain both betacyanins and anthocyanins... plants which have betalains instead of anthocyanins commonly have various other flavonoid compounds, and specifically that anthoxanthins and betaxanthins often occur together. Therefore the betalain taxa evidently have most of the genetic mechanism necessary for the production of anthocyanins" (Cronquist, 1968).

According to Kendrick and Hillman (1971), all three members of the Centrospermae—*Amaranthus caudatus, Mirabilis jalapa* and *Spinacia oleracea*—are devoid of the phytochrome darkreversion process typical of other dicotyledons. In monocotyledons, dark reversion is lacking. Members of the Amaranthaceae, Chenopodiaceae and Portulacaceae resemble certain tropical grasses by virtue of the C_4-dicarboxylic acid photosynthetic route. Though carbon dioxide compensation points are indicative of C_4 plants, they occur only in the Centrospermae and Geraniales. Such close connection with monocots are rather unlikely, but the Centrospermae contains distinct primitive plants which may be linked to the ancestral stock from which both dicots and monocots originated. In fact, Meeuse (1963) conjectured an origin of true Centrospermae from the Bennettitales by the use of morphological evidence.

A comparison of the orthodox circumscription of the Centrospermae, based mainly on morphology and anatomy, with that of families containing betalain pigments is given in Table 5.

TABLE 5. Three classifications of the Centrospermae and its allies

Structural classification	*Chemical classification*	*Compromise classification*
Centrospermae	*Centrospermae*	*Caryophyllidae*
Molluginaceae	Aizoaceae	*Caryophyllales*
Caryophyllaceae	Portulacaceae	Caryophyllaceae
Aizoaceae	Phytolaccaceae	Molluginaceae
Portulacaceae	Chenopodiaceae	*Chenopodiales*
Phytolaccaceae	Amaranthaceae	Phytolaccaceae
Chenopodiaceae	Didiereaceae	Nyctaginaceae
Amaranthaceae	Nyctaginaceae	Didiereaceae
Didiereaceae	Basellaceae	Cactaceac
Nyctaginaceae	Cactaceae	Aizoaceae
Basellaceae	*Excluded*	Portulacaceae
Cactales	Molluginaceae	Basellaccae
Cactaceae	Caryophyllaceae	Chenopodiaceae
		Amaranthaceae

The newer definition of the Centrospermae differs from the old in excluding the Molluginaceae and Caryophyllaceae which possess anthocyanins and lack betalains and in including the Cactaceae which has usually been placed in a separate order but which is found to contain betalains. While the exclusion of the Molluginaceae and Caryophyllaceae directly contradicts the anatomical evidence, the structural data do not argue strongly against the inclusion of the Cactaceae. This clash of orthodox and chemical evidence is augmented by the discovery of the ultrastructural characters of the sieve tube plastids (Behnke, 1969). Sieve tube plastids of the Centrospermae as well as the Molluginaceae and Caryophyllaceae are characterised by peripheral ring-shaped bundles of proteinaceous filaments. These inclusions are absent from the Polygonaceae, Plumbaginaceae and Batidaceae. Though the sieve tube evidence correlate fully with the traditional diagnostic features of the Centrospermae, they do less with the betalain distribution. The Cactaceae, Caryophyllaceae and Molluginaceae are correctly assigned to the Centrospermae. A compromise classification has been proposed by Behnke and Turner (1971), replacing essentially the Centrospermae by two closely related orders—the Caryophyllales *sensu stricto* (without betalains) and the Chenopodiales (with betalains), both coming under the subclass Caryophyllidae.

Rendle suggested that the Centrospermae indicates a passage from the monochlamydeous to the dichlamydeous type of flower. The simpler forms show a floral structure similar to the Urticales and elaborate forms a floral plan similar to the Ranales. The Cactales, according to Mitra (1974), forms a connecting link between the caryophyllaceaous members on the hand and the ranalian members on the other.

Considerable evidence has been adduced favouring a connection between the Caryophyllales and Ranunculales (Takhtajan, 1969). "Among the Caryophyllales the closest links with the Ranunculales are shown by the primitive family Phytolaccaceae which has many features that bring it near the families Menispermaceae and Lardiza balaceae. We find within it forms with apocarpous gynoecia—such as *Anisomeria, Ercilla* and *Phytolacea* subg. *Pircunia*—which at once connect it with the Ranunculales. But affinity with the Ranunculales is shown by the presence not only of apocarpous gynoecia but also of tricolpate pollen grains of the ranunculalean type, with its characteristic granulate surface of the middle part of the aperture membrane. The presence of the curved embryo in the Phytolaccaceae and other Caryophyllales points to an affinity with the Menispermaceae. Stipules are sometimes found in the Caryophyllales, but they are also known in some Menispermaceae and the secondary thickening from successive cambia (concentric type) of the stem of many Caryophyllales is also reminiscent of the Menispermaceae. Affinity with the Ranunculales is further corroborated by biochemical data—the presence of isoquinoline derivatives in the Chenopodiaceae and Cactaceae."

The Centrospermales, according to Rendle, is a natural taxon. Here one encounters diversity in the differentiation of floral envelope coupled with the uniform features of placenta, ovule and embryo. Within the Phytolaccaceae, there are both monochlamydeous (*Microtea*) and dichlamydeous forms (*Limeum*) as well as forms with multiplicity of stamens and carpels. Such a variety has led to concept that the families of the Centrospermales owe their origin to the Phytolaccaceae by development along various lines. On the basis of this concept, the Amaranthaceae and Chenopodiaceae represent a variant of the simple floral plan. The type with two staminal whorls and two carpellary whorls has given rise to the taxa with more highly developed flowers:

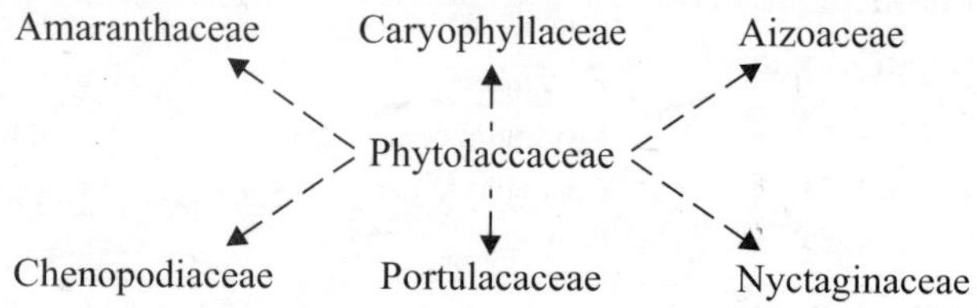

If, from the original type, the inner whorl of stamens and a whorl of carpels disappear, with the consequent reduction in pistil and formation of a petaloid perianth, something like the floral structure of the Nyctaginaceae would result. One can trace the same point of origin for the simpler types of flower of the Aizoaceae like *Mollugo* and *Sesuvium.* In more elaborate type of the Aizoaceae, as in *Mesembryanthemum*, the staminal whorl splits to form a series of petaloid structures and stamens. From the same source, the Portulacaceae may appear by the formation of a pair of bracteoles to constitute the dimerous calyx and by the extinction of the carpellary whorls. A further elaboration is noticed in the floral plan of the Caryophyllaceae where the alternate position of five carpels may be due to the disappearance of one or other of the two whorls in the phytolaccaceous ancestor.

The present author, in line with Core (1955) and Mitra (1974), takes into consideration eight families under the Centrospermales with the characters as follows:

(1) Perianth bract-like	(2)	
(1) Perianth not bract-like	(3)	
(2) Stamens free, staminodes rare	*Chenopodiaceae*	
(2) Stamens united at base, often with alternating staminodes	*Amaranthaceae*	
(3) Flowers haplochlamydeous	(4)	(4)
(3) Flowers diplochlamydeous	(6)	(6)
(4) Perianth segments 5, united; carpel 1	*Nyctaginaceae*	
(4) Perianth segments 4 or 5, free or united; carpels generally more than 1	(5)	(5)
(5) Leaves fleshy or thread-like	*Aizoaceae*	
(5) Leaves not fleshy or thread-like	*Phytolaccaceae*	

(6) Sepals 5; petals 0 or 5 — *Caryophyllaceae*
(6) Sepals 2; petals 4 or 5 — (7) (7)
(7) Twining herbs — *Basellaceae*
(7) Herbs or undershrubs — *Portulacaceae*

Chenopodiaceae
(Beet Family)

GENERAL FEATURES OF THE FAMILY

Habit	—	*Succulent herbs,* with swollen nodes.
Leaves	—	Alternate, simple, entire, membranous or fleshy.
Inflorescence	—	*Cyme or spike without scarious bracts.*
Flowers	—	Monochlamydeous, bisexual or unisexual, regular.
Perianth	—	Small, sepaloid, free or united; segments 3–5, persistent.
Androecium	—	Stamens 5, opposite to perianth segments, free or united.
Gynoecium	—	Carpels 2, united; ovary superior, unilocular; ovule 1, basal, campylotropous; style terminal, short or long.

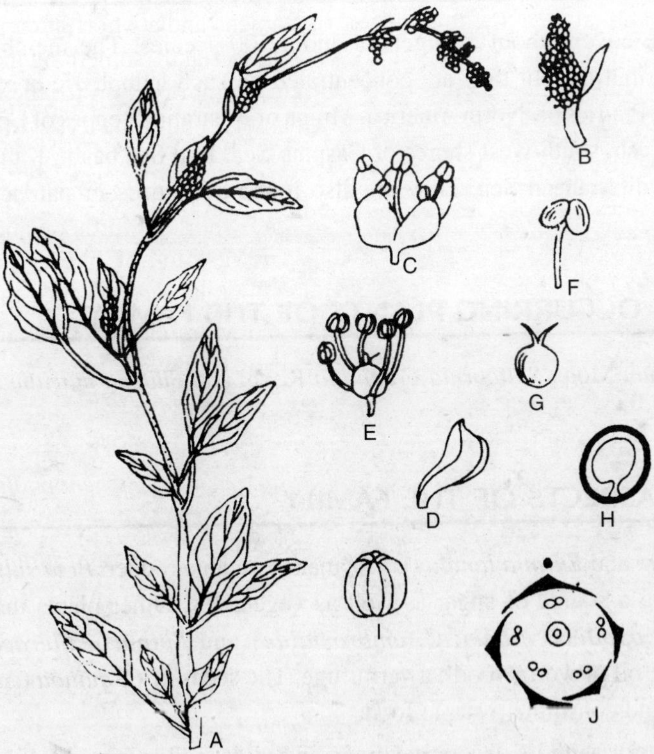

Fig. 14 CHENOPODIACEAE. *Chenopodium album* L. **A**, portion of plant; **B**, inflorescence; **C**, flower; **D**, perianth; **E**, androecium and carpel; **F**, stamen; **G**, carpel; **H**, L. S. of ovary; **I**, fruit; **J**, floral diagram

Fruit — Nut or utricle.
Seeds — Erect, albuminous or exalbuminous; *embryo curved or spirally twisted.*

FLORAL RANGE IN THE FAMILY

The flowers are unisexual in *Atriplex* and *Spinacia.* The male and female flowers are borne on different plants (*Grayia*) or on the same plant (*Sarcobatus*). The flowers may also be irregular, as in *Halocnemum.* The sepals are green and herbaceous (*Beta, Chenopodium*) or hyaline (*Salsola*). In *Monolepis,* the stamen is 1 and in *Salicornia* the number is 2.

DIVISIONS OF THE FAMILY

The Chenopodiaceae is divided into two groups:

Group I. *Cyclolobeae.* Embryo annular, horseshoe shaped or semi circular, partially or completely surrounding perisperm, Examples: *Beta, Chenopodium, Spinacia,* etc.

Group II. *Spirolobeae.* Embryo spirally twisted; perisperm absent or divided into two masses by embryo. Examples: *Anabasis, Haloxylon, Suaeda,* etc.

NUMBER IN AND DISTRIBUTION OF THE FAMILY

The Chenopodiaceae covers about 100 genera and 1,500 species. The members of this family are cosmopolitan in distribution, but they are concentrated on such halophytic or xerophytic areas as the Pampas of Argentina, Prairies of North America, Mulga of Australia, Steppes of Central Asia, Sunderbans of India and Bangladesh, south-west shores of Caspian Sea, Red Sea basin, Karro of South Africa and coast land of the Mediterranean Sea. They are also found as "weeds in salt rich soils around human habitations".

COMMONLY OCCURRING PLANTS OF THE FAMILY

Anthrocnemum indicum Moq., *Salicornia brachiata* Roxb. and *Suaeda maritima* (L.) Dum. are plants of the Sunderbans.

ECONOMIC ASPECTS OF THE FAMILY

The species of *Atriplex* and *Eurotia lanata* (USA) make valuable fodder. *Beta vulgaris* is an example of root economics, being a source of sugar as well as vegetables. Other plants that are used for edible greens include *Chenopodium album, C. ambrosioides* and *Spinacia oleracea.* The seeds of *C. anthelminticum* yield 'oil of wormwood', a vermifuge. The seeds of *C. quinoa* (Andes), called 'quinoa' are cooked like rice and constitute a staple food.

"*C. amaranticolor,* with its decorative green and violet-red coloured foliage, is used as a border plant in gardens." Some species of *Kochia* are grown as ornamentals.

GENERAL FEATURES OF THE FAMILY

Habit	—	Herbs or shrubs.
Leaves	—	Alternate or opposite, simple, exstipulate.
Inflorescence	—	*Congested* (spicate or capitate) *with scarious bracts.*
Flowers	—	Monochlamydeous, regular, bisexual or unisexual, hypogynous.
Perianth	—	*Dry, membranous,* white or coloured; segments 5, persistent.
Androecium	—	Stamens 1–5, opposite to perianth segments; *filaments connate in a cup or tube,* sometimes bearing fimbriate scales; anthers 2-celled.
Gynoecium	—	Carpels 2, united; *ovary superior, unilocular*; ovule 1, campylotropous with a short or long funiculus.
Fruit	—	Nutlet or utricle.
Seeds	—	Lenticular with a rough or polished testa; perisperm mealy; embryo annular or horseshoe shaped.

Amaranthaceae
(Amaranth Family)

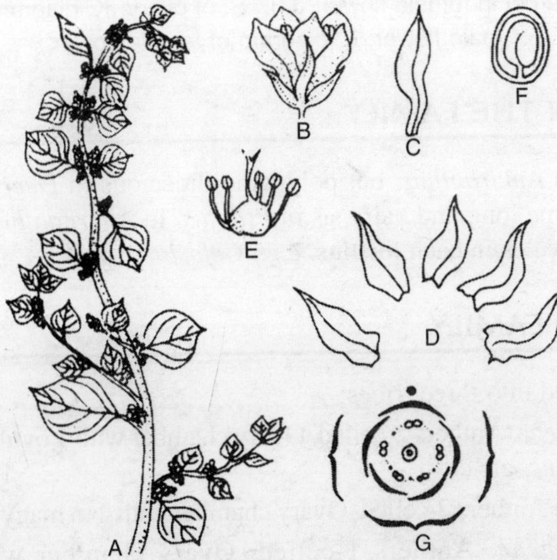

Fig. 15 AMARANTHACEAE. *Aerva lanata* (L.) Juss. **A**, portion of shoot; **B**, flower; **C**, bract; **D**, parts of perianth; **E**, carpel with stamens; **F**, L. S. of ovary; **G**, floral diagram

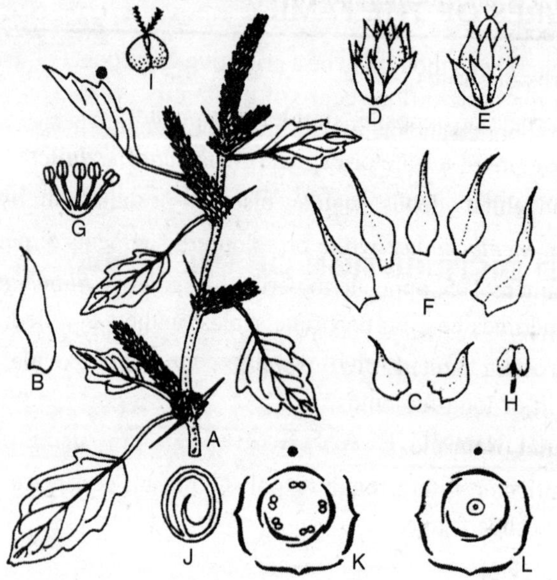

Fig. 16 AMARANTHACEAE. *Amaranthus spinosus* L. **A**, portion of shoot; **B**, bract; **C**, bracteoles; **D**, male flower; **E**, female flower; **F**, parts of perianth; **G**, stamens in male flower; **H**, single stamen; **I**, carpel in female flower; **J**, L. S. of ovary; **K**, diagram of male flower; **L**, diagram of female flower

FLORAL RANGE IN THE FAMILY

The flowers are unisexual in *Amaranthus,* but polygamo-dioecious in *Deeringia amaranthoides.* The perianth segments may become long and hairy, as in *Ptilotus.* In *Alternanthera,* the anther has a single lobe. They may be several ovules in each loculus, *e.g., Celosia.*

DIVISIONS IN THE FAMILY

The Amaranthaceae is divided into three tribes:

Tribe (*i*) *Amarantheae.* Anthers 2-celled. Ovary chamber with 1 ovule. Examples: *Achyranthes, Amaranthus,* etc.,

Tribe (*ii*) *Celosieae.* Anthers 2-celled. Ovary chamber with 2 to many ovules. Example: *Celosia.*

Tribe (*iii*) *Gomphreneae.* Anthers 1-celled. Ovary chamber with 1 ovule. Examples: *Alternanthera, Gomphrena,* etc.

POSITION AND AFFINITY OF THE FAMILY

The Amaranthaceae was included in the series Curvembryeae of the subclass Monochlamydeae by Bentham-Hooker. On the other hand, Engler placed the family in the suborder Chenopodineae of the order Centrospermae within the subclass Archichlamydeae. In Hutchinson's arrangement, the family appeared under the Chenopodiales.

The Amaranthaceae is related to the Chenopodiaceae by the possession of monochlamydeous flowers, uniseriate stamens and a single basal ovule in unilocular ovary. However, the congested inflorescence, scarious bracts and membranous perianth are characters which sharply delimit the amaranths from the chenopods.

The Amaranthaceae has been thought to be a primitive component of the Centrospermales. Recent studies have indicated that the basic inflorescence of the family is a dichasium of three flowers, out of which two generally become aborted and merely the bracteoles remain. In accordance with this suggestion, each bracteated flower is supposed to represent a relic of an ancestral dichasium. Many botanists have felt that the family is not primitive, being derived from caryophyllaceous ancestors.

NUMBER IN AND DISTRIBUTION OF THE FAMILY

The Amaranthaceae contains 65 genera and 900 species. Though the members of the family are inhabitants of the tropical zones, they attain their zenith of development in Africa and America.

COMMONLY OCCURRING PLANTS OF THE FAMILY

Achyranthes aspera L., *Aerva lanata* (L.) Juss., *A. sanguinolenta* (L.) Bl., *Alternathera sessilis* (L.) DC., *Amaranthus spinosus* L., *Cyathula prostrata* (L.) Bl., *Digera muricata* (L.) Mart., *Psilotrichum ferrugineum* (Roxb.) Moq. and *Pupalia lappaceum* (L.) Juss. are weeds.

Allmania nodiflora (L.) R. Br. ex Hook. f. is a dichotomously branched aquatic herb.

Celosia argentea L., a plant growing in barren soils, bears a globose head of pale pink flowers which turn white at maturity.

Cock's Comb (*Celosia cristata* L.) and Globe Amaranth (*Gomph rena globosa* L.) are garden plants.

Deeringia amaranthoides (Lamk.) Merr. is a climbing shrub with purple berries.

ECONOMIC ASPECTS OF THE FAMILY

The Amaranthaceae is of little economic interest. A few plants are of medicinal value, *e.g.*, *Achyranthes aspera*. Some are cultivated as ornamentals, such as some species of *Alternanthera, Celosia, Gomphrena, Iresine* and *Ptilotus. Amaranthus caudatus* (Andes) yields edible grains, while *A. cruentus* and *A. frumentaeeus* are raised as cereals by primitive tribes in tropical Asia. *A. gangeticus* and *A. gracilis* are used as vegetables.

Nyctaginaceae
(Four-O'clock Family)

GENERAL FEATURES OF THE FAMILY

Habit	—	Herbs, shrubs or trees.
Leaves	—	Opposite, simple, entire, exstipulate, *often in unequal pairs*.
Inflorescence	—	Cymose.
Flowers	—	Monochlamydeous, regular, hypogynous, *involucrate*.
Perianth	—	*Petaloid,* united, inflated at base, segments 5, plicate.
Androecium	—	Stamens 8–30, exserted, unequal.
Gynoecium	—	Carpel 1; *ovary superior, unilocular*; ovule 1, basal, campylotropous with an inferior micropyle; stigma peltate.
Fruit	—	*Achene, enclosed within perianth base.*
Seeds	—	Erect, albuminous; perisperm mealy: *embryo curved or folded*; cotyledons unequal.

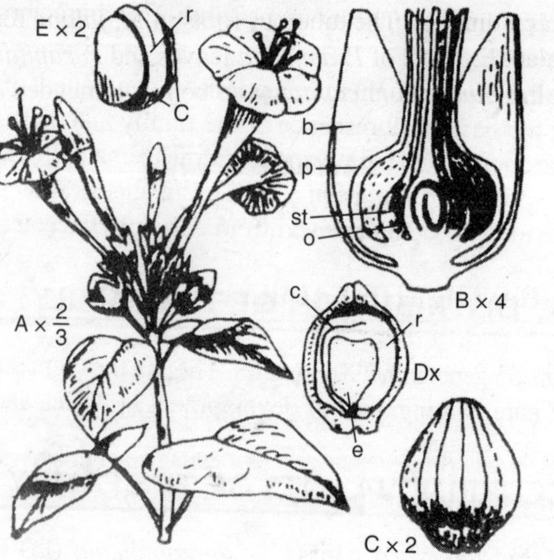

Fig. 17 NYCTAGINACEAE. *Mirabilis jalapa* L. **A**, flowering shoot; **B**, V. S. of lower portion of flower (*p*, perianth; *st*, stamen; *o*, ovary); **C**, anthocarp; **D**, V. S. of anthocarp (*c*, cotyledons; *r*, radicle; *e*, perisperm); **E**, embryo (*c*, outer cotyledons; *r*, radicle). (**A–E**, after Rendle)

FLORAL RANGE IN THE FAMILY

Although *Mirabilis* has a biparous cyme, the inflorescence shows a tendency to become uniparous at the upper part on account of one sided suppression of axes. Each bract is leaf-like and coloured, subtending a flower, *e.g.*, *Bougainvillea*. The bracts may be sepal-like and subtend more than a single flower, as in *Oxybaphus*. The flowers of *Pisonia* are unisexual.

NUMBER IN AND DISTRIBUTION OF THE FAMILY

The Nyctaginaceae embraces about 30 genera and 290 species. The members of this family abound in the warmer parts of the world, but are mainly found in tropical America.

COMMONLY OCCURRING PLANTS OF THE FAMILY

Boerhaavia diffusa L. is a weed.

Bougainvillea spectabilis Willd. is a garden climber with purple bracts.

Four-O'clock Plant or Marvel-of-Peru (*Mirabilis jalapa* L.) is an ornamental shrub.

Pisonia aculeata L., a large straggling shrub, is armed with recurved axillary spines.

ECONOMIC ASPECTS OF THE FAMILY

Economically, the Nyctaginaceae is of little value except as ornamentals. While *Abronia, Bougainvillea* and *Mirabilis* are cultivated for showy flowers, *Pisonia aculeata* is used for hedges. *Boerhavia diffusa* is of medicinal importance as a diuretic. The tuberous roots of *M. jalapa* form the source of a purgative utilised as a substitute for jalap. Extracts of *P. aculeata* leaves and *P. capitata* fruits' are tried on a range of complaints. The leaves of *P. alba* and *P. capitata* can be recommended as a vegetable.

Phytolaccaceae
(Pokeweed Family)

GENERAL FEATURES OF THE FAMILY

Habit	—	Herbs, shrubs or trees.
Leaves	—	Alternate, simple, entire, *exstipulate,* glabrous.
Inflorescence	—	Cymose or racemose.
Flowers	—	Small, bisexual, regular.
Perianth	—	Small, *non-petaloid,* free or united, *segments* 4–5, imbricate.
Androecium	—	Stamens 5–30, inserted on a fleshy disc; anthers 2-celled, dehiscence longitudinal.
Gynoecium	—	*Carpels 1–16*, free or united; *ovary superior,* unilocular; ovule basal, campylotropous.
Fruit	—	Achene, berry, drupe, schizocarp or utricle.
Seeds	—	Embryo large, annular; cotyledons half terete or flat and narrow; endosperm copious and mealy.

FLORAL RANGE IN THE FAMILY

The flowers are invariably hypogynous, but epigynous in *Agdestis.* They are usually actinomorphic, but zygomorphic in *Anisomeria.*

The stamens form a whorl, the members of which are opposite to (*Microtea*) or alternate with (*Rivina*) the perianth segments. There may be two alternating whorls of stamens, the number of which may increase in either of the whorls (*Phytolacca decandra, P. icosandra*). Sometimes the outer whorl of stamens is replaced by a whorl of 5 petals (*Stegnosperma*).

The ovary is superior, except in *Agdestis* where it is inferior.

DIVISIONS OF THE FAMILY

The Phytolaccaceae is divided into five tribes:

Tribe (*i*) *Agdestideae.* Ovary inferior. Example: *Agdestis.*

Tribe (*ii*) *Barbeuieae.* Ovary bicarpellate. Example: *Barbeuia.*

Tribe (*iii*) *Euphytolacceae.* Ovary multicarpellate. Example: *Phytolacca.*

Tribe (*iv*) *Rivineae.* Ovary unicarpellate. Example: *Rivina.*

Tribe (*v*) *Stegnospermaeae*. Staminodes petaloid. Example: *Stegnosperma.*

NUMBER IN AND DISTRIBUTION OF THE FAMILY

The Phytolaccaceae comprises 22 genera and 125 species, being widely distributed in tropical America and the West Indies.

COMMONLY OCCURRING PLANTS OF THE FAMILY

Petiveria alliacea L. is an erect herb with greenish flowers.

Rouge Plant or Bloodberry (*Rivina humilis* L.) is a semiclimbing undershrub, having showy scarlet berries.

ECONOMIC ASPECTS OF THE FAMILY

The Phytolaccaceae is of little economic importance. The species of *Agdestis, Ercilla, Petiveria, Phytolacca, Rivina* and *Trichostigma* are grown for ornamental purpose. *Petiveria alliacea* (South America) is used medicinally. The young shoots of *Phytolacca americana* form a source of edible greens and berries of this plant yield a deep red dye for colouring wines. A red dye is also extracted from the berries of *R. humilis.*

Aizoaceae
(Carpetweed Family)

GENERAL FEATURES OF THE FAMILY

Habit — Herbs or shrubs.

Leaves — Alternate, opposite or whorled, *fleshy* or *thread-like*.

Flowers — Bisexual, regular.

Perianth — Uniseriate, 5- to 8-lobed, free or united.

Androecium — Stamens 3–5 or many, *outermost forming petaloid structures*; anthers 2-celled; dehiscing longitudinally.

Gynoecium — Carpels 3–5; *ovary inferior or superior, 2- to many locular*; placentation apical, axile or basal.

Fruit — Capsule.

DIVISIONS OF THE FAMILY

The Aizoaceae is subdivided on the basis of the characters of the perianth and relative position of the ovary:

Subfamily I. *Ficoideae.* Perianth segments united to form a shorter or longer tube.

Tribe (*i*) *Aizoaeae.* Staminodes absent. Ovary inferior. Capsule loculicidal. Example: *Aizoon.*

Tribe (*ii*) *Mesembryanthemeae.* Staminodes numerous or absent. Examples: *Mesembryanthemum* and *Tetragona.*

Tribe (*iii*) *Sesuvieae.* Staminodes absent. Ovary superior. Capsule opening by a lid. Examples: *Sesuvium* and *Trianthema.*

Subfamily II. *Mollugînoideae.* Perianth segments almost free. Petaloid staminodes generally absent. Ovary superior. Example: *Glinus.*

NUMBER IN AND DISTRIBUTION OF THE FAMILY

The Aizoaceae includes 143 genera and about 2,300 species. The members of this family are highly developed in South Africa.

COMMONLY OCCURRING PLANTS OF THE FAMILY

Glinus lotoides (O. Kuntze) L. is a prostrate herb.

Trianthema portulacastrum L. is a creeping herb on the seashore.

ECONOMIC ASPECTS OF THE FAMILY

Economically, the Aizoaceae is of little significance. The species of *Mesembryanthemum* are suitable for cool green houses. The fruits of *M. acinaciforme* and *M. edule* (South Africa), called 'Hottentot figs', are edible. *Tetragonia tetragonioides* (New Zealand) is a source of edible greens.

Basellaceae
(Basella Family)

GENERAL FEATURES OF THE FAMILY

Habit	—	*Climbing herbs, often succulent.*
Leaves	—	Alternate, petiolate, simple, exstipulate.
Inflorescence	—	Panicles, racemes or spikes.
Flowers	—	Apetalous, bisexual, regular, *often brightly coloured.*
Perianth	—	Uniseriate, free or united into a 5-lobed structure persistent.
Androecium	—	Stamens 5; filaments short; anthers 2-celled, dehiscing lengthwise.
Gynoecium	—	Carpel 1; ovary superior, unilocular; ovule 1, campylotropous; placentation basal.
Fruit	—	*Drupe, enveloped by a persistent perianth.*
Seeds	—	Spherical, with a membranous testa; embryo semiannular or twisted; endosperm copious.

FLORAL RANGE IN THE FAMILY

The stamens are curved in bud (*Anredera*) or straight in bud (*Basella*).

POSITION AND AFFINITY OF THE FAMILY

While Rendle was silent about the Basellaceae, it was mentioned by Lawrence, Core and Hutchinson. It was united with the Chenopodiaceae by Bentham-Hooker. It is probably an advanced terminal offshoot of the Portulacaceae.

NUMBER IN AND DISTRIBUTION OF THE FAMILY

The Basellaceae is composed of four genera and 22 species, being found in tropical Africa, Asia and America.

COMMONLY OCCURRING PLANT OF THE FAMILY

Basella alba L. is a much branched fleshy twining herb.

ECONOMIC ASPECTS OF THE FAMILY

Basella alba and *B. rubra* produce edible shoots. *Boussingaultia gracilis* (tropical America) is cultivated as an ornamental for fragrant white flowers as well as for edible tubers. The tuberous starchy roots of *Ullucus tuberosus* (Andes) are used as a substitute for potatoes.

Portulacaceae
(Purslane Family)

GENERAL FEATURES OF THE FAMILY

Habit — Herbs or undershrubs.
Leaves — Alternate or opposite, simple, *succulent with scarious stipules.*
Inflorescence — Cymose.
Flowers — Bisexual, regular, *hypogynous without a disc.*
Calyx — Sepals 2, imbricate.
Corolla -- Petals 4 or 5, free.
Androecium — Stamens 4 or more, opposite to petals and united with them at base.
Gynoecium — Carpels 3, syncarpous; *ovary semi-inferior,* 1-*celled, containing several campylotropous ovules on basal central placenta.*
Fruit — Capsule with a circumscissile or loculicidal dehiscence.

POSITION AND AFFINITY OF THE FAMILY

Bentham-Hooker included the Portulacaceae in the Caryophyllales.

Rendle placed it under the Centrospermae and Lawrence did like wise.

Like the Caryophyllaceae, the Portulacaeae has come from the same source—the Phytolacceaeae. The floral type of *Portulaca* evolved by the disappearance of the inner whorl of carpels and by the juxtaposition of a pair of bracteoles to the flower to form the dimerous calyx.

The Portullacaceae shows close relationship with the Basellaceae, as the genera *Portulacaria* and *Phillippiamra* stand intermediate betplants belonging to this family are somewhat cosmopolitan. between the two families. It is also related to the Aizoaceae, but botanists do not agree with regard to the placement of some of the genera.

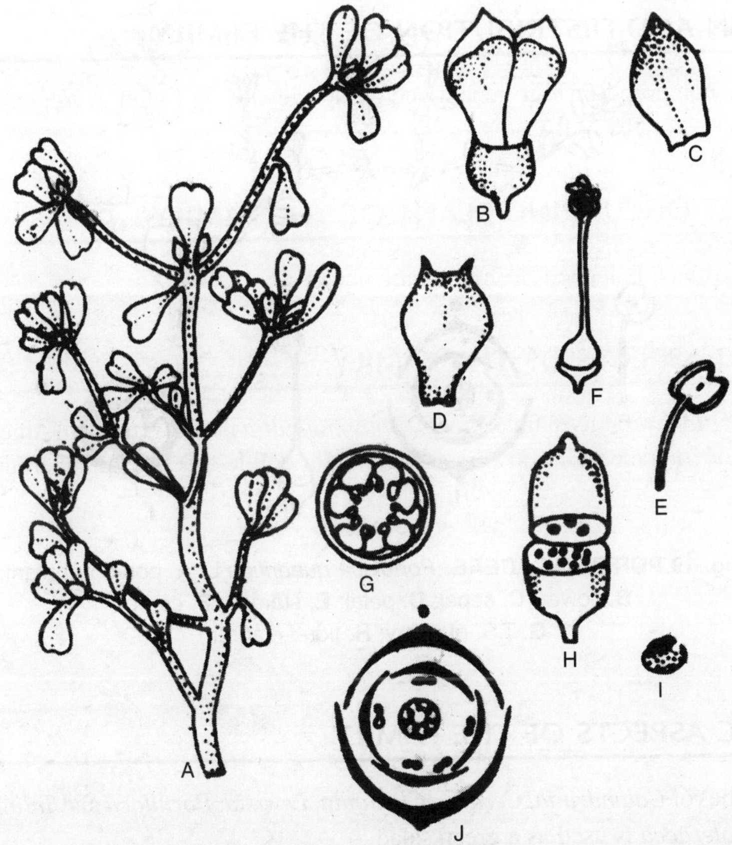

Fig. 18 PORTULACACEAE. *Portulaca oleraceae* L. **A**, portion of plant; **B**, flower; **C**, sepal; **D**, carpel; **E**, stamen; **F**, corpel; **G**, T.S. of ovary; **H**, dehiscence of fruit; **I**, Seed; **J**, floral diagram

NUMBER IN AND DISTRIBUTION OF THE FAMILY

There are 19 genera and about 500 species in the Portulacaceae. The plants belonging to this family are somewhat cosmopolitan, but there is one principal centre of distribution in South America and another in the western part of the United States.

COMMON OCCURRING PLANTS OF THE FAMILY

Rose Moss or Sun Plant (*Portulaca grandiflora* Hook.) is a garden annual with showy red coloured flowers and sensitive stamens.

Purslane (*Portulaca oleracea* L., *P. quadrifida* L., *P. tuberosa* Roxb.), having golden-yellow flowers, are weed in waste places.

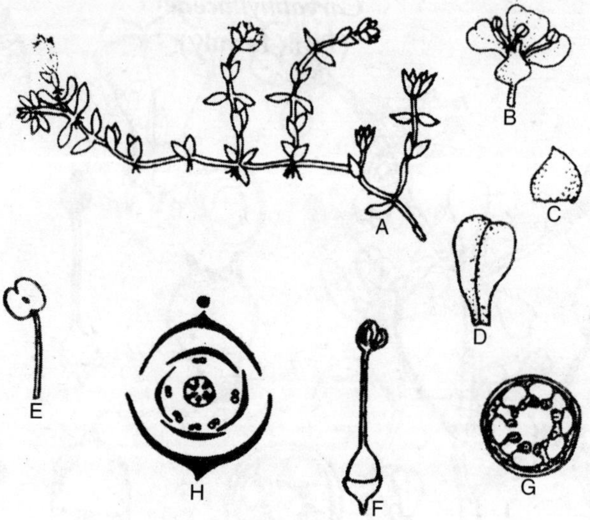

Fig. 19 PORTULACACEAE. *Portulaca quadrifida* L. **A**, portion of plant; **B**, flower; **C**, sepal; **D**, petal; **E**, stamen; **F**, carpel; **G**, T.S. of ovary; **H**, floral diagram

ECONOMIC ASPECTS OF THE FAMILY

Many garden varieties of *Calandrinia, Ceraria, Claytonia, Lewisia, Portulaca* and *Talinum* are cultivated. The herbage of *P. oleracea* is used as a green salad.

GENERAL FEATURES OF THE FAMILY

Habit — Herbs, *having stems with swollen nodes.*

Leaves — Opposite, simple, entire, exstipulate.

Inflorescence — *Dichotomous cyme.*

Flowers — Bisexual, regular, hypogynous or perigynous.

Calyx — Sepals 4 or 5, free, or united.

Corolla — Petals 4 or 5, free, clawed.

Androecium — *Stamens 8–10, in 2 whorls,* free.

Gynoecium — Carpels 2–5, syncarpous; ovary unilocular, ovules many, amphitropous or campylotropous; *placentation free central.*

Fruit — Capsule.

Seeds — Somewhat flattened, albuminous; embryo curved.

Caryophyllaceae (Pink family)

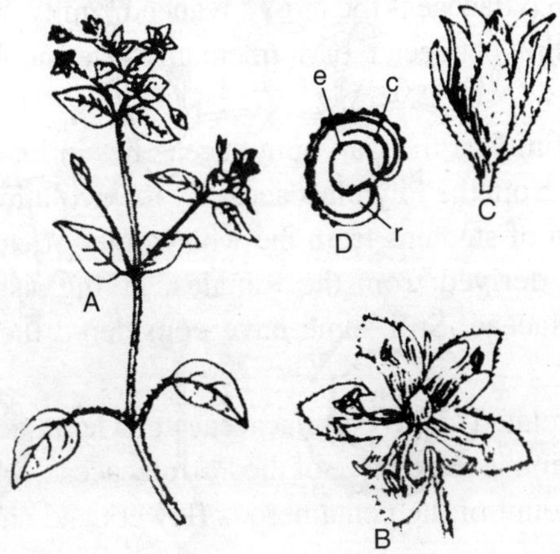

Fig. 20 CARYOPHYLLACEAE. *Stellaria media* (L.) Vill. **A**, flowering shoot; **B**, flower; **C**, capsule open; **D**, L.S. of seed (*c*, cotyledons; *e*, perisperm; *r*, radicle). (**A–D**, after Rendle)

FLORAL RANGE IN THE FAMILY

The flowers vary widely. A typical flower is pentamerous, as *Spergula.* Sometimes the flowers are tetramerous (*Sagina*). In *Stellaria,* the flower is pentamerous with the carpels reduced to three. In one and the same species, we may find considerable variation; a case in point is *Stellaria media.* Here the flower may have the formula $S_5\ P_{5+5}\ A_5\ G_{(3)}$ or $S_5\ P_5\ A_5\ G_{(3)}$.

The sepals are free or may be united into a tube. The petals, may be shorter than the sepals and sometimes totally wanting. They are often red in colour and differentiated into a limb and a claw. At the junction of the limb and claw or the corolla throat, there is an outgrowth or a ligule which constitutes the so called corona. The corona is absent in *Githago* and *Gypsophila,* but present in *Dianthus*, *Lychnis* and *Silene.* An internode is also formed above the calyx on which petals, stamens and carpels are borne. In *Alsine, Arenaria, Holosteum* and others, the petals are more or less notched.

The stamens and carpels differ with regard to their number and arrangement. The inner whorl of stamens is generally absent in *Cerastium.* In some genera, the stamens are slightly perigynous. The styles are united below in *Polycarpon* or are free in *Dianthus.* The carpels are 2 in *Saponaria*, 3 in *Silene* and 5 in *Lychnis.*

DIVISIONS OF THE FAMILY

The Caryophyllaceae is divided into three subfamilies:

Subfamily I. *Alsinoideae.* Sepals distinct. Petals present. Fruit a capsule. Examples: *Arenaria, Cerastium,* etc.

Subfamily II. *Paronychioideae.* Sepals distinct or united. Petals absent. Fruit a utricle. Examples: *Paronychia, Scleranthus,* etc.

Subfamily III. *Silenoideae.* Sepals united. Petals present. Fruit a capsule. Examples: *Dianthus, Lychnis,* etc.

POSITION AND AFFINITY OF THE FAMILY

According to Engler, the Caryophyllaceae is the most advanced family under the Centrospermae. Rendle suggested it to be a connecting link between the Monochlamydeae and Dialypetalae. Hutchinson placed the family in the Caryophyllales.

The ultimate basis of affinity of the Caryophyllaceae lies in its origin. According to one school of thought, the family has come from the Phytolaccaceae by the evolution of petals from the outer whorl of stamens and by the evolution of stamens from the whorl outer of carpels. Others have claimed that the Caryophyllaceae has been derived from the Ranales, giving rise to reduced taxa, such as the Amaranthaceae and Chenopodiaceae. Still some have considered the family as the ancestor of the Primulaceae.

The Caryophyllaceae is related to the Portulacaceae and Geraniaceae. The presence of unilocular ovary with free central placentation is suggestive of the Portulacaceae, while the Caryophyllaceae stands close to the Geraniaceae on account of the pentamerous flowers and nature of stamens.

NUMBER IN AND DISTRIBUTION OF THE FAMILY

The Caryophyllaceae represents about 80 genera and about 2,000 species. The family is primarily available in the north temperate regions, but also distributed in the south temperate lands and tropical mountains; the Mediterranean area is the main centre of distribution.

COMMONLY OCCURRING PLANTS OF THE FAMILY

Sweet William (*Dianthus barbatus* L.), Carnation (*D. chinensis* L.), Baby's Breath (*Gypsophila paniculata* L.) and *G. elegans* Bieb. are season flowers.

Drymaria cordata Willd. and *Spergula arvensis* L. are weeds in cool climates like those of Darjeeling and Shillong.

Polycarpaea corymbosa (L.) Lamk., Chickweed [*Stellaria media* (L.) Vill.] and Cowcockle [*Vaccaria pyramidata* (L.) Medik] are herbaceous plants, generally considered as weeds.

ECONOMIC ASPECTS OF THE FAMILY

Economically, the plants of the Caryophyllaceae are of little or no importance. They are generally unpalatable to livestock. *Spergula arvensis* var. *sativa* is occasionally used as a fodder plant in dry sandy areas. Some are used as vegetables, *e.g., Polycarpon prostratum.* Others are chiefly valuable for their flowers, such as the species of *Agrostemma, Cerastium, Dianthus, Gypsophila, Lychnis, Saponaria* and *Silene.* The roots of *Vaccaria pyramidata* yield 'saponin' which forms lather with water.

POLYPETALAE

The Polypetalae may be regarded as an assortment of orders and families numbering 12 and 82 respectively. Here the petals are free in the flower.

The chief diagnostic characters of the various orders under the Polypetalae are given in the key below (Lawrence, 1951; Core, 1955):

(1) Ovary generally inferior	(2)
(1) Ovary generally superior	(4)
(2) Plants spiny, usually succulent	*Cactales*
(2) Plants not spiny or succulent	(3)
(3) Flowers in umbels; ovules 1–2 in each loculus	*Umbellales*
(3) Flowers not in umbels; ovules several to many in each loculus	*Myrtales*
(4) Floral parts spiral	*Ranales*
(4) Floral parts cyclic	(5)
(5) Plants insectivorous	*Sarraceniales*
(5) Plants not insectivorous	(6)
(6) Placentation parietal; hypanthium absent	*Rhoeadales*
(6) Placentation axile; hypanthium present	(7)
(7) Carpels 1 to many, free or united; ovary surrounded by an androperianth tube	*Rosales*
(7) Carpels united; androperianth tube absent	(8)
(8) Stamens numerous	(9)
(8) Stamens not numerous	(10)
(9) Sepals valvate; placentation axile	*Malvales*
(9) Sepals imbricate; placentation parietal	*Parietales*
(10) Stamens as many as petals and opposite to them	*Rhamnales*
(10) Stamens rarely more than twice as many as petals and alternate with them	(11)
(11) Ovules pendulous with a ventral raphe and micropyle downwards or erect with a ventral raphe and micropyle upwards	*Sapindales*
(11) Ovules pendulous with a ventral raphe and micropyle upwards or erect with a dorsal raphe and micropyle downwards	*Geraniales.*

Order 14. *Cactales*

This is a monotypic order with the characters of its single family, the Cactaceae.

GENERAL FEATURES OF THE FAMILY

Habit	—	*Succulent plants,* with fleshy, globular or cylindrical, angular or ribbed and jointed stems.
Leaves	—	*Reduced to tufts of bristles or spines or glochids arranged in areoles.*
Flowers	—	*Solitary,* bisexual, regular, epigynous.
Perianth	—	*Gradual transition from sepals to petals* and often merging into bracts, many, free or united to form a tube.
Androecium	—	Stamens many, arranged spirally or in clusters.
Gynoecium	—	Carpels many; *ovary inferior, unilocular with many ovules on parietal placentae.*
Fruit	—	Berry.
Seeds	—	Many, exalbuminous; embryo straight or curved.

*Cactaceae**
(Cactus family)

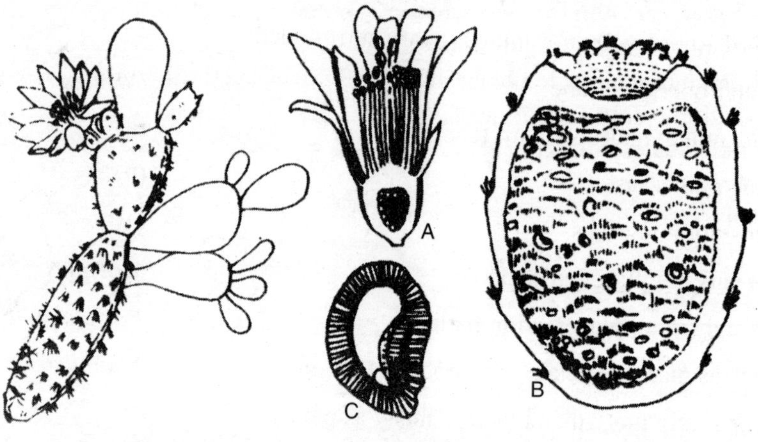

Fig. 21 CACTACEAE. *Opuntia vulgaris* L. **A**, L.S. of flower; **B**, fruit; **C**, L.S. of seed. (**A–C**, after Hutchinson)

DIVISIONS OF THE FAMILY

The Cactaceae is divided into three subfamilies:

Subfamily I. *Cereoideae.* Leaves reduced to minute scales and areoles without glochids. Flowers sessile, usually funnelform or salverform. Examples: *Cereus, Echinocactus, Mammillaria,* etc.

Subfamily II. *Opuntioideae.* Leaves cylindrical to subulate, caducous and areoles with glochids. Flowers sessile, usually rotate. Examples: *Nopalea, Opuntia,* etc.

* For detailed treatment, consult Benson (1957) and Willis (1966).

Subfamily III. *Pereskibideae.* Leaves broad, flat and areoles without glochids. Flowers usually stalked, often in panicles. Examples: *Pereskia.*

POSITION AND AFFINITY OF THE FAMILY

"The Cactaceae is of special interest to botanists for its combination of a primitive, unspecialized flower with highly advanced vegetative organs; to the ecologist for its survival under adverse conditions and drought; and to the evolutionist for its parallel life forms to other unrelated xerophytes, *e.g.*, *Stapelia, Euphorbia* and *Pachypodium.* To the taxonomist it presents great problems, being apparently still in a state of active evolution, and resisting the standard herbarium procedure based on dried specimens" (Rowley, 1978).

Many views have been put forward about the position and relationship of the Cactaceae. Engler retained it in the Opuntiales near the Myrtiflorae and admitted an alliance with the Aizoaceae. Bessey regarded the Cactales to be related to the Cucurbitales. Wettstein included the Cactaceae in the Centrospermae. Bentham-Hooker placed the family in the Ficoidales before the Umbellales and after the Passiflorales. Rendle opined that "the affinity of the family is very doubtful"; he set it up within the Opuntiales, lying between the Myrtiflorae and Umbelliflorae. In Hutchinson's arrangement, the Cactaceae appeared under the Cactales which was again put in between the Cucurbitales and Theales. On the basis of epidermal structure, floral plan and serological data, Mez considered the family as linked to the Loasaceae and the Parietales complex. Buxbaum (1944) derived the Cactaceae from the Phytolaccaceae and took it to exemplify parallel evolution with the Aizoaceae. Evidence pooled from anatomy, embryology and floral morphology led Maheshwari (1945) and others to suggest that this family should be assigned to a position within or near the Centrospermae. Gundersen (1950) treated the family in the Cactales and near the Ranales. Mitra (1974) supported Gundersen's contention of a ranalian relationship as expressed by *Anemarrhena* type of seedling, spirocyclic flowers (simulating *Euryale*), transition from sepals to petals, fasciculate androecium and large persistent antipodals.

NUMBER IN AND DISTRIBUTION OF THE FAMILY

The Cactaceae embraces 87 genera and over 2,000 species. The main centre of distribution of this family "is in the dry districts of the warmer parts of America, especially of Mexico, including the temperate and colder part, as well as the neighbouring portion of the United States".

COMMONLY OCCURRING PLANT OF THE FAMILY

Prickly Pear (*Opuntia elatior* Mill.) is an instance of a cladode.

ECONOMIC ASPECTS OF THE FAMILY

Economically, the members of the Cactaceae are important as ornamentals, *e.g.,* the species of *Cereus, Echinocactus, Epiphyllum, Mammillaria, Schlumbergera,* etc., A few yield edible fruits, such as *C. variabilis, Opuntia elatior, O. ficus-indica and O. megacantha.* Cochineal dye is extracted from cochineal insects which live upon *Nopalea cochenillifera. Ariocarpus, Astrophytum, Aztekium, Dolichothele, Lophophora, Obregonia, Pelecyphora* and *Solisia*, popularly classed as 'peyote' and containing a variety of alkaloids, are used by the American and Mexican Indians for medicinal and hallucinogenic purposes. Some are used as emergency fodder and spines of others are utilised as gramophone needles. "Some are used for hedging, while those with woody skeletons are used for rustic furniture and trinkets."

Order 15. *Ranales*

The Ranales is a natural taxon. Here the floral parts are many and distinct, being spirally arranged. In general, the perianth is not differentiated into calyx and corolla. Both the stamens and carpels are free and numerous.

The Ranales, as suggested by many workers, represents the most primitive dicotyledon. From such a stock, angiospermic orders arose in various directions. The primitiveness of the order is ascribed to the presence of simple and pinnately veined leaves, unilacunar node, vessel-less wood* and spiral to spirocyclic arrangement of floral parts.

That the Ranales is a heterogenous group has also been recognised. The structure of the seed coat is so different in the Annonaceae, Berberidaceae, Magnoliaceae, Myristicaceae, Nymphaeaceae, Ranunculaceae and Lauraceae that the Ranales is considered by Corner (1953) as a grade of families at the level of apocarpy and polypetaly and not a phyletic series. According to Benzing (1967a), the diversity in protoxylem patterns in woody ranalian families is partly a reflection of the extreme evolutionary divergence of the families. This recognition has led to the fragmentation of the group into smaller orders by Hutchinson (1959) and Takhtajan (1969). Some of the newer families arising from this break up have been assigned to orders somewhat different from those of the parent families, as in the Berberidaceae and Nymphaeaceae.

In connection with heterogeneity, another problem to be sorted out is the question of interpretation of the phylogeny of the respective families. Bailey (1954) suggested that the inter-relationships of the families are of a reticulate pattern. Sastry (1969) opined that it would be impossible to arrange them in a linear sequence. Two major factors can possibly be held responsible for the absence of linear relationships; (*a*) most of the connecting links are missing in the existing flora and fossil studies have been of little help and (*b*) the families are polyphyletic. Although the concept of the origin of flowering plants has been gaining momentum, any attempt to define the number and probable course of phyletic lines involved is bound to be speculative at the present state of our knowledge. Nevertheless, it would be worthwhile to examine some efforts in this direction. Modern palynological evidence seems to corroborate Hutchinson's division into woody and herbaceous series, characterised by monocolpate, derived dicolpate or acolpate pollen grains and tricolpate or modified forms respectively. On the basis of nodal anatomy, Money *et al.* (1950), classified the woody families with ethreal oil cells and monocolpate or derived dicolpate, polyporatc or acolpate pollen into two groups (Table 6). A comparison of the floral morphological, embryological and seed characters of the two groups is set in Table 7.

TABLE 6. Families with uni-, tri- or multilacunar nodes

Nodes unilacunar	*Nodes tri- or multilacunar*
Austrobaileyaceae, Trimeniaceae	Winteraceae, Degeneriaceae
Amborellaeeae, Monimiaceae	Himantandraceae, Magnoliaceae
Gomortegaceae, Lauraceae	Annonaceae, Myristicaceae
Hernandiaceae, Chloranthaceae	Eupomatiaceae, Canellaceae
Calycanthaceae, Lactoridaceae	Piperaceae, Sauruaceae

* Found in four families: Amborellaceae, Tetracentraceae, Trochodendraceae and Winteraceae.

TABLE 7. Unilacunar versus tri- or multiacunar families

	Unilacunar families	*Tri- or multilacunar families*
Carpel	Relatively Speicalised	Primitively conduplicate
Stamens	Specialised, usually fasciculate with glandular appendages	Primitive sporophylls, free, glandular appendages absent
Division of PMC	By cell-plate formation	By furrowing
Another tapetum	Plasmodial	Secretory
Archesporium in ovule	Multicellular	Unicellular
Endosperm	Nuclear	Cellular
Seed	Exalbuminous with large embryo	Endospermous with small embryo

Contrary to the view of Bailey (1956), the unilacunar node in more primitive than the tri- and multilacunar and the families of the second group are usually more primitive. Interpretation of palaeobotanical (Pant and Mehra, 1964) and anatomical (Benzing, 1967b) evidence indicate that the older argument of the primitive nature of the trilacunar condition may still hold water. The first seven families belonging to the unilacunar group roughly correspond to the Laurales of Hutchinson (1959). Within this homogenous group, more than one line of evolution is apparent (Shutts, 1960). Though the Calycanthaceae and Chloranthaceae show some resemblances to other members of the group, they differ in several points and may represent independent phyletic lines. Among the ranalian families of the trilacunar group, the Myristicaceae stands out prominently by the possession of unusual floral character, successive division of pollen mother cells and nuclear endosperm. Likewise, the Eupomatiaceae is quite distinct from the others. When the totality of evidence is superimposed on the classificatory scheme based on wood anatomy, some taxa will obviously be left out. This led Sastry (1969) to suggest that neither of the unilacunar and trilacunar groups can be regarded a single phyletic series.

Eames (1961) envisaged five main lines in the group: cabombaceous, magnoliaceous, monimiaceous, ranunculaceous and winteraceous. He also believed that there might be several others represented by single isolated families. In view of the above findings, Sastry (1969) suggested two major lines in each of the woody and herbaceous series (Table 8):

TABLE 8. Concept of woody and herbaceous series

Woody series	*Herbaceous series*
Magnolilian line: Winteraceae, Degeneriaceae, Magnoliaceae. Schisandraceae, Cercidiphyllaceae. Annonaceae, etc.	Nyphaealian line: Nymphaeaceae (wider sense), Ceratophyllaceae.
Lauralian line: Monimiaceae, Gomortegaceae. Hernandiaceae, Gyrocarpaceae, Lauraceae, etc.	Berberidalian line: Lardizabalaceae, Berberidaceae, Menispermaceae, Ranunculaceae, Podophyllaceac, etc.

However, none of the above lines necessarily forms a homogenous group. The question of further splitting them can be decided only when more complete data are forthcoming. At the present moment, the phylogeny of the other families is more hazardous to guess. Whereas some of them might fit well into one of the four major lines, others probably constitute independent lines. Though phylogenetic implications are obvious for a break up into smaller orders, the order Ranales (*sensu lato*) might be retained on grounds of practical expediency.

The present author, following largely the idea of Core (1955), takes into account 11 families in the Ranales with the characters given below:

(1) Aquatic plants	(2)
(1) Terrestrial plants	(3)
(2) Leaves small, finely dissected	*Ceratophyllaceae*
(2) Leaves large, not dissected	*Nymphaeaceae*
(3) Oil cells present	(4)
(3) Oil cells absent	(8)
(4) Leaves opposite	*Monimiaceae*
(4) Leaves alternate	(5)
(5) Flowers trimerous	*Annonaceae*
(5) Flowers not trimerous	(6)
(6) Seeds arillate	*Myristicaceae*
(6) Seeds not arillate	(7)
(7) Fruits partly enclosed in a hollowed axis	*Lauraceae*
(7) Fruits not enclosed in an axis	*Magnoliaceae*
(8) Erect herbs or shrubs	(9)
(8) Climbing shrubs	(10)
(9) Fruit an achene or a follicle	*Ranunculaceae*
(9) Fruit a berry or capsule	*Berberidaceae*
(10) Fruit a drupe	*Menispermaceae*
(10) Fruit a berry	*Lardizabalaceae*

Nymphaeaceae
(Water-lily Family)

GENERAL FEATURES OF THE FAMILY

Habit	— *Aquatic herbs,* with a perennial rhizome.
Leaves	— Floating, *peltate or* falsely so, *long petioled.*
Flowers	— Solitary showy, *long peduncled*, bisexual, regular, hypogynous to perigynous, thalamus cup shaped and fleshy.
Perianth	— Segments many, free spiral, gradually passing from sepals to petals and petals to stamens.
Androecium	— Stamens many, free, adnate to thalamus; *filaments foliaceous.*
Gynoecium	— Carpels many, free enveloped by thalamus; ovary 1- or more-locular with *ovules disposed on lamellate or parietel placentae; stigma sessile*, radiating, often with appendages.
Fruit	— Berry-like fleshy mass.
Seeds	— *Arillate* with both perisperm and endosperm; embryo straight.

FLORAL RANGE IN THE FAMILY

No other family has such variations in the structure of the flower as the Nymphaeaceae. Since this is a distinct morphological feature among certain genera, this has been the basis for the divisions of the family.

DIVISIONS OF THE FAMILY

On the basis of the construction of floral parts, the Nymphaeaceae is divided into three subfamilies:

Subfamily I. *Cabomboideae.* Flowers cyclic, trimerous and hypogynous. Perianth segments in two whorls of 3 each, outer forming sepals and inner petals. Stamens 6. Carpels 3, free. Examples: *Brasenia* and *Cabomba.*

Subfamily II. *Nelumbonoideae.* Flowers acyclic and hypogynous. Perianth segments indefinite, outer whorl of 5 sepals. Stamens indefinite. Carpels many, free. Example: *Nelumbo.*

Subfamily III. *Nymphaeoideae.* Flowers hypogynous, perigynous or epigynous. Perianth segments indefinite whorls of 4–5 each. Stamens indefinite. Carpels many, united. Examples: *Euryale, Nuphar, Nymphaea,* etc.

POSITION AND AFFINITY OF THE FAMILY

The Nymphaeaceae is a primitive family and comes under the herbaceous Ranales. Most botanists have accepted the family as composed of three distinct subfamilies. But Bessey treated each subfamily as a well-defined family, included the Cabombaceae and Nelumbaceae in the Ranales and removed the Nymphaeaceae to the Rhoeadales on the criterion of syncarpous pistil. Hutchinson, while accepting the Cambombaceae, retained the other two subfamilies within the scope of the Nymphaeaceae.

An affinity of the Nymphaeaceae can be traced to the Ranunculaceae through the Cabombaceae. The scattered vascular bundles and arillate seeds of the Nymphaeaceae recall those of the Berberidaceae. Moreover, points favouring the alliance of the Nymphaeaceae with the Papaveraceae include the presence of latex in tissues, large peltate-rayed stigma and laminal placentation.

Some investigators have drawn our attention to a possible relationship between the Nymphaeaceae and Alismataceae. The aquatic habit of the plants and structure of the ovary are essentially similar in the two families, but further connections are difficult to pursue.

NUMBER IN AND DISTRIBUTION OF THE FAMILY

The Nymphaeaceae consists of nine genera and over 90 species. In general, plants of this family are noticed in fresh waters of the world.

COMMONLY OCCURRING PLANTS OF THE FAMILY

Euryale ferox Salisb. is known for the abundant development of prickles on the surface of leaves and fruits.

Lotus (*Nelumbo nucifera* Gaertn.), Yellow Waterlily (*Nuphar luteum* Sibth. & Sm.) and Waterlily (*Nymphaea nouchali* Burm. f.) are found in tanks.

ECONOMIC ASPECTS OF THE FAMILY

The plants of the Nymphaeaceae are mainly important for the beauty of their flowers which exhibit great variety and brilliancy of colours in the tropics. *Cabomba* is a popular oxygenator for aquaria. The fruits and rhizomatous stem of most of the waterlilies are edible. The seeds of *Euryale* and *Nymphaea* are also consumed.

The family is characterised by the single lateral branch produced at a node, dichotomously dissected serrulate leaves, solitary unisexual perianthless flowers and unicarpellate pistil.

Ceratophyllaceae
(Hornwort family)

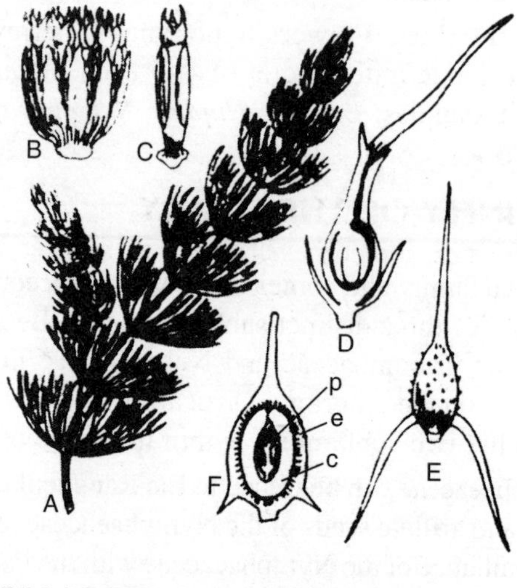

Fig. 22 CERATOPHYLLACEAE. *Ceratophyllum demersum* L. **A**, portion of plant; **B**, male flower; **C**, stamen; **D**, V. S. of female flower; **E**, fruit; **F**, L. S. of fruit (*c*, cotyledons; *e*, endosperm; *p*, plumule). (**A**, after Rendle; **B–E**, after Schumann; **F**, after Baillon)

The family comprises one genus (*Ceratophyllum*) and two or about 30 species of cosmopolitan distribution.

Ceratophyllum demersum L., a fragile herb, is seen in slow moving or shallow stagnant water. "*Ceratophyllum* usually floats in mats just below the surface of water. These mats provide a valuable protection for fish fry but also provide protection for bilharzia carrying snails and malaria or filaria carrying mosquito larvae" (Cook, 1978).

Monimiaceae
(Monimia Family)

The family is distinguished by the monoecious or dioecious plants, exstipulate leaves, valvular dehiscent anthers, achene fruit and copious endosperm.

The Monimiaceae is of great phyletic significance, because it is a combination of magnoliaceous and lauraceous characters. The floral plan is suggestive of the Magnoliaceae with a concave axis in which many stamens and free carpels are depressed. The small embryo embedded in a copious endosperm also speaks for an alliance with the Magnoliaceae. A relationship is indicated with the Lauraceae in terms of perigynous flowers as well as valvular dehiscent anthers and appendaged filaments. The Monimiaceae approaches the Calycanthaceae with regard to the hollowed out receptacle. Rosalian affinities are expressed by apetaly and somewhat urceolate receptacle with valvate stamens. But the biological feature of the flowers, namely the inconspicuous and generally unisexual flowers, is very peculiar to the Monimiaceae.

In the Monimiaceae, there are 30 genera and about 450 species. The members of this family abound in southern hemisphere, especially in Malagasy, Polynesia, Oceania and Australia.

The wood of *Atherosperma moschatum* and *Doryphora sassafras*, both of Australia, is used for making furniture. *Laurelia aromatica* (Peru) yields a type of spice called 'Peruvian nutmeg'. *novae-zelandiae* (New Zealand) yields valuable timber. *Peumus boldus* (Chile) provides edible fruit as well as wood for cabinet work.

Annonaceae (Custard-apple Family)

GENERAL FEATURES OF THE FAMILY

Habit	—	Shrubs or trees, sometimes climbers with naked buds.
Leaves	—	Alternate, *entire, exstipulate, distichous,* gland dotted.
Flowers	—	Bisexual, regular, *hypogynous, spirocyclic, trimerous,* sometimes aromatic.
Calyx	—	Sepals 3, free, persistent, *valvate.*
Corolla	—	Petals 6, in 2 whorls, *valvate.*
Androecium	—	*Stamens many, free, spirally arranged;* filaments short; *anther lobes linear with a prolonged connective.*
Gynoecium	—	Carpels many, free or united; ovary superior; ovules 1 or more, anatropous.
Fruit	—	Aggregate of berries.
Seeds	—	Large, often arillate; embryo small; *endosperm ruminate or marbled.*

FLORAL RANGE IN THE FAMILY

While *Artabotrys* bears recurved hook-like inflorescence, the flowers of *Geanthemum* are borne on subterranean shoots. The flowers are monoecious in *Raimondia* and *Uvariopsis,* but dioecious in *Ephedranthus, Tetrastemma* and *Thonnera.* The flowers of *Sapranthus* are unusually large.

The calyx enlarges, becomes membranous and encloses the fruit in *Sphaerothalamus.*

The corolla is sympetalous in *Asteranthe, Disepalum, Enneastemon* and *Stormia.* The outer petals may be provided with long dorsal appendages, as in *Rollinia* and *Stenanthera.* There are 6 petals in *Haplostichanthus, Monanthotaxis* and *Monocyclanthus*, but 3 petals in *Xylopia.*

The outer stamens are subpetaloid and sterile in *Fusaea.* The anthers are transversely locellate in *Cardiopetalum, Hornschuchia* and *Porcelia.* The connective is not produced beyond the loculi in *Alphonsea.*

The carpel is solitary in *Kingstonia, Mezzetia, Monocarpia* and *Tridimeris.* It may be sunk in the torus, in *Pseudoannona,* but located on an elongated torus in *Mischogyne* and *Toussaintia.* The carpels may be united into a many locular mass, *e.g., Pachypodanthium* or into an unilocular ovary with parietal placentae, *e.g., Monodora.*

DIVISIONS OF THE FAMILY

The Annonaceae is divided into two groups, one with four tribes and the other with a single tribe.

Group I. Pistil apocarpous.

Tribe (*i*) *Hexalobeae.* Perianth members equal. Example: *Hexalobus.*

Tribe (*ii*) *Miliuseae.* Perianth members valvate, usually unequal. Example: *Miliusa.*

Tribe (*iii*) *Uvarieae.* Perianth unjointed or clawed, claws opposite to stamens. Examples: *Unona, Uvaria,* etc.

Tribe (*iv*) *Xylopeae.* Perianth members hollow at base. Examples: *Annona, Xylopia,* etc.

Group II. Pistil syncarpous.

Tribe (*i*) *Monodoreae.* Ovary unilocular with parietal placentae. Example: *Monodora.*

POSITIONS AND AFFINITY OF THE FAMILY

The Annonaceae is a primitive family and comes under the category of the arboreal Ranales. It is regarded as derived from the magnoliaceous stock. Hutchinson treated it, along with the Eupomatiaceae, as comprising the Annonales and as noting it to be related to but more advanced than the Magnoliaceae.

The Annonaceae is allied to the monocotyledons by virtue of its trimerous flowers and small embryo.

NUMBER IN AND DISTRIBUTION OF THE FAMILY

The Annonaceae contains 80 genera and 850 species. The plants of this family are confined to the tropics and are found abundantly in the rain forests of Brazil, West Africa, Sri Lanka, Malaysia and South Burma.

COMMONLY OCCURRING PLANTS OF THE FAMILY

Bastard apple or Bullock's Heart (*Annona reticulata* L.) and Custard apple (*Annona squamosa* L.) are borne on well known trees.

Artabotrys hexapetolus (L.f.) Bhandari is a large scandent shrub with hooked peduncles and glabrous leaves.

Cyathocalyx zeylanicus Champ, ex Hook. f. & Thoms, is a tall tree with deflexed or horizontal branches.

'Ylang-ylang (*Desmos chinensis* Lour.) is a small tree, mostly grown in gardens.

Polyalthia cerasoides (Roxb.) Bedd. is a small tree, often found in thickets.

Mast Tree [*Polyalthia longifolia* (Sonner.) Thw.] is an ornamental tree, having lanceolate leaves with repand margins.

Sageraea laurifolia (Grah.) Blatt. occurs in moist forests of Western Ghats.

Uvaria cordata (Dunal) Alston is a climber noted for its cordate leaves and red flowers.

ECONOMIC ASPECTS OF THE FAMILY

The ripe fruit of *Annona squamosa* is regarded as a dessert and a refreshing drink made out of the sweet pulp. Edible fruits are also obtained from the various species of *Annona* and *Asimina. Bocagea laurifolia, B. virgata, Cyathocalyx zeylanicus, Duguetia quitarensis, Eupomatia laurina* and *Oxandra lanceolata* yield useful timber. The flowers of *Desmos chinensis* furnish 'Macassar oil', a perfume. The perfume of *Mkilua fragrans* is used by Arab women. The bark of *Goniothalamus wightii* produces strong fibres. The seeds of *Monodora myristica* are sometimes utilised as nutmegs. The green foliage of *Polyalthia longifolia* are used for decorations on festivals. The fruits of *Xylopia aethiopica* (West Africa) and *X aromatica* (tropical America) are used as peppers.

Myristicaceae
(Nutmeg Family)

The family is featured by the pellucid dotted leaves, unisexual flowers, monadelphous stamens, arillate seeds and ruminate endosperm.

The Myristicaceae is a taxon of about 16 genera and 380 species, being distributed in the tropical regions of the world and inhabiting lowland rainforests.

Gymnacranthera farquhariana and *Virola surinamensis* possess waxy seeds, utilised as a source of butter for candle making and human consumption. *Myristica fragrans* is the source of 'mace' obtained from the dried aril and of 'nutmeg' from the seed. The pericarp of the fruit is also used to make a jelly preserve and inferior seeds are pressed to make nutmeg butter used in perfumery.

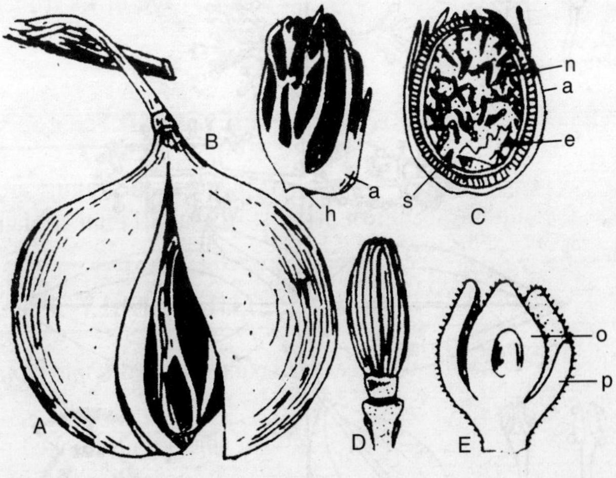

Fig. 23 MYRISTICACEAE. *Myristica fragrans* Houtt. **A**, fruit with pericarp split to expose seed; **B**, seed surrounded by aril *(a)* and showing hilum *(h)*; **C**, L. S. of seed (*a*, aril; *s*, seed coat; n, ruminate endosperm; *e*, embryo); **D**, androecium of male flower; **E**, L. S. of female flower (*p*, perianth; *o*, ovary). (**A**, after Le Maout & Decaisne; **B** & **C**, after Luerssen; **D** & **E**, after Baillon)

Lauraceae
(Laurel Family)

GENERAL FEATURES OF THE FAMILY

Habit	—	Shrubs or trees, with an aromatic bark.
Leaves	—	Alternate or opposite, simple, entire, *exstipulate.*
Inflorescence	—	Panicles, racemes, spikes or umbels.
Flowers	—	Bisexual or unisexual, regular.
Perianth	—	*Small, undifferentiated,* biseriate, of 6 basally connate segments.
Androecium	—	*Stamens in 4 whorls of 3 each,* innermost reduced to staminodes; *anthers* 2- or 4-celled, basifixed; *dehiscing by flap-like valves.*
Gynoecium	—	Carpel 1, ovary superior, unilocular; *ovule solitary pendulous,* anatropous.
Fruit	—	Berry or drupe, surrounded by persistent perianth base seated upon an enlarged receptacle.
Seeds	—	Embryo straight; *endosperm absent.*

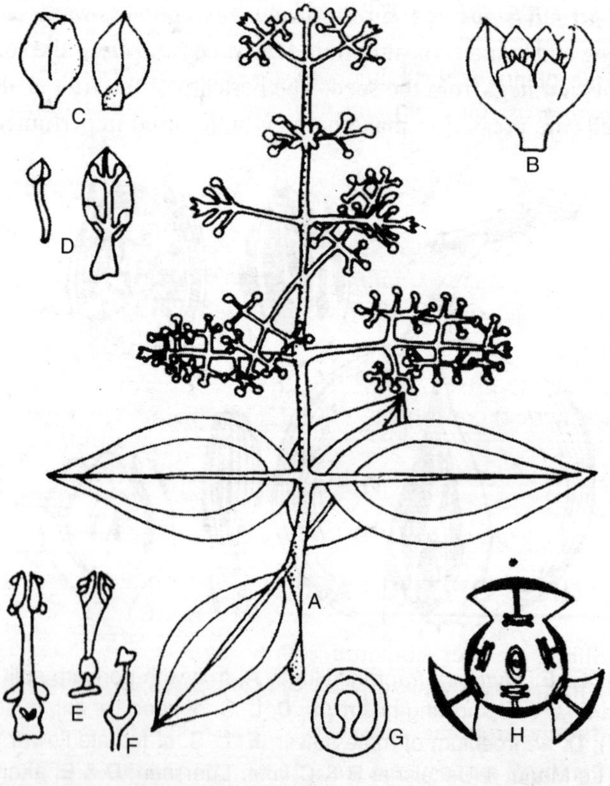

Fig. 24 LAURACEAE. *Cinnamomum zeylanicum* Bl. **A**, portion of shoot; **B**, flower; **C**, perianth; **D**, staminodes; **E**, stamens; **F**, gynoecium; **G**, L. S. of ovary; **H**, floral diagram

DIVISIONS OF THE FAMILY

The Lauraceae is divided into two subfamilies:

Subfamily I. *Lauroideae.* Anthers 4-celled. Examples: *Cinnamomum, Laurus,* etc.

Subfamily II. *Perseoideae.* Anthers 2-celled. Examples: *Cassytha, Persea,* etc.

NUMBER IN AND DISTRIBUTION OF THE FAMILY

The Lauraceae represents about 32 genera and 2,500 species, being prevalent in the tropical and subtropical regions of both hemispheres. The chief centres of distribution include Brazil and South-east Asia.

COMMONLY OCCURRING PLANTS OF THE FAMILY

Cassytha filiformis L. is a thread-like, leafless and greenish parasite.

Cryptocarya floribunda Nees, *Dehaasia kurzii* King, *Machilus villosa* Hook. f. and *Phoebe lanceolata* (Nees) Nees are trees.

ECONOMIC ASPECTS OF THE FAMILY

In an economic sense, the plants of the Lauraceae are important for aromatic oils, edible fruits and useful timber. *Aniba panurensis* (Guyana) and *A. rosaeodora* var. *amazonica* (Brazil) yield 'bois de rose', a perfume. *Cinnamomum burmanni* (Indonesia), *C. cassia* (Burma), *C. massoia* (New Guinea), *C. oliveri* (Australia) and *C. tamala* furnish 'cassia bark', used for flavouring purposes. Camphor is obtained by distillation of the wood of *C. camphora*, *C. loureirii* (Vietnam) and *C. zeylanicum* are cultivated for their aromatic bark called 'Saigon cinnamomum' and 'cinnamomum' respectively. *Beilschmiedia roxburghiana, C. cecidodaphne, Nectandra rodiaei* (Guyana), *Ocotea bullata* (South Africa), *Persea nammu* (China) and *Umbellularia California* (USA) produce valuable timber. *P. gratissima* (tropical America) is the source of edible fruits termed 'alligator pear' or 'avocado'. The leaves of *Laurus nobilis* (Europe) from 'sweet bay', much used in cooking. The root-bark of *Sassafras albidum* (eastern North America) constitutes 'oil of sassafras', a flavouring material.

Magnoliaceae
(Magnolia Family)

GENERAL FEATURES OF THE FAMILY

Habit	—	Shrubs or trees, sometimes climbing.
Leaves	—	Alternate, simple, *entire with bud scales or stipules covering buds.*
Flowers	—	Showy, aromatic, bisexual, regular, hypogynous, *acyclic or spirocyclic.*
Perianth	—	Petaloid, outer whorl sometimes sepaloid, deciduous, *arranged in whorls or in spirals.*
Androecium	—	*Stamens many, free; connective of anthers not projecting.*
Gynoecium	—	*Carpels many, free,* spirally arranged on a thalamus; ovules 2 to many, anatropous on ventral suture.

Fig. 25 MAGNOLIACEAE. *Magnolia campbelli* Hook. f. & Thorns.
A, arrangement of stamens and carpels; **B**, stamen;
C, seed; **D**, L.S. of seed. (**A–D**, after J.D. Hooker)

Fruits — Aggregate of berries or follicles.*

Seeds — *Large, suspended by a silky thread-like funiculus; embryo small; endosperm watery fleshy* and not ruminate.

FLORAL RANGE IN THE FAMILY

The flowers are bisexual in *Magnolia* and *Michelia,* but unisexual in *Drimys* and *Kmeria.*

In *Magnolia,* the thalamus is long and uniformly cylindrical. This can become elongated and narrowed above from a conical base, constituting a gynophore, as in *Michelia.*

The leaves of the floral envelope in *Magnolia* and *Michelia* are in three whorls of three each, all being petaloid or the outer whorl being sepaloid. In *Illicium,* there are many perianth segments; here the inner whorl becomes petaloid and the outer sepaloid. *Liriodendron* has 3 sepals followed by two whorls of 3 petals in each whorl. *Zygogynum* bears a cup like calyx and a few unequal petals.

Elmerrillia has axillary flowers with a sessile gynoecium and introrse anthers, while *Michelia* has axillary flowers with a stipitate gynoecium and latrorse or sublatrorse anthers. In *Aromadendron,* the carpels are concrescent and indehiscent. *Pachylarnax* has concrescent carpels which open entirely along the abaxial suture. But *Talauma* has circumscissile dehiscence of the carpels.

DIVISIONS OF THE FAMILY

Based on the habit, presence or absence of stipules and degree of thalamus development, the Magnoliaceae is separated into three subfamilies:

* The fruit of the Magnoliaceae is usually a spiral multifolliculus" (Takhtajan, 1969).

Subfamily I. *Illicioideae.* Leaves with no stipules. Flowers bisexual or unisexual. Thalamus short. Example: *Illicium.*

Subfamily II. *Magnolioideae.* Leaves with sheathing stipules. Flowers bisexual. Thalamus long. Example: *Magnolia.*

Subfamily III. *Schisandroideae.* Climbing shrub. Leaves with no stipules. Flowers unisexual. Example: *Schisandra.*

POSITION AND AFFINITY OF THE FAMILY

Engler, Rendle and other botanists put the Magnoliaceae under the Ranales. But Hutchinson placed the family within the Magnoliales and considered it as the most primitive among the dicotyledons. The primitiveness of the family is shown by the spiral arrangement of stamens and carpels as well as the apocarpous pistil—characters also shared with the Ranunculaceae. Smith (1945), however, cast some doubt on the assumption of the primitive nature of the Magnoliaceae and thought that the family is relatively specialised from vegetative and reproductive aspects.

From the stand point of construction and orgainisation of flowers, the Magnoliaceae is akin to the Annonaceae. The two families, however, can be set apart from each other (Table 9).

The Magnoliaceae has a great deal of affinity with the gymnosperms. The most striking thing about the wood of *Drimys* is the presence of tracheids with bordered pits in the xylem. This coniferous nature of the wood is "discounted by the fact that the structure of the bast is characteristically dicotyledonous, having sieve tubes with companion cells". Furthermore, the receptacle of *Magnolia* flower bears numerous spirally arranged free sporophylls. This can be compared with the strobilus of *Bennettites,* consisting of a central axis covered with megasporopylls. This has led to a suggestion that the Bennettitales may be ancestral to the modern angiosperms initiated by the Magnoliaceae.

TABLE 9. Demarcation of the Magnoliaceae and Annonaceae.

Magnoliaceae	*Annonaceae*
Stipules present	Stipules absent
Flowers large, showy	Flowers small, showy
Anthers introrse	Anthers extrorse
Endosperm watery fleshy not ruminate	Endosperm ruminate or marbled

NUMBER IN AND DISTRIBUTION OF THE FAMILY

The Magnoliaceae embraces 12 genera and about 220 species. The members of this family belong to the temperate regions of northern hemisphere, with centres of distribution in eastern Asia, Malaysia, Brazil, West Indies and eastern North America.

COMMONLY OCCURRING PLANTS OF THE FAMILY

Magnolia campbellii Hook. f. & Thoms. *M. grandiflora* L., *M. pterocarpa* Roxb and *Michelia champaca* L. are garden plants, producing fragrant and handsome flowers.

ECONOMIC ASPECTS OF THE FAMILY

The bark of *Drimys winteri* (Mexico) and *Illicium verum* (China) is useful in medicine. The bark and flowerbuds of *Magnolia officinalis* (China) yield a valuable drug. *Liriodendron tulipifera* (USA) furnishes us with wood for furniture and cabinet work. The species of *Magnolia* are cultivated domestically for their foliage and aromatic flowers. *M. acuminata, Manglietia hookeri, Michelia baillonii, M. doltsopa* and *Pachylarnax pleiocarpa* produce valuable timber. *Michelia champaca* yields 'champaca oil' from the flowers, camphor from the wood and scented water from the leaves.

Ranunculaceae
(Buttercup Family)

GENERAL FEATURES OF THE FAMILY

Habit	—	Erect or climbing herbs or shrubs.
Leaves	—	Alternate or radical, simple or compound, *often with a twisted sheathing base.*
Inflorescence	—	Cymose or racemose.
Flowers	—	Bisexual, *regular,* acyclic or hemicyclic.
Calyx	—	Sepals 5, petaloid, imbricate.
Corolla	—	Petals 5 or more, sometimes spurred.
Androecium	—	*Stamens many, free*; anthers extrorse.
Gynoecium	—	*Carpels many, free*; ovary superior; ovules few to many.
Fruit	—	1-seeded achene or many-seeded follicle.
Seed	—	Embryo small; endosperm oily.

FLORAL RANGE IN THE FAMILY

Deviations from the normal type of flower occur in the Ranunculaceae. The flowers are frequently regular, but irregular in *Aconitum* and *Delphinium.* Although the calyx and corolla may be generally found, petals are sometimes lacking (*Nigella*). The segments of the perianth are alike in *Helleborus* and differentiated into the calyx and corolla.

The perianth is followed by a varying number of honey leaves which are regarded as metamorphosed stamens or petals. In *Helleborus,* there are 13 rows of stamens and the first component of each row is transformed into a honey leaf; here the honey leaf is the progenitor of the stamens. In *Isopyrum,* the honey leaves are in a whorl of 5 alternating with 5 sepals and hence may be regarded as definite petals. In *Trollius,* the first members of the staminal row become converted into small narrow honey leaves each of which has a short claw and a small flat blade at its base. In *Caltha,* there is no separate nectary and honey is secreted by the carpels. In *Eranthis,* the nectary is absent and honey is secreted by the petals. Definite nectariferous bodies are not even seen in *Paeonia,* but a ring- or disc-like swelling of the receptacle developed at the base of the carpels takes over the function of the nectary.

As regards the number and arrangement of carpels, various types are encountered. Generally, the carpels are numerous, but 5–10 in *Trollius,* 5–12 in *Nigella* and 1 in *Actaea.* The carpels are free in many cases, but spiral in *Myosurus* and *Ranunculus* and somewhat united to form a compound ovary in *Nigella.* In *Coptis,* the carpels are often stalked.

DIVISIONS OF THE FAMILY

The Ranunculaceae is divided into three tribes:

Tribe (*i*) *Anemoneae.* Ovules solitary, arising from base of ventral suture. Examples: *Anemone, Clematis, Myosurus,* etc.

Tribe (*ii*) *Helleboreae.* Ovules arranged in two rows along ventral suture of carpels. Fruit a berry or follicle. Examples: *Aconitum, Caltha,* etc.

Tribe (*iii*) *Paeoniaeae.* Ovules arranged in two rows along ventral suture of carpels, but outer integument well developed. Example: *Paeonia.*

POSITION AND AFFINITY OF THE FAMILY

The Ranunculaceae is usually accepted to be among the most primitive of the dicotyledons. Hutchinson as well as Bentham-Hooker placed the family in the class of early dicots. In Engler's arrangement, the family appeared under the advanced Archichlamydeae. But Rendle and others relegated it to the early Dialypetalae.

The floral parts of the Ranunculaceae are similar in structure and arrangement to those of the Magnoliaceae, but the latter can be separated by the possession of woody habit and distinct stipules. Another relationship can be traced to the Cabomboideae (Nymphaeaceae), the items of common interest being the dimorphic leaves and small hypogynous flowers. On the basis of the structure of essential whorls, such as stamens and carpels, the Ranunculaceae shows resemblance to the Papaveraceae; the latter is characterised by the presence of milky sap, dimerous calyx and syncarpous ovary. Though the family can be linked with the Rosaceae on account of tree and numerous stamens and carpels, the latter stands out prominent by virtue of the general perigynous flowers and cyclic arrangement of stamens. An affinity with the Berberidaceae is indicated by the single carpel of *Actaea,* while the enormous development of the outer integument of the ovule is a characteristic feature of the Berberidaceae. "The Ranunculaceae and Berberidaceae are also phytochemically related in the possession of the alkaloid berberine" (Jury, 1978).

NUMBER IN AND DISTRIBUTION OF THE FAMILY

The Ranunculaceae covers about 50 genera and over 1,800 species. The plants of this family grow largely in the temperate and Arctic regions.

COMMONLY OCCURRING PLANTS OF THE FAMILY

Monkshood (*Aconitum napellus* L.) and Larkspur (*Delphinium ajacis* L.) are garden plants.

Virgin's Bower (*Clematis gouriana* Roxb. ex DC.) is a herbaceous plant which climbs by means of tendrillar petioles.

Coptis teeta Wall., a stemless herb, occurs in temperate regions of Arunachal.

Traveller's Joy [*Naravelia zeylanica* (L.) DC] is a climber with alternate leaves, the terminal leaflet of which turned into a tendril.

Black Cumin or Fennel Flower (*Nigella sativa* L.) is raised by farmers for its seeds.

Indian Buttercup (*Ranunculus scleratus* L.) is an erect herb, thriving in marshes and river banks.

Meadow Rue (*Thalictrum javanicum* Bl.), found mainly in hilly spots, is a slender plant with decompound leaves.

ECONOMIC ASPECTS OF THE FAMILY

Economically, the Ranunculaceae is of domestic importance for the large number of ornamentals. *Aconitum napellus* and *Delphinium ajacis* are commonly cultivated as garden plants.

The plants of this family are rich in alkaloids which have medicinal uses. The notable examples are *Anemone Pulsatilla, Cimicifuga racemosa, Delphinium staphisagria* and *Hydrastis canadensis.*

The seeds of *Nigella sativa* are generally used as a condiment, but also employed as a preservative of woollen garments against the attacks of insects.

Berberidaceae
(Berberry Family)

GENERAL FEATURES OF THE FAMILY

Habit	—	Herbs or shrubs.
Leaves	—	Alternate or basal, simple or compound, exstipulate.
Flowers	—	Bisexual, regular, hypogynous.
Perianth	—	Biseriate, segments free and distinct.
Androecium	—	*Stamens 4–18, in* 2 *whorls; anthers* 2*-celled, opening by flap-like valves.*
Gynoecium	—	*Carpel 1*; ovary superior; ovules many or few.
Fruit	—	Berry or follicle.
Seeds	—	Embryo straight; endosperm copious.

DIVISIONS OF THE FAMILY

The Berberidaceae is divided into two subfamilies:

Subfamily I. *Berberidoideae.* Foliage leaves simple or compound Flowers with honey leaves. Examples: *Berberis, Leontice, Mahonia,* etc.

Subfamily II. *Podophylloideae.* Foliage leaves not pinnately divided. Flowers without honey leaves. Examples: *Achlys, Podophyllum,* etc.

NUMBER IN AND DISTRIBUTION OF THE FAMILY

The Berberidaceae is represented by 13 genera and 550–600 species, chiefly distributed in the temperate zones of northern hemisphere from Central Asia to India as well as from the Andes to the Straits of Magellan.

COMMONLY OCCURRING PLANTS OF THE FAMILY

Berberis asiatica Roxb. ex DC., *B. insignis* Hook. f. & Thoms. and *B. lycium* Royle are shrubs.

Mahonia leschenaultii (Wall. ex Wt.) Takeda is known from Western Ghats to the Nilgiris southwards.

Podophyllum hexandrum Royle is found in western Himalaya at an altitude of 1,843–4,267 m.

ECONOMIC ASPECTS OF THE FAMILY

The Berberidaceae is of importance for ornamentals, *e.g.*, *Berberis buxifolia, B. calliantha, B. thunbergii, Mahonia aquifolium, M. bealei, M. lomariifolia* and *Nandina domestica. B. vulgaris* is known as the obligate host of the aecidial stage of wheat rust. The rhizome of *Podophyllum peltatum* is the source of the drug 'podophyllin'.

Menispermaceae
(Moonseed Family)

The family is recognised by the lianous habit, unisexual and trimerous flowers, drupaceous fruits and curved seeds.

The family includes 65 genera and roughly 350 species, being distributed in the tropical parts of the world.

Cocculus hirsutus (L.) Diels, *Stephania hernandifolia* Walp. and *Tiliacora acuminata* (Lamk.) Miers are climbers. *Coscinium fenestratum* (Gaertn.) Colebr. is found in Western Ghats.

The family is of little economic importance except for ornamentals (*Cissampelos, Cocculus* and *Menispermum*). 'Curare', chiefly obtained from *Chondrodendron tomentosum* is utilised as a muscle relaxant in surgical operations. A tonic is prepared from the roots of *Jateorhiza palmata.* The roots of *Tinospora cordifolia* serve as a febrifuge.

Lardizabalaceae
(Lardizabala Family)

The family is known by the lianas with digitately compound leaves, perulate flower buds, stamens opening lengthwise and baccate fruit.

A family of nine genera and 36 species, the Lardizabalaceae is found in the Himalaya, China, Japan and Chile.

The fruits of all species are edible, particularly *Akebia quinata* and *Decaisnea fargesii. Holboellia coriacea* is cultivated ornamentally.

Order 16. *Rhoeadales*

In Engler's arrangement, the Rhoeadales contained five suborders and seven families: Rhoeadineae (Papaveraceae), Capparidineae (Capparidaceae, Cruciferae, Tovariaceae), Resedineae (Resedaceae), Moringineae (Moringaceae), Bretschneiderineae (Bretschneideraceae). Within this order, Rendle included four families like the Papaveraceae, Capparidaceae, Cruciferae and Resedaceae. Both Lawrence and Core raised the number of taxa to six by also accommodating the Fumariaceae and Moringaceae. Hutchinson split up the whole order into four groups, *viz*. Capparidales (Capparidaceae, Moringaceae, Tovariaceae), Rhoeadales (Papaveraceae, Fumariaceae), Cruciales (Cruciferae) and Resed- (Resedaceae). Like other workers, the present author considers the position of the order after the and before the Sarraceniales.

The Rhoeadales forms a natural group, arising from the Ranales or a ranalian ancestor traceable to *Platystemon* of the Papaveraceae. Hallier suggested the origin of the Cruciferae and Capparidaceae from the Berberidaceae and of the Papaveraceae, along with the Nymphaeaceae, from the Ranunculaceae. Mitra (1974) stated that the evolution of the Rhoeadales took place along different lines from the ranalian plexus.

Among the members of the Rhoeadales, the Resedaceae and Capparidaceae are regarded by Norris (1941) as the most primitive. He further noted that the Cruciferae, Fumariaceae and Papaveraceae arose "by subsequent parallel evolution from a common ancestral group somewhat resembling the existing Resedaceae and Capparidaceae".

The gynoecium of the rhoeadalean flower in general and of the crucifer flower in particular has been a topic for speculation. Some believed it to be bicarpellary. Others accepted it as tetracarpellary. Stilla few took it to be hexacarpellary. Puri (1951) remarked that the problem here "does not concern the number of carpels; rather it is a matter of placentation". He expressed the view that the parietal placentation in the Capparidaceae, Cruciferae, Moringaceae, Papaveraceae, Passifloraceae and possibly other families originated from the axile condition. During the course of evolution, the carpellary margins receded to the periphery. The marginal bundles or their products of fusion retained their inverse orientation, thereby producing inverted placental strands. Hence, there are only two carpels in the crucifer gynoecium.

The presence of isorhamnetin in several members of the Capparidaceae appears to be phylogenetically significant (Daniel and Sabnis, 1977). Isorhamnetin is supposedly a characteristic compound of the Cruciferae, indicating thereby a close affinity between this family and the Capparidaceae. The alkaloid, brassicines, reported from *Brassica, Lepidium* and *Sinapis,* also exists in *Cleome* and *Gynandropsis.* The Capparidaceae is thus grouped with other thio-glucoside containing families like the Cruciferae, Moringaceae and Resedaceae under the order Rhoeadales.

The Rhoeadales is distinguished by the herbaceous habit, bisexual and hypogynous flowers with cyclic perianth segments and androecium, syncarpous pistil with two to many carpels and parietal placentation. The usual six families of the order can be identified by referring to the following synoptic key (Core, 1955):

(1) Flowers actinomorphic	(2)
(1) Flowers zygomorphic	(4)
(2) Gynoecium elevated on a gynophore	*Capparidaceae*
(2) Gynoecium not elevated	(3)
(3) Sepals 4; stamens many; carpels 2 to many	*Papaveraceae*
... Sepals 4; stamens 6; carpels 2	*Cruciferae*
...erigynous	*Moringaceae*
...ogynous	(5)
...arpels not elevated	*Famariaceae*
...els elevated on an androgynophore	(6)
...ort, expanded into a disc	*Resedaceae*
... and slender, not expanded into a disc	*Capparidaceae*

Papaveraceae (Poppy Family)

Habit	—	*Herbs with a coloured or milky sap.*
Leaves	—	Alternate, radical or cauline, lobed, exstipulate.
Flowers	—	Solitary, showy, *bisexual, regular,* hypogynous.
Calyx	—	Sepals 2, free, imbricate, *caducous.*
Corolla	—	Petals 4, in 2 whorls, crumpled, deciduous.
Androecium	—	*Stamens many, in several whorls, free;* anthers basifixed.
Gynoecium	—	Carpels 2 or more, connate; ovary superior, 1- or more-locular; ovules many, anatropous; *placentation parietal;* stigma capitate.
Fruit	—	Capsule, dehiscing by pores or valves.
Seeds	—	Embryo small; endosperm copious, mealy or oily.

FLORAL RANGE IN THE FAMILY

In *Bocconia* and *Macleaya,* the flowers are small and arranged in compound racemes. The flowers are small, yellow and umbellate in *Chelidonium.* In *Meconopsis,* the flowers are blue, purple or yellow and are solitary or in racemes or sometimes forming dichasia or uniparous cymes.

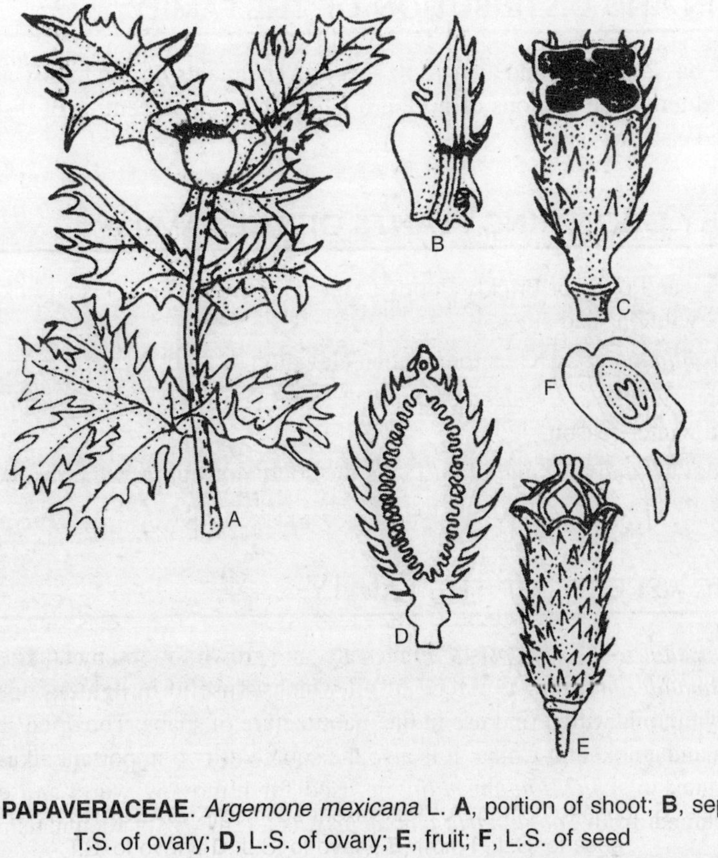

Fig. 26 PAPAVERACEAE. *Argemone mexicana* L. **A**, portion of shoot; **B**, sepal; **C**, T.S. of ovary; **D**, L.S. of ovary; **E**, fruit; **F**, L.S. of seed

In *Argemone,* one can find either trimerous or tetramerous arrangement of perianth segments and carpels. In *Eschscholzia,* the flower is perigynous with a cup-like receptacle. But the receptacle may be somewhat depressed and bear on its rim a trimerous calyx succeeded by a corolla of two alternating trimerous whorls, as in *Platystemon.*

POSITION AND AFFINITY OF THE FAMILY

The Papaveraceae was treated under the Rhoeadales by Engler, Rendle, Lawrence and Core. Bentham-Hooker, however, placed the family under the Parietales. The flowers of *Platystemon* establish a link between this family and the Nymphaeaceae.

There are morphological similarities between the flower of *Hypecoum* (Papaveraceae) and *Leptopyrum fumarioides* (Ranunculaceae).

The floral characters of the Papaveraceae are only slightly altered from those of the Ranunculaceae. The greatest divergence from that family is that the carpels instead of remaining distinct from each other are organised into a compound pistil that ripens in the form of a capsule.

The dimerous calyx and corolla as well as the pod shaped ovary are features of the Papaveraceae shared with the Capparidaceae and Cruciferae. But the seeds of the Papaveraceae are albuminous, whereas those of the other two families are not so. The presence of gynophore in some abnormal flowers of *Argemone* offers a connection between the Papaveraceae and the Capparidaceae.

NUMBER IN AND DISTRIBUTION OF THE FAMILY

The Papaveraceae has 26 genera and about 250 species. The plants of this family are chiefly natives of the subtropical and temperate regions of northern hemisphere, with centres of distribution in western North America and eastern Asia.

COMMONLY OCCURRING PLANTS OF THE FAMILY

Devil's Fig or Mexican Poppy or Prickly Poppy (*Argemone mexicana* L.) is a prickly herb with yellow flowers and grows wild on road sides.

Meconopsis napaulensis DC. is found at an elevation of 2,6000–3,300 m in the Darjeeling area.

Garden Poppy (*Papaver argemone* L.) and Oriental Poppy (*Papaver orientale* L.) are planted in gardens during the winter season.

Opium Poppy (*Papaver somniferum* L.) is the source of opium which is the inspissated sap of unripe capsules.

ECONOMIC ASPECTS OF THE FAMILY

The species of *Eschscholzia, Meconopsis, Papaver,* etc., are grown for ornament. The seeds of *Argemone mexicana* and *P. somniferum* produce a type of oil which is useful in lighting purpose. The seeds of *Glaucium flavum* yield oils which find use in the manufacture of soaps; The dried sap of *P. somniferum* is chewed in India and smoked in China; it is also the source of two important alkaloids, 'coedine' and 'morphine'. The latex of *Chelidonium majus* is used for removing warts and corns. An alkaloid, 'sanguinarine', obtained from *Sanguinaria canadensis*, is a remedy for rheumatism.

GENERAL FEATURES OF THE FAMILY

Habit	—	*Herbs with a pungent juice.*
Leaves	—	Alternate, radical or cauline, simple, auriculate, *often lyrate,* exstipulate.
Inflorescence	—	Raceme without bracts.
Flowers	—	Bisexual, regular, hypogynous, complete.
Calyx	—	Sepals 4, free, in two whorls, *lateral pairs saccate.*
Corolla	—	Petals 4, free, *cruciform with a distinct limb and claw.*
Androecium	—	*Stamens 6, tetradynamous* (2 outer short and 4 inner long), usually with a gland on thalamus.
Gynoecium	—	Carpels 2, syncarpous, *initially unilocular but subsequently biocular by a false partition* (*replum*); ovules many; *placentation parietal.*
Fruit	—	Silicula or siliqua.

Cruciferae
(Mustard family)

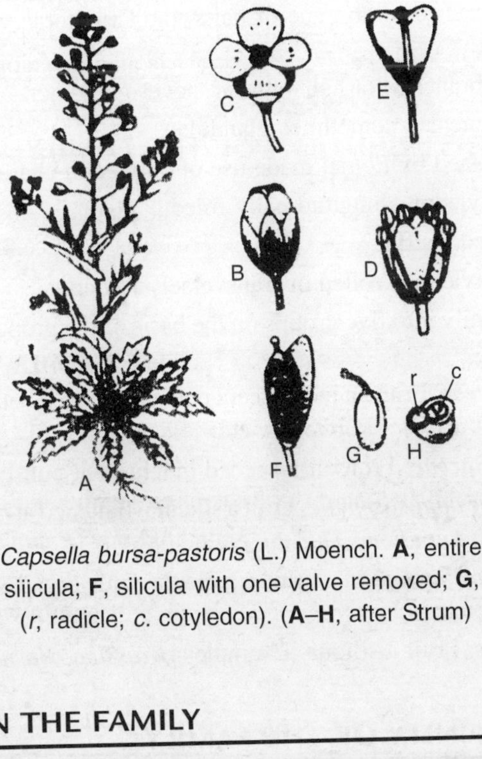

Fig. 27 CRUCIFERAE. *Capsella bursa-pastoris* (L.) Moench. **A**, entire plant; **B**, bud; **C**, flower; **D**, pistil and stamens; **E**, siicula; **F**, silicula with one valve removed; **G**, seed; **H**, seed cut across (*r*, radicle; *c*. cotyledon). (**A–H**, after Strum)

FLORAL RANGE IN THE FAMILY

Deviations from the normal type of floral structure have been noticed in the Cruciferae. In *Iberis* and *Teesdalia,* the flowers are zygomorphic owing to the enlargement of 2 outer petals. Frequently, the petals are very small or the petals may be totally absent, *e.g., Coronopus, Lepidium,* etc. In *Capsella bursa-pastoris,* the petals are sometimes substituted by 4 stamens. An elongation of the thalamus in the form of a gynophore takes place in *Brasssica elongata* and *Diplotaxis hurra.*

The stamens are reduced in number, as in *Cardamine hirsuta* and *Coronopus didymus* with 4 and 2 stamens respectively; here is a failure of the median stamens to double up and so the flowers appear diandrous. Furthermore, the stamens may be numerous (16) and this condition is found in *Megacarpaea.*

More than 2 carpels occur in a number of genera. *Holargidium* and *Tetrapoma* have 4 or more carpels, but *Lepidium sativum* is characterised by 3 carpels. Numerous carpels are reported from plants with abnormal flowers.

In general, the ovary is sessile on the receptacle. But there is a stipe between the ovary and receptacle in *Stanleya.*

The commisural face of the stigma is usually at right angles to the false partition, except in *Matthiola, Moricandia, Streptanthus, Thelypodium* and others.

DIVISIONS OF THE FAMILY

Linnaeus utilised the marked difference between pods of the diffeient genera as the criterion for the classification of his Tetradynamia into two orders:

Order 1. *Siliculosae.* Fruit a silicula.

Order 2. *Siliquosae.* Fruit a siliqua.

Depending on the presence or absence of hairs, the Cruciferae was divided into two series by Prantl:

Series A. Hairs unbranched or hairs absent, never glandular.

Series B. Hairs branched, sometimes glandular.

A classification suggested by Pomel made use of cotyledon characters:

A. *Orthoploceae.* Cotyledons longitudinally foled.

P. *Platylobeae.* Cotyledons flat.

C. *Pleuroploceae.* Cotyledons rolled or transversely folded.

Rendle divided the family into five groups on the basis of the proportion and mode of dehiscence of fruits:

Group I. *Lomentaceae.* Fruit an indehiscent pod, constricted between seeds (like a lomentum). Examples: *Cakile, Raphanus,* etc.,

Group II. *Nucumentaceae.* Fruit an 1-seeded indehiscent nut. Examples: *Bunias* and *Isatis.*

Group III. *Siliculosae angustiseptate.* Fruit a silicula, flattened at right angles to plane of septum. Examples: *Lepidium. Thlaspi,* etc.

Group IV. *Siliculosae latiseptate.* Fruit a silicula with broad septum. Examples: *Alyssum, Camelina,* etc.

Group V. *Siliquosae.* Fruit a siliqua. Examples: *Brassica, Nasturtium, Sisymbrium,* etc.

POSITION AND AFFINITY OF THE FAMILY

The Cruciferae was placed under the order Rhoeadales by Rendle and others and under the cohort Parietales by Bentham-Hooker. On the one hand, it is related to the Papaveraceae and on the other to the Capparidaceae. The three families have in common the features of tetramerous perianth, syncarpous pistil and parietal placentation. This has given rise to the problematic issue as to whether the Cruciferae descended from the Papaveraceae or originated from the Capparidaceae.

The morphology of stamens and carpels of the crucifer flower as well as their anatomy bears testimony to a papaverous ancestry. But the stamens are tetradynamous in the Cruciferae and not so in the Papaveraceae. In the Cruciferae, there is generally no gynophore which is a prominent character of the capparidean flower. A comparison of the floral diagram indicates that the flower of the Cruciferae is closely allied to that of the Capparidaceae. Whereas in the Capparidaceae the stamens exhibit great variation in number and arrangement, in the Cruciferae one stage has become fixed: two lateral shorter stamens and four median longer stamens. Though this phase is generally distinct, it may frequently be seen to spring from a single protuberance and may even remain somewhat united in the adult flower.

NUMBER IN AND DISTRIBUTION OF THE FAMILY

The Cruciferae, covering 380 genera and 3,000 species, is most abundant in northern hemisphere, especially in the cold temperate regions and countries surrounding the Mediterranean basin also in central and south-western Asia.

COMMONLY OCCURRING PLANTS OF THE FAMILY

White Mustard (*Brassica hirta* Moench.), Black Mustard [*B. juncea* (L.) Czern. & Coss. ssp. *juncea*], Indian Rape (*B. napus* L.) and *Eruca sativa* Lamk. are cold season crops.

Shepherd's purse [*Capsella bursa-pastoris* (L.) Moench], *Cardamine hirsuta* Hook. f. & T. And., *Coronopus didymus* (L.) Smith and *Rorippa indica* (L.) Hiern are common weeds.

Cochlearia flava Buch.-Ham. ex Roxb. is an annual herb.

Candy Tuft (*Iberis amara* L.) is a garden annual.

ECONOMIC ASPECTS OF THE FAMILY

Many plants of the Cruciferae are wholesome, being largely used for food: (*a*) for their roots, Horse Radish (*Armoracia rusticana*), Rutabaga (*Brassica napobrassica*), Turnip (*B. rapa*) and Radish (*Raphanus sativus*); (*b*) for their stems, Kohlrabi (*B. caulorapa*); (*c*) for their leaves, Kale (*B. oleracea* var. *acephala*), Cabbage (*B. oleracea* var. *capitata*), Brussels' sprouts (*B. oleracea* var. *gemmiferd*) and Chinese Cabbage (*B. pekinensis*); (*d*) for their inflorescences, Cauliflower (*B. oleracea* var. *botrytis*) and Broccoli (*B. oleracea* var. *italica;* (*e*) for their seeds yielding colza or rape oils, *e.g., B. chinensis, B. napus* var. *napus* and *B. rapa* and for their seeds used as spices, *e.g., B. hirta, B. juncea* and *B. nigra.*

The species of *Arabis, Cheiranthus, Draba, Erysimum, Hesperis, Lunaria, Matthiola,* etc. are grown in garden and rockery for their flowers. Other plants of economic use include *Camelina sativa* which yields a kind of drying oil, *Isatis tinctoria* which is the source of a blue dye and *Lunaria annua* which is utilised for dry bouquets.

Capparidaceae
(Caper Family)

GENERAL FEATURES OF THE FAMILY

Habit	—	Herbs or shrubs, erect or climbing.
Leaves	—	Alternate, simple or compound; stipules, if present, spiny or minute.

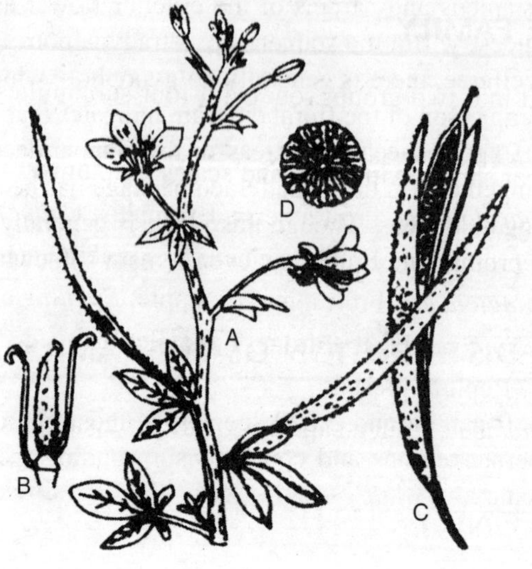

Fig. 28 CAPPARIDACEAE. *Polanisia viscosa* (L.) DC. **A**, portion of flowering shoot; **B**, flower with sepals, petals and most of stamens removed; **C**, ripe fruit; **D**, seed. (**A–D**, after Rendle)

Flowers	—	*Bracteate,* bisexual, regular, hypogynous, *tetramerous.*
Calyx	—	Sepals 4, free or united.
Corolla	—	Petals 4, free.
Androecium	—	Stamens 4 or more, free; filaments long.
Gynoecium	—	Carpels 2 or more, syncarpous; *ovary superior,* 1- or more-locular; ovules many, campylotropous; *placentation parietal with a gynophore.*
Fruit	—	Berry or capsule.
Seeds	—	Many, reniform; embryo folded; endosperm fleshy.

FLORAL RANGE IN THE FAMILY

Departures from the normal type of flower occur in the Capparidaceae. The flowers become medianly zygomorphic due to the unequal development of floral members, as in *Pteropetalum.* In this case, the posterior petals are much larger than the anterior ones. In many species of *Capparis,* the sepals are also unequal and the anterior sepal appears hooded. In some genera, the petals are lacking. In *Emblingia,* the sepals are united to form a four lobed calyx and the two posterior petals form a large hooded corolla.

The stamens vary in number. In some species of *Cleome*, the stamens are 4–6. In *Gynandropsis,* they are 6; in *Capparis* and *Polanisia* numerous. In *Dactylaena micrantha,* 1 stamen is fertile and 3 stamens are reduced to staminodes. In *Haptocarpum brasiliense,* 4 stamens are fertile and 2 sterile.

The flower of the Capparidaceae shows the development of an internode between the petals and stamens (*androphore*) and an internode between the stamens and carpels (*gynophore*). The carpels often project out of the flower. The formation of swellings, discs or tubular structures as lateral outgrowths inside or outside the corolla are also found.

DIVISIONS OF THE FAMILY

The Capparidaceae is divided into two groups, one with four subfamilies and the other with a single subfamily:

Group I. Usually trees or shrubs with hairs and scales. Replum absent.

Subfamily I. *Capparidoideae.* Fruit a berry. Examples: *Capparis, Crataeva,* etc.

Subfamily II. *Dipterygioideae.* Fruit a samara. Example: *Dipterygium.*

Subfamily III. *Emblingioideae.* Fruit a nut. Example: *Emblingia.*

Subfamily IV. *Roydsioideae.* Fruit a drupe. Example: *Roydsia.*

Group II. Usually herbs. Replum present.

Subfamily I. *Clemoideae.* Fruit a capsule or siliqua. Examples: *Cleome, Gynandropsis,* etc.

POSITION AND AFFINITY OF THE FAMILY

Hutchinson included the Capparidaceae under the Capparidales. But Bentham-Hooker placed it under the cohort Parietales and Rendle under the order Rhoeadales.

There is a general agreement among botanists to regard the Capparidaceae and Cruciferae as two closely related families. It has been thought that the two families arose from a common ancestor. According to another idea, the Cruciferae might have been derived from a primitive member of the Capparidaceae. Both families possess an apparently bicarpellate ovary and parietal placentation. Puri (1945) suggested that this condition originated from a quadrilocular ovary which was tetracarpellate and had axile placentation.

NUMBER IN AND DISTRIBUTION OF THE FAMILY

In the Capparidaceae, there are 40–50 genera and approximately 700 species. Generally speaking, the plants belonging to this family are widely distributed in the tropical and subtropical parts of both hemisphere.

COMMONLY OCCURRING PLANTS OF THE FAMILY

Capparis spinosa L. and *C. zeylanica* L. non Hook. f. & Thomas, are way side weeds. *Cleome gynandra* L. is another weed with white or pale purplish flowers.

Capparis sepiaria L. is a hedge plant which climbs by means of stipular hooked spines.

Crataeva nurvala Buch—Ham. is a small tree.

Polanisia viscosa (L.) DC. is a weed with yellow flowers, having an unstalked ovary.

ECONOMIC ASPECTS OF THE FAMILY

Economically, the Capparidaceae is not of much importance to us. The leaves of *Cadaba fruticosa* are used as a poultice for sores and those of *C. trifoliata* in the preparation of medicated oils. *Capparis spinosa* is grown in the Mediterranean area for the unopened flower buds called 'capers' which are useful in seasoning foods. The bark of *Crataeva nurvala* finds application as a remedy for bladder stones; the fruit is edible and the rind is used as a mordant in dyeing. The species of *Capparis, Gynandropsis* and *Polanisia* are sometimes grown as garden plants.

Fumariaceae (Fumitory Family)

GENERAL FEATURES OF THE FAMILY

Habit	—	Delicate smooth herbs or climbers, with a watery sap.
Leaves	—	Alternate, compound, dissected, exstipulate.
Flowers	—	Bisexual, *irregular,* hypogynous, perfect.
Calyx	—	Sepals 2, small, deciduous.
Corolla	—	*Petals 4, in two pairs, somewhat connivent, inner one narrower and coherent and outer basally saccate or spurred.*
Androecium	—	*Stamens 4, free and opposite to petals or 6 and united into two bundles.*
Gynoecium	—	Carpels 2; ovary superior, unilocular; *placentation parietal.*
Fruit	—	Capsule or nut.
Seeds	—	Embryo minute; endosperm fleshy.

POSITION AND AFFINITY OF THE FAMILY

Engler, Wettstein and others accepted the Fumariaceae as a subfamily within the Papaveraceae. Hutchinson recognised the Fumariaceae as a separate family, but included the Hypecoideae in it. The zygomorphic flower as well as the androecial situation led Lawrence, Core and Mitra to treat it as a distinct family under the Rhoeadales.

The Fumariaceae stands close to some extent to the monocotyledonous plants due to the presence of a single cotyledon, subterranean tubers and rhizomes bearing scales in several species of *Corydalis* and *Dicentra.*

NUMBER IN AND DISTRIBUTION OF THE FAMILY

The Fumariaceae contains 16 genera and about 450 species, distributed mainly in the north temperate regions and especially in the Mediterranean basin.

COMMONLY OCCURRING PLANTS OF THE FAMILY

Fumaria indica (Haussk.) Pugsly is a small branched annual herb in waste places.

ECONOMIC ASPECTS OF THE FAMILY

Economically, the Fumariaceae is of little importance except for ornamentals. *Corydalis ramosa* is useful in the treatment of eye diseases. The pea-like tubers of *Dicentra canadensis* form 'corydalis' of commerce.

Resedaceae
(Mignonette Family)

The family is characterised by the glandular stipules, zygomorphic flowers, laciniate petals, open unilocular ovary and capsule with an apical dehiscence.

According to Engler, Bessey and Rendle, the Resedaceae is a phyletically advanced taxon within the Rhoeadales. Following the procedure of Bentham-Hooker, Hutchinson aligned the family with the Violaceae. The family is most closely allied to the Capparidaceae; this is evident by the xerophytic habit, zygomorphic flower, gynophore development and parietel placentation. An additional link between the two families is the presence of a third and innermost seed coat in species of both Capparidaceae and Resedaceae (Orr, 1921). However, the remarkable tetramery of the capparidean flower is absent in the Resedaceae.

The Resedaceae comprises six genera and about 75 species, being developed mainly in the Mediterranean basin and South Africa.

The family is of little economic value. *Reseda luteola* was once used as a source of 'weld', a reddish yellow dye for textile industry. *R. odorata* is grown for fragrant flowers and yields a perfume oil.

Moringaceae
(Moringa Family)

The family is distinguihsed by the alternate pinnately compound leaves, zygomorphic flowers having five stamens and three to five alternating staminodes on a short hypanthium as well as a trivalved elongated fruit with winged seeds.

The position of the Moringaeceae is within the Rhoeadales, as treated by Bessey, Wettstein and Core. Rendle ignored the family. Hutchinson included the family in the Capparidales along with the Capparidaceae and Tovariaceae. Puri (1942) considered the family as distinguished by its complicated gynoecial condition which is also known in the Capparidaceae and Cruciferae. Datta and Mitra (1947) claimed the Moringaceae as most closely allied to the Violaceae. The zygomorphic flower of the Moringaceae having the perianth and stamens borne on a hypanthium, furnishes a link with the Rosaceae.

The family contains one genus *Moringa* and 12 species. It "has a distribution from the Mediterranean and North Africa to the Arabian peninsula and India, and is also represented in south-western Africa and Madagascar" (Heathcote, 1978).

The family is of economic importance for the Horse radish Tree (*Moringa oleifera*), producing fruits. The seeds'yield 'oil of ben', a non-drying oil, which finds application in the lubrication of watches. The roots serve as the source of a condiment.

Order 17. *Sarraceniales*

The order includes generally damp-loving herbaceous plants, having their leaves usually variously modified for trapping insects. Here the flowers are regular with an uniseriate or a biseriate perianth.

The order contains three families which, according to Core (1955), can be distinguished as follows:

(1) Leaves not modified into pitchers, but covered with sensitive or glandular hairs	*Droseraceae*
(1) Leaves modified into pitchers	(2)
(2) Low herbs bogs	*Sarraceniaceae*
(2) Climbing herbs or shrubs	*Nepenthaceae*

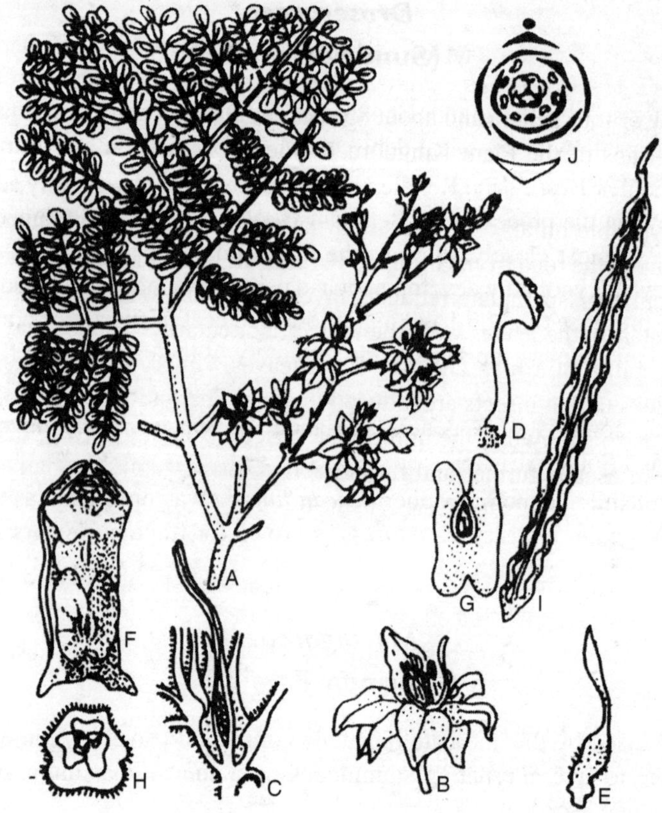

Fig. 29 MORINGACEAE. *Moringa oleifera* Lamk. **A**, Portion of shoot; **B**, flower; **C**, L.S. of flower; **D**, stamen; **E**, gynoecium; **F**, portaion of fruit showing ovary; **I**, fruit; **J**, floral diagram. (**B–J**, after Mitra)

Engler, Rendle, Lawrence and Core considered the three families as forming a natural group, Bentham-Hooker placed the Sarraceniales under the Parietales, the Droseraceae under the Rosales and the Nepenthaceae under the Multiovulate Terrestres. Hallier reduced the Nepenthaceae to the status of a tribe under the Sarraceniaceae. Diels treated the Droseraceae as a component of the Parietales, having an affinity with the Violaceae. Hutchinson shifted the Nepenthaceae to the Aristolochiales, keeping the Sarraceniaceae and Droseraceae within the Sarraceniales. The Droseraceae and Nepenthaceae have very similar pollen (Chanda, 1965), in addition to the obvious exomorphic characters.

"The group", in the words of Porter (1967), "represents an early modification, and a terminal one, from the Ranales by specialization of the leaves and the adoption of a specialized habitat". Despite the admitted differences between the three families, the insect-catching leaves are homologous in all of them (Margraf, 1955). "The insectivorous habit of the Sarraceniales may be presumed to be an evolutionary response to their growth in habitats deficient in available nitrogen. The Sarraceniaceae and Droseraceae commonly grow in water logged soils containing little or no soluble nitrate. The Nepenthaceae occur in wet, tropical forests, which characteristically have nutrient-poor soils. Many other groups of plants have faced similar problems, but very few have learned to meet them by trapping insects" (Cronquist, 1968).

*Droseraceae** (Sundew Family)

This is a small family of four genera and about 83 species. *Drosera* and *Drosophyllum* are nicknamed as the bird-lime or fly-traps of the Plant Kingdom. While *Drosera* is an active trap, *Drosophyllum* is a passive one.

Drosera is available on all continents, but mainly in Australia. It is found in abundance in the wastelands fringing upon the road from Giridih to Parashnath Hills, Jharkhand. Each plant is in possession of a rosette of orbicular or spathulate reddish leaves which are studded with glandular leaves called *tentacles.* From a distance, the leaves wear the look of a circular red spot; it may seem that the spot has been caused by the spittle of a man chewing Betel leaves. "Each hair secretes a drop of fluid which shines so bright in the sun that insects are induced to alight upon it in the hope of getting a sip of water. The fluid, however, is so sticky that the unfortunate insects cannot get away from it. Every effort they make puts the hope of escape further and further away; for gradually the hairs collect, bend over and take a firm hold of them. The insects caught by this trap soon lose their strength and die, and are decomposed and absorbed".

Drosophyllum ranges from Morocco to Portugal and South Spain. In the plant, we notice a short woody stem which passes above into the branched cymose inflorescence. The leaves are long and narrow, being crowded on the upper portion of the stem and branches. "One of the species ... is used in rural districts of Portugal as a substitute for fly paper. The plants are hung in doorways, and by means of their sticky leaf secretions, firmly hold insects which alight on them" (Fuller, 1957).

Dionaea is strictly endemic to the coastal plains of North and South Carolina, USA. The leaf-blade is similar to a steel-trap which is itself an active device. The upper red surface bears six sensitive hairs or triggers. As stimulation is initiated by the movement of an insect, the leaf tends to close abruptly. After 5–10 days, digestion and absorption are complete.

Aldrovanda is widely distributed. It is also prevalent in the saltmarshes south of Calcutta and in the Sunderbans area. It is a modified version of *Dionaea,* as the insect catching mechanism is the same in both. There is an obvious difference: the pointed hairs are numerous in *Aldrovanda* and six in *Dionaea.* Also, the former floats in water and has whorls of leaves.

Roridula has leaves like *Drosera,* without any movement of the tentacles. The former is confined to South Africa. It was once regarded as carnivorous, but has been shown by Lloyd (1942) not to be so. Here the secretion of glands is resinous and not digestive.

Sarraceniaceae (Pitcher-plant Family)

The family is represented by three genera and 17 species. They have limited distribution—*Darlingtonia* in northern California and southern Oregon; *Heliamphora* in Guyana and Venezuela; *Sarracenia* in eastern and north-eastern North America. These genera are linkened to the pitfalls of the Plant Kingdom. Pitfalls are passive traps.

* The material for the Droseraceae and Nepenthaceae have been drawn from an article by the author. See under selected bibliography.

Nepenthaceae
(Nepenthes Family)

The family contains one genus (*Nepenthes*) and about 70 species, inhabiting the tropics of the Old World from China to Australia with the greatest development in Jaintia and Garo Hills of Meghalaya.

Nepenthes conforms to the pitfall type of trap. Here the pitcher is a remarkably intricate device. It attracts insect by virtue of colour or odour and contains a copious amount of liquid (made up of rainwater and dew as well as secreted enzymes and acid). Inside the pitcher, there are a number of downward-pointing stiff hairs which make a precarious foothold. When insects wander around the leaf, they crawl into the pitcher, lose their footing and get drowned in the liquid. Following the death of the insects, their bodies are digested and absorbed by the leaf cells. In contrast with other carnivorous plants, the pitcher plants do not move in consonant with their catch. Moreover, the effectiveness of the insect traps are manifested by the presence of a mass of undigested exoskeletons generally to be noticed in any mature leaf.

The stems of *Nepenthes distillatoria* (Sri Lanka) and *N. reinwardtiana* (Malaysia.) are used as a kind of cordage as well as for making baskets.

Order 18. *Rosales*

Bessey suggested the Rosales as a fairly primitive taxon, being a direct descent from the Ranales. Hallier opined that the Dilleniaceae gave rise to the Rosales directly. Hutchinson supported the origin of woody Rosales from the Magnoliales via the Dilleniales. The Rosales, according to Rende, is a very natural group connected by transitional forms in which there is a passage from actinomorphy to zygomorphy and hypopyny to perigyny. The Byblidaceae and Cephalotaceae are remarkable in that they comprise insectivorous plants. The Podostemonaceae is unique for the striking adaptations to life in rushing water.

In the Rosales, the ovary is surrounded by an androperianth tube. The taxon is large and diversified, as will be evident from the number of families treated under it by the following systematists:

Engler & Diels	:	Saxifragineae (Crassulaceae, Cephalotaceae, Pittosporaceae, Byblidaceae, Brunelliaceae, Cunoniaceae, Myrothamnaceae, Bruniaceae, Roridulaceae, Eucommiaceae), Rosineae, (Platanaceae, Crossosomataceae, Rosaceae, Connaraceae (Leguminosae).
Rendle	:	Crassulaceae, Saxifragaceae, Cunoniaceae, Pittosporaceae, Podostemonaceae, Hydrostachyaceae, Hamamelidaceae, Platanaceae, Rosaceae, Connaraceae, Leguminosae.
Lawrence	:	Crassulaceae, Saxifragaceae, Pittosporaceae, Cunoniaceae, Hamamelidaceae, Eucommiaceae, Platanaceae, Crossosomataceae, Rosaceae, Leguminosae.
Core	:	Podostemonaceae, Crassulaceae, Cephalotaceae, Saxifragaceae, Pittosporaceae, Cunoniaceae, Hamamelidaceae, Platanaceae, Rosaceae, Connaraceae, Leguminosae.
Hutchinson	:	Rasaceae, Dichapetalaceae, Calycanthaceae.
Mitra	:	Rosaceae, Saxifragaceae, Crassulaceae, Cunoniaceae, Pittosporaceae, Connaraceae, Leguminosae.
Cronquist	:	Eucryphiaceae, Cunoniaceae, Davidsoniaceae, Pittosporaceae, Byblidaceae, Columelliaceae, Hydrangeaceae, Grossulariaceae, Bruniaceae, Alseuosmiaceae, Crassulaceae, Cephalotaceae, Saxifragaceae, Rosaceae, Neuradaceae, Chrysobalanaceae, Leguminosae.

The present author largely accepts Mitra's concept of the Rosales, but includes the Podostemonaceae within its scope. The nine families may be distinguished as follows:

(1) Flowers hypogynous	(2)
(1) Flowers perigynous	(7)
(2) Aquatic plants	*Podostemonaceae*
(2) Terrestrial plants	(3)
(3) Succulents	*Crassulaceae*
(3) Non-succulents	(4)
(*A*) Leaves opposite or whorled	*Cunoniaceae*
(4) Leaves alternate	(5)
(5) Leaves pinnate	*Connaraceae*
(5) Leaves simple	(6)
(6) Leaves deciduous	*Saxifragaceae*
(6) Leaves evergreen	*Pittosporaceae*
(7) Fruit a legume	*Leguminosae*
(7) Fruit not a legume	(8)
(8) Stamens many	*Rosaceae*
(8) Stamens 5 or 10	*Saxifragaceae*

Podostemonaceae
(Riverweed Family)

GENERAL FEATURES OF THE FAMILY

Habit	—	*Lichen-like or moss-like herbs, growing in running water and torrential hill streams where they are attached to rocks by haptera.*
Leaves	—	Alternate, simple, linear or subulate.
Inflorescence	—	Solitary or cymose.
Flowers	—	*Minute, bisexual, irregular, borne on long pedicels from a spathe-like structure covering (spathella).*
Perianth	—	Segments 3–5, free or united.
Androecium	—	Stamens 1–4 or many; anthers 2-celled, introrse.
Gynoecium	—	Carpels 2–3, ovary superior; placentation axile; ovules many, anatropous.
Fruit	—	Capsule.
Seeds	—	Testa mucilaginous; embryo straight; endosperm absent.

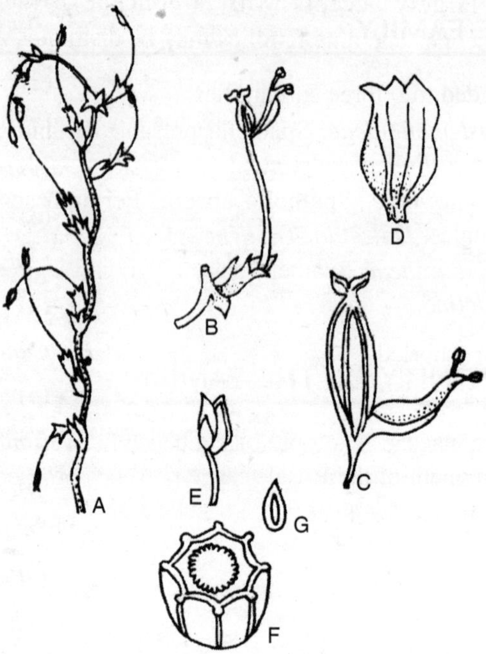

Fig. 30 PODOSTEMONACEAE. *Dicraea dichotoma* (Garden.) Tul. **A**, portion of plant; **B**, infloresence; **C**, flower; **D** spathella (open); **E**, stamen; **F**, T.S. of ovary; **G**, fruit. (**A–G**. after Wight)

POINTS OF BIOLOGICAL INTEREST IN THE FAMILY

The seeds are minute and numerous, having no special contrivance for dissemination. The mucilaginous character of the testa contributes to distribution by the feet of wading birds to which the seeds may cling.

The seeds give rise to primary axes which form thalli and are actually fixed to the substratum. From the thallus, secondary shoots appear as branching leaf bearing axes or rosettes of leaves. The leaves are delicate and change into bracts; they usually terminate in a spathella which encloses the young flower.

The flowers are produced when the streams begin to dry up in summer. When mature, the fruits burst and the seeds are shed on the rocks. With the onset of rains, the seeds germinate and develop into individual plants.

In many cases, silica is deposited in peripheral parts of the plant. This acts as a protection against drying when the level of water recedes and the plants are exposed.

The haptera are formed exogenously from the surface tissues of the thallus. They have growing points like those of roots.

The internal structure consist of parenchyma and collenchyma, without any intercellular spaces. "The absence of intercellular spaces is explained by Goebel as correlated with the habitat in rapidly moving well aerated water; large intercellular spaces to carry oxygen to lower portions in mud or stagnant water being unnecessary" (Rendle 1925). The vascular bundles are closed, collateral and scattered. The xylem is extremely reduced and phloem properly developed. The stomata are absent.

DIVISIONS OF THE FAMILY

The Podostemonaceae is divided into three subfamilies:

Subfamily I. *Podostemonoideae.* Spathella present. Perianth segments minute. Examples: *Dicraea, Podostemon,* etc.

Subfamily II. *Tristichoideae.* Spathella absent. Perianth segments 3, somewhat united. Examples: *Dalzellia, Tristicha,* etc.

Subfamily III. *Weddelinoideae.* Spathella absent. Perianth segments 5, free or united; Example: *Weddelina.*

POSITION AND AFFINITY OF THE FAMILY

The position of the Podostemonaceae is a topic of controversy. Warming, Engler, Rendle and others accepted the family as a component of the Rosales. The presence of free styles, bicarpellate ovary and numerous anatropous ovules, speaks for an alliance with the Saxifragaceae. The study of embryological features led Maheshwari (1945) to conclude that "the Podostemonaceae are much reduced apetalous derivatives of the Crassulaceae". On the basis of chromosome number, Gundersen (1950) placed the Podostemonaceae within the Hamamelidales. Hutchinson (1959) considered the family, along with the Hydrostachyaceae, under the Podostemonales as reduced types of the Saxifragales.

NUMBER IN AND DISTRIBUTION OF THE FAMILY

The Podostemonaceae contains 45 genera and about 130 species. The plants of this family are found mostly in the tropical areas, but are also distributed in the temperate regions.

COMMONLY OCCURRING PLANTS OF THE FAMILY

Subramanyam (1962) listed 13 species of the Podostemonaceae as occurring in India:

Dicraea dichotoma (Gardn.) Tul., an algiform herb, occurs in Western Ghats, Nilgiris and Malabar Hills upto 2,500 m.

Dicraea stylosa (Benth.) Wt., a seaweed-like broad thallus with helmet shaped bracts, is recorded from Western Ghats, South Genara to Travancore and Anamalai Hills upto 1,500 m.

Farmeria indica Willis, a thread-like thallus bearing four seeded ribbed fruits, is found in Western Ghats and in Tambraparni river in Tinnevelly District of Tamil Nadu.

Griffithella hookeriana (Tul.) Warm., a peculiar lichen-like plant with distichous leaves, is prevalent in Western Ghats from Bombay to South Canara at Beltangadi and Sullia.

Hydrobryopsis sessile (Willis) Engl., a small plant with sessile fruits, occurs in Visakhapatnam at Andhra Pradesh, Western Ghats and South Canara at Beltangadi.

Indotristicha ramosissima (Wt.) van Royen, a free floating plant with branched stems and linear scattered leaves, occurs in rivers of South Canara, Malabar, Cochin, Travancore and in the Anamalais upto 1,200 m.

Podostemon barberi Willis, a submerged herb with cleistogamous flowers and ovary with six broad wings, is reported from Western Ghats, the hills of South Canara at Beltangadi and the Bastar District of Madhya Pradesh.

Podostemon subulatus Gardn., a submerged herb with subulate leaves, occurs in Western Ghats at Anamalai Hills.

Terniola zeylanica (Gardn.) Tul., a frondose plant with thallus branching freely, grows in Western Ghats, South Canara at Sullia and Beltangadi, Kerala at Kottayam and Quilon and Tambraparni river in Tinnevelly.

Terniola zeylanica Tul. var. *konkanica* (Willis) Santapau, a thallus with flabelliform lobes and crowded growing points, thrives in and around Khandala, in Thul Ghat near Igatpuri of Western Ghats, in Khadsi river, near Bhorkus, west of Poona and Angul river in Orissa.

Willisia selaginoides (Bedd.) Warm. ex Willis, a tufted herb with shoots resembling *Lycopodium selago,* is found in Western Ghats at Anamalais upto 1,000 m.

Zeylanidium johnsonii (Wt.) Engl., a crustose thallus with very long stamens, occurs in Malabar.

Zeylanidium lichenoides (Kurz) Engl., a ribbon-like branched thallus with stamens of varying length, is reported from South Canara, Travancore, Anamalais at 1,000 m, Western Ghats, near Jeypore in Orissa, Visakhapatnam and Assam.

Zeylanidium lichenoides (Kurz) Engl. var. *bhorense* (Willis) Santapau, a ribbon-like thallus with pedicels 5–8 mm long, is known from Thul Ghat and Khandala.

Zeylanidium lichenoides (Kurz) Engl. var. *khandalense* (Willis) Santapau, a ribbon-like thallus with pedicels 2–5 mm long, is noticed in Thul Ghat near Igatpuri and Khandala, Western Ghats.

Zeylanidium olivaceum (Gardn.) Engl., a crustose thallus bearing erect ribbed fruits, is recorded from Quilon and Kottayam of Kerala. The form *anamalaiense* of this species has been collected from the Anamalais at 1,000 m and the form *griseum* from the Nilgiris in the Pykara river at about the same elevation.

Crassulaceae
(Orpine Family)

The family is distinguished by the succulent character of stems and leaves, stamens and carpels as many as petals and pistil with a scaly gland at the base.

The Crassulaceae shows affinity with the Saxifragaceae (Subramanyam, 1970). *Crassula aquatica,* which has the most reduced endosperm in the Crassulaceae, has a mode of habit similar to the Podostemonaceae. This taxon forms a connecting link between the two families (Subramanyam, 1962) and supports Maheshwari's (1945) contention that the Podostemonaceae are much reduced apetalous derivatives of the Crassulaceae.

In the Crassulaceae, there are 35 genera and 1,500 species. The plants of this family are abundant in the arid regions, especially in south-central Asia, Mediterranean area, South Africa, northern Mexico and south-western United States.

Bryophyllum pinnatum (Lamk.) Oken. and *Kalanchoe laciniata* DC. are fleshy herbs, the former with long tubular pendulous flowers and the latter with erect protandrous flowers.

The members of the family are of importance as ornamentals, such as the species of *Aeonium, Echeveria, Rochea, Sedum,* etc.

Cunoniaceae
(Cunonia Family)

The family is distinguished by the tree-like habit, opposite or whorled leaves and carpellary margins of each placenta distinct or recurved.

The family comprises 26 genera and 250 species, exclusively distributed in the southern hemisphere and largely in the Oceania and Australasia.

The family is of some importance. The species of *Ackama, Callicoma, Ceratopetalum* and *Weinmannia* are of value as ornamentals. *Ceratopetalum apetalum* (New South Wales) yields timber and *Weinmannia pubescens* (northern Andes) tannin.

Connaraceae
(Connarus Family)

The family is characterised by the pinnate leaves, free carpels and arillate seeds.

The family is composed of 16 genera and about 350 species, being prevalent in the tropics of both hemispheres.

The leaves of *Agelaea emetica* (Malagasy) yields an essential oil which promotes vomiting and those of *A. villosa* (West Africa) are used to treat dysentery. The leaves of *Cnestis corniculata* and *C. ferruginea* constitute the source of an astringent and a laxative respectively. The seeds of *Connarus africanus* are converted into a flour which is effective as an anthelmintic. *C. guianensis* (Guyana) yields timber. The bark of *Rourea glabra* (Central America) is used for tanning animal hide and the roots of the same plant yield a strong fibre for making ropes. The fruits and seeds of *R. volubilis* are utilised for poisoning dogs.

Saxifragaceae
(Saxifrage Family)

In this family, the leaves are exstipulate, flowers perigynous or epigynous, stamens and carpels few and endosperm abundant.

The family includes 80 genera and about 1,250 species are cosmopolitan in distribution, but found mostly in the north temperate regions.

The family is important for ornamentals, *e.g., Astilbe, Bergenia, Deutzia, Escallonia, Hydrangea, Philadelphus, Saxifraga,* etc. Some yield edible fruits, *e.g., Ribes americanum,* European Gooseberry (*R. grossularia*), *R. hirtellum,* Black Currant (*R. nigrum*), Red Currant (*R. rubrum*) and White Currant (*R. sativum*).

Pittosporaceae
(Pittosporum Family)

The family is recognised by the coriaceous exstipulate alternate or whorled leaves and by the flowers with short filaments, poricidal anthers, simple style, 2- to-5-cxarpelled ovary ad multiovulate placentae.

A family of nine genera and 200–400 species, the Pittosporaceae is found in the warmer parts of the Old World and is highly developed in Australia.

The family is important for ornamentals, *e.g. Pittosporum* and *Sollya. Pittosporum* also yields timber. *Billardiera longiflora* (Tasmania) is raised for blue edible berries.

Rosaceae
(Rose Family)

GENERAL FEATURES OF THE FAMILY

Habit	—	Herbs, shrubs or trees.
Leaves	—	Alternate, simple or compound, *stipulate* (stipules adnate).
Flowers	—	Bisexual, regular, *perigynous with a disc.*
Calyx	—	Sepals 4 or 5, persistent.

Corolla — Petals 0 or many, imbricate, deciduous.
Androecium — *Stamens many, in whorls,* incurved.
Gynoecium — *Carpels 1 or more,* free; ovary superior; ovules 1 or more in each cell.
Fruit — Achene, berry or drupe.

DIVISIONS OF THE FAMILY

On the basis of floral construction, the Rosaceae is divided into six subfamilies:

Subfamily I. *Chrysobalanoideae.* Thalamus deeply cup shaped. Flowers irregular. Stamens sometimes monadelphous. Carpel 1. Examples: *Acioa, Chrysobalanus, Hirtella,* etc.

Subfamily II. *Neuradoideae.* Thalamus cup shaped. Carpels 5–10, united. Examples: *Grielum* and *Neurada.*

Subfamily III. *Pomoideae.* Thalamus deeply concave. Stamens many. Carpels 5, somewhat united and adnate to thalamus; ovules 2 in each carpel. Fruit a pseudocarp or pome. Examples: *Cotoneaster, Pyrus,* etc.

Subfamily IV. *Prunoideae.* Thalamus cup shaped. Stamens 10, 20 or more. Carpel 1, containing 2 ovules. Fruit a *drupe.* Example: *Prunus.*

Subfamily V. *Rosoideae.* Thalamus concave. Carpels many and free; ovules 2 in each carpel. Examples: *Fragaria, Potentilla, Rosa, Rubus,* etc.

Subfamily VI. *Spiraeoideae.* Thalamus somewhat flat, but never deep or convex. Stamens 10 to many. Carpels 5, free, located in centre. Examples: *Spiraea* and *Quillaja.*

POSITION AND AFFINITY OF THE FAMILY

The Rosaceae comes under the Rosales, although many have included the Leguminosae under the same order and some have put them in separate orders.

The members of the Rosaceae are very much like those of the Ranunculaceae in some respects. The number of floral parts is similar in the two families, but in the Rosaceae the thalamus, instead of taking the form of a cone, is generally cup shaped with the carpels either standing up or embedded in the bottom of the cup.

The Rosaceae is allied to the Leguminosae as regards the perigynous flowers and monocarpellary pistil (in the Chrysobalanoideae and Prunoideae). A connection is also established with the Legume family by the genus *Acioa* where the flowers are zygomorphic with monadelphous stamens and a single carpel.

NUMBER IN AND DISTRIBUTION OF THE FAMILY

The Rosaceae covers 115 genera and 3,200 species. The plants belonging to this family are widely distributed over most of the earth and are abundant in Europe, North America and eastern Asia.

COMMONLY OCCURRING PLANTS OF THE FAMILY

Indian Strawberry (*Fragaria nilgerrensis* Schlecht. ex J. Gay) is a trailing herb with globose and pale pink fruits.

Potentilla supina L. is an annual herb of Darjeeling, Shillong and other hill stations.

Prunus cerasoides D. Don is found in the hills.

Pygeum acuminatum Colebr. is a tree with two seeded fruits.

Roses (*Rosa alba* L., *R. centifolia* L., *R. damascena* Mill., *R. indica* L., *R. involucrata* Roxb.) are armed shrubs.

Bramble (*Rubus hexagynus* Roxb.) is a shrubby plant.

ECONOMIC ASPECTS OF THE FAMILY

The Rosaceae is economically important for edible seeds, such as Almond (*Prunus amygdalus* var. *dulcis*) and edible fruits, such as Coco plum (*Chrysobalanus icaco*), Quince (*Cydonia vulgaris*), Loquat (*Eriobotrya japonica*), Strawberry (*Fragaria chiloensis, F. vesca. F. virginiana*), Apple (*Malus pumila*), Medlar (*Mespilus germanica*), Apricot (*Prunus armeniaca*), Cherry (*P. cerasus*), Plum (*P. domestica*), Peach (*P. persica*), Pear (*Pyrus communis*), Blackberry (*Rubus fruticosus*), Raspberry (*R. idaeus*), etc. *Licania rigida* (Brazil) yields 'oiticica oil' of commerce. The bark of *Moquila utilis* (Brazil) is used in making heat resistant pots. The dried inner bark of *Quillaja saponaria* (Andes) forms the commercial 'soapbark'. *Sorbus aucuparia* yields timber. There are also a number of ornamentals, *e.g.*, *Cotoneaster, Crataegus, Geum, Kerria, Potentilla, Rosa* and *Spiraea*

Leguminosae
(Legume Family)

GENERAL FEATURES OF THE FAMILY

Habit	— Herbs, shrubs or trees, often climbing.
Leaves	— *Alternate, compound, stipulate.*
Flowers	— Regular or irregular, slightly perigynous.
Calyx	— Sepals 4 or 5, free or united.
Corolla	— Petals 4 or 5, *with standard, wings and keel,* free or united.
Androecium	— *Stamens 10 or many,* sometimes less, free or united into a tube.
Gynoecium	— Carpel 1; *ovary superior,* unilocular, ovules 1 to many style long.
Fruit	— *Legume or pod.*
Seeds	— Exalbuminous; embryo large; cotyledons fleshy or leafy.

FLORAL RANGE IN THE FAMILY

The Leguiminosae shows a great deal of variation in the arrangement of floral parts. This fact holds good particularly for the characters of the corolla and androecium.

In the Papilionoideae, the flowers are irregular. They are also papilionaceous. Here the five unequal petals resemble a butterfly. The posterior petal becomes the largest and is called *standard* or *vexilium;* the anterior pair is somewhat united, forming a boat shaped structure termed *carina* or *keel;* the lateral pair constitutes the *alae* or *wings.*

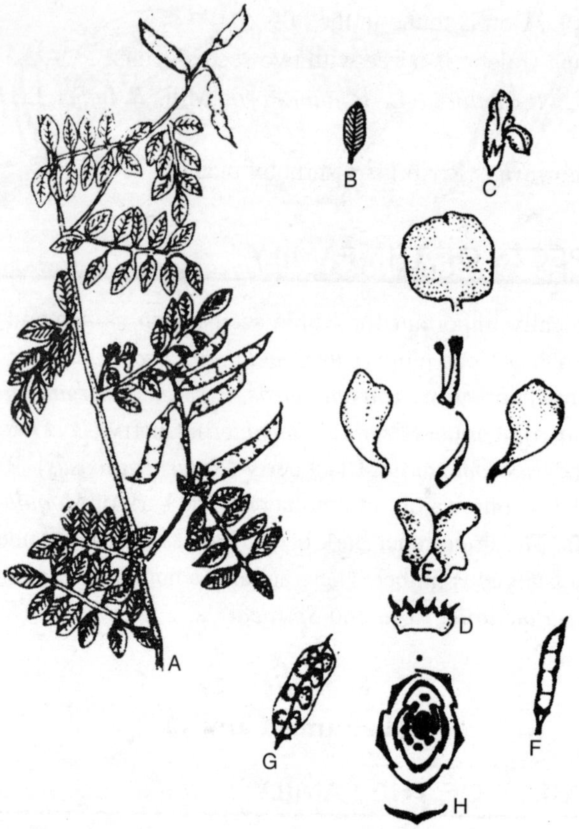

Fig. 31 LEGUMINOSAE. *Tephrosia hamiltonii* J.R. Drumm. **A**, portion of shoot; **B**, leaflet: **C**, flower; **D**, calyx; **E**, parts of corolla; **F**, fruit; **G**, L. S. of fruit; **H**, floral diagram

In general, the stamens are diadelphous in the Papilionoideae. Sometimes, the monadelphous condition is found in *Arachis* and *Crotalaria.* The stamens may be in two bundles of five each, as in *Aeschynomene.* The stamens may also be free, *e.g.*, *Ormosia* and *Sophora.* The posterior stamen is lost in *Chorizema.*

In the Caesalpinioideae, the flowers are tetramerous or pentamerous being medianly zygomorphic. The two upper sepals are connate in *Tamarindus.* In others, the sepals are free and provided with imbricate aestivation. But *Cercis* is characterised by a valvate aestivation. The petals are subequal, as in *Cassia.* The anterior pair of petals becomes reduced in size in *Amherstia* and obliterated in *Tamarindus.* In *Krameria,* however, the petals are represented by glandular scales. The petals are not seen in *Ceratonia* and *Copaifera.*

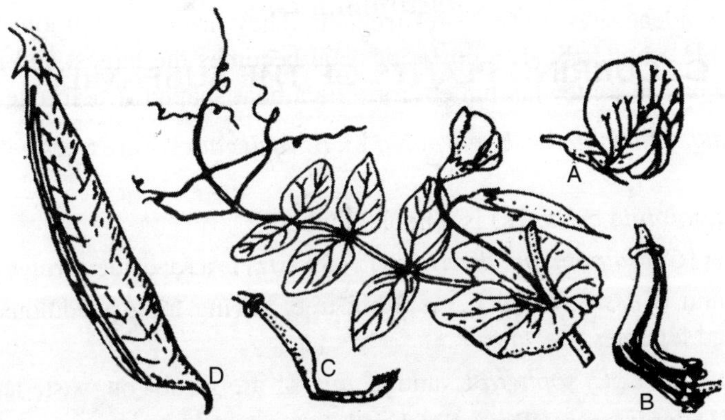

Fig. 32 LEGUMINOSAE. *Pisum sativum* L. **A**, flower; **B**, flower with perianth removed; **C**, ovary; **D**, fruit. (**A–D**, after Baillon)

In the Mimosoideae, the flowers are usually pentamerous. But trimerous, tetramerous and hexamerous conditions are also noticeable. Although valvate aestivation is the general rule, the only exception occurs in *Parkia.* The stamens are many and free in *Acacia,* but many and somewhat united into a bundle in *Inga.*

Generally, the carpel is single. Two carpels have been noted in *Affonsea* and *Tounatea.* Carpels are 10–15 in *Archtdendron.* The styles are rolled and straight, as in *Phaseolus.* Elsewhere, the styles are bending and terminal. The ovules are ascending or pendulous with one or two integuments. The ovules may also be amphitropous or anatropous or campylotropous.

DIVISIONS OF THE FAMILY

The Leguninosae is divided into three subfamilies:

Subfamily I. *Caesalpinioideae.* Inflorescence a raceme. Flowers siightly irregular. Petals imbricate, posterior one innermost and smallest. Stamens 10 or less, some abortive, free. Examples: *Bauhinia, Caesalpinia,* etc.

Subfamily II. *Mimosoideae.* Inflorescence a head or spike. Petals valvate. Stamens 4 to many, free or united. Examples: *Acacia, Mimosa,* etc.

Subfamily III. *Papilionoideae.* Inflorescence a raceme. Petals vexillary, posterior one outermost and biggest. Stamens usually 10, often monadelphous or diadelphous. Examples: *Crotalaria, Desmodium,* etc.

NUMBER IN AND DISTRIBUTION OF THE FAMILY

The Leguminosae includes about 700 genera and 17,000 species and ranks as the second or third largest family of flowering plants, being exceeded in size only by the Compositae and perhaps by the Orchidaceae. Its members form a part of every major vegetation type in the world, although they seem to be most important in semi-arid regions.

Caesalpinioideae

COMMONLY OCCURRING PLANTS OF THE SUBFAMILY

Camel's Foot (*Bauhinia acuminata* L., *B. purpurea* L., *B. variegata* L.) are trees with deeply emarginate leaves.

Fever Nut (*Caesalpinia crista* L.) is a hook climber.

Peacock Flower [*Caesalpinia pulcherrima* (L.) Swartz] is a roadside shrub with crimson flowers.

Indian Laburnum (*Cassia fistula* L.) is a big tree, having long pendulous racemes of yellow flowers and rod shaped legumes.

Cassia occidentalis, L. *C. sophera* L. and *C. tora* L. are shrubs on waste lands.

Gold Mohur [*Delonix regia* (Boj.) Raf.] and *Peltophorum pterocarpum* (DC.) Backer ex K. Heyne are trees, the former with deep crimson to salmon flowers and the latter with yellow flowers.

Kingiodendron pinnatum (Roxb.) Harms is a small tree.

Parkinsonia aculeata L. is a large shrub.

ECONOMIC ASPECTS OF THE SUBFAMILY

The wood of *Acrocarpus fraxinifolius* is used for furniture, boards, shingles and tea boxes. From the trunks of *Bauhinia purpurea,* we get the gum called 'semi-gond'. The bark of *B. vahlii* yields the fibres termed 'selu' which is utilised for making ropes. The pods of *Caesalpinia coriaria, C. digyna* and *C. sappan* are used in the tanning of leather. Valuable resins are obtained from *Copaifera mopane* (Africa), *Hymenaea courbaril* (West Indies) and *Trachylobium verrucosum* (Zanzibar). The microscopic stain, 'haematoxylin', comes from the wood of *Haematoxylon campechianum.* The gum obtained from *Kingiodendron pinnatum* is valued as a wood preservative.

Cercis angustifolia (Arabia) yields a cathartic. The infusion of the leaves of *Cassia auriculata* is slightly aromatic and finds use as a substitute for tea. The leaves of *C. acutifolia, C. angustifolia* and *C. obovata,* when dried, form the drug 'senna'. The bark decoction of *Saraca indica* is used as a remedy for defective menstruation. *Tamarindus indica* is used for culinary purposes.

Mimosoideae

COMMONLY OCCURRING PLANTS OF THE SUBFAMILY

Acacia auriculaeformis A. Cunn. ex Benth. is a tree which bears phyllodes.

Bead Tree or Coral Pea (*Adenanthera pavonina* L.) is a large deciduous unarmed tree.

Silk Flower [*Albizia lebbeck* (L.) Benth.] is a road side tree.

Entada phaseoloides (L.) Merr. is a woody climber, having snake-like branches.

Sensitive Plant (*Mimosa pudica* L.) is a straggling herb.

Neptunia plena (L.) Benth. and *N. prostrata* (Lamk.) Baill. are aquatic weeds with sensitive leaves.

Manila Tamarind [*Pithecellobium dulce* (Roxb.) Benth.] is a medium sized tree.

Prosopis chilennis (Molina) Stuntz is a desert shrub.

Rain Tree [*Samanea saman* (Jacq.) Merr.] is a tree, being planted on road sides.

ECONOMIC ASPECTS OF THE FAMILY

A number of trees furnish valuable timber, *e.g.*, *Acacia melanoxylon* (Australia), *A. visco, Albizia lebbeck, A. procera, Enterolobium cyclocarpum* (tropical America) and *Xylia xylocarpa* (Burma). *Acacia armata* and *A. karroo* form good sand binders or hedges. The seeds of *Adenanthera pavonina* are employed as beads for rosaries; the powdered wood of this tree is made into a paste and used by Brahmins for marking their foreheads. *Albizia moluccana* and *A. chinensis* are used as a shade for Tea in India, while *Inga edulis* is a shade for Coffee in Colombia and Central America. The seeds of *Entada phaseoloides* are utilised for curling garments. The seeds of *Prosopis juliflora* are ground up and utilised as an animal feed.

From the wood of *Acacia catechu* are obtained three articles, namely, 'cutch', 'kath' and 'kheersal'. *A. decurrens* (Australia) and *A. dealbata* (Australia) are useful in tanning. The flowers of *A. farnesiana* on distillation, yield the well known 'Cassie perfume'. Strong fibres, extracted from the stems of *A. leucophloea* and *A. pennata,* are used for making ropes and fishing nets. The trunks of *A. nilotica, A. scorpioides* and *A. senegal* yield 'gum acacia', 'gum ghati' and 'gum arabic' respectively. The pods of *A. rugata* are in demand as a detergent for washing silk and woolen fabrics.

Papilionoideae

COMMONLY OCCURRING PLANTS O[illegible]UBFAMILY

Crab's eye or Indian Liquorice (*Abrus precatorius* L.) is [illegible]ial wiry climber with pink flowers and usually scarlet seeds having a black end.

Flame-of-the-forest [*Butea monosperma* (Lamk.) Taub.], Butterfly Pea or Mussel shell Creeper (*Clitoria ternatea* L.), Rattlewort (*Crotolaria sericea* L.) and Coral Tree [*Erythrina variegata* L. var. *orientalis* (L.) Merr.] are plants with very showy flowers.

Butea parviflora Roxb. is a woody climber.

Sword Bean [*Canavalia gladiata* (Jacq.) DC.] is a large climber.

Crotalaria retusa L. is a garden plant.

Dalhousiea bracteata Grah. ex Benth. is a scandent shrub.

Telegraph Plant [*Desmodium gangeticum* (L.) DC.] is a weed of waste places and is noted for the variation movements of two lateral leaflets of the trifoliate leaves.

Melilotus alba Medik. and *M. indica* (L.) All are field weeds in the winter season.

Cowage (*Mucuna prurita* Hook.) is a climber whose pods are covered with stinging hairs.

Ormosia robusta Wt. is an erect tree, having imparipinnate leaves.

Pongamia pinnata (L.) Pierre is a tree, being planted on avenues.

Psophocarpus tetragonolobus (L.) DC. is a slender climber.

Psoralea corylifolia L. is an erect annual.

Pueraria tuberosa (Roxb. ex Willd). DC. is a shrubby climber, having tuberous roots.

Smithia abyssinica (A. Rich) Verdc. is a common herb (of Pareshnath Hills, Bihar) where the leaves are sensitive.

Hoary Pea or Wild Indigo [*Tephrosia purpurea* (L.) Pers.] is found on road sides.

Teramnus labialis Spreng. is a slender twiner.

Uraria rufescens (DC.) Schindl. occurs in dry forest undergrowth.

ECONOMIC ASPECTS OF THE SUBFAMILY

The species of *Astragalus, Glycyrrhiza, Medicago, Melilotus, Psoralea* and others are useful in medicine. *Lupinus alba, Sesbania cannabina* and others form natural fertiliser. The species of *Onobrychis, Trifolium* and many others are fodder plants. Ornamentals include the species of *Baptisia, Cytisus, Genistea, Wisteria,* etc.

Arachis hypogaea is cultivated for the fruits, the seeds of which are eaten and also yield oil. The pulses, such as Pigeon Pea (*Cajanus cajan*), Gram (*Cicer arietinum*), Lentil (*Lens culinaris*), Green Gram (*Phaseolus aureus*), Black Gram (*P. mungo*), Pea (*Pisum arvense, P. sativum*), *Lathyrus sativus,* Soybean (*Glycine max*), etc. are important food grains. The pods of Horse Gram (*Dolichos biflorus*), Black Bean (*D. lablab*), Lima Bean (*Phaseolus lunatus*), Scarlet Runner (*P. coccineus*), Broad Bean (*Vicia faba*), and Asparagus Bean (*Vigna sinensis*) as well as the flowers of *Sesbania grandiflora* are common table vegetables. *Pachyrhizus erosus* and *Psoralea esculenta* have edible tuberous roots. The seeds of *Trigonella corniculata* are used as a flavouring agent.

The seeds of *Abrus precatorius* are used by goldsmiths as small weights called 'ratty'. The pith-like stems of *Aeschynomene aspera* and *A. indica* are utilised for making shola hats, fishing floats, nets, baskets and rafts. Many trees yield timber, e.g. *Baphia nitida* (West Africa), *Castanospermum australe* (Australia), *Dalbergia latifolia, D. sissoo, Erythrina caffra* (South Africa), *Pterocarpus indicus* and others. The gum exuding from the bark of *Butea monosperma* is the so called 'Bengal kino'. The main produce of *Crotalaria burhia* and *C. juncea* is their fibres, used for making cordage, coarse gunny cloth and canvas. The bark of *Erythrina suberosa* is suitable for cork plugs and insulation boards. The powdered wood of *E. variegata* is made into face powder. Dyes like 'indigo' and 'santalin' are obtained from *Indigofera tinctoria* and *Pterocarpus santalinus* respectively. The leaves, flowers and twigs of *Genistea tinctoria* are the source of a yellow dye used for colouring fabrics. *Myroxylon balsamum* var. *balsamum* (Venezuela) and *M. pereirae* var. *pereirae* (Peru) produce balsam. *Pterocarpus marsupium* is the source of the red gum 'kino'; *P. officinalis* (Central America) yields a red resin.

Order 19. *Parietales*

Usually regarded as an artificial assemblage, the Parietales is a large taxon. Here the families are variously redistributed by systematists: 31 by Engler and Diels, 22 by Lawrence and Core and 10 by Rendle. The presence of parietal placentation is almost a constant character.

The present author considers 12 families which may be characterised as follows:

(1) Endosperm starchy	(2)
(1) Endosperm oily	(4)
(2) Leaves opposite or whorled	*Elatinaceae*
(2) Leaves alternate	(3)
(3) Leaves scale-like; sap watery	*Tamaricaceae*
(3) Leaves not scale-like; sap coloured	*Bixaceae*
(4) Leaves opposite or whorled	*Guttiferae*
(4) Leaves alternate	(5)
(5) Petals somewhat united	(6)

(5) Petals free	(8)
(6) Fruits winged	*Dipterocarpaceae*
(6) Fruits not winged	(7)
(7) Plants with latex	*Caricaceae*
(7) Plants without latex	*Theaceae*
(8) Pistil apocarpous	*Dilleniaceae*
(8) Pistil syncarpous	(9)
(9) Flowers with a gynophore	*Passifloraceae*
(9) Flowers without a gynophore	(10)
(10) Sap coloured	*Cochlospermaceae*
(10) Sap watery	(11)
(11) Flowers often imperfect	*Flacourtiaceae*
(11) Flowers perfect	(12)
(12) Placentation parietal	*Violaceae*
(12) Placentation axile	*Theaceae*

Bixaceae
(Bixa Family)

The family is recognised by the reddish sap, palminerved leaves, horseshoe shaped anthers and seeds with a red fleshy testa.

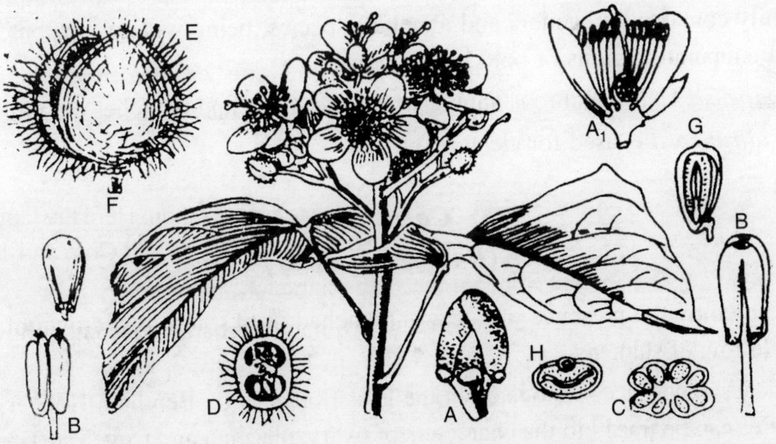

Fig. 33 BIXACEAE. *Bixa orellana* L. **A**, flower bud and glands; **A_1** V.S.of flower; **B**. anthers; **C**, T. S. of anther; **D**, T.S. of ovary; **E**, fruit; **F**, seed; **G**, L. S. of seed; **H**, T.S. of seed. (**A–H**, after Le Maout & Decaisne)

A family of one genus (*Bixa*) and one or four species, the Bixaceae is found in South America. An orange dye, 'annatto', produced from the seed testa of *Bixa orellana,* is used to colour butter, cheese and other foods.

Tamaricaceae
(Tamarisk Family)

The family is characterised by the sinuous branchlets, scale-like leaves and hairy seeds.

The family is divided into two tribes:

Tribe (*i*) *Reaumurieae.* Flowers solitary. Seeds with endosperm. Examples: *Hololachne* and *Reaumuria.*

Tribe (*ii*) *Tamarisceae.* Flowers borne in spike-like or racemose inflorescence. Seeds without endosperm. Examples: *Myricaria* and *Tamarix.*

There are four genera and about 120 species in the Tamaricaceae; these are desert, steppe and sea shore plants of the Mediterranean area and Central Asia.

Tamarix dioica Roxb. and *T. troupii* Hole are found in the islands and sandy banks of Hooghly, West Bengal.

Several species of *Tamarix* are important as ornamentals due to their feathery foliage, catkin-like inflorescence and pink flowers. *T. aphylla, T. dioica* and *T. troupii* are employed as fireweed; the galls borne on these trees are rich in tannin and are used for dyeing. The wood of *T. articulata* (North Africa) is used for house construction. *T. mannifera* (Egypt to Afghanistan) yields 'manna'.

Elatinaceae
(Waterwort Family)

The family is characterised by the aquatic habit, opposite or whorled leaves, paired stipules, axillary cymose trimerous to tetramerous flowers and axile placentation.

The family contains two genera and about 33 species, being widely distributed throughout the subtropical and temperate regions of both hemispheres.

Bergia capensis L. is a glabrous annual where the stems and branches are pink and translucent.

Bergia suffruticosa is used for cleaning teeth.

Caricaceae
(Papaya Family)

The family is known by the presence of latex, unbranched trunks, terminal crown of leaves, union of petals and double row of stamens.

The Caricaceae was accommodated in the Passifloraceae by Bentham-Hooker; the relationship of the two families can be traced to the characters of ovary, placenta and fruit. Van Tieghem considered the Caricaceae as related to the Cucurbitaceae due to the ovular structure which has two thick integuments and a long persistent nucellus, Hutchinson placed the Caricaceae in the Cucurbitales.

The family comprises four genera and 30 species which abound in tropical America.

Papaw (*Carica papaya*) is well known for the large edible melon-like fruits—"the melon that grows on a tree"; the latex of the unripe fruits is used to make meat tender when cooking; a medicinal compound, 'papain' is obtained from the fruits. *C. candicans* (Peru), *C. chrysophila* (Colombia) and *C. pentagyna* (Ecuador) are cultivated for their edible pericarp or sweet juicy seed envelopes.

C. cundinamarcensis is raised successfully at higher altitudes in the tropics where *C. papaya* would fail. The fruits of *Jacaratia mexicana* and *Jarilla caudata* are eaten in the Andes.

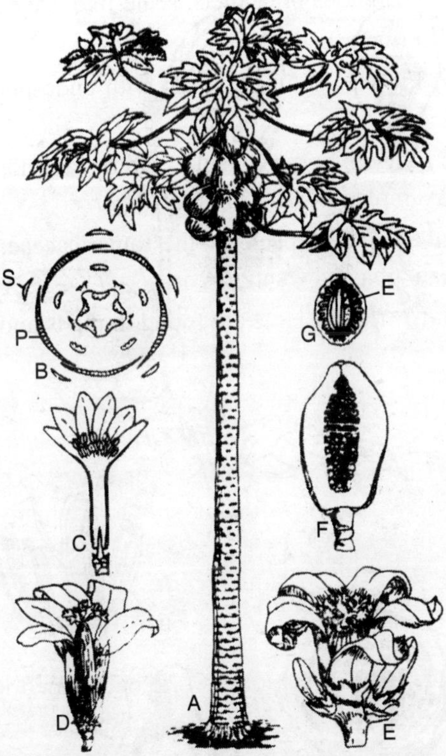

Fig. 34 CARICACEAE. *Carica papaya* **L. A**, female tree bearing fruits; **B**, diagram of bisexual flower (*s*, sepal; *p*, petal; *st*, stamen; *o*, ovary); **C**, bisexual flower cut open; **D** & **E**, female flower; **F**, young fruit cut open; **G**, L. S. of seed (*e*, endosperm). (**A**, **B** & **F**, after Martius; **C**, **D** & **E**, after Kohier; **G**, after Engler)

In the Theaceae, we find exstipulate leaves, spiral arrangement of perianth segments, several whorls of numerous stamens and axile placentation.

The Theaceae is divided into two tribes:

Tribe (*i*) *Camellieae.* Anthers mostly versatile. Fruit a loculicidal capsule, rarely a drupe. Embryo straight or radicle bent round. Example: *Camellia.*

Tribe (*ii*) *Ternstroemieae.* Anthers mostly basifixed. Fruit a berry or dry and indehiscent. Embryo horseshoe shaped, rarely almost straight. Example: *Eurya.*

The Theaceae embraces about 29 genera and 1,100 species. The plants of this family are distributed in the tropical and subtropical parts of the world.

Wild Tea (*Camellia kissi* Wall.) is found in Assam, Khasi Hills and Eastern Himalaya.

The family provides a good number of ornamentals, *e.g.*, *Camellia, Cleyera, Eurya, Franklinia, Gordonia* and *Stewartia. Camellia sasanqua* is the source of 'tea seed oil', a non-drying oil. *C. sinensis* is the Tea plant of commerce. *Gordonia lasianthus* produces tannin. *Haploclathra paniculata* yields valuable timber.

Theaceae
(Tea Family)

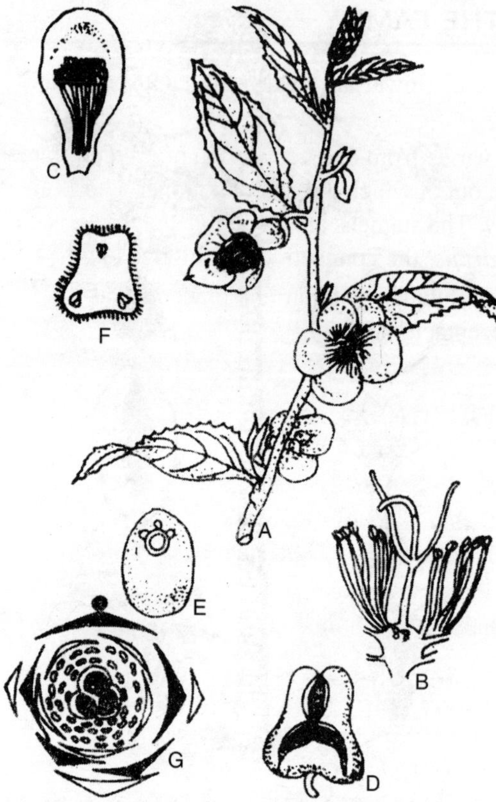

Fig. 35 THEACEAE. *Camellia sinensis* (L.) O. Kuntze. **A**, portion of shoot; **B**, V, S. of flower; **C**, petal with numerous stamens; **D**, fruit ;**E**, seed; **F**, T.S. of fruit; **G**, floral diagram. (**A–E**, after Melchior; **G**, Baillon)

Dipterocarpaceae
(Dipterocarp Family)

GENERAL FEATURES OF THE FAMILY

Habit	—	Gigantic trees, with an *abundant resin.*
Leaves	—	Alternate, simple, entire, *stipulate, pubescent.*
Inflorescence	—	Racemes.
Flowers	—	Bisexual, regular, hypogynous.
Calyx	—	Sepals 5, united, *persistent.*
Androecium	—	Stamens 5 or more, *free or slightly polyadelphous;* connective produced.
Gynoecium	—	Carpels 3, united; *ovary superior; trilocular with 2 pendulous ovules in each loculus.*

Fruit — *Samara.*
Seeds — *Exalbuminous.*

FLORAL RANGE IN THE FAMILY

Deviations from the normal type of flower are few in the Dipterocarpaceae. In general, the calyx and corolla are pentamerous.

The number of stamens varies from one genus to another. The stamens are 5 in *Monoporandra.* They may be 15 in three whorls, out of which the outer and inner whorls are antisepalous and the middle antipetalous. e.g., *Hopea pierrei.* The stamens are also monadelphous, as in *H. recopei.* Except in a few species of *Shore* and *Stemonocarpus,* the connective is prolonged at the tip of anther.

Sometimes the ovary is unilocular with parietal placentae, as in *Marquesia.* It may be adnate to the calyx tube or sunk in the receptacle (*Vatica*).

DIVISIONS OF THE FAMILY

The Dipterocarpaceae is divided into two subfamilies:

Subfamily I. *Dipterocarpoideae.* Androecium without a gynophore; anthers basifixed. Wood with intercellular resin canals and multiseriate rays. Example: *Dipterocarpus.*

Subfamily II. *Monotoideae.* Androecium with a gynophore; anthers dorsifixed. Wood without resin canals and with uniseriate rays. Example: *Monotes.*

POSITION AND AFFINITY OF THE FAMILY

The Dipterocarpaceae is a family under the Guttiferales of some authors and the Parietales of others. Bentham-Hooker included the genera *Ancistrocladus* and *Luphira* in this family, but Gilg transferred them to the Ancistrocladaceae and Ochnaceae respectively. Hence, the Dipterocarpaceae has relationship with these families.

The Dipterocarpaceae agrees with the Magnoliaceae in the possession of large stipules. While the calyx persists in the Dipterocarpaceae, it is definitely not so in the Magnoliaceae. The Dipterocarpaceae resembles the Guttiferae by virtue of the resinous content, but differs from it in having stipules and alternate leaves. The Dipterocarpaceae stands very close to the Theaceae; whereas resin canals are present in some members of the former, they are altogether absent in the latter.

NUMBER IN AND DISTRIBUTION OF THE FAMILY

The Dipterocarpaceae consists of about 15 genera and 580 species. This is chiefly a tropical family, extending from India to New Guinea and the Philippines.

COMMONLY OCCURRING PLANTS OF THE FAMILY

Anisoptera scaphula (Roxb.) Pierre, Garjan balsam (*Dipterocarpus turbinatus* Gaertn. f.), *Isauxis lanceaefolia* King and Sal Tree (*Shorea robusta* Gaertn. f.) are huge trees.

ECONOMIC ASPECTS OF THE FAMILY

The plants of the Dipterocarpaceae are of great value for resins and timber, while the seeds of some species are meant for human consumption. '*Damar penak*" is derived from *Balanocarpus heimii* and *Shorea hypochra.* 'Eng oil' is the product of *Dipterocarpus tuberculatus*. The stems of *D. turbinatus* are exploited for 'garjan oil', a balsam or resin, which is useful in making varnishes and lithographic inks; the timber of this tree is used for boat building and structural work. *Drybalanops aromatica* (East Indies) is the source of 'Borneo camphor' or 'Sumatra camphor'. *Hopea micrantha* produces the valuable dammar called 'mata kuching'. *H. parviflora* yields excellent timber which is particularly useful in pile and bridge construction. A dark strong smelling thick balsam termed 'chooa' is obtained by distilling the resinous juice which exudes from the bark wound of *Isauxis lanceaefolia* (Bangladesh). *Shorea aptera* furnishes fats and oils. *S. koodarsii* provides a dammar resin, used as an incense. *S. robusta* is a timber tree and yields from the stem a resin known as 'dhoona' which is burnt as an incense; the seeds produce 'sal butter' which is used as an adulterant for ghee and instead of cocoa butter in the manufacture of chocolates. The wood of *Vateria indica* is used as plank timber; oleoresin obtained from the trunk is 'piney varnish' when soft and 'Indian copal' or 'Copal varnish' when hard, being used for varnishing pictures and anatomical preparations, for making candles and ornaments; the seeds produce vegetable butter, 'Malabar tallow' or'Vateria fat' or 'piney tallow', which finds use as a luminant, in confectionary and as a substitute for ghee; the bark is in demand in the manufacture of arak and jaggery. The starchy seeds of *Doona*, *Vatica* and *Vateria acuminata* are edible.

Dilleniaceae
(Dillenia Family)

The presence of leaves with parallel lateral nerves, persistent calyx, numerous stamens in bundles, gynoecium of separate carpels, parietal placentation, seeds with a crested or laciniate aril and minute straight embryo are characteristic features of the Dilleniaceae.

The family is made up of 18 genera and 530 species of tropical regions, reaching the highest development in America and Australasia.

The fruits of *Actnidia chinensis* (eastern Asia) are edible. The leaves of *Curatella americana* (Central America) are sometimes used as sand paper. The fleshy sepals enclosing the fruits of *Dillenia indica* and *D. pentagyna* are used as vegetables. The species of *Dillenia, Hibbertia* and *Tetracera* are important as ornamentals.

Passifloraceae
(Passion-flower Family)

GENERAL FEATURES OF THE FAMILY

Habit	—	Herbs or shrubs, *climbing by tendrils.*
Leaves	—	Alternate, simple, lobed, stipulate, *often with glands on petiole.*
Flowers	—	Bisexual or unisexual, regular; receptacle large, like a basin or cup and *bears variously modified corona.*
Calyx	—	Sepals 5, imbricate.
Corolla	—	Petals 5, sometimes absent, free, imbricate.
Androecium	—	Stamens 5, elevated by a gynandrophore; anthers 2-celled.

Gynoecium — Carpels 3–5; style free or united; *ovary superior, unilocular;* ovules several to many, anatropous; *placentation parietal;* stigma capitate.

Fruit — Berry or capsule.

Seeds — Many, *often with a mucilaginously pulpy aril;* embryo with a fleshy endosperm.

NUMBER IN AND DISTRIBUTION OF THE FAMILY

The Passifloraceae covers 20 genera and about 600 species. Though the members of the family spread over the warmer parts of the world, the chief centre of distribution lies in tropical America.

COMMONLY OCCURRING PLANTS OF THE FAMILY

Adenia hondala (Gaertn.) de Wild. is found in Western Ghats and other moist hilly tracts in Peninsular India.

Passion-flower (*Passiflora foetida* L., *P. lunata* Juss. ex DC., *P. suberosa* L.) is a showy garden climber.

ECONOMIC ASPECTS OF THE FAMILY

Economically, the Passifloraceae is important for ornamentals and for edible fruits. The latter is produced by *Passiflora edulis, P. laurifolia, P. ligularis, P. mollissima* and *P. quadrangularis.*

Cochlospermaceae (Cochlospermum family)

The family is akin to the Bixaceae, but differs from it in having palmately lobed leaves, three- to five-valved fruits and an oily endosperm.

The family represents two genera and about 38 species of tropical regions.

Cochlospermum religiosum bears large yellow flowers.

Cochlospermum gillivraei yields a useful 'kapok'. *C. religiosum* is the source of 'Kuteera gum'. The bark of *C. vitifolium* is used to make cordage.

Flacourtiaceae (Flacourtia Family)

The family can be diagnosed by the pellucid dotted leaves, polygamous flowers, undifferentiated perianth, enlarged receptacle, variously modified disc and numerous stamens.

The position of the Flacourtiaceae is uncertain. Hallier placed it in the Passionales, Hutchinson in the Bixales and others in the Parietales or Guttiferales.

The family takes into account 89 genera and about 1,250 species of the tropical and subtropical regions.

Flacourtia indica (Burm. f.) Merr. and *F. jangomas* (Lour.) Raeusch. are found throughout the greater part of tropical India.

The fruits of *Aberia caffra* (South Africa) and *Flacourtia indica* as well as *F. ramontchi* (Malagasy) are edible. *Berberidopsis corallina, Carrierea calycina, Idesia polycarpa* and *Oncoba spinosa* are noteworthy ornamentals. The wood of *Casearia elliptica* is used for carving and making combs. *Gossypiospermum praecox* (Venezuela) yields valuable timber. *Hydnocarpus wightiana* and *Taraktogenos kurzii* are the source of 'chalmoogra oil', used in the treatment of leprosy and other cutanous diseases. The seeds of *Pangium edule* (Malaya) are eaten after prolonged boiling in water to eliminate hydrocyanic acid.

Violaceae
(Violet Family)

GENERAL FEATURES OF THE FAMILY

Habit	—	Herbs or shrubs, rarely climbing.
Leaves	—	Alternate, simple, *stipulate.*
Inflorescence	—	Solitary or terminal spikes with bracts and bracteoles.
Flowers	—	Bisexual, regular or irregular.
Calyx	—	Sepals 5, imbricate, persistent.
Corolla	—	Petals 5, anterior one spurred; imbricate or twisted.
Androecium	—	Stamens 5; filaments short; *connective of anther dilated or prolonged, connivent over pistil.*
Gynoecium	—	*Carpels 3; ovary superior unilocular;* ovules 1 or more, campylotropous; *placentation parietal; stigma curved or s-shaped.*
Fruit	—	Capsule.
Seeds	—	Subglobose or ovoid, albuminous; embryo straight.

FLORAL RANGE IN THE FAMILY

The flowers of the Violaceae are almost regular, as in *Allexis, Gloeospermum* and *Rinorea.* They may be slightly irregular, *e.g., Amphirrhox* or highly irregular, *e.g., Hybanthus* and *Viola.*

The members of each calyx are generally free, but somewhat united in *Leonia* and *Melicytus.* The sepals are also uniform in size and green in colour. They may be appendiculate, as in *Viola.* The petals are also large and coloured.

The stamens and carpels vary in number and arrangement. The stamens may be monadelphous as in *Agatea* or diadelphous as in *Anchietea.* The carpels are 4 or 5 (*Leonia*) and 2 or 5 (*Melicytus*). In *Viola,* the connective of the anterior pair of anthers is spurred or it may be appendaged.

DIVISIONS OF THE FAMILY

The Violaceae is divided into two subfamilies:

Subfamily I. *Leonioideae.* Stamens united into a column. Fruit nut-like with a woody pericarp. Example: *Leonia.*

Subfamily II. *Violoideae.* Stamens somewhat united. Fruit a capsule. Example: *Viola.*

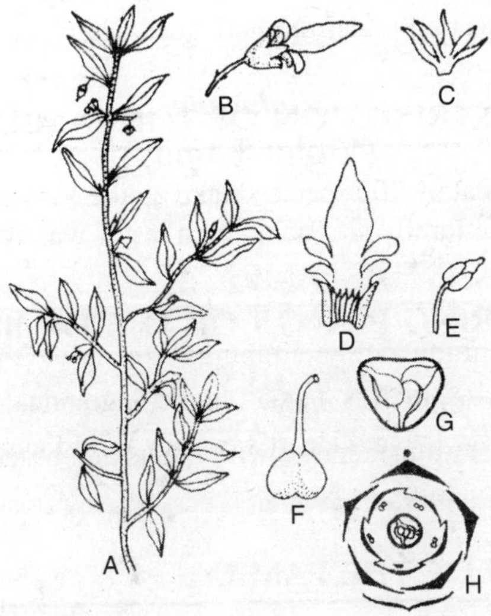

Fig. 36 VIOLACEAE. *Hybanthus enneaspermus* F. v. Muell. **A**, portion of plant; **B**, flower; **C**, sepal; **D**, corolla; **E**, stamen; **F**, carpel; **G**, T. S. of ovary; **H**, floral diagram

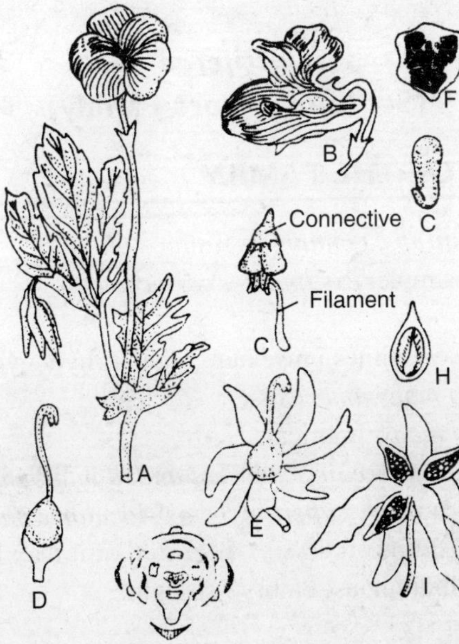

Fig. 37 VIOLACEAE. *Viola tricolor* L. **A**, portion of shoot; **B**, V.S. of flower; **C**, stamen; **D**, young carpel; **E**, mature carpel; **F**, T. S. of ovary; **G**, seed; **H**, L.S. of seed; **I**, dehiscence of fruit; **J**, floral diagram. (**B** & **I**, after Oliver)

POSITION AND AFFINITY OF THE FAMILY

The phylogenetic position of the Violaceae is uncertain. The family indicates a relationship with the Bixaceae and hence brought close to it by Bentham-Hooker. Hallier opined that the Violaceae is akin to the Balsaminaceae, Flacourtiaceae and Turneraceae due to the spurred corolla and thus included it as a primitive member of his Polygalines. Bessey retained it in the Guttales. In the arrangement of Rendle as well as Wettstein, the family appeared under the Parietales. Hutchinson placed it in the Violales.

NUMBER IN AND DISTRIBUTION OF THE FAMILY

Twenty two genera and approximately 900 species make up the Violaceae. Although widely distributed in all continents, the plants of the family are mainly natives of warmer climate.

COMMONLY OCCURRING PLANTS OF THE FAMILY

Hybanthus enneaspermus (L.) F.V. Muell is a small branching annual herb.

Sweet Violet (*Viola odorata* L.) and Heart's ease or Wild Pansy (*V. tricolor* L.) are ornamental plants.

ECONOMIC ASPECTS OF THE FAMILY

The plants of the Violaceae are chiefly important as garden plants. *Anchietea salutaris, Cornostylis hybanthus*, *Hybanthus ipecacuanha, Viola canescens* and *V. cinerea* are of medicinal value. *Leonia glycycarpa* yields edible fruits. A perfume is extracted from the flowers of *V. odorata,* while the leaves of the same plant produce an essential oil which is used for scenting soaps.

Guttiferae
(St. John's-wort Family)

GENERAL FEATURES OF THE FAMILY

Habit — *Herbs or shrubs, containing resin or oil.*
Leaves — *Opposite,* simple, *exstipulate, coriaceous.*
Inflorescence — Cymose.
Flowers — Bisexual, sometimes unisexual, regular, hypogynous.
Calyx — *Sepals 4 to many, imbricate.*
Corolla — *Petals 4 to many,* imbricate.
Androecium — *Stamens many, free or somewhat united into bundles.*
Gynoecium — Carpels 3–5; *ovary superior, 3- to 5-locular with 1 to many ovules in each cell.*
Fruit — Capsule or drupe.
Seeds — Large, exalbuminous, embryo straight.

FLORAL RANGE IN THE FAMILY

The flowers of the Guttiferae show considerable variation in the number and arrangement. Frequently, bracteoles develop beneath the calyx or corolla to make a sharp distinction between the sepals and petals. In *Tovomita*, the thalamus is concave or convex. A cruciform pattern characterises *Havetiopsis;* here two pairs of sepals are succeeded by two pairs of petals and two pairs of stamens by a whorl of 4 carpels. In other cases, the dimerous calyx and corolla are followed by two or three whorls of stamens and a whorl of carpels or by indefinite spirally arranged stamens. The cruciform arrangement also occurs in some species of *Clusia* with respect to the bracteoles and calyx, but the petals and stamens are all arranged in spirals.

The androecium shows remarkable diversity. The stamens are free or united to various degrees. They may form a cup, a lobed synandrium or are arranged in four or live bundles (*Hypericum*). In *Oedematopus tetrameros,* there arc 8–12 stamens in two or three whorls. Staminodes are found either variously united or converted into secretory organs.

The carpels may be equal in number with the petals or twice or thrice as many as or still fewer in number. They are united into a single whorl. The styles are free or partly connate, sometimes very short The stigmas are generally broad. The ovary is unilocular in *Ascyrum.* The ovules are anatropous or amphitropous and are fixed by a short funiculus to the ventral wall of the carpel or to the inner angle. The ovules may be basal in *Calophyllum, Mammea* and other genera, they have two thick integuments.

DIVISIONS OF THE FAMILY

The Guttiferae is divided into five subfamilies:

Subfamily I. *Calophylloideae.* Leaves opposite. Flowers bisexual or unisexual. Stamens many, free or united at apex or base. Carpels 1–4, each with 1–2 ovules. Fruit indehiscent. Cotyledons thick. Examples: *Calophyllum, Mesua,* etc.

Subfamily II. *Clusioideae.* Leaves opposite. Flowers bisexual or unisexual. Stamens many, free or in groups. Fruit various. Cotyledons small. Examples: *Clusia* and *Garcinia.*

Subfamily III. *Hypericoideae.* Leaves opposite. Flowers bisexual. Stamens usually many, usually in 5 or 3–8 bundles before petals. Styles 3–5, usually free. Fruit a 1- or 3- to 5-locular septicidal capsule or indehiscent. Embryo usually straight with not very thick cotyledons. Examples: *Hypericum,* etc.

Subfamily IV. *Kielmeyeroideae.* Leaves alternate or opposite. Flowers usually bisexual. Stamens many, Carpels 3- to 5-locular. Example: *Kielmeyera.*

Subfamily V. *Moronobeoideae.* Leaves opposite. Flowers bisexual. Stamens many, in five groups or united into a tube. Style long, 5-lobed. Fruit a berry. Seeds without an aril; embryo undifferentiated. Example: *Pentadesma.*

NUMBER IN AND DISTRIBUTION OF THE FAMILY

The Guttiferae, with about 40 genera and 1,000 species, is almost exclusively tropical in distribution.

COMMONLY OCCURRING PLANTS OF THE FAMILY

Alexandrian Laurel or Dilo Tree (*Calophyllum inophyllum* L.) is a handsome evergreen glabrous tree and is planted for its beautiful coriaceous leaves.

Beauty Leaf or Poon Spar Tree or Sirpoon Tree (*Calophyllum elatun* Bedd.) is a very magnificent evergreen tree whose bark is furrowed with longitudinal fissures.

Brindonia Tallow Tree or Mangosteen Oil Tree [*Garcinia indica* (Dupetit-Thours) Choisy] is a tall graceful slender tree with dropping branches.

Ceylon Iron wood Tree (*Mesua ferrea* L.) is an elegant tree, having white fragrant flowers with curled petals and a big patch of yellow coloured anthers in the middle.

ECONOMIC ASPECTS OF THE FAMILY

The tropical species of the Guttiferae are economically important as a source of edible fruits, dyes, timber, oils and resins. The wood of *Calophyllum brasiliense, C. elatum* and *Mammea longifolia* are in great demand for cabinet making and construction work. *Garcinia cowa, G. mangostana, G. pedunculata, G. speciosa, G. xanthochymus, Mammea americana* (tropical America) and *Mesua ferrea* (West Africa) arc cultivated for their edible fruits. A yellow pigment called 'gamboge' is derived from *G. hanburyi* (Thailand) and is used for yellow silk fabrics. From the seeds of *G. indica,* we get the well known 'kokum butter' or 'mangosteen oil' which is used in cosmetics textile and other industries. *G. kola* (West Africa) is the source of 'bitter kola'. *Hypericum connatum, H. laxiusculum* and *H. teretiusculum* are of medicinal value. The thick sap of *Pentadesma butryacea* (West Africa) is a substitute for butter.

Order 20. *Malvales*

The Malvales is a natural group. This is indicated by the presence of stellate and mucilage cells or canals. The flowers are generally bisexual, regular, hypogynous and pentacyclic. The calyx is valvate. The stamens are usually many and united into a column; hence the taxon was named Columniferae by Eichler. The placentation is axile.

Bessey assumed that the Malvales originated from the Ranales via the Rosales. Hallier opined that the Ranales was the direct ancestor of the Malvales. Some botanists also believed that the Guttiferales gave rise to the Malvales.

The order is related to the Guttiferae by the regular hypogynous flowers with pentamerous calyx, corolla and by the androecium which indicates a tendency to connation of parts. The Malvales forges a link with the Euphorbiaceae due to the frequent occurrence of branched multicellular hairs and ovary structure which often separates the one seeded parts in the fruit; an endotropic course of the pollen tube is common to both groups. The Bombacaceae resembles the Paeoniaceae (Ranunculaceae) as far as centrifugal development and trunk-vascular supply of stamens are concerned.

Engler and Diels conceived the Malvales as made up of four suborders and seven families: Elaeocarpineae (Elaeocarpaceae), Chlaeneae (Chalaenaceae), Malvineae (Tiliaceae, Malvaceae, Bombacaceae, Sterculiaceae), Scytopetalineae (Scytopetalaceae). Gunderson (1950) and Benson (1957) followed the Englerian concept with reference to the order. Bessey expanded the order to incorporate into it also the Balanopsidaceae, Ulmaceae, Moraceae and Urticaceae. Bentham-Hooker included the Elaeocarpaceae Tiliaceae and Bombacaceae in the Malvaceae. The Malvales formed Rendle's seventh order in the Dialypetalae after the Guttiferales and before the Tricoccae. Both Lawrence and Core treated the Malvales as containing five families (Elaeocarpaceae, Tiliaceae, Malvaceae, Sterculiaceae, Bombaceae). Hutchinson (1959) split up the order into the Tiliales (which comprises the Dirachmaceae, Scytopetalaceae, Tiliaceae, Sterculiaceae, Peridiscaceae, Bombacaceae) and Malvales (which embraces the Malvaceae). The present author takes into account four families which are characterised as follows:

(1) Sepals united	*Sterculiaceae*
(1) Sepals free	(2)
(2) Stamens free or in separate bundles	*Tiliaceae*
(2) Stamens united into a column	(3)
(3) Pollen grains spiny; carpels 5 to many	*Malvaceae*
(3) Pollen grain smooth; carpels 2–5	*Bombacaceae*

Tiliaceae
(Linden Family)

GENERAL FEATURES OF THE FAMILY

Habit — Shrubs or trees, rarely herbs; inner bark fibrous; juice mucilaginous.

Leaves — Alternate, simple, entire or toothed, stipulate or exstipulate.

Inflorescence — *Cymose.*

Flowers — Bisexual or unisexual, regular.

Calyx — Sepals 5, free or united.

Corolla — Petals 5, free, imbricate.

Androecium — *Stamens many, inserted on a disc, free or polyadelphous; anthers, 2-celled.*

Gynoecium — Carpels 2–5, syncarpous; *ovary superior, 2-locular to l0-locular with 1 or more ovules in each cell.*

Fruit — Capsule or drupe.

Seeds — Albuminous; embryo curved with leafy cotyledons.

FLORAL RANGE IN THE FAMILY

The flowers of the Tiliaceae are bisexual, but unisex ... *ptera* and *Vasivaea.*

Although pentamerous calyx and corolla are ... most of the flowers, *Sparmannia* is characterised by tetramery. The petals may be absent, as in *P* ... and *Solmsia.*

The carpels indicate much variation. In *Belotia,* the placentation is basally axile and distally parietal. The ovary may remain unilocular and the placentation parietal, as in *Goethalsia* and *Mollia.* The ovary can also be initially unilocular with five to six deeply intruded placentae that tardily cohere or unite at the centre.

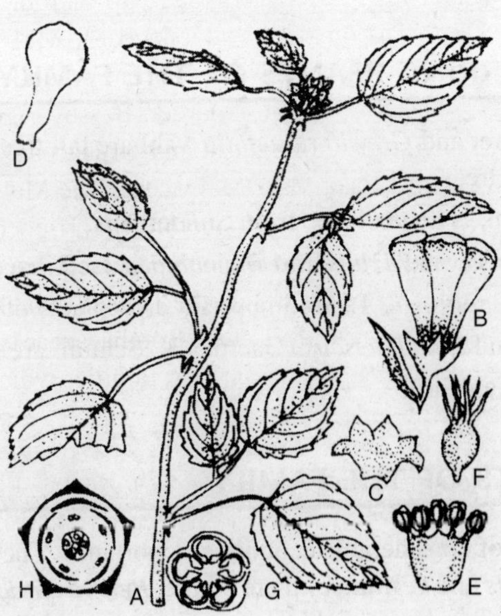

Fig. 38 TILLACEAE. *Corchorus* sp. **A**, portion of shoot; **B**, flower; **C**, Calyx; **D**, petal; **E**, stamen; **F**, carpel; **G**, T.S. of ovary; **H**, floral diagram

DIVISIONS OF THE FAMILY

The Tiliaceae is divided into four tribes:

Tribe (*i*) *Apeibeae.* Sepals free. Anther lobes not confluent. Stamens with apical appendages; androphore absent. Carpels 6 to many. Examples: *Ancistrocarpus, Apeiba,* etc.

Tribe (*ii*) *Brownloweae.* Sepals connate at base. Anther lobes confluent at apex. Examples: *Berrya, Brownlowia,* etc.

Tribe (*iii*) *Greweae.* Petals with basal glands. Androphore present. Examples: *Duboscia, Grewia,* etc.

Tribe (*iv*) *Tilieae.* Sepals connate at base. Stamens without appendages. Carpels 2–5. Example: *Tilia.*

POSITION AND AFFINITY OF THE FAMILY

Rendle included the Tiliaceae in his seventh order Malvales after the Guttiferales and before the Tricoccae. Hutchinson separated the Tiliaceae from the Malvales and placed it under the Tiliales.

The Tiliaceae is similar to the Sterculiaceae in habit, leaves and stipules. It is akin to the Elaeocarpaceae and hence the latter was incorporated by Bentham-Hooker into the Tiliaceae. Engler set the Elaeocarpaceae apart from the Tiliaceae due to the mucilage as well as the presence of imbricate sepals and porous dehiscence of anthers. The Tiliaceae is allied to some members of the Euphorbiaceae which has free stamens inserted upon a gynophore.

NUMBER IN AND DISTRIBUTION OF THE FAMILY

The family is composed of about 41 genera and 400 species. Although mostly found in the tropics, some plants of the Tiliaceae extend even to the temperate regions.

COMMONLY OCCURRING PLANTS OF THE FAMILY

Berrya cordifolia (Willd.) Burret and *Grewia tiliaefolia* Vahl are tall trees, while *G. elastica* Royle and *G. subinequalis* DC. are small trees.

Brownlowia tersa (L.) Kosterm. is a small tree from Sunderbans.

Corchorus aestuans L. non Forsk. and *Triumfetta rhomboidea* Jacq. are weeds of waste places.

Corchorus capsularis L. and C. *olitorius* L. are important crop plants of India.

Elaeocarpus floribundus Bl. and *E. sphaericus* (Gaertn.) K. Schum. are found in North Bengal.

Muntigia calabura L. is a garden shrub.

ECONOMIC ASPECTS OF THE FAMILY

The plants of the Tiliaceae are of considerable economic importance. The berries of *Aristotelia macqui, Grewia elastica, G. subinequalis* and *G. tiliaefolia* are edible. *Berrya cordata, B. cordifolia, Erinocarpus nimmonii, Tilia americana* and *T. vulgaris* yield timber. The leaves of *Corchorus aestuans* are good for stomach disorders. The stem bark of *C. capsularis* and *C.olitorius* produces the 'jute' of commerce. The seeds of *Elaeocarpus sphaericus* are made into beds. *Sparmannia africana* (South Africa) is grown for its handsome clusters of white flowers.

Malvaceae
(Mallow Family)

GENERAL FEATURES OF THE FAMILY

Habit	— Herbs, shrubs or trees; inner bark fibrous; juice mucilaginous.
Leaves	— Alternate, simple, *palminerved, stipulate* (stipules free lateral).
Flowers	— Bisexual, regular, hypogynous, *subtended by a whorl of bracts and bracteoles called epicalyx.*
Calyx	— Sepals 5, united, valvate.
Corolla	— Petals 5, adnate below to a staminal column, twisted.
Androecium	— *Stamens many, united into a bundle, epipetalous; anthers 1-celled, reniform, extrorse; pollen grains large, spherical and spiny.*
Gynoecium	— Carpels 5 to many, syncarpous; *ovary superior multilocular,* with 1 or more ovules in each loculus; placentation axile; *style 1, passing through a hollow staminal column and divided at top into as many branches as there are carpels, each branch ending in a stigma.*
Fruit	— Capsule or schizocarp.
Seeds	— Naked or hairy, round; embryo curved with folded cotyledons.

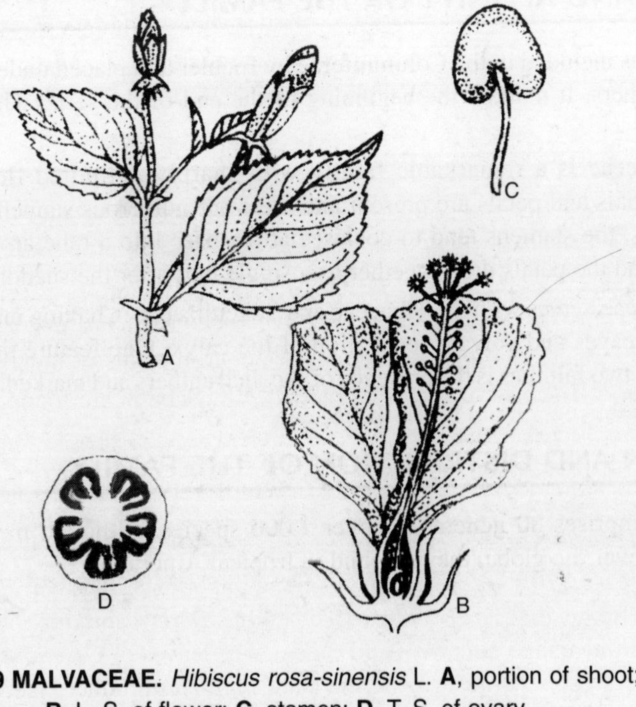

Fig. 39 MALVACEAE. *Hibiscus rosa-sinensis* L. **A**, portion of shoot; **B**, L. S. of flower; **C**, stamen; **D**, T. S. of ovary

FLORAL RANGE IN THE FAMILY

In the Malvaceae, there is nor much deviation in the floral structure from the normal type. The flowers may be solitary and showy or arranged in a racemose manner or more complicated arrangements of the flower occur.

The epicalyx is absent in *Abutilon*. Its number may be large, *e.g.*, *Althaea* or 3, *e.g.*, *Gossypium*. The position of the median carpel may vary. It is either posterior or anterior.

The number of carpels may differ. It may be 1 or 2 in *Plagianthus* and 15–20 in *Abutilon*. Still, sometimes the carpels are indefinite and arranged one above the other in somewhat vertical rows, as in *Malope trifida*.

The ovules are anatropous and fixed to the inner angle of the carpel. Their position may vary from ascending to horizontal to pendulous. This sort of variation may be found in one and the same cell, as in *Abutilon* and allied genera.

The Malvaceae is divided into five tribes:

Tribe (*i*) *Abutileae*. Stigmas apical or nearly so. Examples: *Abutilon, Sida,* etc.

Tribe (*ii*) *Hibisceae*. Carpels persistent. Fruit a capsule. Examples: *Gossypium* and *Hibiscus*.

Tribe (*iii*) *Malopeae*. Carpels in two or more spirally arranged or superposed whorls. Example: *Malope*.

Tribe (*iv*) *Malveae*. Stigmas decurrent on style branches. Examples: *Althaea, Malva,* etc.

Tribe (*v*) *Ureneae*. Stigmas and style branches twice as many as carpels. Example: *Urena,* etc.

POSITION AND AFFINITY OF THE FAMILY

The Malvaceae was included in the Columniferae by Eichler and placed under the Malvales by Rendle, Hutchinson and others. It marked the beginning of the end of the series Thalamiflorae by Bentham-Hooker.

The Malvaceae is a remarkable family with sharply delimited floral characters. Like the Ranunculaceae, sepals and petals are present and there are numerous stamens and carpels. In contrast with the buttercups, the stamens tend to coalesce at the base into a tube around the carpels in such a fashion that they and the petals drop together as a single unit after the shedding of pollen grains.

The Malvaceae agrees with the Tiliaceae and Sterculiaceae in having mucilaginous plants, simple alternate stipulate leaves and valvate aestivation of the calyx. One feature that has set the Malvaceae apart from the other two families is the presence of 1-celled anthers and markedly monadelphous stamens.

NUMBER IN AND DISTRIBUTION OF THE FAMILY

The Malvaceae comprises 80 genera and over 1,000 species. Although members of the family are widely distributed over the globe, they Abound in tropical America.

COMMONLY OCCURRING PLANTS OF THE FAMILY

Indian Mallow (*Abutilon theophrastii* Medic.) is an annual herbaceous shrub found in bushes.

Changeable Rose (*Hibiscus mutabilis* L.) is often grown in gardens for the flowers, the petals of which change from white to red during the course of day.

Chinese Rose or Shoe Flower (*Hibiscus rosa-sinensis* L.) is a garden shrub.

Kydia calycina Roxb. is a small deciduous tree.

Anoda hastata L., *Malachra capitata* (L.) L., *Malvastrum coromandelianum* (L.) Garcke, *Pavonia odorata* Willd., *Sida acuta* Bruma. f., *S. cordifolia* L., *S. glutinosa* Cav., *S. rhombifolia* L., *S. veronicaefolia* Lamk., *Urena lobata* L. and *U. sinuata* L. are weeds of waste places.

Portia Tree [*Thespesia populnea* (L.) Sol. ex Corr.] is an avenue tree which flowers in all seasons.

ECONOMIC ASPECTS OF THE FAMILY

Several plants of the Malvaceae are economically important, because of the presence of fibres and mucilaginous juice.

Abutilon theophrastii (China) is cultivated for the fibres which are utilized in rug making. The seed hairs of *Gossypium arboreum*, *G. barbadense*, *G. herbaceum*, *G. hirsutum*, *G. peruvianum* and *G. punctatum* form the 'cotton' of commerce. The seeds of *Gossypium* produce an oil for lighting purpose; the oil cake, obtained from pressed seeds, is used as a kind of fodder. 'Madras hemp' is derived from the fibres of the bark of *Hibiscus cannabinus*. Strong fibres are also obtained from *H. elatus, Kydia calycina, Sida rhombifolia* and *Urena lobata*.

The plants of the Malvaceae have some edible properties. The half ripe fruits of *Abelmoschus esculentus* are widely used in soups and as vegetables. The roots of *Althaea officinalis* are used in confectionery. The leaves of *Hibiscus cannabinus* serve as a source of the edible greens. The calyx and epicalyx of *H. sabdariffa* have an acid taste, being used for making pickles and jellies. The mucilaginous root of *Pavonia hirsuta* (Zimbabwe) is put into milk to hasten butter production

The Malvaceae includes plants of ornamental value, *e.g. Althaea rosea*, *Hibiscus mutabilis* and *H. rosa-sinensis*.

Bombacaceae
(Bombax Family)

GENERAL FEATURES OF THE FAMILY

Habit	—	*Tall trees, with thick trunks and spreading branches.*
Leaves	—	Alternate, simple or palmately compound, stipulate.
Flowers	—	Showy, bisexual, regular, hypogynous.
Calyx	—	Sepals 5, united, valvate.
Corolla	—	Petals 5, free, imbricate.
Androecium	—	Stamens 5 to many, distinct or monadelphous; staminodes often present; anthers 1-, 2- or more-celled; *pollen grains smooth.*
Gynoecium	—	Carpels 2–5; ovary superior, 2- to 5-locular; *placentation axile.*
Fruit	—	*Capsule with a woolly or pithy pericarp.*

NUMBER IN AND DISTRIBUTION OF THE FAMILY

The Bombacaceae contains 20 genera and about 140 species, being prevalent in the tropics.

COMMONLY OCCURRING PLANTS OF THE FAMILY

Baobab Tree (*Adansonia digitata* L.) is a "fantastic looking tree with a short trunk of considerable size close to the ground and tapering gradually upwards".

Red Silk Cotton Tree (*Bombax ceiba* L.) is a tree with "a very striking feature in any landscape where it occurs, in the months of December, when it loses its foliage, and January, when it bursts into a blaze of scarlet flowers upon the naked branches".

While Cotton Tree [*Ceiba pentandra* (L.) Gaertn.] is a tall tree with horizontally spreading branches.

Cullenia exarillata Robyns is a tall tree, found in Western Ghats.

ECONOMIC ASPECTS OF THE FAMILY

The acid pulp of the fruits of *Adansonia digitata* makes a good beverage. The seed hairs of *Bombax ceiba* yield floss for stuffing pillows, while those of *Ceiba pentandra* give us the so-called 'kapok'. *C. acuminata*, *C. aesculifolia* and *Chorisia insignis* produce floss which can be used as an alternative for kapok. Strong fibres are also obtained from the pods of *C. speciosa* (Brazil). The fruits of *Durio zibethinus* (Malaya) are edible, but taste like "French custard passed through a sewer pipe". *Ochroma pyramidala* (South America) is the source of the world's lightest commercial wood which has a specific gravity of 0.12.

Sterculiaceae
(Sterculia Family)

GENERAL FEATURES OF THE FAMILY

Habit	—	Shrubs or trees, sometimes herbs with mucilage.
Leaves	—	Alternate, simple, entire or lobed or digitately compound, stipulate.
Inflorescence	—	*Cymose.*
Flowers	—	*Bisexual, sometimes unisexual, regular* or irregular, hypogynous.
Calyx	—	Sepals 3–5, somewhat partially united, valvate.
Corolla	—	Petals 5, occasionally 0, imbricate.
Androecium	—	*Stamens many, free or monadelphous; anthers 2-celled with intervening staminodes.*
Gynoecium	—	Carpels 2–5; *ovary syncarpous, often on a gynophore*; ovules few to many.
Fruit	—	Capsule or follicle.

Seeds — Arillate, often winged; embryo with flat, folded or rolled, leafy cotyledons; endosperm fleshy.

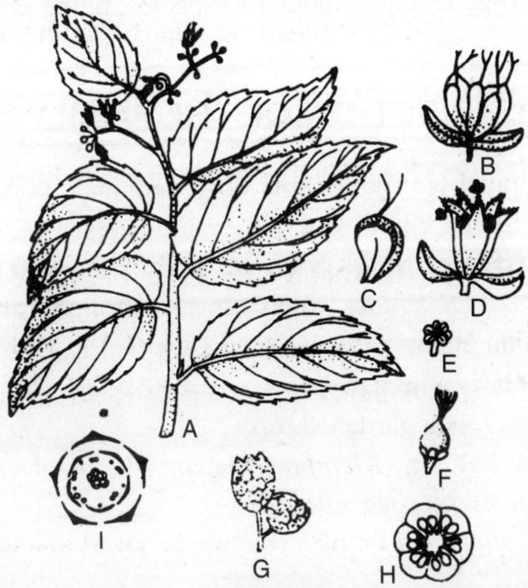

Fig. 40 STERCULIACEAE. *Guazuma ulmifolia* Lamk. **A**, portion of shoot **B**, flower; **C**, petal; **D**, staminal whorl; **E**, fertile stamen; **F**, carpel; **G**, fruit; **H**, T. S. of ovary; **I**, floral diagram

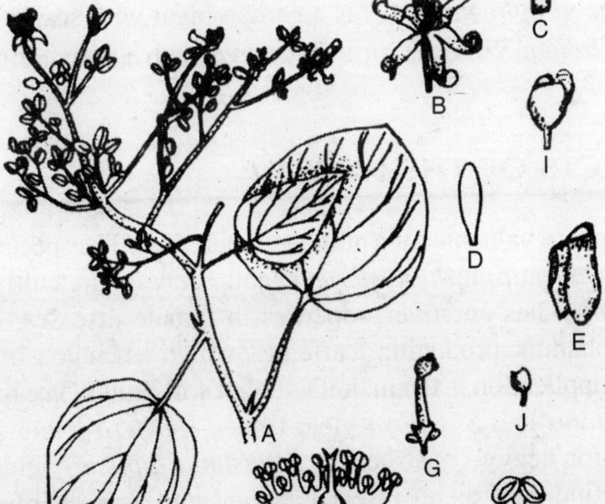

Fig. 41 STERCULIACEAE. *Kleinhovia hospita*.. L. **A**, portion of shoot; **B**, flower; **C**, sepal; **D**, petal; **E**, modified petal; **F**, androgynophore; **G**, staminal column; **H**, stamen; **I**, fruit

FLORAL RANGE IN THE FAMILY

The flowers are generally bisexual, but may be unisexual (*Cola, Heritera, Sterculia,* etc.). They are regular or irregular and pentamerous.

Frequently, the petals are reduced in size and sometimes absent. The stamens are in two whorls; those of the inner (antipetalous) are fertile and those of the outer (antisepalous) sterile. Many flowers show the presence of staminodes, but they are wanting in *Eriolaena.* Though the androgynophore is present in a number of plants, it is lacking in *Helicteres*, *Mansonia* and other genera.

NUMBER IN AND DISTRIBUTION OF THE FAMILY

This is a pantropical and subtropical family of about 60 genera and 700 species.

COMMONLY OCCURRING PLANTS OF THE FAMILY

Devil's Cotton or Perennial Indian Hemp [*Abroma augustum* (L.) L. f.] is a shrub.

Buettneria herbacea Roxb. is a herb.

Dombeya mastersii Hook. f. is a garden shrub.

Firmiana colorata (Roxb.) R. Br., *Kleinhovia hospita* L., *Pterygota alata* (Roxb.) R. Br. and *Sterculia foetida* L. are trees planted on road sides.

Bastard Cedar (*Guazuma ulmifolia* Lamk.) is a middle sized tree with woody tubercled globose capsule.

East Indian Screw Tree (*Helicteres isora* L.) is a shrub, the follicles of which on dehiscing twist spirally and expel the seeds.

Heritera fomes Buch-Ham. is a tree from which the Sunderbans takes its name.

Melhamia magnifolia Blatt. & Hallb. occurs in Rajasthan.

Melochia corchorifolia L. and *Waltheria indica* L. are weeds of waste places.

Noon Flower (*Pentapetes phoenicia* L.) is a garden plant with scarlet flowers.

Pterospermum acerifolium Willd. is an avenue tree with white odorous bisexual flowers and long fleshy sepals.

ECONOMIC ASPECTS OF THE FAMILY

The plants of the Sterculiaceae are valuable for a number of purposes. The species of *Abroma*, *Bracychiton*, *Dombeya*, *Firmiana*, *Fremonodendron*, *Pterospermum* and *Reevesia* are cultivated as ornamentals. The root-bark of *Abroma augustum* has curative properties in female diseases. *Cola acuminata* (tropical America) is the source of cola nuts, producing 'caffeine', which when consumed, confers the power to endure fatigue and thus find application in the manufacture of soft drinks. The bark of *Guazuma ulmifolia, Helicteres isora, Sterculia urens* and *S. villosa* yield fibres. *Heritera fomes* provides us with the best firewood as well as wood for heavy construction. *Mansonia dipikae* yields timber. The flowers of *Pterospermum acerifolium* find uses as an insect repellant and as a disinfectant. From the seeds of *Theobroma, cacao*, yielding the alkaloids 'thein' and 'theobromin' and widely cultivated in South India and Sri Lanka, we get chocolate, cocoa and cocoa butter.

Order 21. *Geraniales*

The order is closely related to the Malvales, from which is differentiated by the typically obdiplostemonous androecium and prevalence of herbaceous forms. Generally, the stamens are twice as many as sepals. The ovules are pendulous with an upwardly-directing micropyle and a dorsal raphe. The styles are often persistent in the fruit. The seeds are without any endosperm.

Engler and Diels treated the order as composed of six suborders and 21 families: Geraniieae (Oxalidaceae, Geraniaceae, Tropaeolaceae, Linaceae, Erythroxylaceae, Zygophyllaceae, Cneoraceae,

Rutaceae, Simaroubaceae, Burseraceae, Meliaceae, Akariaceae), Malpighiineae (Malpighiaceae, Trigoniaceae, Vochysiaceae), Polygalineae (Tremandraceae, Polygalaceae), Dichapetalineae, (Dichapetalaceae) Tricoccae (Euphorbiaceae, Daphniphyllaceae), Callitrichineae Callitrichaceae). Bessey accommodated all these families in the Geraniales. Hallier, Wettstein and Rendle regarded them to cover three orders. Hutchinson distributed some of the families of the Geraniales among four orders (Geraniales, Rutales, Meliales, Malpighiales) and several other families among five other orders. Evidence from different branches of botany show that the Geraniales, as constructed by Engler, is not a natural group. On the other hand, the order is accepted as composed of somewhat four or five disjunctive taxa. The present author takes into account Core's concept of the Geraniales which embraces 14 families as follows:

(1) Flowers unisexual	(2)
(1) Flowers bisexual	(5)
(2) Flowers reduced	(3)
(2) Flowers not reduced	(4)
(3) Aquatic plants; carpels 2	*Callitrichaceae*
(3) Terrestrial plants; carpels 3	*Euphorbiaceae*
(4) Ovules 2 in each loculus	*Burseraceae*
(4) Ovule 1 in each loculus	*Simaroubaceae*
(5) Secretory cells present	(6)
(5) Secretory cells absent	(7)
(6) Leaves gland-dotted	*Rutaceae*
(6) Leaves not gland-dotted	*Meliaceae*
(7) Flowers actinomorphic	(8)
(7) Flowers zygomorphic	(12)
(8) Carpels splitting apart at maturity	(9)
(8) Carpels not splitting apart at maturity	(10)
(9) Carpels splitting as mericarps	*Geraniaceae*
(9) Carpels splitting as nutlets	*Zygophyllaceae*
(10) Sap acidic	*Oxalidaceae*
(10) Sap not acidic	(11)
(11) Loculi 4–5, all fertile	*Linaceae*
(11) Loculi 3–4, only 1 fertile	*Erythroxylaceae*
(12) Anthers with a poricidal dehiscence	*Polygalaceae*
(12) Anthers with a lengthwise dehiscence	(13)
(13) Leaves alternate	*Tropaeolaceae*
(13) Leaves opposite	*Malpighiaceae*

Callitrichaceae
(Water-starwort Family)

In this family, the plants are aquatic herbs with opposite leaves. Here the flowers are unisexual and naked, the male consisting of a single stamen and the female of two carpels.

The genus *Callitriche* is of doubtful affinity and the position of the family is rendered obscure by the extremity in the reduction of vegetative and floral structure as well as vasculature. Hallier treated it as a primitive component of the Guttales, being derived from the Linaceae. According to Eichler, Baillon,

Wettstein, Pax and Rendle, the Callitrichaceae is an advanced member of the Euphorbiales. Clarke suggested that the family should be placed in the Caryophyllaceae. Robert Brown, de Candolle, Hegelmaier, Bentham-Hooker and others considered the family as related to the Haloragaceae. Hutchinson placed the family in his Lythrales together with the Lythraceae, Onagraceae, Trapaceae and Holarrhagaceae. Others went to the extent of suggesting a relationship with the Sympetalae and this was based on the presence of unitegmic ovules.

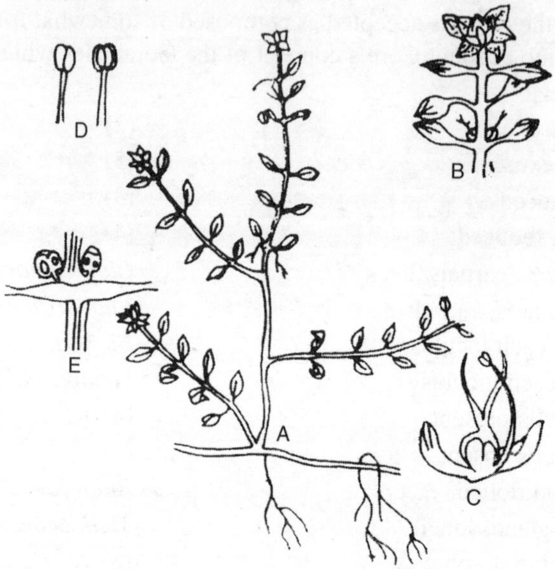

Fig. 42 CALLITRICHACEAE. *Callitriche stagnalis* Scop. **A**, portion of plant; **B**, portion of enlarged branch; **C**, flower; **D**, stamens; **E**, fruits at a node. (**A–E**, after Subramanyam)

The family has one genus (*Callitriche*) and about 17 species which are distributed from the arctic to subantarctic regions minus South Africa.

Callitriche stagnalis Scop. is a small flaccid herb, rooting at the nodes and with carpels narrowly winged at the margins.

In southern Germany, "it is possible to predict the abundance of certain pollutants by the species composition and state of *Callitriche* plants" (Cook, 1978).

Euphorbiaceae (Spurge Family)

GENERAL FEATURES OF THE FAMILY

Habit	—	Herbs, shrubs or trees, *often with a milky sap.*
Leaves	—	Simple, entire or lobed, with caducous or persistent stipules.
Inflorescence	—	Racemose, cymose, mixed or cyathium.
Flowers	—	*Unisexual*, regular, hypogynous, bracteate.
Perianth	—	Simple, sepaloid, 5-merous, sometimes double or absent.

Fig. 43 EUPHORBIACEAE. *Ricinus communis* L. **A**, upper portion of flowering shoot; **B**, unopened male flowers; **C**, open male flower; **D**, anthers; **E**, female flower; **F**, T.S. of ovary; **G**, L. S. of ovary; **H**, capsule; **I**, coccus split open; **J**, seed; **K**, V. S. of seed. (**A–K**, after Rendle)

Androecium	—	Stamens 1 to many (in male flowers), free or united; anthers 2-celled.
Gynoecium	—	Carpels 3 (in female flowers,) syncarpous; *ovary superior*, *3-locular*; *ovules collateral, pedulous with a ventral raphe*; *placentation axile*; *micropyle carunculate.*
Fruit	—	Regma, sometimes drupe.
Seeds	—	Albuminous; embryo bent or straight; cotyledons flat or narrow.

FLORAL RANGE IN THE FAMILY

The Euphorbiaceae shows a wide range of variations in the floral structure. The flowers may be bisexual, pentamerous and diplochlamydeous. The male and female are heteromorphic, *e.g., Aporusa.*

The perianth segments are either free or united, their numbers being 3, 4, 5 or 6.

The stamens may be 1 in *Euphorbia,* 2 or 3 in *Sapium*, 4–8 in *Baccaurea* and several hundred in *Ricinus*, *Trewia,* etc. In *Breynia*, *Jatropha*, *Wielandia*, etc., the stamens are monadelphous. Sometimes, out of 10–15 stamens arranged in two to three whorls, the inner whorl is monadelphous and the outer remains free, as in *Chrozophora.* The stamens may be polyadelphous in *Lasiococca.*

The carpels are more than 3 in some species of *Phyllanthus.* There are 2–15 carpels in *Glochidion.* In *Antidesma*, the ovary becomes unilocular due to the abortion of other carpels. The ovary may be 2- to 5- celled, as in *Baccaurea.*

In many genera, the rudiments of stamens and carpels are present; a case in point is *Pridetia.* Sometimes androphore and gynophore may be seen, as in *Bridelia.*

DIVISION OF THE FAMILY

The Euphorbiaceae is divided into two groups, each with two subfamilies; one subfamily contains eight tribes and the other two tribes.

A. *Platylobeae.* Cotyledons much broader than radicle.

Subfamily I. *Crotonoideae.* Latex present. Ovule 1 in each loculus.

Tribe (*i*) *Acalypheae.* Inflorescence a raceme, spike, axillary or terminal. Flowers usually apetalous. Calyx of male flowers valvate. Stamens erect in bud. Examples: *Acalypha*, *Ricinus*, etc.

Tribe (*ii*) *Adrianeae.* Inflorescence a simple terminal spike or raceme. Flowers usually apetalous. Calyx of male flowers valvate. Stamens erect in bud. Example: *Manihot.*

Tribe (*iii*) *Clutieae.* Male flowers with petals, in groups or cymes, these partial inflorescence axillary or in complex inflorescence. Examples: *Codiaeum* and *Clutia.*

Tribe (*iv*) *Crotoneae.* Stamens bent inwards in bud. Example: *Croton.*

Tribe (*v*) *Euphorbieae.* Inflorescence a cyathium. Example: *Euphorbia.*

Tribe (*vi*) *Gelonieae.* Male flowers apetalous, in groups or cymes, these partial inflorescence axillary or complex inflorescence which are leaf opposed. Example: *Gelonium.*

Tribe (*vii*) *Hippomaneae.* Inflorescence axillary terminal, spike-like; partial inflorescence cymes. Male flowers apetalous. Examples: *Excoecaria*, *Sapium*, etc.

Tribe (*viii*) *Jatropheae.* Inflorescence a dichasial thyrse. Male flowers usually apetalous. Calyx of male flowers valvate. Stamens erect in bud. Examples: *Hevea* and *Jatropha.*

Subfamily II. *Phyllanthoideae.* Latex absent. Ovules 2 in each loculus.

Tribe (*i*) *Bridelieae.* Calyx of male flower valvate. Embryo larger, little shorter than endosperm. Example: *Bridelia.*

Tribe (*ii*) *Phyllantheae.* Calyx of male flowers imbricate. Embryo larger, little shorter than endosperm. Examples: *Baccaurea*, *Phyllanthus*, etc.

B. *Stenolobeae.* Cotyledons as wide as radicle.

Subfamily I. *Porantheroideae.* Latex absent. Ovules 2 in each loculus. Example: *Poranthera.*

Subfamily II. *Ricinocarpoideae.* Latex present. Ovule 1 in each loculus. Example: *Ricinocarpos.*

POSITION AND AFFINITY OF THE FAMILY

There is much difference of opinion as to the systematic position of the family. Hallier treated the Euphorbiaceae in his Passionales. Engler included the family in the Geraniales being close to the Malvales. Wettstein as well as Rendle created an order (Tricoccae) for it. Bentham-Hooker placed the family in the Unisexuales just before the Urticaceae. In Hutchinson's arrangement, the family appeared under the Euphorbiales.

The Euphorbiaceae is closely related to the Malvales in general and to the Sterculiaceae in particular due to the presence of monadelphous stamens (in about 25 percent of the genera), rudiments of stamens and carpels, androphore or gynophore occasionally, caruncule and the nature of embryo. The family bears a relationship with the Geraniales and Sapindales on account of the nature of ovules.

It has been suggested that the Euphorbiaceae arose from the Malvales. This was made possible by the reduction of the inner whorl of perianth and abortion of one of the components of the essential whorls during the course of evolution.

NUMBER IN AND DISTRIBUTION OF THE FAMILY

The Euphorbiaceae contains about 300 genera and over 5,000 species. The members of the family are world wide in distribution with two centres of abundance—one in Brazil and the other in Indo-Malaysia.

COMMONLY OCCURRING PLANTS OF THE FAMILY

Acalypha fallax Muell-Arg., *A. indica* L., *Chrozophora rottleri* (Geis.) Juss. ex Spreng., *Croton bonplandianum* Baill., *Euphorbia bombaiensis* Santapau, *E. hirta* L., *E. nivulia* Buch-Ham., *Phyllanthus asperulatus* Hutch., *P. simplex* Retz. and Indian Stinging Nettle (*Tragia involucrata* L.) are weeds.

Acalypha hispida Burm. f., *Codiaeum variegatum* (L.). Bl., Coral Plant (*Jatropha multifida* L.) and *Poinsettia pulcherrima* R. Garh are garden plants.

Agrostistachys indica Dalz., known from Western Ghats, occurs in evergreen forests and usually along the banks of streams.

Antidesma ghesaembilla Gaertn. is a village shrub.

Bishop Wood or Vinegar Wood (*Bischofia javanica* Bl.), *Bridelia squamosa* Gehrm., Monkey-face Tree [*Mollotus philippinensis* (Lamk.) Muell.-Arg.], Child-life tree (*Putranjiva roxburghii* Wall.) and *Trewia nudiflora* L. are trees.

Cicca acida (L.) Merr. and Embelic Myrobalan (*Emblica officinalis* Gaertn.) are fruit trees.

Euphorbia antiquorum L. is a hedge plant with a fleshy leafless spinous three angled irregularly narrowed stem.

Euphorbia heterophylla L. is an annual herb with leaves embracing the cymes spotted scarlet near the base but green towards the apex.

Euphorbia neriifolia auct. pl. non L. is a shrub, bearing stipulary spines and obovate succulent glabrous leaves.

Milk-hedge (*Euphorbia tirucalli* L.) and Jew's Slipper [*Pedilanthus tithymaloides* (L.) Poit.] are hedge plants.

Excoecaria agallocha L. and *Sapium indicum* Willd. are met with in the Sunderbans.

ECONOMIC ASPECTS OF THE FAMILY

Economically, the members of the Euphorbiaceae are of considerable importance. They provide us with food, drugs, rubber and other products.

Some plants yield edible fruits and roots. The fruits of *Aleurites moluccana*, *Baccaurea sapida*, *Bridelia squamosa*, *Hemicyclia andamanica*, *H. sepiaria* and *Trewia nudiflora* are used for culinary purposes. *Baliospermum micranthum* is used as a source of the edible greens. The fruits of *Cicca acida* and *Emblica officinalis* are made into preserves. *Manihot esculenta* is cultivated for the starchy roots which produce 'tapioca' by a special preparation.

Some plants are medicinally valuable, *e.g.*, [*Claoxylon indicum Hura crepitans*, *Jatropha curcas*, *Mallotus philippinensis* etc. *Croton eluteria* (Bahamas) yields the so called 'cascarilla bark'.] From

Croton tiglium and Ricinus communis are derived 'croton oil' and 'castor oil' respectively, both of which are purgatives. Others are used for a number of ailments, such as *C. alanosana* in toothache, *C. cortesiana* and *C. humilis* in skin diseases, *C. echinocarpa* in eye troubles and *C. nivea* in fevers.

Aleurites fordii (China) constitutes 'Tung oil' which is useful in varnishes. *A. moluccana* yields 'candlenut oil' which finds use as a preservative for the hulls of vessels. The bark of *Bischofia javanica* and *Bridelia retusa* are useful in tanning. *Aporusa lindleyana, Chaetocarpus castanocarpus, C. coriaceous, Coelodepas calycinum, Drypetes griffithi* and *Gelonium lanceolatum* furnish us with useful timber. Gum exuded from the trunk of *Croton xalapensis* (Mexico) is used as a substitute of toothpaste. The coloured wood of *Daphniphyllum himalayense* is used by the hill people for marking the forehead. 'Candelilla wax' is extracted from the stems of *Euphorbia antisyphylitica* and *Pedilanthus pavonis*, both being plants of Mexico and Texas (USA). The seeds of *Givotia rottleriformis* give oil, suitable for lubricating machinery. The latex of *Hevea brasiliensis* form the 'Para rubber' of commerce and *Manihot glaziovii* the 'Ceara rubber'. Ink is prepared from the ripe fruits of *Kirganelia reticulata* (Philippines). *Sapium sebiferum* has a waxy seed covering utilised for candles.

There are some ornamentals in the Euphorbiaceae, *e.g.* certain species of *Acalypha Codiaeum, Dalechampia, Euphorbia, Jatropha* and others.

Burseraceae
(Incense-tree Family)

The family is characterised by the presence of resin ducts, compound leaves, unisexual flowers, distinct stamens, a short single style and two ovules per loculus.

The Burseraceae is divided into three tribes:

Tribe (*i*) *Bursereae.* Drupe with an endocarp of fused parts; exocarp always dehiscing by valves. Examples: *Bursera, Commiphora*, etc.

Tribe (*ii*) *Canarieae.* Drupe with an endocarp of fused parts; exocarp not dehiscing by valves. Example: *Canarium.*

Tribe (*iii*) *Protieae.* Drupe with 2–5, free or adhering parts but not fused in stony endocarp; exocarp occasionally dehiscing by valves. Examples: *Protium*, etc.

There are about 17 genera and 500 species in the Burseraceae. The members of this family reach their peak of development in north-eastern Africa and tropical America.

Boswellia serrata Roxb. ex Colebr., *Bursera penicillata* (Sesse & Moc. ex DC.) Engl., *Garuga pinnata* Roxb. and *Protium serratum* (Wall. ex Colebr.) Engl. are trees.

Several plants of the family are of value as a source of balsams, gums, resins, nuts and timbers. *Boswellia carteri* produces 'gum olibanum' and *B. frereana* 'African elemi'. Indian frankincense is the product of *B. serrata*, 'Mexican linaloe' of *Bursera glabrifolia* and *B. penicillata* and 'chibou' of *B. gummifera. Canarium commune* provides 'Java almond' and *C. ovatum* 'Pili nut'. The wood of *C. littorale, C. schweinfurthii, Dacrycodes costata, Santiria laevigata* and *S. tomentosa* is used for carpentary and building construction. *C. luzonicum* supplies 'Manila elemi' and *C. strictum* 'black dammar'. *Commiphora africanum* yields 'African bdellium', *C. erythraea* 'herbol myrrh' and *C. opobalsamum* 'Mecca balsam'. 'Carana gum' is derived from *Protium carana* and 'Brazilian elemi' from *P. heterophyllum.* Some species of *Bursera* and *Garuga* are cultivated as ornamentals.

Simaroubaceae (Quassia Family)

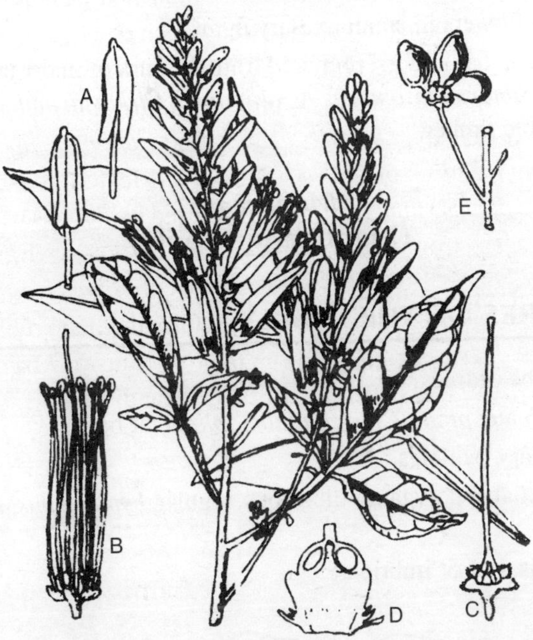

Fig. 44 SIMAROUBACEAE. *Quassia amara L.* **A**, stamens; **B**, stamens and ovary; **C**, ovary; **D**, V.S. of lower portion of ovary; **E**, fruits. (**A–E**, after Kohler)

The family is distinguished by the presence of a bitter bark, pinnate leaves, unisexual flowers, filament bases with scales and a single ovule per loculus.

The position of the genus *Balanites* is debatable. According to Record (1921), *Balanites* differs from zygophyllaceous taxa like *Bulnesia*, *Guaiacum* and *Porlieria* in xylotomy. However, it is similar to these taxa in having thick-walled fibres with bordered pits, abundance of fusiform parenchyma cells, small vessel pitting and vasicentric tracheids (Metcalfe and Chalk, 1950). Heimsch (1942) considered that *Balanites* shows greater relationship to the Zygophyllaceae than the Simaroubaceae, although the genus can be set aside from the former family on the basis of the width and height of medullary rays. In floral anatomical features, the genus resembles the Zygophyllaceae. In the basic chromosome number (x=9), *B. aegyptiaca* simulates *Fagonia cretica* of the Zygophyllaceae and *Quassia amara* of the Simaroubaceae (Darlingtoni and Wylie, 1961). *Balanites* resembles the Zygophyllaceae in essential embryological characters and differs from the Simaroubaceae in the smaller number of parietal layers, presence of endothecium and absence of nucellar cap (Kapil and Ahluwalia, 1963; Nair and Jain, 1965; Davis, 1966). Erdtman (1971) stated that the pollen grains of the Balanitoideae are akin to those of *Harrisonia* of the Simaroubaceae. Furthermore, *B. roxhurghii* exhibits the probable existence of saponins and steroids. In the absence of leucoanthocyanins, *Balanites* is more closer to the members of the Zygophyllaceae than those of the Simaroubaceae. However, it differs from both taxa in the positive reaction for alkaloids. Hence, taking the totality of evidence, it has been tentatively suggested by Parvati and Narayana (1978) that *Balanites* be retained in the Zygophyllaceae but in subfamily Balanitoideae. But similarities with the Simaroubaceae do not rule out a probable common ancestral stock for the two taxa.

A family of about 20 genera and 120 species, the Simaroubaceae is found in the tropics and subtropics.

Ailanthus excelsa Roxb. is a tall tree with much branched panicles. *Balanites aegyptiaca* (L.) Del. is a spiny tree, having flowers in small axillary cymes.

The family is important for 'bitters' prepared from the bark of many trees, e.g., *Ailanthus excelsa, Balanites aegyptiaca* and *Simarouba amara. A. altissima, Kirkia acuminata, Picrasma excelsa* and *Simaba cedron* yield valuable timber.

Meliaceae (Mahogany Family)

GENERAL FEATURES OF THE FAMILY

Habit — Shrubs or trees.

Leaves — Alternate, *pinnate, exstipulate, oblique* at base.

Inflorescence — Axillary panicles.

Flowers — Bisexual or polygamo-dioecious, regular, *hypogynous with an annular or a tubular disc.*

Calyx — Sepals 4 or 5, imbricate.

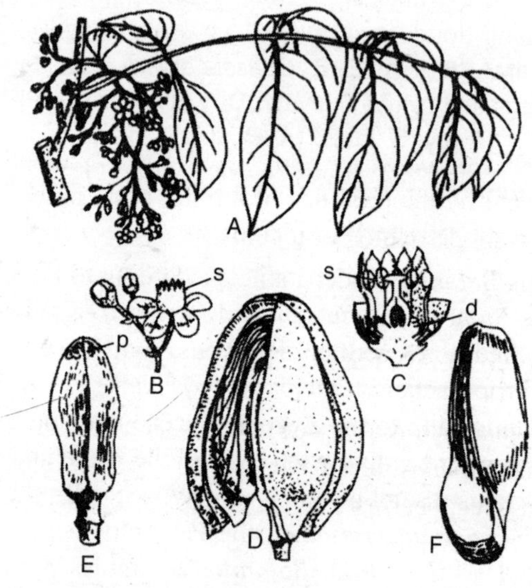

Fig. 45 MELIACEAE. *Swietenia mahagoni* (L.) Jacq. **A**, leaf and inflorescence; **B**, buds and open flower; **C**, V.S. of flower; **D**, dehiscence of fruit with valve removed to expose seeds; **E**, central axis of fruit (*p*, points of seed attachment); **F**, seed. (**A** – **F**, after Rendle)

Corolla — Petals 4 or 5, imbricate or twisted, *adnate to staminal* tube.

Androecium — Stamens 4 or more, in 2 whorls, sometimes turned into staminodes; *filaments cohere to form a columnar tube to whose mouth anthers are inserted.*

Gyaoecium — Carpels 3–5; *ovary superior, 3- to 5- locular with 1 or 2 ovules in each loculus.*

Fruit — Berry, capsule or drupe.

Seeds — Winged, albuminous or exalbuminous.

DIVISIONS OF THE FAMILY

The Meliaceae is divided into three subfamilies:

Subfamily I. *Cedreloideae.* Stamens free. Ovules not more than 2 in each loculus. Examples: *Cedrela, Pteroxylon*, etc.

Subfamily II. *Melioideae.* Stamens monadelphous Ovules not more than 2 in each loculus. Examples: *Azadirachta, Melia*, etc.

Subfamily III. *Swietenioideae.* Stamens monadelphous. Ovules more than 2 in each loculus. Examples: *Soymida, Swietenia, etc.*

POSITION AND AFFINITY OF THE FAMILY

The Meliaceae was placed in the Terebinthales by Hallier who regarded it as being derived from the Rutaceae. Bessey retained it in the Geraniales and Bentham-Hooker followed his practice. Wettstein treated it together with the Burseraceae and Polygalaceae. Hutchinson segregated it as the sole family of his Meliales. In Rendle's arrangement, the Meliaceae was the last family under the Rutales.

The family stands very close to the Rutaceae, Sapindaceae, Burseraceae and Simaroubaceae. In the Meliaceae, the leaves are not gland-dotted and the flowers are characterised by the presence of a staminal tube.

NUMBER IN AND DISTRIBUTION OF THE FAMILY

The Meliaceae, with about 50 genera and 550 species, is primarily distributed in the tropics and common as understory trees in rainforests.

COMMONLY OCCURRING PLANTS OF THE FAMILY

Amoora cucullata Roxb. and *A. rohituka* Wt. & Arn. are planted on road sides.

Margosa or Neem (*Azadirachta indica* A. Juss.), Persian Lilac (*Melia azedarach* L.) and Mahogany [*Swietenia mahagoni* (L.) Jacq.] are trees, bearing winged fruits.

Indian Satinwood (*Chloroxylon swietenia* DC.) is a tall tree and so is Moulmein Cedar (*Toona ciliata* M.J. Roem.).

Chickrassy (*Chickrassia tabularis* A. Juss.) is a large deciduous tree.

Dysoxylum binectariferum (Roxb.) Hook. f. ex Bedd. is an evergreen tree with a cylindrical stem.

Xylocarpus granatum Koen., a small tree with large round fruits, is found in the Sunderbans.

ECONOMIC ASPECTS OF THE FAMILY

Cedrela odorata is the source of 'cedar wood' which is used for making cigar boxes. *Khaya senegalensis* and *Swietenia mahogani* form cabinet wood. Other timber yielding plants include *Chloroxylon swietenia*,

Dysoxylum malabaricum, *Entandrophragma candollei*, *Swietenia macrophylla*, *Toona ciliata* and *Walsura robusta*.

The fruits of *Aglaia odorata* are used in flavouring tea. The leaves of *Azadirachta indica* are used as a pot herb and an insect repellant; the twigs of this plant are widely used for tooth brushes; the seeds yield the 'margosa oil' of commerce. *Carapa guianensis* and *C. moluccensis* produce 'carapa oil', an illuminant. Oil from *Chisocheton macrophyllus* is used also an illuminant. The flowers of *Chickrassia tabularis* yield a red dye; the trunk provides a yellow transparent gum; the astringent bark is of medicinal value. The fruits of *Lansium domesticum* and *Sandoricum koetjape* are edible. Other medicinal plants include *Melia composita*, *Soymida febrifuga* and *Trichilia emetica*.

Some plants of the Meliaceae are grown as ornamentals, such as the species of *Amoora*, *Melia*, *Swietenia*, *Turraea*, etc.

Rutaceae
(Citrus Family)

GENERAL FEATURES OF THE FAMILY

Habit	—	Shrubs or trees.
Leaves	—	Simple or compound, exstipulate, *gland-dotted.*
Flowers	—	Bisexual, regular, hypogynous, with a *cushion-like or an annular disc.*
Calyx	—	Sepals 4 or 5, free or united, imbricate.
Corolla	—	Petals 4 or 5, imbricate or valvate.
Androecium	—	Stamens 4 or more, free or polyadelphous, obdiplostemonous; anthers introrse.
Gynoecium	—	Carpels 4 or more, syncarpous; *ovary superior, elevated on a dise, multilocular with axile placentae*; ovules 2 or more in each loculus; stigma capitate.
Fruit	—	Amphisarca or hesperidium, sometimes capsule.
Seeds	—	Embryo large, straight or curved; endosperm little or none.

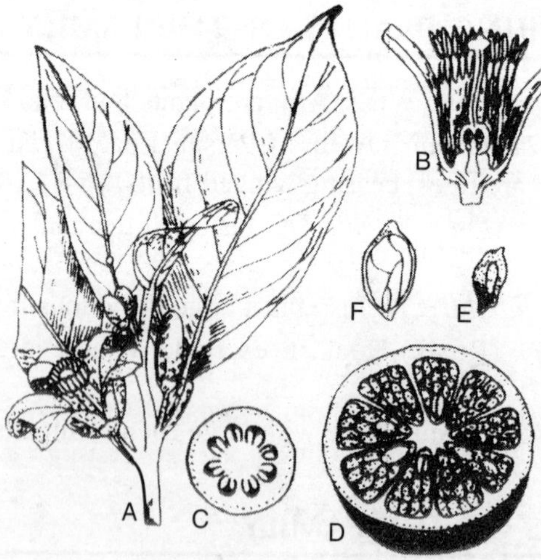

Fig. 46 RUTACEAE. *Citrus aurantium* L. **A**, portion of flowering shoot; **B**, V.S. of flower, **C**, T.S. of ovary; **D**, T.S. of fruit; **E**, seed; **F**, L.S. of seed. (**A–F**, after Rendle)

FLORAL RANGE IN THE FAMILY

The flowers of the Rutaceae are usually pentamerous, but tetramerous (*Acronychia*, *Boenninghausenia*) and trimerous (*Lunasia*) flowers are also found. In *Ruta*, the lateral flowers are tetramerous and the terminal flower is pentamerous. The flowers of *Empleurum*, *Toddalia* and *Xanthoxylum* are unisexual. In *Almeidea* and *Dictamnus*, the flowers are zygomorphic.

Although the sepals are free in some cases, they are often united into a tube (*Cusparia*). The sepals are obsolete in *Asterolasia* and very unequal in *Moniera*.

As a rule, the petals are free. However, they are united and form a campanulate corolla in *Correa*.

The stamens are numerous in *Peltostigma*, but monadelphous in *Drummondita* and *Philotheca*. They are inserted above the disc in *Zieria* and provided with a hairy scale at the base in *Chorilaena* and *Nematolepis*.

In some flowers, the ovary is many-locular and entire or slightly grooved with a terminal undivided style. In others, the ovary is deeply lobed and styles spring from the centre. In each carpel, several ovules are arranged in two rows, but only 2 are superposed or collateral. Only 1 ovule may be present in an ovary. In *Feronia*, one can see a syncarpous ovary with many placentae upon which many ovules are arranged in many rows. A gynophore is noticed in *Bosistoa* and *Xanthoxylum*.

DIVISIONS OF THE FAMILY

The Rutaceae is divided into five subfamilies based on fruit characters:

Subfamily I. *Aurantioideae*. Fruit a berry. Examples: *Citrus*, *Feronia*, *Murraya*, etc.

Subfamily II. *Flindersioideae*. Fruit a loculicidal or septicidal capsule. Example. *Flindersia*.

Subfamily III. *Rutoideae*. Carpels some what distinct in fruit, opening on inside with a separating endocarp. Examples: *Boronia*, *Correa*, *Cusparia*, *Fagara*, *Ruta*, etc.

Subfamily IV. *Spathelioideae*. Fruit a winged drupe. Example: *Spathelia*.

Subfamily V. *Toddalioideae*. Fruit a drupe or dry and winged. Examples: *Ptelea*, *Skimmia*, *Toddalia*, etc.

POSITION AND AFFINITY OF THE FAMILY

Hallier attributed particular significance to the Rutaceae. He derived it from stocks ancestral to the Berberidaceae and regarded it as the precursor of the Terebinthales—an order in which he also included all the amentiferous families. Engler placed this family under the Geraniales. In Rendle's arrangement, the Rutaceae appeared within a separate order (Rutales) after the Geraniales and before the Sapindales. This family, along with the Simaroubaceae and Burseraceae, was segregated by Hutchinson into his concept of the Rutales.

The Rutaceae is allied to the Meliaceae, Sapindaceae and Anacardiaceae due to the heath-like habit, leaf structure, disc formation and obdiplostemonous androecium. A relationship of this family has been traced to the Euphorbiaceae on account of the ventral raphe of ovule in some genera.

NUMBER IN AND DISTRIBUTION OF THE FAMILY

The Rutaceae includes 190 genera and about 900 species. Though the members of the family are widely distributed in the tropical and temperate zones, they are predominant in Australia and South Africa.

COMMONLY OCCURRING PLANTS OF THE FAMILY

Acronychia pedunculata (L.) Miq. is seen in scrubby thickets near Canning, West Bengal.

Wood Apple [*Aegle marmelos* (L.) Corr.] and Elephant Apple [*Limonia elephantum* (Corr.) Panigrahi] are fruit trees, usually armed with axillary straight strong spines.

Atlantia monophylla (Roxb.) DC. is a garden shrub.

Boenninghausenia albiflora (Hook.) Reichb. ex Meissn. occurs in Khasi Hills and temperate Himalaya.

Lemon [*Citrus limon* (L.) Burm. f.], Shaddock [*C. maxima* (J. Burm.) Merr.] and Orange [*C. sinensis* (L.) Osbeck] plants are characterised by winged jointed petioles.

Clausena heptaphylla Wt. & Arn. is prevalent in country side shrub berries.

Tooth brush Plant [*Glycosmis pentaphylla* (Retz.) DC] is an odorous shrub found all over northern India.

Merope angulata (Willd.) Swingle and *Paramignya citrifolia* Hook, f. are thorny undershrubs of the Sunderbans.

Curry leaf Plant [*Murraya koenigii* (L.) Spreng.] is extensively cultivated in South India.

Chinese Box [*Murraya paniculata* (L.) Jack] is a garden plant.

Skimmia arborescens T. And. ex Gamble is a glabrous shrub found in Khasi Hills and temperate Himalaya.

Toddalia asiatica (L.) Lamk. is a shrub, armed with peculiar spines and noticed in Khasi Hills and Orissa.

ECONOMIC ASPECTS OF THE FAMILY

The fruit pulp of *Aegle marmelos* is very nutritious and is used for making sherbets. *Casmiroa edulis* yields white sapote', an apple like fruit with a custard-like pulp. Some of the world's important fruits come from different species of *Citrus*; such as *C. aurantifolia* (lime), *C. aurantium*(sour orange), *C. grandis* (pumelo), *C. limon*(lemon), *C. medica* (citron), *C. paradisi* (grape fruit), *C. reticulata* (tangerine), and *C. sinensis* (orange). The fruits of *Evodia fraxinifolia* are used in chutney. The fruits of *Limonia elephantum* are edible and are used for making preserves. The leaves of *Murraya koenigii* are widely used as a condiment. *Xanthoxylum piperitum* provides 'Japan pepper'.

The dried leaves of *Barosma betulina*, *B. crenulata* and *B. serratifolia* form the drug 'buchu'. The violet scented 'oil of boronia' used in perfumery, is derived from *Boronia megastigma. Cusparia febrifuga* furnishes us with 'cusparia bark', a substitute for quinine. Other medicinal plants include *Pilocarpus pennatifolius* and *Ruta graveolens.*

'Mexican elemi', an oleoresin, is derived from *Amyris balsamifera* and *A. elemifera.* The twigs of *Glycosmis pentaphylla* and *Xanthoxylum alatum* are utilised as chew sticks or tooth brushes. The bark and flowers of *M. paniculata* are used for making cosmetics; the wood of the root is good for making kris handles. The seeds of *Peganum harmala* produce a red dye and the root-bark of *Toddalia* a yellow dye.

Some species of *Atlantia, Barosma, Calodendrum, Dictamnus, Diosma, Murraya, Phellodendron, Poncirus*, etc. are of value as ornamentals.

Oxalidaceae
(Wood-sorrel Family)

The family is recognised by the exstipulate leaves, actinomorphic flowers, imbricate calyx, contorted corolla, stamens connate at base, free styles, capsular fruit and seeds with an elastic testa.

Fig. 47 OXALIDACEAE. *Averrhoa carambola* L. **A**, flower; **B**, flower with petals removed; **C**, stamens and staminodes; **D**, anther; **E**, fruit; **F**, T.S. of fruit. (**A–F**, after Hutchinson)

The family comprises three genera and about 900 species, being prevalent in the tropics and subtropics and having richest representation in the Southern Hemisphere.

Biophytum sensitivum (L.) DC. is a weed of rice fields and road sides with a rosette of sensitive pinnate leaves and dimorphic flowers. Wood-sorrel (*Oxalis corniculata* L.) is a weed with ternate leaves, conistuting a leaf mosaic.

Averrhoa bilimbi and *A. carambola* are grown for their acid fruits. The leaves of *Oxalis acetosella* are sometimes used in salads and stems of *O. pescaprae* used as a vegetable. *O. crenata*, *O. deppei* and *O. tuberosa* yield edible tubers.

Linaceae
(Flax Family)

The family is known by the pentamerous actinomorphic flowers, contorted corolla of distinct and fugaceous petals, basally connate filaments, small staminodes and septicidally dehiscent capsule.

The phyletic position of the Linaceae is a matter of dispute. Bessey, Wettstein, Rendle, Lawrence and Core placed it in the Geraniales. Hallier shifted it to the Guttales. Hutchinson included it in the Malpighiales and Cronquist in the Linales.

The family represents 13 genera and about 300 species of the temperate and subtropical regions of both hemispheres.

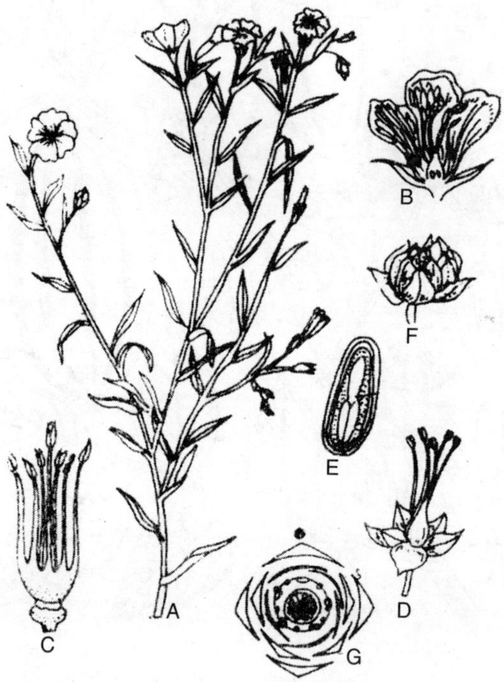

Fig. 48 LINACEAE. *Linum usitatissimum* L. **A**, portion of plant; **B**, V.S of flower; **C**, androecium; **D**, pistil with intact sepals; **E**, L.S. of seed; **F**, fruit: **G**, floral diagram. (**A–G**, after Baillon)

The family is of importance for the Flax Plant (*Linum usitatissinm*), the bark of which yields a fine fibre for weaving into linen cloth. The seeds of this plant produce 'linseed oil' and the resulting oil-cake is a cattle-feed. *Ctenolophon parviflorum* (Malaya) yields a hard durable timber. *Hugonia obtusifolia* and *H. platysephala*, both of Africa, provide edible fruits. A few species of *Linum* and *Reinwardtia* are cultivated ornamentally.

Erythroxylaceae
(Coca Family)

The family is closely related to the Linaceae from which it is differentiated by the presence of monadelphous stamens, appendaged petals and drupaceous fruits.

The family was placed in the Linaceae by Bentham-Hooker and Hallier. Hutchinson transferred it to the Malpighiales. Most other workers treated it as a distinct family and as a member of the Geraniales.

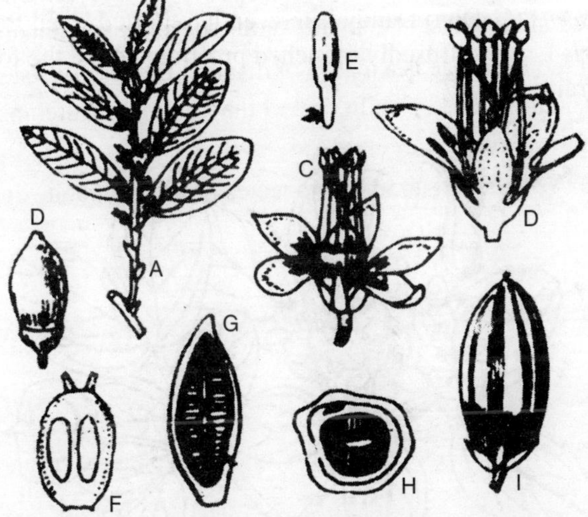

Fig. 49 ERYTHROXYLACEAE. *Erythroxylum coca* L. **A**, shoot with flowers, **B**, flower bud; **C**, flower; **D**, V.S. of flower; **E**, anther; **F**, V.S. of ovary; **G**, V.S. of seed; **H**, T.S. of seed; **I**, fruit. (**A–I**, after Le Maout & Decaisne)

The family covers four genera and about 260 species, being distributed in tropical America.

The leaves of *Erythroxylum coca* and *E. truxillense* are chewed by the people of Bolivia and Peru; this practice enables them to live without food or drink for several days. *E. coca* is also the source of 'cocaine'; an extract of this alkaloid, together with cola, finds use in the manufacture of soft drinks.

Geraniaceae
(Geranium Family)

The family is characterised by the pentamerous actinomorphic flowers, carpels splitting apart at maturity as mericarps and absence of endosperm.

The Geraniaceae embraces 11 genera and about 750 species. The members of this family are widely distributed in the subtropical and temperate regions.

Geranium ocellatum Jacq., found in the hills, is a straggling herb with rose coloured flowers.

In an economic sense, the family is of some importance. *Erodium cicutarium* is esteemed as a range forage plant. The roots of *Geranium nepalense* and *G. wallichianum* have astringent properties and are used as tanning material. The leaves of *Pelargonium graveolens* and *P. odoratissimum* yield, on distillation, 'geranium oil', an essential oil.

Zygophyllaceae
(Caltrop Family)

The family is characterised by the pinnate or two foliolate leaves, persistent paired stipules, flowers with a prominent disc, free stamens bearing basal scales and carpels splitting apart at maturity as spiny or tuberculate nutlets.

The Zygophyllaceae takes into account 25 genera and 240 species. The members of this family are found generally in the drier areas of the world and particularly in the salt deserts.

Kallstroemia pubescens (G. Don) Dandy is a recently reported weed from West Bengal. *Tribulus cistoides* L. and *T. terrestris* L. are diffusedly branched prostrate herbs, the former without spiny fruits and the latter with spiny fruits.

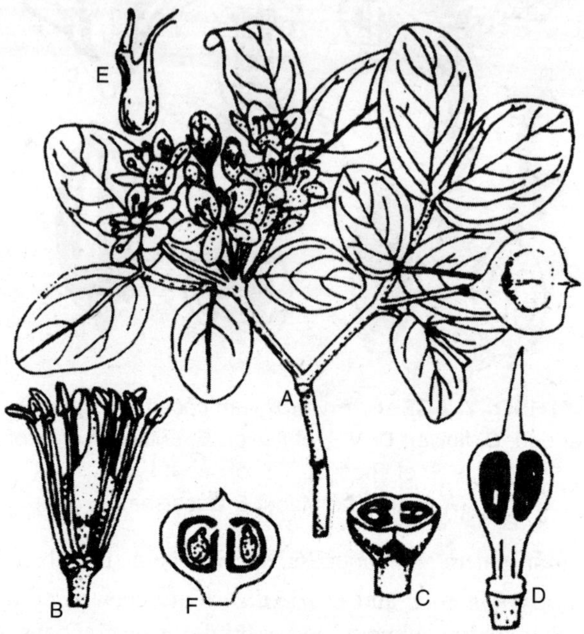

Fig. 50 ZYGOPHYLLACEAE. *Guaiacum officinale* L. **A**, small shoot bearing inflorescence and one fruit; **B**, flower with sepals and petals removed; **C**, T.S. of ovary; **D**, L.S. of ovary; **E**, ovule; **F**, L.S. of fruit. (**A–F**, after Berg *A* Schmidt)

The family is of economic value due to *Guaiacum officinale* which yields 'lignum vitae', one of the heaviest and hardest of commercial wood with a specific gravity of 0.95–1.25; it is also the source of 'guaiacum', a bitter resin, useful in medicine. Other timber yielding plants include *Bulnesia arborea*, *B. sarmientii* and *G. sanctum*. *Neoschroetera tridentata* and *Zygophyllum fabago* are used medicinally. A few species are of ornamental value, *i.e.* those of *Larrea*, *Tribulus*, *Zygophyllum*, etc. Some species of *Nitraria* yield edible fruits.

Tropaeolaceae
(Nasturtium Family)

The presence of long petioled alternate succulent leaves, zygomorphic flowers, distinct stamens and schizocarpic fruits are the characteristics of the Tropaeolaceae.

The family is composed of two genera and about 90 species extending from Mexico to Central Chile and Argentina.

Some species of the family are cultivated as ornamentals, *e.g.*, *Tropaeolum majus* and *T. peregrinum.* The unripe seeds of *T. majus* are occasionally pickled. *T. tuberosum* (Andes) yields 'cubios', a type of edible tuber.

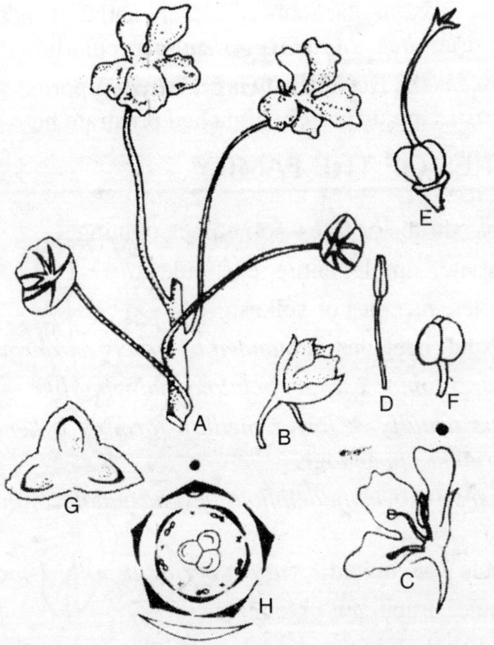

Fig. 51 TROPAEOLACEAE. *Tropaeolum majus* L. **A**, portion of plant; **B**, flower; **C**, V.S. of .flower; **D**, stamen; **E**, pistil; **F**, fruit; **G**, T.S. of ovary; **H**, floral diagram

Malpighiaceae
(Malpighia Family)

The family is diagnosed by the lianous habit, opposite leaves, appressed medifixed unicellular hairs, glandular calyx, clawed petals, peculiar anthers and lobed or winged fruits.

The Malpighiaceae is divided into two groups:

Group I. *Planitorae.* Tours concave or flat. Mericarps not winged. Examples: *Byrsonima*, *Malpighia*, etc.

Group II. *Pyramidotorae.* Torus pyramidal. Mericarps usually winged. Examples: *Tetrapteris*, etc.

The family is mostly treated as being akin to the Geraniaceae. Hallier placed it in the Polygalales. Hutchinson included it in the Malpighiaceae as derived from the tiliaceous stocks.

The family is represented by 60 genera and 800 species of the tropics, particularly in South America.

The family is of some economic significance. The bark of *Banisteriopsis caapi*, *B. inebrians*, *B. quitensis* and the species of *Tetrapteris* is used as the source of a narcotic beverage by the Indian tribes of the north-western part of the Amazon basin. *Byrsonima spicata* (West Indies) supplies 'shoemaker's bark'. The leaves of *Hiptage bengalensis* are used to treat skin diseases. The fruits of *Malpighia glabra* is good for making preserves. *Acridocarpus alternifolius*, *Banisteria laevifolia*, *Heteropterys beecheyana*, *Stigmaphyllon ciliatum*, etc. make fine ornamentals.

Polygalaceae
(Milkwort Family)

GENERAL FEATURES OF THE FAMILY

Habit	— Herbs, shrubs or trees, sometimes twining.
Leaves	— Alternate, simple, entire, exstipulate.
Inflorescence	— Panicles, racemes or spikes.
Flowers	— Bisexual, irregular, *subtended by bracts and bracteoles.*
Calyx	— *Sepals 5, inner 2 large, petaloid and wing-like.*
Corolla	— *Petals usually 3, lower median forming a keel and often bearing a dorsal fimbriated appendage.*
Androecium	— *Stamens 3–10, monadelphous beyond middle*; *anthers 1–celled, opening by apical pores.*
Gynoecium	— Carpels 2–5, united; *ovary superior, usually 2-locular*; placentation axile.
Fruit	— Capsule, drupe, nut or samara.
Seeds	— Pilose with a conspicuous strophiole; embryo axial, straight, endosperm soft, fleshy.

POSITION AND AFFINITY OF THE FAMILY

There is little agreement with respect to the phylogenetic position of the Polygalaceae. Warming included it in the series Sapindales and after the Aceraceae. Engler and Bessey placed it in the Geraniales. Wettstein put it in theTerebinthales. In the arrangement of Bentham-Hooker, it appeared between the Parietales (Papaveraceae, Cruciferae etc.) and Caryophyllinae. It was treated within the Polygalales by Hallier as well as Hutchinson. Chodat suggested that the Polygalaceae is "a very natural family, not closely allied with any other". Rendle regarded the family as of doubtful position.

NUMBER IN AND DISTRIBUTION OF THE FAMILY

The family is made up of 17 genera and 1,000 species of cosmopolitan distribution, except for the arctic zone, Polynesia and New Zealand.

COMMONLY OCCURRING PLANTS OF THE FAMILY

Polygala arillata Buch.-Ham. ex D. Don, found in Shillong, is a graceful shrub.

Polygala chinensis L. is a well-known weed in pastures.

Xanthophyllum flavescens Roxb., a large tree of Darjeeling, bears pink and yellow panicles.

ECONOMIC ASPECTS OF THE FAMILY

Economically, the Polygalaceae is of slight importance. The species of *Comesperma*, *Polygala* and *Securidaca* are grown as ornamentals. A few species of *Polygala* yield dyes. *P. butyracea* (tropical Africa) produces a fibre. The dried roots of *Polygala senega* (USA) form the source of a glucosidal drug called 'senega'.

Order 22. *Sapindales*

The order is closely related to the Geraniales from which it differs in having ovules pendulous with a ventral raphe and downward microphyle or ovules erect with a ventral raphe and upward microphyle.

According to Engler and Diels, the Sapindales is made up of 11 suborders and 23 families: Buxineae (Buxaceae), Empetrineae (Empetraceae), Coriarineae (Coriariaceae), Limnanthineae (Limnanthaceae), Anacardiineae (Anacardiaceae), Celastrineae (Cyrillaceae, Pentaphyllaceae, Corynocarpaceae, Aquifoliaceae, Hippocrataceae, Salvadoraceae, Stackhousiaceae, Staphyleaceae), Icacineae (Icacinaceae, Celastraceae, Atoxiaceae, Hippocastanaceae, Sapindaceae), Sabiineae (Sabiaceae), Melianthineae (Melianthaceae), Didiereineae (Didiereaceae) and Balsamineae (Balsaminaceae). All other systematists have considered the families of the Sapindales to belong in from three to six orders. There is a general agreement that the Sapindales should contain such families as the Sapindaceae, Hippocastanaceae, Sabiaceae, Melianthaceae, Aceraceae and Anacardiaceae, while the Celastraceae, Aquifoliaceae and Hippocrataceae should be accommodated in the Celastrales. The present author, following Core (1955), takes into account six families with the characters given below:

(1) Flowers zygomorphic	(2)
(1) Flowers actinomorphic	(4)
(2) Herbaceous plants	*Balsaminaceae*
(2) Woody plants	(3)
(3) Leaves opposite	*Hippocastanaceae*
(3) Leaves alternate	*Sapindaceae*
(4) Stems with resin ducts	*Anacardiaceae*
(4) Stems without resin ducts	(5)
(5) Fruit a berry or capsule	*Celastraceae*
(5) Fruit a samara	*Aceraceae*

Balsaminaceae
(Jewelweed Family)

The family is recognised by the herbs with watery stems, irregular flowers, anthers connate around ovary and capsule with an explosive dehiscence.

The presence of nectariferous spur in the flower had led some botanists to link the Balsaminaceae with the Geraniaceae or Tropaeolaceae. While the spur is strictly an outgrowth of the calyx in the

Balsaminaceae, receptacular tissues are involved in spur formation in the Geraniaceae and Tropaeolaceae. In the past, there was a general tendency to interpret the Balsaminaceae as a member of the Geraniales. The current view is to treat the family in the Sapindales.

The Balsaminaceae is composed of four genera and 500–600 species. The members of this family are widely distributed, especially in the tropics of Africa and Asia.

Hydrocera triflora (L.) Wt. & Arn. is an aquatic weed with floating fistular stem. Balsam (*Impatiens balsamina* L.) is a garden herb with showy variegated flowers.

The species of *Impatiens* are raised for their ornamental value, as the flowers are abundant and handsome.

Hippocastanaceae
(Horse-chestnut Family)

The family is characterised by the pinnately compound leaves, thyrsoid inflorescence, irregular flowers, simple stigma, coriaceous capsule and cotyledons adherent face-to-face.

A family of two genera and 25 species, the Hippocastanaceae abounds in the north temperate zone and South America.

Aesculus indica Colebr. ex Camb. is a very handsome tree from the Punjab and Uttar Pradesh.

Several species of the family are cultivated as ornamentals, e.g., *Aesculus californica*, *A. glabra*, *A. hippocastanum* and *A. parviflora.* The roots are crushed and used for washing woollen clothes. The seeds are utilised as fodder.

Sapindaceae
(Soapberry Family)

GENERAL FEATURES OF THE FAMILY

Habit	—	Shrubs, trees or tendril climbers.
Leaves	—	Alternate, *palmately or pinnately compound*, exstipulate.
Inflorescence	—	*Panicles or dichotomous cymes.*
Flowers	—	Small, regular or irregular, *polygamous with an annular or oblique disc.*
Calyx	—	Sepals 4 or 5, free or united, often equal.
Corolla	—	Petals 4 or 5, often bearded with a basal scale.
Androecium	—	*Stamens 5–10, free, inserted inside or outside disc.*
Gynoecium	—	Carpels 3; ovary lobed or entire, 3-locular; *ovule ascending with a ventral raphe*; style terminal.
Fruit	—	Berry or capsule.
Seeds	—	Exalbuminous, often arillate; embryo convoluted or curved.

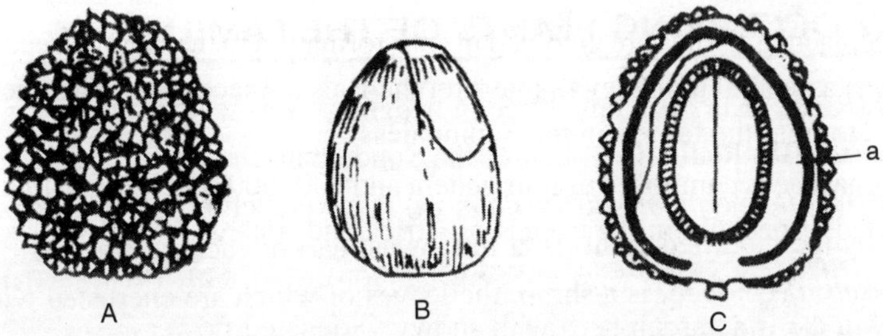

Fig. 52 SAPINDACEAE. *Litchi chinensis* (Gaertn.) Sonner. **A**, nut; **B**, aril enclosing seed; **C**, V.S. of nut (*a*, aril). (**A–C**, after Rendle)

FLORAL RANGE IN THE FAMILY

Both regular and irregular flowers are met with in the Sapindaceae. The flowers are regular in *Litchi*, *Nephelium* and *Sapindus*, but irregular in *Allophylus*, *Cardiospermum* and *Erioglossum.*

Both unisexual and bisexual flowers occur in the same plant; these flowers are polygamous. Sometimes, bisexual and staminate flowers or bisexual and pistillate flowers are found in separate plants; these are polygamo-dioecious.

The disc is produced into five elongated horns in *Xanthoceras.* The sepals are valvate in *Koelreuteria.* The petals are absent in *Dodonaea*, *Schleichera* and others.

The stamens are frequently 6, but 5 in *Turpinia* and 12 or more in *Deinbollia.*

The style is mostly 1, but rarely 2–4 and 1- to 3-lobed. The ovary is usually superior; rarely, the loculus number is 1, 2 or 4.

DIVISIONS OF THE FAMILY

Based on ovule characters, the Sapindaceae is divided into two subfamilies:

Subfamily I. *Dodonaeoideae.* Ovules 2 or more in each loculus, erect or pendulous with an upwardly micropyle. Examples: *Dodonaca*, *Turpinia*, etc.

Subfamily II. *Sapindoideae.* Ovule 1 in each loculus, erect with a downwardly micropyle. Examples: *Nephelium*, *Sapindus*, etc.

POSITION AND AFFINITY OF THE FAMILY

In Rendle's arrangement, the Sapindaceae appeared under the Sapindales between the Rutales and Rhamnales. Bentham-Hooker placed the family under the Sapindales after the Celastrales and before the Rosales. Lawrence included the family in the Sapindales between the Geraniales and Rhamnales.

In habit and general floral structure, the Sapindaceae is related to the Anacardiaceae. In the former, the flowers are mostly irregular and in the latter regular.

The presence of ovules with a ventral raphe, trimerous pistil and arillate seeds make the Sapindaceae as being akin to the Euphorbiaceae.

NUMBER IN AND DISTRIBUTION OF THE FAMILY

The Sapindaceae is chiefly a tropical family with about 150 genera and 2,000 species.

COMMONLY OCCURRING PLANTS OF THE FAMILY

Allophylus serratus (Roxb.) Kurz is a small tree in hedges.

Aphania danura (Bl.) Radlk. is a plant of the Sunderbans area.

Baloon Vine (*Cardiospermum halicacabum* L.) is a weed climbing by means of tendrils, a pair of which is formed by the modified pedicels at the lowest part of each raceme.

Dodonaea viscosa (L.) Jacq. is a shrub, the leaves of which are encrusted with viscid secretion and the fruits of which are three winged.

Erioglossum rubiginosum (Roxb.) Bl. is a plant prevalent in village shrubberies.

Longan [*Euphoria longan* (Lour.) Steud.] and Litchi [*Litchi chinensis* (Gaertn.) Sonner.] are fruit trees.

Lac Tree [*Schleichera oleosa* (Lour.) Oken] is a large deciduous tree which is considered to be one of the best hosts for the culture of lac insects.

ECONOMIC ASPECTS OF THE FAMILY

Many useful fruits are derived from the Sapindaceae, such as *Blighia sapida* (tropical Africa), *Euphoria longan*, *Litchi chinensis*, *Melicoccus bijugatus* (tropical America), *Nephalium lappaceum* (Malaya) and *Pappea capensis* (South Africa). In these plants, the fleshy aril which enlarges enormously and surrounds the seed completely forms the edible portion of the fruit.

Other economically important plants of the family include the species of *Dodonaea*, *Paullinia*, *Sapindus* and *Schleichera*. The leaves of *D. viscosa* are chewed as a stimulant; the twigs are utilised for manuring fields; the wood is employed for turnery, engraving, tool handles, walking sticks and as fire-wood. *P. cupana* (Brazil) produces the stimulating substance 'quarana' and is thus cultivated as a beverage plant. 'Yoco', another beverage, is the product of *P. yoco*. The pericarp of the fruits of *Sapindus laurifolius* and *S. mukorossi* makes lather with water and is largely used as a substitute for soap, especially for washing silk and woollen fabrics. The seeds of *Schleichera oleosa* give the 'kusum oil' of commerce; the leaves and twigs are lopped for fodder; the flowers yield a dye.

The ornamentals in the family comprise *Koelreuteria paniculata*, *Xanthoceras sorbifolia*, etc.

Anacardiaceae
(Cashew Family)

GENERAL FEATURES OF THE FAMILY

Habit — Shrubs or trees, with a *resinous juice*.

Leaves — Alternate, simple or compound, exstipulate.

Inflorescence — *Panicles*.

Flowers — Small, regular, sometimes unisexual or *polygamous with a cup-shaped or glandular disc.*

Calyx — Sepals 3–5, connate.

Corolla — Petals 3–5, free.

Androecium — *Stamens 5–10*, free.

Gynoecium — Carpels 1–5; *ovary unilocular, sometimes 5-locular; ovule ascending or pendulous with a dorsal raphe.*

Fruit — Drupe or nut.

Seeds — Exalbuminous; embryo large with fleshy cotyledons.

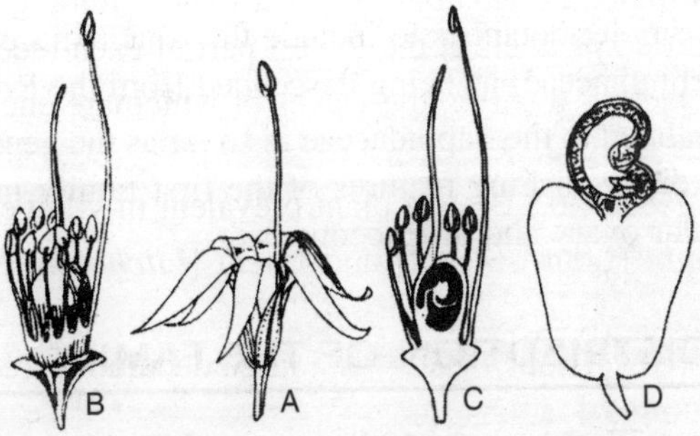

Fig. 53 ANACARDIACEAE. *Anacardium occidentale* L. **A**, flower; **B**, flower with petals removed; **C**, V.S. of flower with petals removed; **D**, V.S. of fruit. (**A–D**, after Baillon)

FLORAL RANGE IN THE FAMILY

Deviations from the normal type of flower are relatively few in the Anacardiaceae. The flowers often become unisexual, as in *Lannea*, *Pistacia*, *Rhus*, *Semecarpus*, etc. by the abortion of stamens or carpels. The inflorescence is comb-like and fasciated in *Laurophyllus.* The female bracts are concrescent into large fig-like receptacles in *Blepharocarya.*

The thalamus may be convex, bearing nectar secreting outgrowths between the stamens. A cup-like disc exists between the stamens and carpels, as in *Rhus.* The disc may be located outside the stamens, as in *Mangifera.* A gynophore is seen in *Melanorrhoea.*

The calyx is spathaceous is *Gluta.* The petals are persistent and enlarged in the fruit in *Melanorrhoea.*

The number of stamens may vary. There are 5 stamens in *Mangifera,* but only 1 stamen is found to be fertile. In *Rhus*, the stamens are also 5. In the case of *Anacardium*, the stamens may be 7–10 with only 1 becoming fertile. *Buchanania* has 10 stamens, all of which are fertile. The connective is bilobed and dilated in *Androtium.*

Though unilocular ovary is the rule in the family, exception is found in *Buchanania* where the ovary is 5-locular. The ovary is inferior in *Drimycarpus.*

DIVISIONS OF THE FAMILY

The Anacardiaceae is divided into five tribes:

Tribe (*i*) *Dobineae.* Female flowers naked. Carpel 1. Example: *Dobinea.*

Tribe (*ii*) *Mangifereae.* Leaves simple and entire. Carpels 1–5 Examples: *Buchanania*, *Mangifera*, etc.

Tribe (*iii*) *Rhoideae.* Carpels 4–5; ovary free from floral axis with a solitary ovule. Examples: *Pistacia* and *Rhus.*

Tribe (*iv*) *Semecarpeae.* Ovary adnate to and sunken in thalamus. Example: *Semecarpus.*

Tribe (*v*) *Spondieae*. Leaves pinnately compound. Carpels 4–5, connate; ovules as many as carpels. Example: *Spondias*.

POSITION AND AFFINITY OF THE FAMILY

There is a general agreement among botanists to include the Anacardiaceae in the Sapindales. Hallier regarded the family (sub-Terebinthaceae) as being descended from the Rutaceae.

The Anacardiaceae is related to the Sapindaceae in so far as the general habit and floral structure are concerned. However, the distinguishing features of the first family include the presence of intra-staminal disc, usually unilocular ovary and drupaceous fruit.

NUMBER IN AND DISTRIBUTION OF THE FAMILY

The Anacardiaceae covers about 77 genera and 600 species. Although the family is primarily a tropical one, it extends into eastern Asia, Mediterranean area and North America.

COMMONLY OCCURRING PLANTS OF THE FAMILY

Cashew-apple or Cashew-nut (*Anacardium occidentale* L.) is small or medium-sized evergreen tree with an interesting fruit structure where a reniform nut is perched upon a pyriform fleshy peduncle.

Bouea oppositifolia (Roxb.) Meissn is a tree from the Andamans and Sunderbans.

Cuddapah Almond (*Buchanania lanzan* Spreng.), Indian Ash Tree or Wodier Wood [*Lannea coromandelica* (Houtt.) Merr.], Mango (*Mangifera indica* L.), Hogplum (*Spondias dulcis* Forst. f.) and Indian Hogplum [*Spondias pinnata* (L. f.) Kurz] are well known fruit trees.

Sumac (*Rhus chinensis Mill*), a tree of Khasi and Naga Hills, has pinnately compound leaves.

Marking Nut Tree (*Semecarpus anacardium* L. f.) is a remarkable plant, the fruit of which has a roundish nut seated on the top of a pyriform fleshy peduncle.

ECONOMIC ASPECTS OF THE FAMILY

Economically, the members of the Anacardiaceae are of tremendous importance. The seed of *Anacardium occidentale* is edible after roasting; the enlarged stalk makes a fine preserve; the juice of the stalk forms a pleasant beverage and the well known 'kaju wine'. The seed of *Buchanania lanzan* is a substitute for almond and is used in confectionary; the seed also yields an aromatic oil called 'cheroonji oil'. The wood of *Cotinus coggygria* is used for making picture frames and inlaid cabinet work. The fruits of *Dracontomelon mangiferum*, *Harpephyllum caffrum*, *Mangifera indica*, *Pistacia vera*, *Spondias dulcis* and *S. pinnata* are edible. The wood of *Drimycarpus racemosus* is used for canoes and planking. The sap of *Gluta* sp. (Malaya) is used as a good lacquer. The acrid juice from the trunk and rind of the fruit of *Holigarna arnottiana* is used as varnish by painters. The gum exuding from the trunk of *Lannea coromandelica* is used for calico printing, plastering, etc. *Melanorrhoea usitata* yields a natural varnish called 'Burmese lacquer'. 'Mastic', a resin, is the product of *Pistacia cabulica* and *P. lentiscus*. *P. integerrima*, *Rhus coriaria* (Sicily) and *Schinopsia lorentzii* (Paraguay) are cultivated for tannin. *P. terebinthus* (eastern Asia) produces 'Chian turpentine'. The fruit of *R. succedanea* is the source of a natural wax which is used for making candles. *R. vernicifera* (China) yields a juice which forms the source of a natural lacquer. *S. quebracho-colorado* yields useful timber.

The Anacardiaceae includes a few ornamental plants, *e.g. Cotinus coggygria, Rhus typhina* and *Schinus molle.*

Celastraceae
(Staff-tree Family)

The family is known by the fasciculate or cymose flowers, ovary with a disc and berry or capsule in which the seeds are enveloped by a brightly-coloured aril.

The family embraces 55 genera and 850 species of wide distribution.

Cassine glauca (Rottb.) O. Kuntze is a graceful deciduous tree.

Economically, the family is of considerable importance. *Cassine glauca* produces wood for cabinet work; the root of this plant is used as an emetic and as a cure for snake bite. The buds and leaves of *Catha edulis* are used to prepare 'Khat', a beverage of the natives of Arabia. The seeds of *Kokoona zeylanica* (Sri Lanka) act as a source of oil. The species of *Celastrus, Euonymus, Pachystima*, etc. are cultivated as ornamentals. *Elaeodendron americanum, E.purpureus, Euonymus tingens, Hippocratea acapulcensis, Maytenus boania, M. ilicifolia* and *M. senegalensis*, are of medicinal value.

Aceraceae
(Maple Family)

The family is closely allied to the Sapindaceae from which it is distinguished by the opposite and palmate leaves, regular flowers and samaroid fruits.

The family comprises two genera and 102–152 species, being found in the temperate regions and tropical mountains.

Acer caesium Wall, ex Brandis is a tall deciduous tree from Himalayas.

A. macrophyllum, A. pseudoplatanus, A. rubrum and *A. saccharinum* produce valuable timber. *A. nigrum* and *A. saccharophorum* are tapped for the sweet sap which is used for making maple candy, maple sugar and maple syrup.

Order 23. *Rhamnales*

In the Rhamnales, the plants are usually woody. Here the flowers are unisexual or apetalous or both.

The order differs from other polypetalous taxa in having a low stamen number and ovary subtended or surrounded by a disc. It is distinguished from the Geraniales and Sapindales in the presence of stamens in a single whorl, as many as the sepals and opposite to the petals.

Engler and Diels treated the Rhamnales as compound of two families: Rhamnaceae and Vitaceae. Bessey included these families in the Celastrales. Wettstein and Hallier placed these families under a separate order (Rhamnales), which indicated a relationship with the Rosales. According to Rendle, the Celastrales and Rhamnales constitute a parallel series derived from some diplostemonous type resembling the Rutales. Gundersen supported Bessey's contention from cytological aspects. Hutchinson accepted the order Rhamnales, but added two more families like the Elaeagnaceae and Heteropyxidaceae. Benson agreed with the concept of the Rhamnales as put forward by Engler.

The families of the Rhamnales, according to Core (1955), can be recognised by the following characters:

(1) Trees, shrubs or lianas; fruit a drupe *Rhamnaceae*

(1) Climbing shrubs; fruit a berry *Vitaceae*

Rhamnaceae
(Buckthorn Family)

GENERAL FEATURES OF THE FAMILY

Habit	—	Shrubs or trees, *often spiny.*
Leaves	—	Alternate, *simple unlobed*, coriaceous, stipulate.
Inflorescence	—	*Axillary cymes.*
Flowers	—	Bisexual, regular, *perigynous with a disc filling calyx tube.*
Calyx	—	Sepals 4 or 5, united, *valvate.*
Corolla	—	Petals 4 or 5, clawed and horned.
Androecium	—	Stamens 4 or 5, free, *antipetalous.*
Gynoecium	—	Carpels 3; *ovary* superior or inferior, *3-locular with 1 erect basal ovule in each loculus.*
Fruit	—	Berry, capsule, drupe or samara.
Seeds	—	Albuminous or exalbuminous; embryo large and straight.

POSITION AND AFFINITY OF THE FAMILY

In Bentham-Hooker's arrangement, the Rhamnaceae appeared under the Celastrales after the Geraniales and before the Sapindales. Rendle placed the family under the Rhamnales lying between the Celastrales and Rosales. Lawrence included it in the Rhamnales between the Sapindales and Malvales. Hutchinson accommodated it in the Rhamnales between the Santalales and Myrsinales.

The Rhamnaceae bears an affinity with the Vitaceae from the standpoint of floral construction, but is distinguished from it by having simple leaves and receptacular cup.

NUMBER IN AND DISTRIBUTION OF THE FAMILY

The Rhamnaceae comprises 58 genera and about 900 species. The plants of this family are cosmopolitan in distribution.

COMMONLY OCCURRING PLANTS OF THE FAMILY

Berchemia floribunda Wall. is a pretty shrub found in the sub-himalayan tracts.

Gouania tiliaefolia Heyne is a climber with inferior three-winged fruits.

Helinus lanceolatus Brand. is a cirrhose scandent shrub.

Ventilago denticulata Willd. is a strong climber.

Jujube Tree or Indian Plum (*Ziziphus mauritiana* Lamk.) is a small tree.

Ziziphus oenoplia (L.) Mill. and *Z. xylopyrus* Willd. are straggling shrubs.

ECONOMIC ASPECTS OF THE FAMILY

Many members of the Rhamnaceae are of value as drugs, for fruits, for timber and for other purposes. Some species of *Berchemia, Ceanothus, Discaria, Hovenia, Noltea, Pomaderris, Reynosia, Ziziphus,* etc. are grown ornamentally. *C.reclinatus* (tropical America) supplies 'Mabee bark'.The bark of *Colubrina asiatica* (Samoa) is used for cleaning mats. *Frangula alnus, Maeosopis eminii* and *Z. chloroxylon* produce valuable timber. The wood of *Gouania domingensis* is utilised for making chewsticks. The root extract of *G. tiliaefolia* (Philippines) is used as a substitute for soap. While *Hovenia dulcis* yields and edible fruit axis, the fruits of *Z. lotus, Z. mauratiana, Z. vulgaris* and *Z. xylopyrus* are edible. 'Black ironwood', one of the heaviest of woods with a specific gravity of 1.3, is derived from *Krugiodendron ferreum* (West Indies). 'Sap green' pigment is the product of the berries of *Rhamnus cathartica, R. frangula* and *R. purshiana* (USA) are cultivated for their bark providing the purgatives, 'frangula' and 'cascara sagrada' respectively. 'Lokao' or 'Chinese green', a green powdery substance, is made from the bark of *R. globosa* and *R.utilis.* The dried unripe fruits of *R. infectoria* furnish yellow and green dyes. The bark of *Ventilago denticulata* is used for making cordage; the root gives a red dye which is utilised for dyeing cotton fabrics and tassar silk. Extracts of *V. oblongifolia* (Malaya) is used as a poultice to cure cholera. The fruits of *Z. joazeiro* (Brazil) are used as fodder.

Vitaceae
(Grape Family)

GENERAL FEATURES OF THE FAMILY

Habit	—	Shrubs, *climbing by leaf opposed tendrils.*
Leaves	—	Alternate, simple or digitate, stipulate.
Inflorescence	—	Cymes, panicles, racemes or spikes, *arising opposite to a leaf at node.*
Flowers	—	Small, greenish, sometimes unisexual, *hypogynous with a prominent disc.*
Calyx	—	Sepals 4 or 5, toothed or entire, connate.
Corolla	—	Petals 4 or 5, caducous, sometimes connate.
Androecium	—	Stamens 4 or 5, *antipetalous.*
Gynoecium	—	Carpels 2, syncarpous; *ovary partially sunk in disc*, 2-to 6-locular; *ovules 1–2 in each loculus, ascending with a ventral raphe*; *placentation axile*; *stigma capitate or discoid.*
Fruit	—	Berry.
Seeds	—	Embryo small; *endosperm cartilaginous.*

POSITION AND AFFINITY OF THE FAMILY

Bentham-Hooker included the Vitaceae under the Celastrales after the Geraniales and before the Sapindales. In Rendle's arrangement, this family appeared under the Rhamnales between the Celastrales and Rosales. Hutchinson placed the family in the Rhamnales lying between the Santalales and Myrsinales.

The Vitaceae descended from the Rutaceae by the disappearance of the antipetalous whorl of stamens. It stands very close to the Rhamnaceae with a similar floral structure. The Vitaceae, however, differs from the Rhamnaceae by the climbing habit, baccate fruit, albuminous seeds and small embryo.

NUMBER IN AND DISTRIBUTION OF THE FAMILY

The Vitaceae embraces 12 genera and approximately 700 species. Although plants belonging to this family are widely distributed in the tropical and subtropical areas of the world, some often extend into the temperate regions.

COMMONLY OCCURRING PLANTS OF THE FAMILY

Cayratia carnosa (Lamk.) Gagnep. is a short herb without tendrils and with lower leaves about 60 cm across and the upper ones 15–30 cm across.

Cayratia pedata (Lamk.) Juss. ex Gagnep. and *Cissus setosa* Roxb. are found in the country sides.

Cissus quadrangularis L. is a climber with a jointed quadrangular herbaceous sympodial stem.

Cissus woodrowii (Stapf ex Cooke) Santapau is an erect shrub without tendrils.

Leea aequata L. is shrub which is 1.2–3.0 m in height.

L. herbacea Buch–Ham. is another shrub, attaining a height of 3.6–4.8 m.

ECONOMIC ASPECTS OF THE FAMILY

The family is of economic significance due to the various types of grape which arise from the species of *Vitis*, such as *V. aestivalis* (summer grape), *V. labrusca* (fox grape), *V. riparia* (forest grape), *V. rotundifolia* (muscadine), *V. vinifera* (wine grape) and *V. vulpina* (winter grape). "Apart from having edible fruits, the dried Grapes become Raisins and a seedless form provides Sultanas. Currants come from a Corinthian variety. In addition the seeds are the source of Grape seed oil used in the manufacture of soap, paints and (after refining) in certain foods. The waste products are used as cattle food, tannin and cream of Tartar" (Perry, 1972).

The stems of *V. papillosa* (Java) and *V. sicyoides* (Mexico) are utilised as cordage.

The Vitaceae has some medicinal importance. The herbaceous leaves of *Cayratia pedata* and *Cissus setosa* are roasted and oiled and then applied to boils to bring about suppuration.

The ornamental value of the family is enhanced by such plants as *Parthenocissus tricuspidata* and *P. quinquefolia.* Some species of *Cissus* and *Rhoicissus* are cultivated as desirable house plants.

Order 24. *Myrtales*

The order is closely allied to the Rosales and Umbellales. This is a natural taxon with a transition from the perigyny to epigyny. Here the flowers are bisexual and polypetalous, perianth tetramerous, placentation axile and style simple. Another characteristic feature is the presence of intraxylary phloem in the stem.

The Myrtiflorae of Engler and Diels was made up of four suborders and 23 families: Thymelaeineae (Geissolomataceae, Peneaceae, Oliniaceae, Thymelaeaceae, Elaeagnaceae), Myrtineae (Lythraceae, Heteropyxidceae, Sonneratiaceae, Crypteroniaceae, Punicaceae, Lecythidaceae, Rhizophoraceae, Nyssaceae, Alangiaceae, Combretaceae, Myrtaceae, Melastomaceae, Hydrocaryaceae, Onagraceae, Haloragaceae), Hippuridineae (Hippuridaceae, Thelygonaceae), Cynomorineae (Cynomoriaceae). Wettstein accepted the order with a slight modification by merging the Hydrocaryaceae with the Onagraceae and separating the Gunneraceae as distinct from the Haloragaceae. Bessey also accepted the concept, but transferred the Thymelaeineae to the Celastrales and retained the Nyssaceae in the Cornaceae (under the Umbelliferae). Rendle adopted the Engler treatment, but interpreted the Nyssaceae and Alangiaceae as separate taxa of the Umbelliflorae. Hallier removed the Thymelaeineae to the Daphnales, the Alangiaceae to the Santalales and the Onagraceae to the Ranales. He restricted the order Myrtines for such families as the Lythraceae, Lecythidaceae, Rhizophoraceae, Combretaceae, Myrtaceae and Melastomaceae. Hutchinson distributed the various families among five orders: the Aquilariaceae and Thymelaeaceae in the Thymelaeales, the Elaeagnaceae and Heteropyxidaceae in the Rhamnales, the Lythrales in the Araliales and the Myrtaceae, Lecythidaceae, Sonneratiaceae, Punicaceae, Combretaceae and Melastomaceae in the Myrtales. After a study of *Geissoloma marginatum*, Dahlgren and Rao (1959) questioned the inclusion of the Geissolomataceae in the Myrtales. They emphasised its similarity with the Cleaceae and Salvadoraceae. Bhatnagar (1963) suggested the separation of *Duabanga* and *Sonneratia* from the Lythraceae. He felt that the Lythraceae and Onagraceae did not resemble much in their floral vasculature and that they could only tentatively be retained in the Myrtales. On the basis of floral anatomy, Gupta (1962) supported the transference of *Barringtonia*, *Careya* and *Couroupita* to a separate family Lecythidaceae.

While Lawrence considered 14 families in the Myrtiflorae and Cronquist 13 in the Myrtales, Core mentioned 12 families in the order. Core's notion is followed here, but the sequence of the families has been changed.

(1) Flowers perigynous	(2)
(1) Flowers epigynous	(5)
(2) Petals usually present	(3)
(2) Petals usually none	(4)
(3) Leaves with 3–9 longitudinal veins	*Melastomaceae*
(3) Leaves not so veined	*Lythraceae*
(4) Plants covered with silvery scales	*Elaeagnaceae*
(4) Plants not covered with scales	*Thymelaeaceae*
(5) Aquatic plants	(6)
(5) Terrestrial plants	(7)
(6) Petals present	*Trapaceae*
(6) Petals reduced or none	*Haloragaceae*
(7) Loculi of ovary superposed	*Punicaceae*
(7) Loculi of ovary not superposed	(8)
(8) Leaves opposite	(9)
(8) Leaves alternate	(11)
(9) Generally lianas	*Combretaceae*
(9) Not lianas	(10)
(10) Seashore plants with vivipary	*Rhizophoraceae*
(10) Not seashore plants	*Onagraceae*
(11) Fruit often woody	*Lecythidaceae*
(11) Fruit not woody	*Myrtaceae*

Thymelaeaceae (Mezereum Family)

The family is characterised by the simple leaves, perigynous flowers, uniseriate gamosepalous perianth, petals reduced to hypanthial appendages, definite member of stamens and superior unilocular ovary.

The phyletic position of the Thymelaeaceae is debatable. Bessey treated the family in the Celastrales. Hallier interpreted it as the primitive group of his Daphnales—an order originating from the Linaceae. Engler and Rendle placed the Thymelaeaceae in the Myrtiflorae. Lawrence and Core included it in the Myrtales. Hutchinson once suggested that the affinity of the family lay with the Caryophyllales through the Geraniales. This idea was later discarded with the statement appended as "I now consider the Thymelaeaceae to be the apetalous relatives of the Bixaceae, especially of the exotic family Flacourtiaceae".

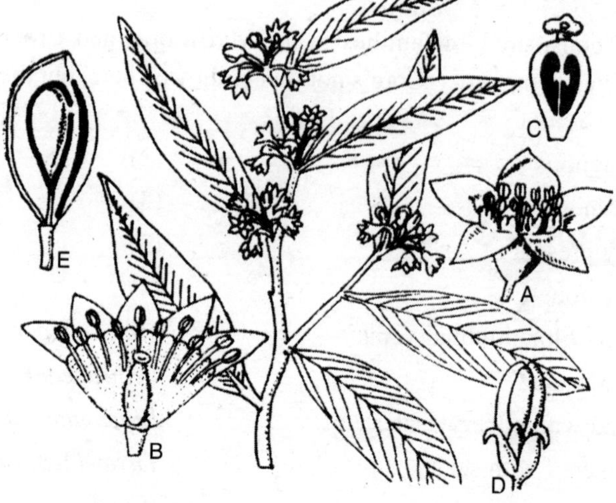

Fig. 54 THYMELAEACEAE. *Aquilaria malaccensis* Lamk. **A**, flower; **B**, flower spread open; **C**, V.S. of ovary; **D**, fruit; **E**, V.S. of fruit. (**A–E**, after W. Hooker)

The family includes 45 genera and perhaps 500 species of cosmopolitan distribution, excepting the polar regions and concentrating in Australia, South Africa, Mediterranean area and the Steppes of central and western Asia.

The family is of economic importance. The resin impregnated wood of *Aquilaria malaccensis* is the source of 'agaru' of commerce. The fibrous inner bark of *Daphne cannabina* is utilised for preparing 'Nepal paper'. The specie of *Daphne, Dirca, Edgeworthia* and *Pimelea* are cultivated ornamentally. The well known 'rice paper' is made from *E. tomentosa* and *Wikstroemia canescens.* The lace-like inner bark of *Lagetta lintearia* is suitable for making dress.

Elaeagnaceae
(Oleaster Family)

The family is recognised by the silvery or golden-brown stellate or lepidote scales that densely cover the leaves and achenes enveloped by a fleshy persistent perianth.

Many systematists have treated the Elaeagnaceae as being an ally of the Thymelaeaceae with the androperianth tube indicating a relationship between the two families and hence they are included under the Myrtales or Myrtiflorae. Hutchinson placed the Elaeagnaceae in the Rhamnales, suggesting it to be "a more advanced group then and related to the Rhamnaceae".

The family contains three genera and about 50 species of largely steppe and rock plants with the centres of abundance in southern Asia, Europe and North America.

Economically, the family is of some importance. The species of *Elaeagnus* are cultivated as deciduous or evergreen shrubs. The fruit of *E.angustifolia* and *Shepherdia argentea* are edible. The fruits of *Hippohae rhamnoides* (France) are sometimes made into jelly or as a sauce eaten with fish and meat; the hard wood has been used for turnery.

Melastomaceae
(Melastoma Family)

GENERAL FEATURES OF THE FAMILY

Habit — Herbs or shrubs.

Leaves — Opposite or verticillate, simple, *entire with* 3–9 *equal conspicuous longitudinal nerves.*

Flowers — Bisexual, regular, perigynous.

Calyx — Sepals 4 or 5, free and persistent, sometimes reduced to a rim on receptacle or united to form a cap.

Corolla — *Petals 4 or* 5, free, twisted or *imbricate.*

Androecium — *Stamens as many as or twice as many as petals*; *filaments* unequal and curved inwards; anthers dehisce by apical pores and bear a sickle form appendage on connective.

Gynoecium — Carpel 1; *ovary inferior or half-inferior, 4- to 5-locular* with many ovules in each loculus.

Fruit — Berry or capsule.

Seeds — Exalbuminous; embryo small; cotyledons fleshy.

DIVISIONS OF THE FAMILY

The Melastomaceae is divided into three subfamilies:

Subfamily I. *Astronioideae.* Placentation basal or parietal. Fruit many-seeded. Embryo minute. Examples: *Astronia*, *Kibbessia*, etc.

Subfamily II. *Melastomatoideae.* Placentation axile. Fruit many-seeded. Embryo minute. Examples: *Melastoma*, *Osbeckia*, etc.

Subfamily III. *Memecyloideae.* Fruit 1- to 5-seeded. Embryo large. Example: *Memecylon.*

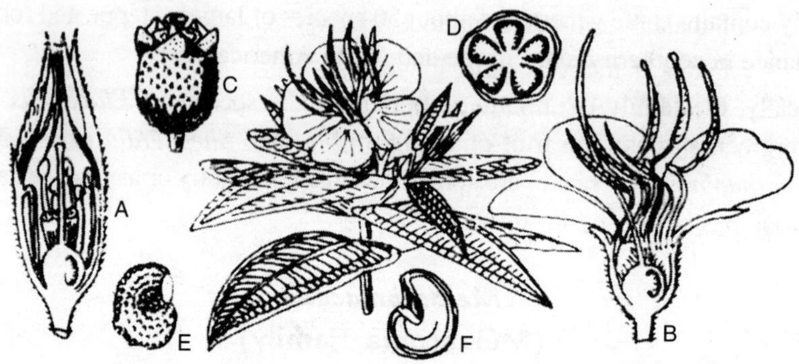

Fig. 55 MELASTOMACEAE. *Melastoma malabathricum* L. **A**, V.S. of closed flower; **B**, V.S. of open flower; **C**, fruit; **D**, T.S. of ovary; **E**, seed; **F** V.S. of seed. (**A–F**, after Baillon)

NUMBER IN AND DISTRIBUTION OF THE FAMILY

This family, of about 240 genera and 3,000 species, is distributed in the warmer parts of the earth, the chief centre of concentration being South America and forming a characteristic of the flora of Brazil.

COMMONLY OCCURRING PLANTS OF THE FAMILY

Melastoma malabathricum L. occurs from the base of the Himalaya to Darjeeling.

Memecylon capitellatum L. and *M. umbellatum* Burm. f. are small trees.

Osbeckia chinensis L. and *O. crinita* Benth. are found in nurseries.

Sonerila tenera Royle is a herb.

Tibouchina semidecandra Cogn. is a shrub with deep violet flowers and orange anthers.

ECONOMIC ASPECTS OF THE FAMILY

The species of *Bertolonia*, *Gravesia* and *Sonerila* are grown for their variegated leaves and those of *Centradenia*, *Heterocentron*, *Medinilla* and *Rhexia* for their flowers. The leaves and fruits of *M. rubicunda* are edible. The wood of *Memecylon umbellatum* makes excellent firewood and charcoal; the pulp of the fruits is edible; the leaves and flowers contain dyes and are employed to dye mats and cotton fabrics.

Lythraceae
(Loosestrife Family)

GENERAL FEATURES OF THE FAMILY

Habit	—	Herbs, shrubs or trees, *often with 4-angled branches.*
Leaves	—	*Opposite or whorled*, simple, entire.
Flowers	—	Bisexual, regular, perigynous.
Calyx	—	Sepals 3- to 6-lobed, united, *valvate.*
Corolla	—	Petals as many as calyx lobes, *crumpled.*
Audroecium	—	Stamens 4–8, *inflexed in bud.*
Gynoecium	—	Carpels 2–6; *ovary superior*, *2- to 6-locular*; ovules many, anatropous and ascending; placentation axile; style simple; stigma capitate.
Fruit	—	Capsule.
Seeds	—	*Exalbuminous*; embryo straight; cotyledons flat.

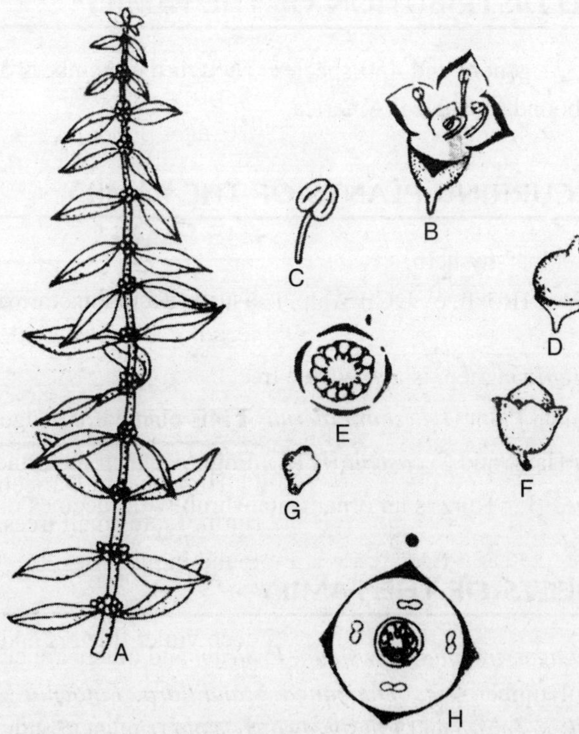

Fig. 56 LYTHRACEAE. *Ammania baccifera* L. **A**, portion of plant; **B**, flower; **C**, stamen; D, carpel; **E**, T.S. of ovary; **F**, fruit; **G**, seed; **H**, flora diagram.

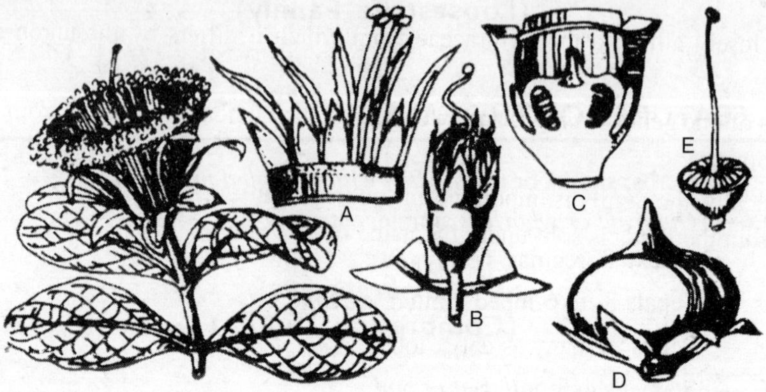

Fig. 57 LYTHRACEAE. *Sonneratia acida.* Roxb. **A**, perianth and stamens; **B**, young fruit; **C**, V.S. of ovary; **D**, mature fruit; **E**, ovary. (**A–E**, after Hutchinson)

NUMBER IN AND DISTRIBUTION OF THE FAMILY

The Lythraceae has about 22 genera and 450 species. Though the members of this family are widely distributed, they mostly abound in tropical America.

COMMONLY OCCURRING PLANTS OF THE FAMILY

Ammannia baccifera L. is a marshy herb.

Duabanga grandiflora (Roxb. ex DC.) Walp. is a large deciduous tree of Nepal, Bhutan, Assam and Sikkim.

Lagerstroemia thorelii Gagnep. is a way side tree.

Indian Privet of Henna Plant (*Lawsonia inermis* L.) is planted in hedges.

Sonneratia apetala Ham. and *S. caseolaris* (L.) Engl. occur in the Sunderbans.

Woodfordia fruticosa (L.) Kurz is an ornamental shrub with racemes of red flowers.

ECONOMIC ASPECTS OF THE FAMILY

Some species of *Cuphea*, *Lagerstroemia*, *Lawsonia*, *Lythrum* and others are cultivated ornamentally. A few plants produce useful timber, *e.g.*, *Duabanga grandiflora*, *Lafoenia speciosa*, *Lagerstroemia lanceolata*, *L. parviflora*, *L. speciosa* and *Physocalymma scaberrimum* (South America). The leaves of *Lafoenia pacari* yields a yellow dye. *Lawsonia inermis* is the source of 'henna', an orange dye used for personal adornment; the flowers of this plant, on distillation, yield an aromatic oil, useful in perfumery and embalming. A liquid, called 'keora water', is obtained from *Sonneratia caseolaris* by distillation. The wood *of Woodfordia fruticosa* is used as fire wood.

Punicaceae
(Pomegranate Family)

The family is closely allied to the Lythraceae from which it differs by the union of ovary with the receptacle.

A family of one genus (*Punica*) and two species, the Punicaceae extends from the Balkans to the Himalaya and Socotra.

Economically, the family is important for 'pomegranate', which is the fruit of *Punica granatum*; a syrup made from the seeds is sold under the name of 'grenadine'.

Combretaceae
(Combretum Family)

GENERAL FEATURES OF THE FAMILY

Habit	—	Shrubs or trees, often climbing.
Leaves	—	Alternate or opposite, simple, exstipulate.
Inflorescence	—	Panicles, racemes or spikes.
Flowers	—	Bisexual or polygamo-dioecious, regular; floral axis forms a tube.
Calyx	—	Sepals 4 or 5, united, *valvate.*
Corolla	—	Petals 4 or 5, alternate with sepals.
Androecium	—	*Stamens 4–5, in 1 whorl, or 6–10, in two whorls,* inserted on tube or limbs of calyx.
Gynoecium	—	Carpel 1; *ovary inferior, unilocular*; *ovules 2–6 in each loculus,* pendulous and anatropous; style long, fili form, stigma pointed.
Fruit	—	Drupe, winged at angles.
Seeds	—	Exalbuminous; cotyledons irregularly folded or spirally rolled.

NUMBER IN AND DISTRIBUTION OF THE FAMILY

The Combretaceae comprises 20 genera and about 475 species. The plants belonging to this family are found in the tropical and subtropical regions of the world.

COMMONLY OCCURRING PLANTS OF THE FAMILY

Anogeissus latifolia (DC.) Wall. ex Bedd. is a big tree, having a cluster of winged fruits.

Calycopteris floribunda (Roxb.) Lamk. is a diffuse or scandent shrub.

Combretum C. ovalifolium Roxb. and *roxburghii* Spreng. are large climbing shrubs.

Gyrocarpus jacquini Gaertn. is a large tree.

Lumnitzera racemosa Willd. is a small tree of the Sunderbans.

Drunken Sailor or Rangoon Creeper (*Quisqualis indica* L.) is a garden climber.

Arjun Tree or White Murdah [*Terminalia arjuna* (Roxb. ex DC.) Wt. & Arn.], Bedda Nut or Beleric Myrobalan [*T. bellerica* (Gaertn.) Roxb. ex Fleming], Indian Silver Grey Wood or White Chuglam (*T. bialata* Steud.), Indian Almond or Malabar Almond (*T. catappa* L.), Black Myrobalan or Chebulic Myrobalan (*T. chebula* Retz.) and Laurel or Sain (*T. tomentosa* Wt. & Arn.) are trees with angled fruits.

ECONOMIC ASPECTS OF THE FAMILY

The Combretaceae has some economic value. The gum, exuding from the trunk of *Anogeissus latifolia*, is used for calico printing, paper sizing and confectionery. The fruits of *Calycopteris floribunda* are used in the treatment of jaundice. The species of *Combretum* and *Quisqualis* are of ornamental significance. The young fruits of *Terminalia bellirica* and *T. chebula* are rich in tannin. The fruits of *T. catappa* are edible. *T. ivorensis* and *T. superba*, both of West Africa, are important sources of timber.

Rhizophoraceae
(Mangrove Family)

GENERAL FEATURES OF THE FAMILY

Habit	—	Shrubs or trees.
Leaves	—	Opposite, simple, stipulate, coriaceous.
Flowers	—	Bisexual, legular, epigynous.
Calyx	—	Sepals 4–8, valvate.
Corolla	—	Petals 4–8, fringed, *convolute or inflexed in bud.*
Androecium	—	*Stamens 8 to many*, located in outer edge of a lobed disc; *anthers 4-celled* with an introrse longitudinal dehiscence.
Gynoecium	—	Carpels 2–5, united; *ovary superior, inferior or half-inferior*; ovules 2 in each loculus, pendulous and anatropous; style 1, with a small unlobed stigma.
Fruit	—	Berry or capsule.

FLORAL RANGE IN THE FAMILY

The flowers are polygamo-dioecious or dioecious in *Blepharistemma* and *Sterigmapetalum.* In the latter, the petals are much divided. In *Kandelia*, the petals are divided into thread-like segments. Only in *Anopyxis*, *Cassipourea* and *Gynotroches*, the ovary is superior. In the remainder of the genera, the ovary is semi-inferior or inferior.

NUMBER IN AND DISTRIBUTION OF THE FAMILY

The Rhizophoraceae is composed of 16 genera and about 120 species. Many plants of this family form the characteristic vegetation of muddy tidal flats and shore lines throughout the palaeotropical regions of the world.

COMMONLY OCCURRING PLANTS OF THE FAMILY

Blepharistemma membranifolia (Miq.) Ding Hou is a handsome straight stemmed tree.

Bruguiera gymnorrhiza (L.) Lamk is a large tree.

Carallia brachiata (Lour.) Merr. is an evergreen tree, having shiny leaves.

Ceriops tagal (Perr.) C.B. Robins is a large shrub of the Sunderbans.

Kandelia candel (L.) Druce is a small tree.

Rhizophora apiculata Bl. and *R. mucronata* Lamk. are small trees of the Sunderbans.

ECONOMIC ASPECTS OF THE FAMILY

Economically, the Rhizophoraceae is of some importance. The wood of *Carallia integerrima* is used for house building, furniture, cabinet making, ornamental panelling, picture frames, rice pounders, etc; the leaf infusion is a substitute for tea in Malacca. The bark-decoction of *Ceriops tagal* stops haemorrhage; tannin is extracted from the roots and fruits and a dye from the bark. *Poga oleosa* yields 'poga nuts'. The wood of *Rhizophora mucronata* is used as fuel and for making charcoal; the bark is used for tanning and dyeing; the fruit is edible and a light wine is prepared from the juice by fermentation.

Onagraceae
(Evening-primrose Family)

GENERAL FEATURES OF THE FAMILY

Habit	—	Herbs, sometimes aquatic.
Leaves	—	Alternate or opposite, simple, exstipulate.
Flowers	—	Bisexual, regular, *epigynous*, *tetramerous*.
Calyx	—	Sepals with 4 limbs, united, *valvate*.
Corolla	—	Petals 4, *convolute*.
Androecium	—	*Stamens in 2 whorls of 4 each, outer whorl larger than inner*; pollen grains spherical, porous with protruding stoppers.
Gyroecium	—	Carpels 4; *ovary 4-locular*; *ovules many*, anatropous; placentation axile: style simple, long and filiform; *stigma capitate*, *notched or 4-rayed*.
Fruit	—	Berry or capsule.
Seeds	—	Exalbuminous; embryo straight or nearly so.

POSITION AND AFFINITY OF THE FAMILY

Hutchinson placed the Onagraceae in the Lythrales and considered the latter to be derived from the Caryophyllales. Hallier treated it as one of the primitive taxa in his Polygalines—an order regarded by him as descended from the Linaceae. It was included in the Myrtiflorae or Myrtales by Bessey, Wettstein, Lawrence and Core.

Fig. 58 ONAGRACEAE. *Ludwigia adscendens* (L.) Hara **A**, portion of shoot; **B**, flower; **C**, sepal; **D**, petal; **E**, stamen; **F**, carpel; **G**, T.S. of ovary; **H**, seed; **I**, floral diagram

The Onagraceae is closely related to the Myrtaceae, Melastomaceae and Lythraceae.

NUMBER IN AND DISTRIBUTION OF THE FAMILY

The Onagraceae embraces 18 genera and about 640 species. Although the members of the family are cosmopolitan in distribution, they are abundant in the temperate regions of North America and in South America. Most of them occur in dry or moist open habitats.

COMMONLY OCCURRING PLANTS OF THE FAMILY

Ludwigia adscendens (L.) Hara is a herb, partly creeping in the mud and partly floating on the surface of tanks.

Ludwigia perennis L. and *L. prostrata* Roxb. are herbs occurring in the rice fields.

ECONOMIC ASPECTS OF THE FAMILY

The Onagraceae is of little economic value. The species of *Clarkia*, *Fuchsia*, *Godetia* and *Oenothera* are cultivated for their ornamental flowers. From the perianth of *Fuchsia* is extracted 'fuchsine', a red dye used as a staining reagent. The leaves and bark of *F. magellanica* have medicinal uses.

Myrtaceae
(Myrtle Family)

GENERAL FEATURES OF THE FAMILY

Habit	—	Shrubs or trees.
Leaves	—	Opposite, simple, entire, exstipulate, *gland-dotted.*
Flowers	—	Bisexual, regular, *epigynous with a disc lining calyx tube.*
Calyx	—	Sepals 4 or 5, often united, sometimes thrown off as a lid.
Corolla	—	Petals 4 or 5, free, sometimes thrown off.
Androecium	—	*Stamens many, bent inwards in bud,* free or polyadelphous; anthers dorsifixed or versatile; *connective gland tipped.*
Gynoecium	—	Carpels 2–5, syncarpous; *ovary inferior, 1- to 2-locular; ovules 2 to many in each loculus,* pendulous, anatropous or campylotropous; placentation axile; style long, kneed or flexuose.
Fruit	—	Berry or capsule, tipped by calyx limbs.
Seeds	—	Exalbuminous; testa hard.

FLORAL RANGE IN THE FAMILY

Departures from the normal type of floral structure are few in the Myrtaceae. The number of sepals is 4 or 5 in the same genus; the same can be said about the number of petals. The petals may form a cap-like structure which is thrown off after the fruits attain maturity, as in *Eucalyptus.*

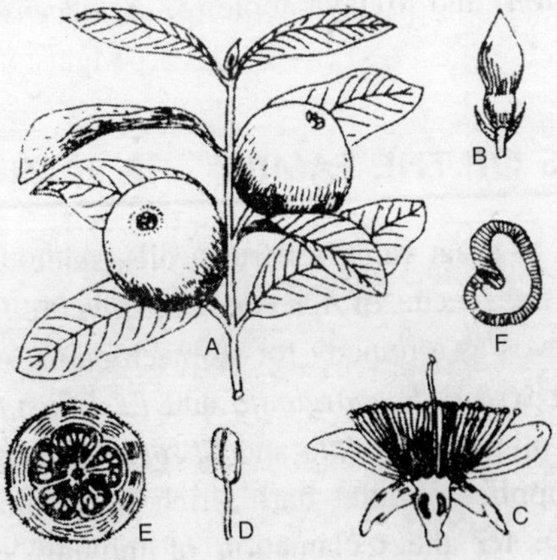

Fig. 59 MYRTACEAE. *Psidium guajava* L. **A**, branch with young fruit; **B**, flower bud; **C**, V.S. of flower; **D**, stamen; **E**, T.S. of ovary; **F**, L. S. of seed to show embryo. (**A–F**, after Nidenzu)

The stamens are generally free. Sometimes, they are united into bundles (*Melaleuca*). The anthers are usually versatile, but basifixed in *Calothamnus.*

The number of carpels is subjected to some variation. In *Pimenta* and *Syzygium,* the carpels are 2–3. The ovary is unilocular in *Calycothrix, Verticordia* and related genera; elsewhere, the ovary has more than 1 loculus.

DIVISIONS OF THE FAMILY

The Myrtaceae is divided into two subfamilies, one with two tribes and the other with a single tribe:

Subfamily I. *Leptospermoideae.* Leaves opposite or alternate. Fruit dry, a capsule or nut-like.

Tribe (*i*) *Chamaelaucieae.* Ovary 1-locular. Examples: *Calycothrix*, *Verticordia*, etc.

Tribe (*ii*) *Leptospermeae.* Ovary many-locular. Examples: *Callistemon*, *Eucalyptus*, *Melaleuca*, etc.

Subfamily II. *Myrtoideae.* Leaves always opposite. Fruit fleshy, typically a berry and rarely a drupe.

Tribe (*i*) *Myrteae.* Ovary 2- to 5-locular. Examples: *Myrtus*, *Psidium*, *Syzygium*, etc.

NUMBER IN AND DISTRIBUTION OF THE FAMILY

The Myrtaceae covers about 100 genera and 3,000 species. The plants of this family belong to warmer climates, with two centres of distribution—one in Australia and another in America.

COMMONLY OCCURRING PLANTS OF THE FAMILY

Acmena acuminatissima (Bl.) Merr. & Perry is a tree from the Andamans.

Decaspermum fruticosum J.R. & G. Forst. is a slender tree of Khasi Hills, Meghalaya.

Eucalyptus globulus Labill. is cultivated throughout India.

Myrtle (*Myrtus communis* L.) is a shrub encountered in hedges.

Guava (*Psidium guajava* L.), Black Plum or Indian Blackberry [*Syzygium cumini* (L.) Skeels], Rose apple [*S. jambos* (L.) Alston] and Malaya apple [*S. malaccense* (L.) Merr. face Perry] are well known fruit trees.

ECONOMIC ASPECTS OF THE FAMILY

Economically, the Myrtaceae is of great value for fruits, oils, spices, timbers and ornamentals. Those cultivated ornamentally include the species of *Acmena*, *Callistemon*, *Feijoa*, *Tristania*, etc. The berries of *Decaspermum fruticosum* serves as a remedy for stomach pains. A few yield valuable timber, *e.g.* *Eucalyptus diversicolor*, *E. leucoxylon*, *E. marginata* and *E. robusta.* The leaves of *E. globulus*, on distillation, give us 'eucalyptus oil'. *E. maculata* and *E. rostrata* provide 'citron gum' and 'red gum' respectively. *E. occidentalis* supplies 'Mallet bark', rich in tannin. *Leptospermum laevigatum* is extensively planted in Australia for the reclamation of moving sand. The leaves of *Melaleuca leucadendron* (Malacca) furnish 'cajeput oil'. Spices like 'allspice' and 'cloves', are derived from *Pimenta dioica* (tropical America) and *Syzygium aromaticum* (Zanzibar) which are dried unripe fruits and unopened flower buds respectively. *P. racemosa* yields 'oil of bay rum'. The species of *Eugenia*, *Feijoa* and *Syzygium* produce edible fruits; the fruits of *Myricaria cauliflora*, *Psidium guajava* and *P. sellowiana* are also edible.

Lecythidaceae
(Brazilnut Family)

GENERAL FEATURES OF THE FAMILY

Habit	—	Shrubs or trees.
Leaves	—	Alternate, simple, *not gland-dotted.*
Inflorescence	—	Racemose or solitary.
Flowers	—	Large, bisexual, regular or irregular.
Calyx	—	Sepals 4–6, valvate or slightly imbricate.
Corolla	—	Petals 4–6, free or united into a tube.
Androecium	—	*Stamens many*, in several series, sometimes outer ones become staminodes and resemble a corona; anthers basifixed.
Gynoecium	—	Carpels 2–6 or more, united; *ovary inferior.*
Fruit	—	*Berry or woody capsule.*

NUMBER IN AND DISTRIBUTION OF THE FAMILY

The Lecythidaceae, with about 20 genera and 450 species, is centred in the wet areas of tropical South America with some representatives in Africa and Asia.

COMMONLY OCCURRING PLANTS OF THE FAMILY

Barringtonia acutangula (L.) Gaertn. occurs along the banks of rivers and canals.

B. racemosa (L.) Spreng. is found in the Sunderbans.

Careya arborea Roxb. is a small tree.

Cannon ball Tree (*Couroupita guianensis* Aubl.) is a garden tree, being cauliflorous.

ECONOMIC ASPECTS OF THE FAMILY

Economically, the Lecythidaceae is of some importance. *Barringtonia acutangula*, *Couroupita guianensis*, *Lecythis grandiflora* and *L. ollaria* produce good timber. Edible nuts are furnished by *Bertholletia excelsa, L. ollaria, L. usitata* and *L. zabucajo.* The leaves of *Careya arborea* are used for making 'bidis' and for covering 'Burmese cherroot'. 'Bark cloth' is made from the fibres, yielded by *Couaratari tauari. Grias cauliflora* is cultivated for the fruits called 'anchovy pear', useful in making pickles.

Haloragaceae
(Water-milfoil Family)

The family is known by the monoecious plants, unisexual flowers, stamens four or eight and inferior ovary with a solitary ovule in each loculus.

The family is divided into two subfamilies, one containing two tribes and the other without any:

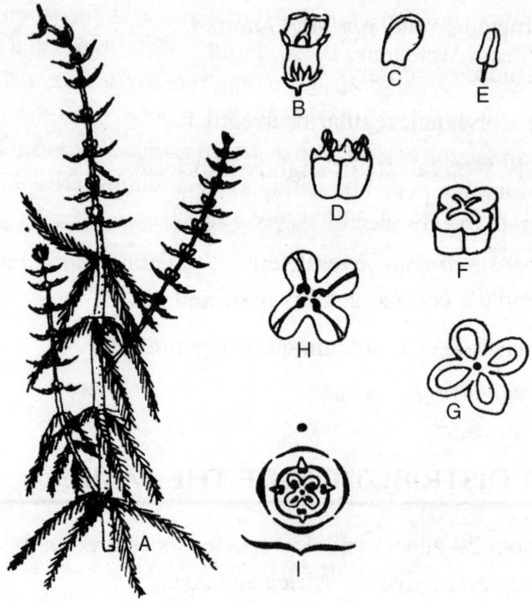

Fig. 60 HALORAGACEAE. *Myriophyllum indicum* Willd. **A**, portion of plant; **B**, flower; **C**, corolla; **D**, corolla with stamens; **E**, stamen; **F**, carpel; **G**, T.S. of ovary; **H**, fruit; **I**, floral diagram

Subfamily I. *Haloragoideae.* Leaves exstipulate. Inflorescence small. Petals present or absent. Ovary 2- to 4-locular. Fruit dry.

Tribe (*i*) *Halorageae.* Fruit indehiscent. Examples: *Haloragis*, *Proserpinaca*, etc.

Tribe (*ii*) *Myriophylleae.* Fruit dehiscing into two or four nutlets. Example: *Myriophyllum.*

Subfamily II. *Gunneroideae.* Leaves stipulate. Inflorescence large. Petals absent Ovary unilocular. Fruit succulent. Example: *Gunnera.*

The family has been interpreted as a reduced form of the Onagraceae. Hallier considered it as derived from the Nymphaeaceae and hence placed it in the Ranales. Hutchinson included it in the Lythrales, suggesting it as advanced over the Onagraceae.

There are seven genera and about 180 species in the Holoragaceae. The members of this family are widely distributed, but abundant in Australia.

Myriophyllum indicum Willd. and *M. tuberculatum* Roxb. are seer in the border of the salt lakes and other moist places near Calcutta.

Trapaceae
(Water-chestnut Family)

The family is often included in the Onagraceae from which it differ by the branched assimilatory roots, uniseriate multicellular hairs or petioles, pedicels and lower surface of leaves, floating leaves with inflated petioles, single whorl of stamens and one seeded drupe with two to four upwardly directed spines.

As early as 1898, Raimann created a family Hydrocaryaceae for accommodating the genus *Trapa.* This view received support from Wettstein (1935). Pulle (1938) stated that the family name should be changed to the Trapaceae in the absence of any genus like *Hydrocarya.* However, Bentham-Hooker (1883), Rendle (1925) and Hutchinson (1959) included the taxon in the Onagraceae. Eames (1953) expressed the opinion that on anatomical evidence *Trapa* is not allied to the Onagraceae and does not belong to it; the vascular bundles here are without a cambium and sphaeraphides are of common occurrence. Gibbs (1958) cited the absence of raphides. in *Trapa* and their presence in the Onagraceae. The embryological features of *Trapa*, such as semi-inferior and bilocular ovary, eight-nucleate embryosac, lack of endosperm, Solanad type of embryo, extremely well developed and haustorial suspensor and extremely unequal cotyledons, justify its removal from the Onagraceae and its assignment to the Trapaceae (Ram, 1956).

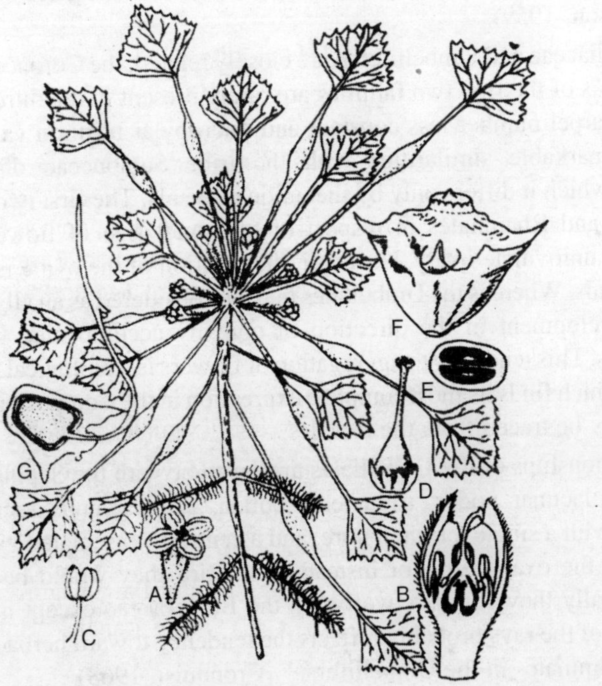

Fig. 61 TRAPACEAE. *Trapa nutans* L. var. *bispinosa* (Roxb.) Makino. **A**, flower; **B**, V.S. of flower; **C**, stamen; **D**, disk and pistil; **E**, T.S. of ovary; **F**, fruit; **G**, V.S. of germinating seed

A family of one genus (*Trapa*) and about 30 species, the Trapaceae belongs to the Old World.

Water chestnut [*Trapa natans* L.var. *bispinosa* (Roxb.) Makino] is an aquatic herb with the floating leaves rhomboidal and submerged leaves pinnatipartite as well as root-like.

The fruits of *Trapa* are eaten either raw or cooked, the starchy cotyledon forming the edible part.

Order 25. *Umbellales*

The Umbellales is a natural taxon. It is characterised by a tendency of the flowers to be in umbels, great reduction in the calyx associated with epigyny, single whorl of stamens and presence of two uniovulate carpels.

Engler and associates conceived the order as composed of three families: Araliaceae, Umbelliferae, Cornaceae. Though the circumscription of the order was adopted by Rendle, it was not accepted in toto by Bessey and Wettstein. Bessey treated both *Garrya* and *Nyssa* within the concept of the Cornaceae. Wettstein expanded the order to accommodate the Garryaceae. However, Hallier recognised two families: Cornaceae, Umbelliferae; here the Araliaceae was merged with the Umbelliferae. Rickett (1945) created two orders, the Umbellales for the Araliaceae and Umbelliferae and the Cornales for the Cornaceae and Nyssaceae. Revising his earlier views, Hutchinson (1948) constructed the Umbellales for the Umbelliferae only, considered the Cornaceae within the Cunoniales (which is a derivative of the Rosales) and set up the Araliaceae as a distinct taxon (Araliales) originating from the Cornaceae. He also regarded the Umbellales as a climax group, possibly derived from the Saxifragales stock and parallel with the Araliaceae (Hutchinson, 1959).

While the Araliaceae and Umbelliferae are closely related, the Cornaceae stands somewhat alone. The secretory passages of the first two families are rarely present in the third. Moreover, the leaves are generally opposite, carpel number less constant and micropylar position varies in the Cornaceae. The family indicates remarkable similarities with the tribe Sambuceae of the sympetalous family Caprifoliaceae from which it differs only by the union of petals. The first two families bear relationship with the Celastrales and Rhamnales in respect of the small size of flowers, single staminal whorl conspicuous disc and uniovulate ovary. However, the position of the ovule is different and the stamens are opposite to the petals. Whereas the Umbellales may be considered as an ally of these taxa, it represents a further line of development in the direction of epigyny accompanied by the aggregation of the inconspicuous flowers. This tendency to aggregation of flowers is a biological factor of prime importance in insect pollination which finds its most complete expression in the Compositae. The probable phylogeny of the Umbellales may be traced from the Rosales.

"The true relationships of the Umbellales appear to be with the Sapindales, in which compound leaves, stipules, multilacunar nodes, trinucleate pollen, schizogenous secretory canals, vasicentric parenchyma, vessels with a single terminal pore, and a centrally vasculated gynoecium are well known. If the Araliaceae had the ovary superior instead of inferior they would be perfectly at home in the Sapindales. Anatomically they compare well with the Burseraceae except in having somewhat wider wood rays. The width of the rays probably reflects the tendency toward herbaceousness which begins in the Araliaceae and culminates in the Umbelliferae" (Cronquist, 1968).

The usual three families of Umbellales, according to Core (1955), may be recognised as follows:

(l) Flowers 4-merous	*Cornaceae*
(1) Flowers 5-merous	(2)
(2) Fruit a berry or drupe	*Araliaceae*
(2) Fruit a cremocarp	*Umbelliferae*

Araliaceae
(Ginseng Family)

The family is characterised by the leaves with a stellate indumentum, umbellate inflorescence, pentamerous flowers, inferior uniovulate ovary and fruit a berry or drupe.

The family is divided into three tribes:

Tribe (*i*) *Aralieae*. Petals imbricate, sessile with a broad base. Examples: *Aralia*, *Panax*, etc.

Tribe (*ii*) *Mackinlayeae*. Petals valvate, shortly clawed. Example: *Mackinlaya.*

Tribe (*iii*) *Scheffiereae*. Petals valvate, not clawed. Examples: *Hedera*, *Schefflera*, etc.

The family contains about 65 genera and 700 species, having centres of distribution in tropical America and Indo-Malaysia.

Helwingia himalaica Hook. f. & Thoms, ex Clarke, common in Darjeeling, is an undershrub with unisexual umbels about the middle of the upper surface of the leaves as well as near the apex of the phyllodes. *Heteropanax fragrans* (D. Don) Seem is a small unarmed tree. *Panax fruticosum* L. is a garden shrub with decompound or dissected sheathing leaves. *P. pseudo-ginseng* Wall. is a herb met with in Khasi Hills, Meghalaya. *Schefflera venulosa* (Wt. & Arn.) Harms. is a climbing shrub.

The family is of some economic importance. The species of *Acanthopanax*, *Aralia*, *Dizygotheca*, *Fatsia*, *Polyscias* and *Schefflera* as well as *Hedera helix* are grown as ornamentals. The wood of *Acanthopanax ricinifolium* (Japan) is used for making hats. Medicinal extracts are made from *Aralia cordata* and *A. racemosa.* 'Virginian sarsaparilla' and 'Ginseng' are obtained from *A. nudicaulis* and *Panax schinseng* (Korea) respectively. The roots of *P. quinquefolium* (USA) serve as a stomachic and stimulant. The pith of *Tetrapanax papyriferus* (China) is used for making 'rice paper'. *Tieghemopanax murrayi* produces valuable timber.

Umbelliferae
(Carrot Family)

GENERAL FEATURES OF THE FAMILY

Habit — *Aromatic herbs with fistular stems.*

Leaves — Alternate, dissected, exstipulate with a *sheathing base.*

Inflorescence — *Simple or compound umbels.*

Flowers — Bisexual, regular, *epigynous* with a disc.

Calyx — *Sepals* 5, united, toothed.

Corolla — *Petals 5*, free, *unequal.*

Androecium — *Stamens 5, free*; anthers versatile.

Gynoecium — *Carpels* 2, syncarpous; *ovary inferior,* 2-locular; ovule 1 in each loculus, pendulous with a single integument and ventral raphe; styles 2; stigmas capitate.

Fruit — *Cremocarp*, dehiscing into two dry segments *(mericarps)* which are suspended from a slender biforked axis (*carpophore*); each mericarp is provided with longitudinal ridges (*costae* or *jugae*) and between two ridges are furrows (*valleculae*) which are traversed by oil ducts (*vittae*).

Seeds — Albuminous; embryo small.

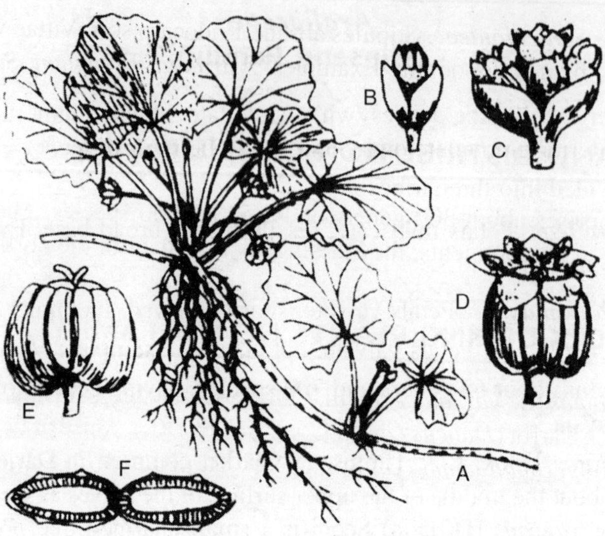

Fig. 62 UMBELLIFERAE. *Centella asiatica* (L.) Urban. **A**, portion of plant bearing flower and fruit; **B**, young inflorescence; **C**, older inflorescence; **D**, flower; **E**, fruit; **F**, T.S. of fruit. (**A–F**, after Rendle)

FLORAL RANGE IN THE FAMILY

The most distinguishing character of the Umbelliferae is the umbellate inflorescence. Although compound umbel is the commonest form of inflorescence, simple umbels are found in *Astrantia* and *Eryngium.* In the species of *Azorella*, the umbel is reduced to a single flower. The flowers may be aggregated into dense heads, as in *Eryngium.* The terminal flower is seen in *Daucus.*

The flowers are generally bisexual. But unisexual flowers are noted in *Echinophora* where a central female flower is surrounded by male umbels. A dioecious condition is noticed in *Arctopus. Torilis anthriscus* is in possession of polygamous flowers. The flowers are usually white, sometimes yellow or pink and seldom blue.

The calyx is smaller than the corolla and is represented by inconspicuous teeth. But these structures are well developed in *Astrantia*, *Eryngium*, *Sanicula*, etc. The calyx has a slight imbricate aestivation, while the corolla has an open, valvate or imbricate.

DIVISIONS OF THE FAMILY

Based on stipule and fruit characters, the Umbelliferae is divided into three subfamilies:

Subfamily I. *Apioideae.* Stipules absent. Endocarp soft or fibrous. Vittae various, style at tip of a stylopodium. Examples: *Apium*, *Coriandrum*, *Echinophore*, *Ferula*, etc.

Subfamily II. *Hydrocotyloideae.* Stipules present. Fruit with a woody endocarp and without a carpophore. Vittae absent or sunk in primary ridges only. Examples: *Azorella* and *Centella.*

Subfamily III. *Saniculoideae*. Stipules absent. Endocarp soft. Vittae various. Style surrounded by a stylopodium. Examples: *Astrantia*, *Eryngium*, *Sanicula*, etc.

NUMBER IN AND DISTRIBUTION OF THE FAMILY

The Umbelliferae comprises about 300 genera and 2,500–3,000 species. Though the members of the family are represented in all continents, they are mostly abundant in the north temperate zone.

COMMONLY OCCURRING PLANTS OF THE FAMILY

Celery [*Apium graveolens* L. *var dulce* (Mill.) DC.], Coriander (*Coriandrum sativum* L.), Cumin (*Cuminum cyminum* L.), Carrot (*Daucus carota* L. var. *sativa* DC.), Anise or Fennel (*Foeniculum vulgare* Mill.), Ajowan [*Trachyspermum ammi* (L.) Sprague ex Turrill] and *T. roxburghianum* (DC.) Craib are grown in India.

Bunium persicum (Boiss.) Fedts. is found in Kashmir.

Bupleurum mucronatum Wt. & Arn., occurring in Hazaribagh, Ranchi and Dehradun, is a weed with undivided leaves and yellow flowers.

Pennywort [*Centella asiatica* (L.) Urban] is a weed with simple undivided reniform crenate leaves.

Water Hemlock (*Cicuta virosa* L.) is a herb with straight hollow stems.

Oenanthe javanica (Bl.) DC. is found in shady places near ditches and ponds.

ECONOMIC ASPECTS OF THE FAMILY

The Umbelliferae is of great economic importance. *Angelica archangelica* (USA) is the source of an aromatic oil, 'angelica oil', useful in confectionary. Some are used as vegetables, *e.g.*, *Anthriscus cerefolium* (France), *Apium graveolens*, *Ligusticum scoticum* (Chile & New Zealand) and *Pteroselinum crispum* (southern Europe). *Arracacia xanthorhiza* (Peru to Mexico), *Chaerophyllum bulbosum* (Europe), *Daucus carota* var. *sativa*, *Pastinaca sativa* and *Sium sisarum* (south-eastern Europe) are cultivated for their edible roots and *Bunium bulbocastanum* and *Conopodium majus* for their edible tubers. *Carum carvi*, *Coriandrum sativum*, *Cuminum cyminum*, *Foeniculum vulgare*, *Levisticum officinale* (southern Europe), *Pimpinella anisum* (Mediterranean), *Trachyspermum ammi* and *T. burghianum* are raised for their fruits which are used as spices.

Some plants are useful in medicine, such as some species of *Apium*, *Centella* and *Eryngium*. *Dorema ammoniacum* (Iran) is the source of 'gum ammoniac'. The roots of *Bupleurum jucundum* are used in liver troubles. *Ferula foetida* and *F. narthex* (Afghanistan) yield 'asafoetida', an ill-scented drug. *F. galbaniflua* and *F. rubicaulis* are the sources of 'gum galbanum'.

A few plants are grown ornamentally, *e.g.* the species of *Aegopodium*, *Astrantia*, *Eryngium*, *Heracleum*, *Myrrhis* and others.

Cornaceae
(Dogwood Family)

The family is recognised by the woody or subligneous plants, umbellate inflorescence, inferior ovary and fleshy indehiscent drupe or berry.

The family includes 13 genera and over 100 species. While most of the species are found in the temperate regions of Asia and North America, some occur in the tropics and subtropics of South America and New Zealand.

Benthamidia capitata (Wall.) Hara, found in the hills near Dehradun and Darjeeling, is a tree with decussate simple leaves and heads (umbels) of scentless yellow flowers, each head having an involucre of four large bracts. *Mastixia rostrata* Bl. occurs in Assam hills.

The family is of importance for ornamentals, *e.g. Aucuba, Cornus, Davidia* and *Griselinia. C. florida, Cynoxylon nuttallii, Nyssa aquatica, N. sylvatica* and *Swida alternifolia* yield timber. The fruits of *Cornus mas* are used for making preserves and for the alcoholic beverage 'vin de cornouille'.

SYMPETALAE

The Sympetalae is a combination of orders and families numbering eight and 45 respectively. Here the flowers show a greater or less union of the petals.

The diagnostic characters of the eight orders of the Sympetalae are indicated by the analytical synopsis provided below (Lawrence, 1951; Core, 1955):

(1) Ovary superior	(2)
(1) Ovary inferior	(7)
(2) Stamens free from corolla	*Ericales*
(2) Stamens borne on corolla	(3)
(3) Stamens opposite to corolla lobes	(4)
(3) Stamens alternate with corolla lobes	(5)
(4) Stamens usually in one whorl	*Primulales*
(4) Stamens usually in two to three whorls	*Ebenales*
(5) Corolla scarious; fruit a circumscissile capsule	*Plantaginales*
(5) Corolla herbaceous; fruit not a circumscissile capsule	(6)
(6) Stamens inserted near corolla base; ovary with 2 carpels	*Gentianales*
(6) Stamens adnate to corolla tube; ovary with 2–5 carpels	*Tubiflorales*
(7) Stamens distinct, rarely coherent	*Rubiales*
(7) Stamens coherent of variously connate	*Campanulales*

Order 26. *Ericales*

According to many systematists, the Ericales is a primitive component of the Sympetalae and forms a connecting link between the Archichlamydeae and Metachlamydeae. The primitiveness of the group is indicated by the retention of the polypetalous characters in some of its members. Bessey viewed the origin of the order from the Geraniales. Hallier derived it from the Guttiferales. Hutchinson interpreted it as arising from his Theales.

The Ericales is a natural taxon. It is characterised by the usually pentamerous and actinomorphic flowers. Here the petals are both united and free. The androecium is obdiplostemonous and inserted at the edge of a hypogynous nectariferous disc. The pollen grains are often in tetrads. The placentation is axile and the ovules are unitegmic and numerous.

Engler and Diels conceived the Ericales to be made up of two suborders and four families: Ericineae (Clethraceae, Pyrolaceae, Ericaceae), Epacridineae (Epacridaceae). Rendle considered five families (Clethraceae, Ericaceae, Pyrolaceae, Epacridaceae, Diapensiaceae) in the order. Lawrence adopted Engler's concept of the order and Mitra supported Rendle's idea. In the order, Core accommodated six such families as the Clethraceae, Pyrolaceae, Lennoaceae, Ericaceae, Epacridaceae and Diapensiaceae. Cronquist included seven families in the order: Cyrillaceae, Clethraceae, Ericaceae, Epacridaceae, Empetraceae, Pyrolaceae, Monotropaceae. Hutchinson raised the number of families in the order to eight by creating more families like the Monotropaceae and Vacciniaceae. The present author takes into account five families which are characterised below:

(1) Stamens free from corolla	(2)
(1) Stamens borne on corolla	(4)
(2) Corolla sympetalous	*Ericaceae*
(2) Corolla polypetalous	(3)
(3) Shrubs	*Clethraceae*
(3) Low herbs	*Pyrolaceae*
(4) Carpels 3	*Diapensiaceae*
(4) Carpels 5	*Epacridaceae*

Clethraceae (Pepperbush Family)

The family is characterised by the shrubby habit, polypetalous corolla, anthers inflexed in bud, pollen grains single (and not in tetrads) and generally trilocular ovary.

Asa Gray accepted the family as a tribe bf the Ericaceae. This view has been followed by Bentham-Hooker and other systematists. Klotzsch's contention that the taxon deserves the rank of the family was accepted by Drude and others. Hallier considered it as the primitive family of his Ericales, being derived from the Ochnaceae.

There is one genus (*Clethra*) and about 120 species in the Clethraceae which belongs to tropical America.

Several species of *Clethra* are grown ornamentally for their aromatic summer bloom, *e.g.*, *C. acuminatum*, *C. alnifolia*, *C. arborea*, *C. monostachya* and *C. tomentosa.*

Pyrolaceae (Pyrola Family)

The family is recognised by the herbaceous habit, polypetalous corolla and uniformly loculicidal capsule. The family is separated into two subfamilies:

Subfamily I. *Monotropoideae.* Parasitic or saprophytic herbs without green colour, leaves being reduced to scales. Anthers opening by longitudinal slits and pollen grains simple. Examples: *Allotropa*, *Monotropa*, *Pterospora*, *Sarcodes*, etc.

Subfamily II. *Pyroloideae.* Green leafy herbs with somewhat scapose habit. Anthers opening by terminal pores and pollen grains in tetrads. Examples: *Chimaphila*, *Moneses*, *Pyrola*, etc.

Several botanists have expressed the view that the Pyrolaceae is not sufficiently distinct from the Ericaceae as to deserve the status of a separate family. Rydberg (1914) and Small (1914) recognised the Pyrolaceae as a distinct taxon, but segregated the saprophytic chlorophyll deficient Monotropoideae as the Monotropaceae. Copeland (1947) discarded the idea that the Pyrolaceae and Monotropaceae constitute two different families apart from the Ericaceae. This idea was previously shared by Henderson (1920) who considered the Pyrolaceae as a subfamily sprung from the Andromedeae. Wodehouse (1935) suggested that the single pollen grains of the Monotropoideae are unlike any other ericaceous pollen. According to him, the Monotropoideae is distinct from the Ericaceae and unlike the Pyroloideae. The Pyrolaceae is closely related to the Ericaceae from which it differs by the incomplete septation of the ovary and the undifferentiated embryo. It may be derived from an ericaceous stock with poly petalous corolla, but the shrubby habit was lost in the course of evolution and the herbaceous habit gradually became saprophytic.

A family of ten genera and about 70 species, the Pyrolaceae abounds in the north temperate and arctic regions.

Monotropa uniflora L. and *Pyrola rotundifolia* L. are found in temperate Himalaya.

Ericaceae
(Heath Family)

GENERAL FEATURES OF THE FAMILY

Habit	—	*Shrubs* or trees, rarely herbs.
Leaves	—	Alternate, opposite or whorled, simple, exstipulate.
Inflorescence	—	Racemose or solitary.
Flowers	—	Bisexual, regular or irregular, hypogynous or epigynous.
Calyx	—	Sepals 4–5, free or united, persistent.
Corolla	—	*Petals 4–7, united.*
Androecium	—	*Stamens 8–10, free, arising from a nectariferous disc*; anthers 2-celled.
Gynoecium	—	Carpels 4–5, joined; ovary inferior or superior; placetation axile; style long; stigma capitate or lobed.
Fruit	—	Berry or capsule.
Seeds	—	Embryo small; endosperm fleshy.

DIVISIONS OF THE FAMILY

The Ericaceae is divided into four subfamilies:

Subfamily I. *Arbutoideae.* Petals united, deciduous. Anthers often with bristle-like appendages. Ovary superior. Fruit a berry or loculicidal capsule. Seeds triangular ovate, not winged. Examples: *Arbutus*, *Epigaea*, etc.

Subfamily II. *Ericoideae.* Corolla persistent. Anthers appendaged. Fruit a loculicidal capsule. Seeds round, not winged. Examples: *Calluna*, *Erica*, etc.

Subfamily III. *Rhododendroideae.* Petals sometimes free, deciduous. Anthers without appendages. Fruit a septicidal capsule. Seeds with a strongly ribbed coat, often winged. Examples: *Daboecia*, *Ledum*, *Rhododendron*, etc.

Subfamily IV. *Vaccinioideae.* Petals united, deciduous. Ovary inferior. Fruit a berry. Examples: *Thibaudia*, *Vaccinium*, etc.

NUMBER IN AND DISTRIBUTION OF THE FAMILY

The Ericaceae contains about 100 genera and 3,000 species. The members of this family form the characteristic vegetation of many regions of acid soils, especially in moors, swamps and mountain slopes throughout the temperate areas of the world. They are also found to a limited extent in the arctic region and tropical mountains.

COMMONLY OCCURRING PLANTS OF THE FAMILY

Gaultheria fragrantissima Wall. and *Lyonia ovalifolia* (Wall.) Drude are found in Khasi Hills, Meghalaya.

Rhododendron hookeri Nutt., a shrub of Darjeeling, Sikkim and Himalaya, is noted for its cluster of blazing red flowers.

ECONOMIC ASPECTS OF THE FAMILY

The Ericaceae is important for ornamentals which is contributed by the species of *Arctostaphylos*, *Calluna*, *Erica*, *Pieris*, *Rhododendron*, etc. The leaves of *Agapetes saligna* are used as tea substitutes. The foliage and twigs of *Andromeda polifolia* (Russia) are used for tanning purposes. Fruits of *Arbutus unedo* are used for making preserves and alcoholic drinks. *A. menziesii* produces valuable timber. *E. arborea* (Mediterranean) provides 'briar root wood' which is used to make tobacco pipes. The leaves of *Gaultheria procumbens* furnishes 'oil of wintergreen', an essential oil, used for medicinal preparations. *G. shallon* forms 'lemon leafs, *Ledum latifolium* 'Labrador tea' and *Vaccinium arctostaphylos* 'Broussa tea'. Many species yield edible fruits, *e.g.*, Huckleberry (*Gaylussacia baccata*), Cranberry (*Oxycoccus palustris*), Blueberry (*V. angustifolium*, *V. corymbosum*, *V. membranaceum*, *V. myrtilloides*, *V. ovatum*, *V. vacillans)* and Cowberry (*V. vitis-idaea*).

Epacridaceae
(Epacris Family)

The family is closely allied to the Ericaceae from which it differs by the single whorl of epipetalous stamens, one-celled anthers and tricarpellate gynoecium.

The Epacridaceae is separated into two subfamilies:

Subfamily I. *Epacridoideae.* Stems without annular scars. Leaves rarely sheathing. Bracteoles usually persistent. Examples: *Epacris*, *Leucopogon*, etc.

Subfamily II. *Richeoideae.* Stems with annular scars. Leaves sheathing. Bracteoles often caducous. Example: *Richea.*

The family includes 30 genera and about 400 species. They are chiefly found in Australia and Tasmania, but are also distributed in New Zealand, Hawaii, South America and India.

The family is important for ornamentals like the species of *Cyathodes*, *Dracophyllum*, *Epacris*, *Leucopogon* and *Richea.* The leaves and roots of *Styphelia malayana* (Malaysia) are used medicinally; the inner bark of stems is used for making canoes waterproof.

Diapensiaceae
(Diapensia Family)

The families characterised by the pentamerous perianth, absence of a disc, stamens borne on the corolla and pollen grains not in tetrads.

The family comprises seven genera and 20 species, inhabiting the cooler and temperate regions of the northern hemisphere.

The family is important for ornamentals, *e.g.*, the species of *Diapensia*, *Pyxidanthera*, *Schizocodon* and *Shortia.* The foliage of *Galax aphylla* are collected and preserved for decorative purposes.

Order 27. *Primulales*

The Primulales is a natural taxon. It is characterised by the pentamerous flowers, single whorl of stamens opposite to the petals, ovary superior with free central placentation and ovules with two integuments.

The prevalence of pentamery, unilocular ovary with free central placentation and seeds with a copious endosperm indicate a connection of the Primulales with the Centrospermales. It has been assumed that the order originated from a dialypetalous stock allied to that from which the Centrospermales sprang.

Bessey placed the Plumbaginaceae and Plantaginaceae in his Primulales. Hallier set up the Primulales near his Bicornes, deriving it from the Ochnaceae of his Guttales. Similar circumscription was made by Wettstein who regarded the Primulales as related to the Ericales (Bicornes). The order, according to Rendle, Lawrence and Cronquist, contains three families: Theophrastaceae, Myrsinaceae, Primulaceae. Core grouped the Myrsinaceae, Plumbaginaceae and Primulaceae into two orders: the Primulales comprising the Primulaceae and Plumbaginaceae and the Myrsinales covering the Myrsinaceae. The present author, in agreement with Core, takes into account three families which may be distinguished as follows:

(1) Styles 5; fruit an achene or utricle	*Plumbaginaceae*
(1) Style 1; fruit a berry, capsule or drupe	(2)
(2) Shrubs or trees; fruit a berry or drupe	*Myrsinaceae*
(2) Herbs; fruit a capsule	*Primulaceae*

Myrsinaceae
(Myrsine Family)

GENERAL FEATURES OF THE FAMILY

Habit	—	*Shrubs or trees.*
Leaves	—	Alternate, simple, *coriaceous with resin ducts*, exstipulate.
Inflorescence	—	Racemose.
Flowers	—	Bisexual or unisexual, regular.
Calyx	—	Sepals 4–6, free or united, persistent.
Corolla	—	Petals 4–6, rotate or salverform.
Androecium	—	Stamens as many as and opposite to corolla lobes; anthers 2-celled.
Gynoecium	—	Carpels 4–6; ovary usually superior; *ovules generally buried in proliferating placentae*; *style simple.*
Fruit	—	*Drupe.*
Seeds	—	Embryo cylindrical; endosperm copious.

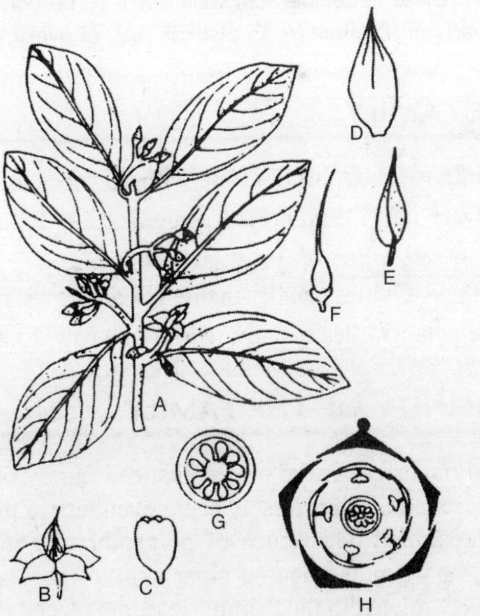

Fig. 63 MYRSINACEAE. *Ardisia solanacea* Roxb. **A**, portion of shoot; **B**, flower; **C**, calyx; **D**, petal; **E**, stamen; **F**, pistil; **G**, T.S. of ovary; **H**, floral diagram

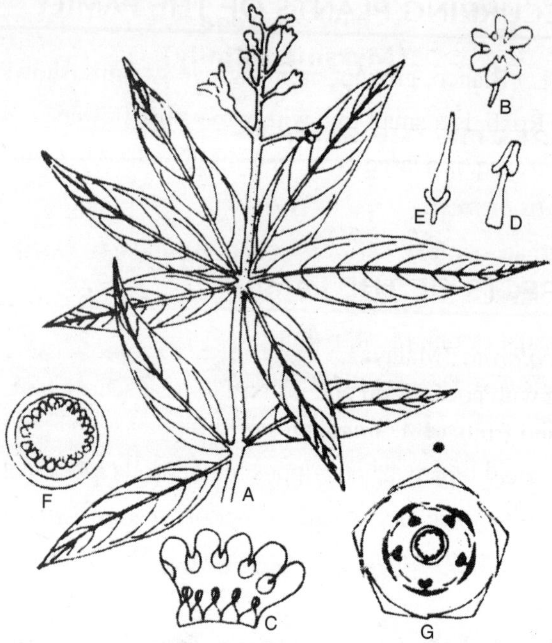

Fig. 64 MYRSINACEAE. *Jacquinia ruscifolia* Jacq. **A**, portion of shoot; **B**, flower; **C**, corolla spread out; **D**, stamen; **E**, pistil; **F**, T.S. of ovary; **G**, floral diagram

DIVISIONS OF THE FAMILY

The Myrsinaceae is divided into two subfamilies, one with no tribes and the other with two tribes:

Subfamily I. *Maesoideae.* Ovary half inferior. Fruit many seeded. Example: *Maesa.*

Subfamily II. *Myrsinoideae.* Ovary superior. Fruit one seeded.

- **Tribe** (*i*) *Ardisieae.* Ovules usually many in many rows, rarely few. Example: *Aegiceras.*
- **Tribe** (*ii*) *Myrsineae.* Ovules usually few in one row. Example: *Rapanea.*

POSITION AND AFFINITY OF THE FAMILY

According to Wettstein, the Myrsinaceae is the most advanced family of the Primulales. This is due to the frequency of dioecism, presence of transversely septate anthers in many genera, prevalence of half inferior ovary in the Maesoideae and occurrence of polyembryony in *Ardisia.* Hallier regarded the family as less advanced than the Primulaceae and more akin to the Theophrastaceae. Hutchinson did not consider the family as closely related to the Primulaceae and placed the two taxa at widely divergent points in his classification. He treated the Myrsinaceae, together with the Theophrastaceae and Aegicerataceae, under the Myrsinales—an order probably derived from the same stock as his Rhamnales and Celastrales.

NUMBER IN AND DISTRIBUTION OF THE FAMILY

There are 32 genera and 1,000 species in the Myrsinaceae. The members of this family are widely distributed in the tropical and subtropical areas, with the northern limits in Japan, Mexico and Florida as well as the southern limits in South Africa and New Zealand.

COMMONLY OCCURRING PLANTS OF THE FAMILY

Aegiceras corniculatum (L.) Blanco, growing in mangrove swamps, shows vivipary.

Ardisia solanacea Roxb. is a small tree with waxy pink flowers.

Jacquinia ruscifolia Jacq. is a large evergreen shrub with spinescent acuminate leaves and ovoid orange red fruits.

ECONOMIC ASPECTS OF THE FAMILY

Leaf extract of *Ardisia colorata* (Malaya) is used to treat stomach complaints, while the sap of *A. fulginosa* (Java) is boiled with coconut oil and utilised to treat scurvy. The fruits of *A. crispa* (Malaya) are edible. The flowers and fruits of *A. squamulosa* (Philippines) are employed to flavour fish dishes. Some species of *Myrsine* yield timber which can be employed for furniture and turnery.

Primulaceae
(Primrose Family)

GENERAL FEATURES OF THE FAMILY

Habit — *Herbs.*

Leaves — Opposite or whorled, simple.

Flowers — Bisexual, regular, hypogynous.

Calyx — Sepals 5, united, generally persistent.

Corolla — *Petals* 5, *united.*

Androecium — *Stamens as many as corolla lobes and opposite to them*, arranged in a single whorl; anthers 2-celled, introrse.

Gynoecium — Carpels 5; ovary superior; *placentation free central.*

Fruit — *Capsule, dehiscent by five teeth or valves.*

Seeds — Embryo straight; endosperm copious.

FLORAL RANGE IN THE FAMILY

The flowers may be solitary and axillary (*Anagallis*) or racemes on lateral or terminal axes to form an umbel (*Primula*) or racemes or spikes (*Lysimachia*).

The flowers are zygomorphic in *Coris.*

Deviations from the typical pentamerous plan take place in *Centunculus* where the flower is 4- to 5-merous, in *Lysimachia* where the flower is 5- to 6-merous and in *Trientalis* where the flower is 5- to 9-merous.

The corolla is absent in *Glaux.* It is generally gamopetalous, but polypetalous in *Pelletiera.* It has a shorter tube in *Anagallis*, but the tube is longer in *Primula.* The limb may be spreading or connivent into a cup or reflexed (*Cyclamen*).

Though the stamens are arranged in a single whorl, the missing whorl is occasionally represented by scales in *Samolus* and *Soldanella.*

The ovary may be superior or half inferior. In *Hottonia* and *Samolus*, the ovules are anatropous and the hilum basal.

DIVISIONS OF THE FAMILY

The Primulaceae is divided into five tribes:

Tribe (*i*) *Corideae.* Flowers medianly zygomorphic. Calyx spiny. Example: *Coris.*

Tribe (*ii*) *Cyclamineae.* Flowers with sharply reflexed corolla limbs. Ovary superior. Capsule with a valvular dehiscence. Example: *Cyclamen.*

Tribe (*iii*) *Lysimachieae.* Corolla lobes contorted in bud. Ovary superior. Capsule valvular or a pyxidium. Example: *Anagallis*, *Lysimachia*, etc.

Tribe (*iv*) *Primuleae.* Corolla lobes imbricate or quincuncial. Ovary superior. Capsule valvular. Examples: *Androsace*, *Dodecatheon*, etc.

Tribe (*v*) *Samoleae.* Ovary half inferior. Example: *Samolus.*

NUMBER IN AND DISTRIBUTION OF THE FAMILY

The Primulaceae contains about 28 genera and nearly 700 species. The members of this family are represented in all continents, but are most abundant in the north temperate zone.

COMMONLY OCCURRING PLANTS OF THE FAMILY

Anagallis arvensis L. var. *caerulea* Gren. & Godr. is a low annual herb with gland-dotted leaves and blue flowers.

Centunculus tenellus Duby is a very small annual.

Primula umbellata (Lour.) Bentv., an annual herb, bears small red or white flowers.

ECONOMIC ASPECTS OF THE FAMILY

The family is of importance for ornamentals, *e.g.*, *Anagallis*, *Cortusa*, *Cyclamen*, *Dodecatheon*, *Omphalogramma*, *Primula* and *Soldanella. Lysimachia vulgaris* is used as a febrifuge and yields a yellow dye. The flowers of *P. veris* are utilised for homemade wine.

Plumbaginaceae (Leadwort Family)

GENERAL FEATURES OF THE FAMILY

Habit	—	Herbs or shrubs.
Leaves	—	Alternate, often rosulate, sometimes with chalk or water glands on surface.
Flowers	—	Bisexual, regular.
Calyx	—	Sepals 5, united, plicate, *often ribbed.*
Corolla	—	Petals 5, united, contorted or imbricate.
Androecium	—	Stamens 5, opposite to corolla lobes; anthers 2-celled.
Gynoecium	—	Carpels 5; *ovary superior, unilocular; styles 5*, glandular or hairy; stigma filiform.
Fruit	—	*Achene* or *utricle.*
Seeds	—	Embryo straight; endosperm crystalline granular.

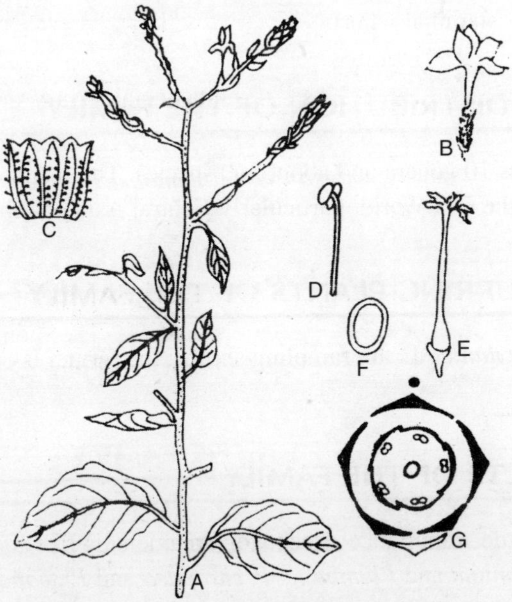

Fig. 65 PLUMBAGINACEAE. *Plumbago zeylanica* L. **A**, portion of shoot; **B**, flower; **C**, calyx; **D**, stamen; **E**, pistil; **F**, T.S. of ovary; **G**, floral diagram

DIVISIONS OF THE FAMILY

The Plumbaginaceae is divided into two tribes:

Tribe (*i*) *Plumbagineae.* Inflorescence simple, both bracteoles being sterile. Stamens hypogynous. Styles free only in upper portion. Example: *Plumbago.*

Tribe (*ii*) *Staticeae.* Inflorescence compound, one bracteole being sterile. Stamens perigynous. Styles free almost to base. Examples: *Acantholimon*, *Limonium*, *Statice*, etc.

POSITION AND AFFINITY OF THE FAMILY

Hallier placed the Plumbaginaceae in the Centrospermae. Wettstein, Lawrence and Rendle treated the family in an order by itself (Plumbaginales). Hutchinson considered it as one of the two families under the Primulales.

The relationship of the Plumbaginaceae to the Centrospermales can be traced to the ovary being unilocular with a basal ovule borne on a long funiculus and embryo surrounded by a mealy endosperm. The Basellaceae speaks for an alliance with the Plumbaginaceae on account of the flowers having five petals united below and one series of antipetalous stamens. But the campylotropous ovule and curved embryo of the Centrospermales offer contrast with the anatropous ovule and straight embryo of the Plumbaginaceae. Since anatropous ovule and straight embryo also occur in the Centrospermales, it does preclude the suggestion that the Plumbaginaceae is derived from the Centrospermales or from a form closely related to that taxon.

The Plumbaginaceae is sometimes grouped under the Primulales. This is due to a common floral plan with antipetalous stamens, uniloculur ovary and bitegmic ovule. Exceptions, however, are found in the ovary structure with five styles, attachment of a solitary ovule, mealy endosperm as well as the absence of any trace of outer staminal whorl.

NUMBER IN AND DISTRIBUTION OF THE FAMILY

The Plumbaginaceae includes 10 genera and about 560 species. The members of this family abound in the dry or saline habitats of the Old World, particularly Central Asia and the Mediterranean.

COMMONLY OCCURRING PLANTS OF THE FAMILY

Plumbago indica L. and *P. zeylanica* L. are rambling shrubs; the farmer is cultivated in gardens and the latter is a weed.

ECONOMIC ASPECTS OF THE FAMILY

The Plumbaginaceae is of little importance except for ornamentals like the species of *Acantholimon*, *Armeria*, *Ceratostigma*, *Limonium* and *Plumbago*. *P. europaea* and *P. scandens* are used to treat dental ailments. Extracts from the roots and leaves of *P. zeylanica* are used to treat skin diseases.

Order 28. *Ebenales*

The Ebenales, according to Hallier, is not a natural taxon. It bears no relationship with the pentacyclic orders. It may be that the order had a distinct origin from the Dialypetalae (Rendle, 1925). Bessey suggested that the Geraniales gave rise to both Ericales and Ebenales along parallel lines. Wettstein sought a possible link with the Malvales, Geraniales and Rutales which are characterised by the diplostemonous androecium having a tendency to indefinite stamens and septate isomerous gynoecium with an axile placentation. Copeland (1938) stated that "the Theaceae … most nearly represent the ancestry of the Ebenales". Hutchinson opined that the order evolved from the Magnoliales through the Dilleniales and Bixales.

The Ebenales is distinguished by the woody habit, actinomorphic and hypogynous flowers, gamopetalous corolla, epipetalous stamens in two to three whorls and ovary basically septate with one to two ovules in each loculus.

Engler and Diels considered the order as composed of two suborders and seven families: Sapotineae (Sapotaceae, Hoplestigmataceae), Diospyrineae (Ebenaceae, Diclindantheraceae, Symplocaceae, Styracaceae, Lissocarpaceae). Hallier treated the Sapotaceae in his Sapotales and the Ebenaceae and Styracaceae in his Santalales. Hutchinson followed a different view by placing the families of the Styracales and Ebenales apart from each other. Cronquist included five families in the order, viz. Sapotaceae, Ebenaceae, Styracaceae, Lissocarpaceae, Symplocaceae. Both Lawrence and Core accepted four families under the Ebenales and the present author adopts their plan in the following manner:

(1) Ovary superior	(2)
(1) Ovary inferior	(4)
(2) Stamens as many as corolla lobes	*Sapotaceae*
(2) Stamens twice as many as corolla lobes	(3)
(3) Flowers unisexual; styles 2–8	*Ebenaceae*
(3) Flowers bisexual; style 1	*Styracaceae*
(4) Ovary incompletely septate	*Styracaceae*
(4) Ovary completely septate	*Symplocaceae*

Sapotaceae (Sapodilla Family)

GENERAL FEATURES OF THE FAMILY

Habit	—	*Shrubs or trees bearing laticiferous ducts*; young portions rusty and tomentose.
Leaves	—	Alternate, simple, entire, coriaceous, *usually exstipulate.*
Flowers	—	Bisexual, regular.
Calyx	—	Sepals 4–8, sometimes in two whorls, free or slightly united.
Corolla	—	Petals as many as or 2–4 times as many as sepals, somewhat united.
Androecium	—	*Stamens epipetalous, in 2 or 3 whorls, outer antisepalous whorl reduced to staminodes or wanting.*
Gynoecium	—	Carpels as many as or twice as many as stamens; *ovary superior, completely septate with many chambers, each chamber with a single ascending ovule.*
Fruit	—	Berry.
Seeds	—	Exalbuminous; embryo large with a small radicle and broad foliaceous cotyledons.

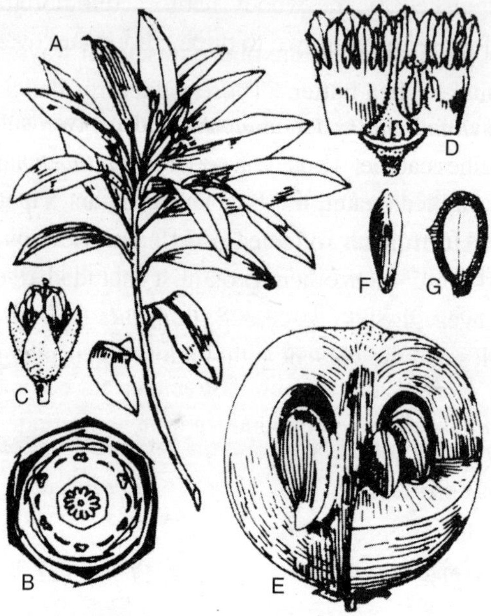

Fig. 66 SAPOTACEAE. *Manilkara achras* (Mill.) Fosberg **A**, flowering shoot, **B**, floral diagram; **C**, flower in female condition; **D**, corolla spread out; **E**, fruit with left side in L.S. and right in T.S. and L.S; **F**, seed to show long hilum; **G**, L.S. of seed to show embryo. (**A–G**, after Engler)

DIVISIONS OF THE FAMILY

The Sapotaceae is separated into two tribes:

Tribe (*i*) *Mimusopeae.* Petals with appendages. Examples: *Manilkara* and *Mimusops.*

Tribe (*ii*) *Palaquieae.* Petals without appendages. Examples: *Achras*, *Butyrospermum*, *Chrysophyllum*, etc.

NUMBER IN AND DISTRIBUTION OF THE FAMILY

The Sapotaceae includes 35–75 genera and about 800 species. The members of this family are widely distributed throughout the tropics and chiefly in low land and lower montane rain forest.

COMMONLY OCCURRING PLANTS OF THE FAMILY

Dichopsis pentaphylla Hook, f., Indian Butter Tree [*Madhuca butyracea* (Roxb.) Macbride], Mohwa Tree (*M. indica* Gmel.), Wild Sapote Tree [*M. longifolia* (L.) Macbride], American Bully or Sapodilla Plum [*Manilkara achras* (Mill.) Fosberg], *M. hexandra* (Roxb.) Dub., Bullet-wood Tree or Indian Medlar (*Mimusops elengi L.*), *Pouteria grandifolia* (Wall.) Baehni and *Sarcosperma arboreum* Hook, f. are trees.

ECONOMIC ASPECTS OF THE FAMILY

Economically, the members of the Sapotaceae are of use in several ways. The seeds of *Butyrospermum paradoxum* (tropical Africa) make 'shea butter'. Those valued for delicious and edible fruits include *Calocarpum sapota, Chrysophyllum cainito, Lucuma mammosa, Manilkara achras,* etc. *Chrysophyllum olivaeforme* (West Indies) furnishes cabinet wood. The seeds of *Madhuca butyracea* produce the vegetable butter, called 'phulwa', used as a cold cream, lip salve and luminant. The flowers of *M. indica* are used for distilling spirit and the seeds form 'mohwa butter', employed for cooking, for burning lamps and for adulterating ghee. From the seeds of *M. longifolia* is extracted the 'phulwa oil' of commerce, utilised as a substitute for ghee and for soap making. The latex of *Manilkara achras* yields 'chickle', used for making chewing gum. An aromatic oil, obtained from the flowers of *Mimusops elengi,* finds use in the manufacture of perfumes. *M. globosa* is the source of 'balata', a commercially important latex. The latex of *Palaquium ellipticum* is used as cement and for water proofing. *P. gutta* (Indo-Malaysia) was once considered as the main source of 'gutta percha'. Some species of *Sideroxylon* yield good timber.

Ebenaceae
(Ebony Family)

The family is characterised by the absence of milky sap and presence of unisexual flowers, stamens twice as many as corolla lobes, multilocular ovary, ovules bitegmic and typically in pairs, styles two to eight as well as seeds with a small hilum.

Ths Ebenaceae contains five genera and 400–500 species. The members of this family are distributed in tropical and subtropical regions with the greatest development in India and Malaya.

Diospyros peregrina (Gaertn.) Gurke and *Maba buxifolia* Pers. are trees.

The family is of economic importance as a source of timber, *e.g., Diospyros ebenum, D.kurzii, D. melanoxylon, D. montana, D. quaesita, D. reticulata* and *Euclea pseudoebenus. D. kaki, D. lotus* and *D. virginiana* produce edible fruits. The leaves of *D. tomentosa* are used for rolling 'bidis'. The fruits of *D. peregrina* is utilised for smearing the bottoms of boats and for steeping fish-nets in order to make them waterproof.

Styracaceae
(Storax Family)

The family is recognised by the stellate or lepidote indumentum, linear anthers, imperfectly septate ovary and drupe.

A family of 12 genera and 180 species, the Styracaceae abounds in the warmer regions of eastern Asia, Mediterranean area, south-eastern United States as well as Central and South America.

Bruinsmia polysperma (Clarke) van Steenis occurs in Khasi Hills, Meghalaya.

The family is economically important for resins like 'Siam benzoin' obtained from *Styrax benzoides* and *S. tonkinensis*, 'Sumatra benzoin, from *S. benzoin* and 'storax' from *S. officinalis. S. japonica* yields finegrained wood used in making umbrella handles; the seeds are the source of an oil.

Symplocaceae (Symplocos Family)

The family is known by the inferior and completely septate ovary as well as berry-like fruit with a persistent calyx.

The Symplocaceae, with one genus (*Symplocos*) and about 300 species, is found in tropical and subtropical Asia and America.

Symplocos paniculata Miq. is found in Assam hills.

The hard seeds of *Symplocos laurina* are strung as beads and put around the necks of children. The bark of *S. racemosa* is used for dyeing; sometimes, it is powdered for 'abir', a colouring matter.

Order 29. *Gentianales*

The Gentianales (which is equivalent to the Contortae of some and the Loganiales of others) is a natural taxon. This is reflected in the nature of aestivation, structure of calyx and number of carpels. As a matter of fact, the plants bear opposite or pinnately compound leaves. The flowers are actinomorphic and have two carpels. The corolla is gamopetalous, the lobes being usually convolute. The stamens are adnate to or near the corolla base.

Engler and Diels conceived the Contortae to contain two suborders and six families: Oleineae (Oleaceae), Gentianeae (Desfontaineaceae, Loganiaceae, Gentianaceae, Apocynaceae, Asclepiadaceae). Bessey suggested that the Gentianales arose from the Geraniales which separated into two phyletic lines, one giving rise to the Loganiales and the other to the Polemoniales. Hallier interpreted the origin of the Loganiales from the Linaceae of his Guttales and placed all the families of the Contortae under the Tubiflorae. Wettstein as well as Rendle "restricted the Contortae to contain only the suborder Gentianineae and segregated the Oleineae as the Oleales, considering the latter to have been derived from stocks allied to the Staphyleaceae and the former to be advanced over the Tubiflorae". Hutchinson split up the Gentianales into three large taxa: Loganiales (Potaliaceae, Loganiaceae, Buddleiaceae, Antoniaceae, Spigeliaceae, Strychnaceae, Oleaceae), Apocynales (Plocospermaceae, Apocynaceae, Periplocaceae, Asclepiadaceae), Gentianales (Gentianaceae, Menyanthaceae). Following Lawrence and Core, the present author treats five families with their characters given below:

(1) Milky sap absent	(2)
(1) Milky sap present	(5)
(2) Shrubs or trees	(3)
(2) Herbs	(4)
(3) Stamens fewer than corolla lobes	*Oleaceae*
(3) Stamens as many as corolla lobes	*Loganiaceae*
(4) Style two-cleft	*Loganiaceae*
(4) Style not cleft	*Gentianaceae*
(5) Filaments distinct; follicles linear	*Apocynaceae*
(5) Filaments united; follicles lanceolate	*Asclepiadaceae*

Oleaceae
(Olive Family)

GENERAL FEATURES OF THE FAMILY

Habit — Shrubs or trees, climbing or erect.

Leaves — *Opposite*, simple or imparipinnate.

Inflorescence — Axillary or terminal, compound racemes.

Flowers — Polygamous or dioecious, regular, *dimerous.*

Calyx — Sepals 4, small, united.

Corolla — Petals 4–6, *free or connate.*

Androecium — Stamens 2, epipetalous; *anthers with cells back to back.*

Gynoecium — Carpels 2, united; *ovary superior, 2-locular*; *ovules 2 in each loculus, anatropous*, ascending or pendulous; style simple; stigma bilobed.

Fruit — Berry, capsule or drupe.

Seeds — Albuminous; embryo straight.

FLORAL RANGE IN THE FAMILY

The flowers are arranged in a cyme, as in *Jasminum.* In *Syringa*, the inflorescence is a compound raceme.

In general, the flowers are bisexual. Unisexual flowers, however, are found in *Fraxinus* as well as in some species of *Olea.*

The calyx is absent in *Fraxinus excelsior.* In some species of *Jasminum*, more than 5 sepals are seen.

The corolla is usually tubular or rotate in *Jasminum*, *Ligustrum* and *Olea.* The petals may be free or united. They are 6 in number in *J. nudiflorum.* The aestivation is imbricate in *Jasminum* and valvate in *Syringa.*

The stamens are usually 2 and transverse, but may be median. The number of stamens is 4 in *Hesperelaea* and *Tessarandra.*

The ovary is generally bilocular, but sometimes tri- or tetralocular.

DIVISIONS OF THE FAMILY

The Oleaceae is divided into two subfamilies:

Subfamily I. *Jasminoideae.* Fruit vertically constricted and divided into two parts. Seeds erect. Example: *Jasminum.*

Subfamily II. *Oleoideae.* Fruit not vertically constricted. Seeds pendulous. Examples: *Fraxinus*, *Syringa*, etc.

NUMBER IN AND DISTRIBUTION OF THE FAMILY

The Oleaceae consists of 29 genera and about 600 species. The members of this family are found in the tropical, subtropical and temperate zones of the world, although they are concentrated in Australasia and South-east Asia.

COMMONLY OCCURRING PLANTS OF THE FAMILY

Jasminum angustifolium Vahl, *J. multiflorum* (Burm. f.) Andr. and *J. sambac* (L.) Ait. are garden plants.

Privet (*Ligustrum robustum* Bl.), *Linociera ramiflora* (Roxb.) Wall., Coral Jasmine or Sorrowful Tree (*Nyctanthes arbor-tristis* L.) and *Olea dioica* Roxb. are trees.

ECONOMIC ASPECTS OF THE FAMILY

The Oleaceae is of importance on account of fruits, timber, dyes and ornamentals. The plants which are cultivated ornamentally include *Chionanthus, Forsythia*, *Jasminum*, *Ligustrum*, *Osmonthus*, *Phillyrea*, *Syringa*, etc. *Fraxinus americana*, *F. excelsior*, *F. nigra*, *F. oregona*, *F. ornus*, *F. quadrangulata. Notelaea ligustrina*, *Olea europaea* and *O. laurifolia* yield useful timber. The corolla of *Nyctanthus arbortristis* is rich in colouring matter and is utilised as an auxiliary to other dyes; the leaves are often used to polish wood instead of sand paper.

Loganiaceae
(Logania Family)

GENERAL FEATURES OF THE FAMILY

Habit	—	*Herbs, shrubs or trees.*
Leaves	—	Opposite, alternate or whorled, simple, *stipulate.*
Inflorescence	—	Cymose.
Flower	—	Bisexual, regular.
Calyx	—	Sepals 4 or 5.
Corolla	—	Petals 4–5, united.
Androecium	—	*Stamens as many as corolla lobes*, epipetalous and alternate with them.
Gynoecium	—	Carpels 2, united; *ovary superior*, *2-locular*; *placentation axile*; *style two-cleft.*
Fruit	—	Berry, capsule or drupe.
Seed	—	Winged; embryo straight; endosperm bony or fleshy.

FLORAL RANGE IN THE FAMILY

The stamens are reduced to 1 in *Usteria.* The ovary is half inferior in *Mitreola;* it may be unilocular in some species of *Strychnos* and incompletely bilocular in *Fagraea.*

DIVISIONS OF THE FAMILY

The Loganiaceae is divided into two subfamilies:

Subfamily I. *Buddlejoideae.* Trichomes multicellular. Internal phloem absent. Leaves usually lobed or toothed. Flowers 4-merous. Endosperm cellular. Examples: *Buddleja*, *Nuxia*, etc.

Subfamily II. *Loganioideal.* Trichomes unicellular, rarely multicellular. Internal phloem present. Leaves entire. Flowers 4-merous or 5-merous. Endosperm nuclear. Examples: *Logania*, *Strychnos*, etc.

NUMBER IN AND DISTRIBUTION OF THE FAMILY

The Loganiaceae is a family of about 30 genera and 600 species, being represented in the tropical, subtropical and temperate regions.

COMMONLY OCCURRING PLANTS OF THE FAMILY

Buddleja asiatica Lour. is a large shrub.

Gelsemium elegans (Gardn. & Champ.) Benth. is a large climber with golden yellow flowers.

Mitrasacme alsinoides R. Br., a small weak annual herb, is found in waste places.

ECONOMIC ASPECTS OF THE FAMILY

Various species of *Buddleja*, *Gelsemium* and *Spigelia* are found in cultivation. *Fagraea crenulata*, *F. elliptica* and *F. fragrans* yield good timber. The roots of *Gelsemium sempervirens* constitute 'false jasmine root'. The seeds of *Strychnos ignatii* are employed as a remedy for cholera. The seeds of *S. nux-vomica* yield the alkaloids 'brucine' and 'strychnine'. The seeds of *S. potatorum* are rubbed into a paste which, when added to dirty water, causes the impurities to settle to the bottom. *S.spinosa* produces edible fruits. The bark of *S. toxifera* produces 'curare' to poison arrows.

Gentianaceae
(Gentian Family)

GENERAL FEATURES OF THE FAMILY

Habit — Herbs.

Leaves — *Opposite*, simple, entire, *exstipulate.*

Inflorescence — *Cymose.*

Flowers — Bisexual, regular, hypogynous.

Calyx Sepals 4 or 5, united, imbricate.

Corolla — Petals 4 or 5, united below into a tube, twisted.

Androecium — *Stamens as many as corolla lobes* and alternating with them; anthers 2-celled, introrse, basifixed or dorsifixed.

Gynoecium — Carpels 2; *ovary unilocular with a glandular disc*; *ovules many with parietal placentation*; style simple; stigma entire or lobed.

Fruit — Leathery or membranous *bivalved capsule.*

Seeds — Embryo small; endosperm copious.

FLORAL RANGE IN THE FAMILY

In general, the inflorescence is a biparous cyme. But the lateral branches may form uniparous cymes.

The bracteoles are connate at the base and adnate to the calyx tube in *Zonanthus.* The calyx is inflated and as long as the corolla tube (*Prepusa*) or the calyx lobes are broadly winged (*Canscora*). The corolla is infundibuliform in *Crawfurdia* and *Hockinia.* It is 3-lobed in *Pycnosphaera* and the lobes are 5–12 in *Sabbatia* and 7–10 in *Lapithea.* The corolla lobes are dorsally spurred or saccate in *Halenia.* At the base of the corolla lobes, there are nectar secreting pits, *e.g.*, *Swertia.* The corolla throat is scaly or fimbriate. The aestivation of one corolla is twisted or induplicate valvate.

The stamens are inserted at different levels. In *Leiphaimos* and *Voyria*, the androecium is syngenesious. The dehiscence of the anther is apical and porous, as in *Exacum.* The anthers are connate and introrsely dehiscent in *Tapeinostemon*, spirally twisted in *Orphium* and appendaged at the apex by a gland in *Sebaea.*

The position of the carpels is median or transverse. Although parietal placentation is generally found, it develops into a bilocular ovary with axile placentation in advanced members of the family.

DIVISIONS OF THE FAMILY

The Gentianaceae is divided into two subfamilies:

Subfamily I. *Gentianoideae.* Plants terrestrial. Leaves opposite. Corolla lobes twisted in bud. Examples: *Gentiana*, etc.

Subfamily II. *Menyanthoideae.* Plants aquatic or marshy. Leaves alternate. Corolla lobes induplicate valvate. Examples: *Menyanthes*, *Nymphoides*, etc.

POSITION AND AFFINITY OF THE FAMILY

The Gentianaceae was included in the Gentianales near the Polemoniales by Bentham-Hooker. In Engler's arrangement, this family appeared under the suborder Gentianineae of the Contortae. Hutchi-nson placed the family in the Gentianales after the Asterales and before the Primulales.

The distribution of various secondary metabolites definitely keeps the two subfamilies of the Gentianaceae chemically very distinct (Daniel and Sabnis, 1978). The Gentianoideae possess flavones (as *o*- and *c*-glycosides), xanthones, syringic acid, protocatechuic acid and *p*-coumaric acid; while flavonols, leucoanthocyanins and saponins are present in the Menyanthoideae. These chemical evidence strongly justify the elevation of these two subfamilies to family level, Gentianaceae *sensu lato* and Menyanthaceae, as already practised by Engler and Hutchinson.

The Gentianaceae is related to the Caryophyllaceae in having cymose inflorescence and dichotomous branching. It also bears a relationship with the Apocynaceae and Loganiaceae, but differs from them by the absence of latex, presence of completely united carpels and parietel placentation.

NUMBER IN AND DISTRIBUTION OF THE FAMILY

The Gentianaceae contains about 80 genera and 900 species. Though the members of the family are world wide in distribution, they are mostly abundant in the temperate zone and even extend into the arctic or alpine region.

COMMONLY OCCURRING PLANTS OF THE FAMILY

Canscora decussata Roem. & Schult. is a herb with amplexicaul leaves.

Canscora diffusa (Vahl) R. Br. is a dichotomously branched herb, having the upper leaves connate.

Canscora grandiflora Wt. is a herb, bearing perfoliate leaves.

Centaurium roxburghii (G. Don) Druce, a herb with pink flowers, is found in dry rice fields.

Enicostema hyssopifolium (Willd.) Verdoorn is a glabrous herb with white flowers borne in axillary clusters.

Hoppea dichotoma Willd. is a much branched glabrous herb.

Snowflake [*Nymphoides cristatum* (Roxb.) O. Kuntze] is a floating tank herb with white flowers.

Swertiachirayita (Roxb. ex Flem.) Karst. is a shrub that grows in the Himalaya.

ECONOMIC ASPECTS OF THE FAMILY

The seeds of *Blackstonia perfoliata* (Europe) yield a yellow dye. *Chirona baccifera* (South Africa) is a purgative. *Erythraea centaurium* (Europe) and *Menyanthes trifoliata* (Europe) are also of medicinal value. Some species of *Exacum*, *Gentiana*, *Menyanthes*, *Nymphoides*, *Sabatia,* etc. are grown as ornamentals. *Gentiana lutea* is the source of 'Gentian root tonic'. The leaves of *Swertia chirayita* form the 'chireta' of commerce, being used as a stomachic.

Apocynaceae
(Dogbane Family)

GENERAL FEATURES OF THE FAMILY

Habit	—	Herbs or shrubs, erect or climbing, *often with latex.*
Leaves	—	Opposite or whorled, simple, entire, *exstipulate.*
Inflorescence	—	Panicles, sometimes cymose.
Flowers	—	Bisexual, regular, hypogynous with bracteoles.
Calyx	—	Sepals 5, united, *imbricate.*
Corolla	—	Petals 5, united, hypocrateriform or rotate, twisted; throat provided with appendages.
Androecium	—	*Stamens 5, epipetalous, free from stigma*; filaments short, appearing from corolla throat; *pollen grains not in pollinia,* but distinct or in tetrads.
Gynoecium	—	Carpels 2; *ovary superior, free in region of ovary*, but united near style and stigma; ovules many, anatropous, pendulous.

Fruit — Baccate, drupaceous *or follicular.*

Seeds — *Flat, winged* or *comose*; embryo large, straight; endosperm fleshy or cartilaginous.

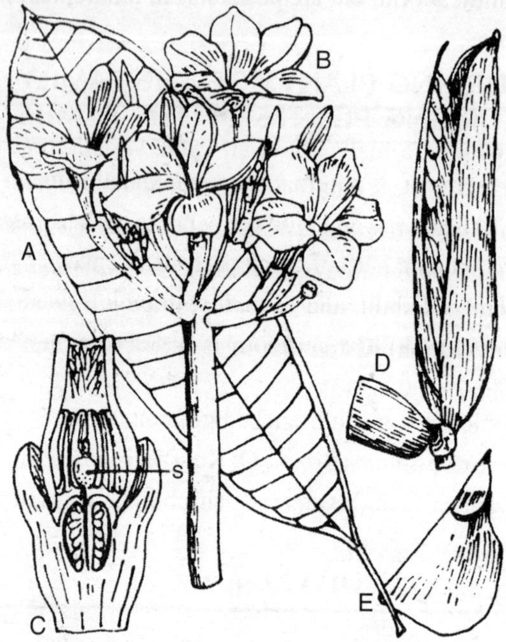

Fig. 67 APOCYNACEAE. *Plumieria rubra* L. **A**, leaf; **B**, inflorescence; **C**, L.S. of flower (*s*, swollen receptive part of stigma); **D**, fruit; **E**, winged seed. (**A–E**, after Rendle)

DIVISIONS OF THE FAMILY

The Apocynaceae is divided into two subfamilies:

Subfamily I. *Echitoideae.* Corolla lobes twisted to right. Stamens closely adhere to stigma; anther tails distinct. Seeds hairy. Examples: *Nerium*, *Vallaris*, etc.

Subfamily II. *Plumieroideae.* Corolla lobes twisted to left. Stamens free or slightly attached to stigma; anther tails absent. Seeds usually hairless, Examples: *Plumieria*, *Vinca*, etc.

POSITION AND AFFINITY OF THE FAMILY

In Bentham-Hooker's arrangement, the Apocynaceae appeared under the Gentianales. It was placed in the Apocynales between the Loganiales and Rubiales by Hutchinson. Engler included the family in the suborder Gentianineae of the Contortae. Hallier considered the Asclepiadaceae together with the Apocynaceae and sought their origin from the Linaceae. Bessey accepted the two families as distinct and treated both in his concept of the Gentianales which he derived from the Geraniaceae.

The Apocynaceae is closely related to the Asclepiadaceae from which it differs by the presence of a single style, pollen grains not in pollinia, stamens free from the stigma and presence of translators connecting the contents of the adjoining anther sacs.

NUMBER IN AND DISTRIBUTION OF THE FAMILY

The Apocynaceae comprises about 180 genera and 1,500 species. The members of this family are widely distributed throughout the world, but are abundant in the tropical regions.

COMMONLY OCCURRING PLANTS OF THE FAMILY

Aganosma caryophyllata (Roxb. ex Sims) G. Don, *Allamanda cathartica* L., *Beaumontia grandiflora* Wall., *Ichnocarpus frutescens* (L.) Ait. & Ait. f. and *Strophanthus wallichii* A. DC. are climbing shrubs.

Devil Tree or Shaitan Wood [*Alstonia scholaris* (L.) R. Br.] is a tall tree with whorled leaves.

Bengal Currants (*Carissa congesta* Wt.), Oleander (*Nerium indicum* Mill.), *Tabernaemontana divaricata* (L.) R. Br. ex Roem. & Schult. and *Thevetia peruviana* (Pers.) K. Schum. are shrubs.

Chonemorpha fragrans (Moon) Alston is a great climber, bearing sweet smelling white flowers.

Holarrhena anti-dysenterica (L.) Wall. ex A. DC. is a wild tree.

Pagoda Tree or Temple Tree (*Plumieria rubra* forma *acuminata* Santapau & Irani ex Shah) is a small fleshy deciduous tree.

Willughbeia edulis Roxb. is a climber, having peduncles converted into tendrils.

Wrightia coccinea Sims, *W. tinctoria* R. Br. and *W. tomentosa* (Roxb.) Roem. & Schult are small trees.

ECONOMIC ASPECTS OF THE FAMILY

The members of the Apocynaceae are mainly important as ornamentals, whereas some yield fruits, drugs, latex and other articles. Some species of *Aganosma*, *Allamanda*, *Beaumontia*, *Plumieria*, *Thevetia*, etc. are cultivated for their showy flowers. The bark of *Alstonia scholaris* is astringent and forms the 'ditta bark' of commerce. *Aspidosperma cruentum* (Central America) supplies valuable timber. *A. quebracho-blanco* (Argentina) furnishes 'quebracho-blanco bark'. The fruits of *Carissa congesta* and *Couma macrocarpa* are edible. Rubber is derived from *Carpodinus lanceolatus*, *Dyera costulata*, *Funtumia elastica* (West Africa), *Hancornia speciosa* (Brazil), *Landolphia kirkii* (tropical Africa) and others. The bark of *Chonemorpha fragrans* yields fibres of good quality. The bark of *Holarrhena anti-dysenterica* is useful in amoebic dysentery. The latex of *Hunteria zeylanica* (Malaya) is used for treating yaws. The bark and wood of *Nerium indicum* are utilised for the manufacture of a rat poison. The wood of *Plumieria rubra* forma *acuminata* is used for making Indian drum, tom-tom and other musical instruments. The roots of *Rauvolfia nitida* and *R. serpentina* give us the 'wonder drug' for lowering blood pressure. The dried fruits of *Rhazya stricta* are utilised as a coagulant in cheese-making. The seeds of *Strophanthus combe*, *S. hispidus*, etc. provide 'strophanthin', a heart stimulant and precursor of cortisone, a specific for arthritis. *S. sarmentosus* is a source of commercial cortisone. The roots of *Tabernaemontana divaricata* serve as an antidote to scorpion bite. The leaves of *Wrightia tinctoria* furnish wrappers for 'bidis'; the seeds contain an indigo-like blue dye called 'goda indrajao' which is used as an anthelmintic. The root of *W. tomentosa* is used in the treatment of snake bite; the seeds yield a red-coloured oil which has medicinal applications.

Asclepiadaceae
(Milkweed Family)

GENERAL FEATURES OF THE FAMILY

Habit	—	Herbs or shrubs, *with latex.*
Leaves	—	Opposite, simple, entire, exstipulate.
Flowers	—	Bisexual, regular, hypogynous.
Calyx	—	Sepals 5, united, quincuncial.
Corolla	—	Petals 5, united, valvate, provided with horn-like appendages (*corona*).
Androecium	—	Stamens 5, epipetalous; *anthers adnate to stigma*; *pollen grains cohere to form pollinia* and attached to a black horny structure called *translator* or *caudicle*; pollinia united in pairs or tetrads to a gland (*retinaculum*) on stigma.
Gynoecium	—	Carpels 2, free below; ovules many on marginal placentae; *stigma five-angled to which anthers adhere* (*gynostegium*), each angle of stigma being accompanied by a sticky disc (*corpusculum*) on groove.
Fruit	—	*Pair of follicles.*
Seeds	—	*Comose*; embryo large, straight; endosperm scanty, cartilaginous.

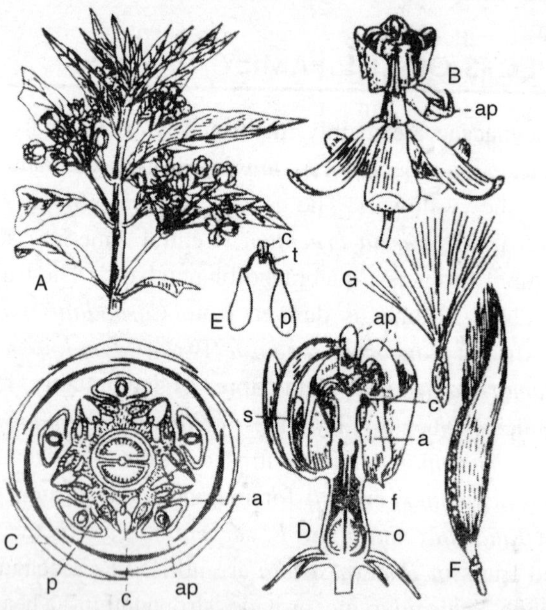

Fig. 68 ASCLEPIADACEAE. *Asclepias curassavica* L. **A**, flowering shoot; **B**, flower (*c*, corolla; *ap*, corolla appendage); **C**, floral diagram; **D**, gynostegium (*ap*, appendage from back of stamen; *c*, corpusculum; *a*, anther where pollenmasses are embedded; *f*, filament; *o*, ovary; *s*, stigmatic surface); **E**, pollinium (*c*, corpusculum; *t*, translator; *p*, pollinia); **F**, dehiscing follicle; **G**, L.S. of seed (**A–G**, after Rendle)

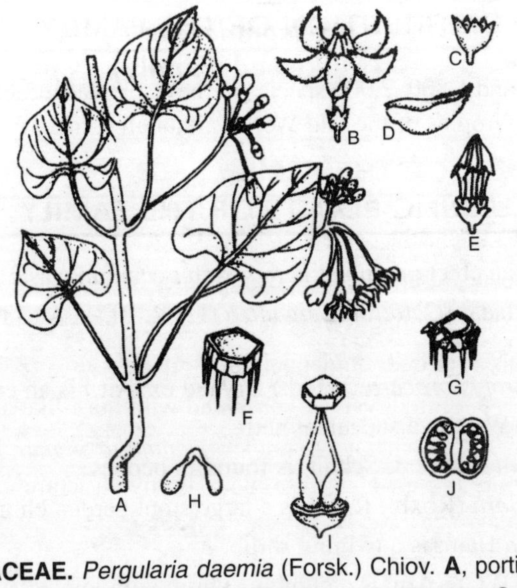

Fig. 69 ASCLEPIADACEAE. *Pergularia daemia* (Forsk.) Chiov. **A**, portion of shoot; **B**, flower; **C**, calyx; **D**, petal; **E**, outer corona at top; **F**, inner corona at top; **G**, disc with pollinia; **H**, pollinia; **I**, carpel; **J**, T.S. of ovary

FLORAL RANGE IN THE FAMILY

In general, the flowers are small. But large flowers are found in *Ceropegia*, *Stapelia* and *Stephanotis.*

The flowers are usually regular, but irregular flowers are sometimes met with. The corolla is divided and forms a rotate structure, but the tube may be longer and constitute a pitcher-like corolla (*Ceropegia*) or salverform corolla (*Stephanotis*). One can find the occurrence of corona-like appendages which spring from the corolla itself or from the back of the stamens.

The ovary is usually apocarpous, but the styles are united to form a swollen head or may become beaked or somewhat conical or flattened.

DIVISIONS OF THE FAMILY

Based on pollen grain characters, the Asclepiadaceae is divided into two subfamilies with four tribes in one subfamily and none in the other subfamily:

Subfamily I. *Cynanchoideae.* Pollen aggregated into two to four waxy bodies called pollinia.

- **Tribe** (*i*) *Asclepiadeae.* Pollinia 2 in each anthers, *i.e.*, 10 in all. Examples: *Asclepias*, *Calotropis*, etc.
- **Tribe** (*ii*) *Gonolobeae.* Pollinia in pairs in each anther but transverse. Example: *Gonolobus.*
- **Tribe** (*iii*) *Secamoneae.* Pollinia 4 in each anther, *i.e.*, 20 in all. Examples: *Genianthus*, *Secamone*, etc.
- **Tribe** (*iv*) *Tylophoreae.* Pollinia in pairs in each anther, but erect. Examples: *Marsdenia*, *Tylophora*, etc.

Subfamily II. *Periplocoideae.* Pollen granular and in tetrads; translators spoon-shaped and provided with an adhesive disc. Examples: *Cryptolepis*, *Hemidesmus*, etc.

NUMBER IN AND DISTRIBUTION OF THE FAMILY

There are about 250 genera and 1,800–2,000 species in the Asclepiadaceae. The members of the family are mainly distributed in the tropics of the Old World, although some are found in tropical America.

COMMONLY OCCURRING PLANTS OF THE FAMILY

Asclepias curassavica L. is an erect perennial herb which occurs in waste places.

Giant Milkweed or Madar [*Calotropis gigantea* (L.) R. Br. ex Ait.] is a large shrub covered with a woolly tomentum.

Swallow-wart [*Calotropis procera* (Willd.) Dryand ex. Ait.] is an erect white downy shrub.

Caralluma fimbriata Wall. is a succulent herb.

Cryptolepis buchanani Roem. & Schult. is found in hedges.

Cryptostegia grandiflora (Roxb.) R. Br. is a large stout garden climber.

Cynanchum callialata Ham. is a twining shrub.

Decalepis hamiltonii Wt. & Arn. is a climbing shrub with jointed stems.

Dischidia rafflesiana Wall. is a stout twiner with pitchers whose cavity is filled with rain water and adventitious roots.

Dregea volubilis (L. f.) Benth. ex Hook. f. is a tall stout climber.

Finlaysonia maritima (Bl.) Backer ex K. Heyne is a large climber of the Sunderbans.

Indian Sarsaparilla [*Hemidesmus indicus* (L.) Schult.] is a thin twining shrub.

Wax plant (*Hoya parasitica* Wall.) is a twining epiphyte with waxy leaves and flowers.

Rajmahal Hemp [*Marsdenia tenacissima* (Roxb.) Moon] is a climbing shrub with a very short stem.

Oxystelma secamone (L.) Karst. is a perennial climbing herb.

Pentatropis capensis (L.f.) Bullock is a twining slender herb of the Sunderbans.

Pergularia daemia (Forsk.) Chiov. is a climbing undershrub, bearing spinous follicles.

Sarcolobus carinatus Wall. is a climber which occurs in the Sunderbans.

Sarcostemma acidum (Roxb.) Voigt commonly occurs in dry areas of Peninsular India.

Telosma minor (Andr.) Craib is a twining glabrous undershrub.

Tylophora indica (Burm. f.) Merr. is a twining slender herb.

ECONOMIC ASPECTS OF THE FAMILY

Economically, the members of the Asclepiadaceae are of some importance. Some species of *Asclepias* act as livestock poisons. The seed hairs of *A.curassavica, A. incarnata* and *A.syriaca* are used as a substitute for 'kapok'. The fleshy stems of *Caralluma fimbriata* and tubers of *Ceropegia pusilla* are edible. A few plants are used medicinally, *e.g. Calotropis, Caralluma* and *Hemidesmus indicus.* From the bark of *Calotropis gigantea, C. procera, Leptadenia pyrotechnica* and *Marsdenia tenacissima* fibres are extracted. *Ceropegia, Hoya, Oxypetalum, Periploca, Stapelia,* etc. are planted for ornament. *Cryptostegia grandiflora* and *C. madagascariensis* form good sources of rubber. The latex of *Gymnema lactiferum* (Sri Lanka) is used as a substitute for milk. *Marsdenia cundurango* supplies 'cundurango bark'. *M. tinctoria* yields an indigo-like dye. The latex of *Matelea* is utilised as an arrow poison. *Oxystelma secamone* has drinkable milk.

Order 30. *Tubiflorales*

The order is a natural taxon. Here the plants are herbaceous with regular or irregular tetracyclic flowers, gamopetalous corolla, hypogynous and epipetalous stamens and unitegmic ovules.

Engler and Diels set up the order to contain eight suborders and 22 families: Convolvulineae (Convolvulaceae, Polemoniaceae, Fouqueriaceae), Lennonineae (Lennoaceae), Boragineae (Hydrophyllaceae, Boraginaceae), Verbenineae (Verbenaceae, Labiatae), Solanineae (Nolanaceae, Solanaceae, Scrophulariaceae, Bignoniaceae, Pedaliaceae, Martyniaceae, Orobanchaceae, Gesneriaceae, Columelliaceae, Lentibulariaceae, Globulariaceae), Acanthineae (Acanthaceae), Myoporineae (Myoporaceae), Phrymineae (Phrymaceae). Wettstein did likewise except that the Fouqueriaceae was removed to the Parietales and that the Plantaginaceae was accommodated in the order. Rendle also followed Engler's circumscription, but the Convolvulaceae was segregated as the Convolvulales, the Fouqueriaceae and Lennoaceae omitted and the Selaginaceae created. Lawrence treated all the families in the order, but excluded the Columelliaceae. Core considered all these families under the order, but placed the Fouqueriaceae within the Parietales and Lennoaceae within the Ericales. Hutchinson split up the entire order into seven taxa; under the Lignosae, Bignoniales (Cobaeaceae, Bignoniaceae, Pedaliaceae, Martyniaceae) and Verbenales (Ehretiaceae, Verbenaceae, Stilbeaceae, Chloanthaceae, Phrymaceae) and under the Herbacae, Solanales (Solanaceae, Convolvulaceae, Nolanaceae), Personales (Scrophulariaceae, Acanthaceae, Gesneriaceae, Orobanchaceae, Lentibulariaceae, Columelliaceae), Polemoniales (Polemoniaceae, Hydrophyllaceae, Cuscutaceae), Boraginales (Boraginaceae) and Lamiales (Myoporaceae, Selaginaceae, Globulariaceae, Labiatae).

To trace a point of origin for the Tubiflorales among polypetalous dicots is fraught with difficulty. The order, according to Rendle, arose from a stock characterised by pentamerous flowers, oligomerous gynoecium, numerous marginal ovules and disc development below the ovary. Wettstein envisaged a type close to the Rosales in which sympetaly is of occasional incidence. For example, the Saxifragaceae resembles the Solanaceae or Scrophulariaceae on the basis of bilocular ovary with numerous anatropous ovules on large placentae.

The Tubiflorales is a large group, illustrating several lines of floral development. In the Convolvulaceae, the flowers are actinomorphic with ovary bilocular having two ovules in each loculus or tetralocular having one ovule in each loculus. It is related to the Polemoniaceae, but differs from it by the terminal style and ovary structure. The flowers of the Polemoniaceae tend to be slightly zygomorphic, but the carpels may be two, three or five and ovules few to many in each loculus. The floral plan of the Hydrophyllaceae is similar to that of the Polemoniaceae, except that the ovary is bicarpellary and bilocular or unilocular due to the failure of the large marginal placentae to meet in the centre. Although in the Boraginaceae the flowers are zygomorphic, the originally bilocular ovary is divided into four one-ovuled portions by a later formed septum. Zygomorphy does not extend to the calyx in the Verbenaceae where the corolla is somewhat two-lipped. The calyx is sometimes regular in the Labiatae, but it is prominently two-lipped. The two carpels of the Verbenaceae unite to form an originally bilocular ovary; each loculus contains two ovules and is later divided by a partition into one-ovuled parts. The reduction and specialisation of the ovary is complete in the Labiatae where the bilocular ovary is divided into four one ovuled chambers. The Solanaceae is recognised by the usually regular isostemonous flowers and bicarpellary generally bilocular ovary with numerous ovules. With four or two fertile stamens and often somewhat zygomorphic flowers, the tribe Salpiglossideae forms a connecting link with the Scrophulariaceae where the zygomorphy of the corolla and androecium is ordinarily conspicuous. The

Gesneriaceae shows a tendency towards epigyny; here the ovary is unilocular and contains numerous ovules. The Orobanchaceae also contains an unilocular many-ovuled ovary. The Lentibulariaceae has a markedly bilabiate corolla and shows a reduction in the number of stamens as well as a free central placentation in the ovary. The floral plan of the Bignoniaceae is similar to that of the Scrophulariaceae. The Acanthaceae indicates a specialisation of the flower and fruit associated with a great variety of stamen development and a remarkable mechanism for seed dispersal.

The present author, following Core (1955), takes into account 17 families in the Tubiflorales with their characters given below:

(1) Corolla regular	(2)
(1) Corolla irregular	(8)
(2) Climbing herbs or shrubs	*Convolvulaceae*
(2) Erect herbs or shrubs	(3)
(3) Ovary 3-celled	*Polemoniaceae*
(3) Ovary not 3-celled	(4)
(4) Ovary separating into nutlets at maturity	(5)
(4) Ovary not separating into nutlets at maturity	(6)
(5) Style usually deeply two-cleft	*Hydrophyllaceae*
(5) Style not deeply two-cleft	*Solanaceae*
(6) Stamens 2 or 4	*Verbenaceae*
(6) Stamens 5	(7)
(7) Carpels 2	*Boraginaceae*
(7) Carpels 5	*Nolanaceae*
(8) Ovary separating into nutlets at maturity	(9)
(8) Ovary not separating into nutlets at maturity	(11)
(9) Stamens 4	*Boraginaceae*
(9) Stamens 2 or 4	(10)
(10) Style terminal	*Verbenaceae*
(10) Style gynobasic	*Labiatae*
(11) Ovary inferior	*Gesneriaceae*
(11) Ovary superior	(12)
(12) Parasitic or insectivorous plants	(13)
(12) Green plants, hardly parasites and not insectivorous	(14)
(13) Root parasites	*Orobanchaceae*
(13) Insectivorous aquatics	*Lentibulariaceae*
(14) Fruit commonly spiny	(15)
(14) Fruit not spiny	(16)
(15) Placentation axile	*Pedaliaceae*
(15) Placentation parietal	*Martyniaceae*
(16) Anther united	*Acanthaceae*
(16) Anther usually free	(17)
(17) Herbaceous plants	*Scrophulariaceae*
(17) Woody plants	*Bignoniaceae*

Convolvulaceae
(Morning-glory Family)

GENERAL FEATURES OF THE FAMILY

Habit	—	Herbs or shrubs, *twining.*
Leaves	—	Alternate, simple, often lobed, *exstipulate.*
Inflorescence	—	Cymose.
Flowers	—	Large, showy, bisexual, regular, hypogynous.
Calyx	—	Sepals 5, free, imbricate, persistent.
Corolla	—	Petals 5, united, infundibuliform, *twisted.*
Androecium	—	Stamens 5, epipetalous; anthers 2-celled, introrse; *intrastaminal disc annular or cupular.*
Gynoecium	—	Carpels 2, syncarpous; *ovary superior, 2-locular with 2 ovules or 4-locular with 1 ovule in each loculus; placentation axile*; stigma bifid.
Fruit	—	Berry or capsule.
Seeds	—	Exalbuminous; *embryo with folded cotyledons.*

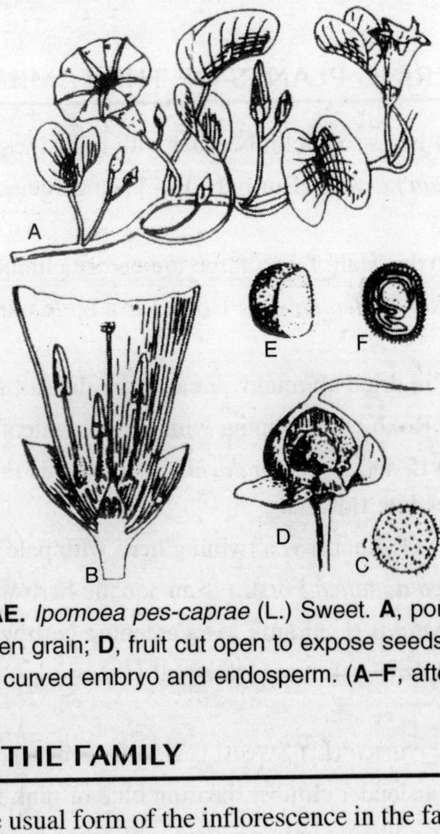

Fig. 70 CONVOLVULACEAE. *Ipomoea pes-caprae* (L.) Sweet. **A**, portion of plant; **B**, L.S. of lower portion of flower; **C**, pollen grain; **D**, fruit cut open to expose seeds; **E**, seed; **F**, L.S. of seed showing curved embryo and endosperm. (**A–F**, after Rendle)

FLORAL RANGE IN THE FAMILY

Although biparous cyme is the usual form of the inflorescence in the family, *Ipomoea pes-tigridis* has a monochasial cyme. The inflorescence is paniculate in *Porana.* In *Hildebrandtia*, the flowers are unisexual and tetramerous. The flowers are cleistogamous in *Dichondra.*

The calyx is gamosepalous in *Wilsonia* and the corolla lobes are imbricate in *Cressa*.

The loculi of anthers are separated from the broad connective in *Lysiostyles*.

In some species of *Ipomoea*, there are 3 carpels in place of 2. In *Erycibe*, there is no style; here the stigma is large and sessile. The style branches are 4 in *Evolvulus* and many in *Nephrophyllum*. The ovary is deeply 2-lobed in *Dichondra*.

DIVISIONS OF THE FAMILY

The Convolvulaceae is divided into two subfamilies:

Subfamily I. *Convolvuloideae.* Autotropic plants with green leaves. Interstaminal scale lacking. Cotyledons folded. Examples: *Cressa*, *Evolvulus*, *Ipomoea*, etc.

Subfamily II. *Cuscutoideae.* Parasites without green leaves. Interstaminal scale present. Cotyledons very reduced or lacking. Example: *Cuscuta.*

NUMBER IN AND DISTRIBUTION OF THE FAMILY

The Convolvulaceae covers about 50 genera and 1,800 species. The members of this family abound in tropical Asia and tropical America.

COMMONLY OCCURRING PLANTS OF THE FAMILY

Argyreia capitata (Vahl) Choisy is a strong climber met with in hedges.

Elephant Creeper [*Argyreia nervosa* (Burm. f.) Boj.] is a garden climber with cordate leaves, the lower surface of which is silvery.

Bonamia semidigyna (Roxb.) Hall. f. is a large tomentose climber, bearing white flowers.

Moon Flower [*Calonyction aculeatum* (L.) House] is a large climber with white flowers which open at sunset.

Cressa cretica L. is seen in dried up muddy areas nears the salt flates of Calcutta.

Dodder (*Cuscuta reflexa* Roxb.) is a twining whitish yellow leafless thread-like parasitic herb.

Evolvulus alsinoides (L.) L. and *E. nummularius* L. are weeds in grassy places, the former with blue flowers and the latter with white flowers.

Hewittia sublobata (L. f.) O. Kuntze is a twining herb with pale yellow flowers.

Water Bindweed (*Ipomoea aquatica* Forsk.) is an aquatic herb with sagittate leaves.

Sweet potato [*Ipomoea batatas* (L.) Lamk.] is a creeping herb with tuberous roots.

Goat's foot (*Ipomoea biloba* Forsk.) and Blue-bell [*I. pes-caprae* (L.) Sweet] are sand binders on sea coast.

Morning Glory [*Ipomoea cairica* (L.) Sweet] is a garden twiner.

Ipomoea nil (L.) Roth is a slender climber, bearing blue or pink flowers.

Ipomoea pes-tigridis L. is a hirsute twining plant with three-to nine-lobed leaves.

Needle Creeper (*Ipomoea quamoclit* L.) is a garden twiner with pinnatisect leaves and crimson flowers.

Jacquemontia pentantha (Jacq.) G. Don is a garden plant.

Merremia emarginata (Burm. f.) Hall. f. is a prostrate herb which strikes roots at the nodes.

Bridal Creeper (*Porana paniculata* Roxb.), *Rivea hypocrateriformis* (Lamk.) Choisy and *Stictocardia tiliifolia* (Desr.) Hall. f. are climbers.

ECONOMIC ASPECTS OF THE FAMILY

The Convolvulaceae is significant for numerous ornamentals, *e.g. Calonyction aculetum, Convolvulus cneorum, C. japnicus, Ipomoea purpurea, I. quamoclit, I. tricolor*, etc.

The family is important for *I. batatas* which produces starchy tuberous roots. *I. aquatica* is used as vegetables. The rootstock of *Calystegia sepium* (New Zealand) is cooked and eaten.

A few plants are of medicinal value, such as *Convolvulus scammonia, Exogonium purga, Ipomoea paniculata*, etc.

Polemoniaceae (Phlox family)

The family is distinguished by the absence of milky sap and presence of gamosepalous calyx, stamens inserted at different levels on the corolla tube, tricarpellate ovary and indefinite ovules.

The flowers may be borne singly or arranged in dichasia. They are zygomorphic in *Bonplandia* and *Loeselia.* The corolla may be campanulate (*Cobaea*), infundibuliform to rotate (*Polemonium*) and salverform (*Phlox*). The corolla tube is filiform in some species of *Gilia.*

The family contains 18 genera and about 300 species, being mainly distributed in the western United States.

Economically, the family is important for a number of ornamentals contributed by the species of *Cobaea, Collomia, Gilia, Linanthus, Phlox* and *Polemonium.*

Hydrophyllaceae (Waterleaf family)

The family is recognised by the tendency towards helicold cymes, imbricate aestivation of corolla, stamens without scales between the filaments, style usually deeply two-cleft and numerous ovules borne on parietal placentae.

The Hydrophyllaceae represents about 18 genera and 250 species. The members of this family are widely distributed, but abundant in western North America.

Hydrolea zeylanica (L.) Vahl, an annual herb, bears lanceolate leaves and bright blue flowers.

Hydrophyllum canadensis serves as a remedy for skin diseases. The species of *Nemophila, Phacelia* and *Wiganda* are cultivated as ornamentals.

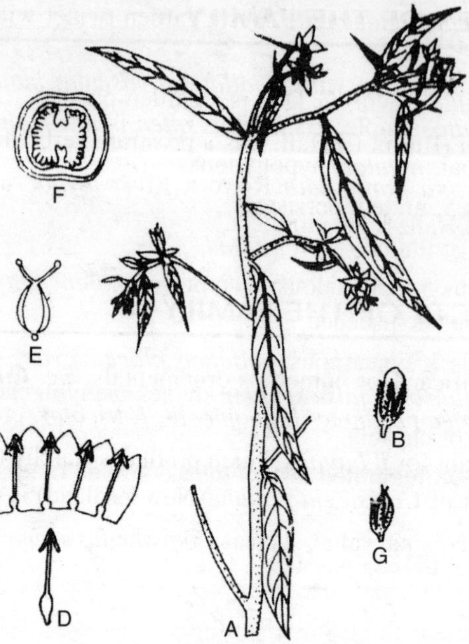

Fig. 71 HYDROPHYLLACEAE. *Hydrolea zeylanica* (L.) Vahl. **A**, portion of plant; **B**, flower bud; **C**, stamen with persistent calyx; **D**, stamen; **E**, pistil; **F**, T.S. of ovary

Solanaceae (Potato Family)

Fig. 72 SOLANACEAE. *Solanum torvum* Swartz. **A**, portion of flowering shoot; **B**, V.S. of flower; **C**, upper portion of anther; **D**, berry; **E**, T.S. of ovary. (**A–E**, after Rendle).

GENERAL FEATURES OF THE FAMILY

Habit	—	Herbs or shrubs, *with bicollateral vascular bundles.*
Leaves	—	*Alternate*, simple, *exstipulate*, *often in unequal pairs.*
Flowers	—	Bisexual, *regular*, hypogynous.
Calyx	—	Sepals 5, united, persistent.
Corolla	—	Petals 3, united, *often plaited.*
Androecium	—	Stamens 5, epipetalous; disc present: *anthers apparently connate*, 2-celled with a porous dehiscence.
Gynoecium	—	Carpels 2, syncarpous, *obliquely placed*; *ovary superior* 2-*locular or imperfectly* 4-*locular by a false septum*; ovules many in each loculus; placentation axile.
Fruit	—	Berry or capsule.
Seeds	—	Many with a pitted testa; embryo curved or straight.

FLORAL RANGE IN THE FAMILY

In *Solanum*, the inflorescence is a monochasial cyme with the suppression of secondary branches on the same side. The flowers are generally regular, but irregular in *Brunfelsia*, *Hyoscyamus* and a few other genera.

The corolla may assume various forms. It is campanulate in *Atropa*, infundibuliform in *Datura*, rotate in *Solanum* and bilabiate in *Schizanthus*.

The development of a false septum in the ovary is a frequent character in the Solanaceae. This results in the ovary being 3- to 5-celled in *Nicandra* and 4-celled in *Datura.* The vary may be 1-celled in the upper part, as in *Capsicum.* The number of ovules is usually many, but a few ovules are seen in *Cestrum.*

DIVISIONS OF THE FAMILY

The Solanaceae is divided into two groups, one with three tribes and the other with two tribes:

Group A. All stamens fertile. Embryo very much curved.

- **Tribe** (*i*) *Datureae.* Ovary 4-celled, septum dividing placenta equally. Example: *Datura.*
- **Tribe** (*ii*) *Nicandreae.* Ovary 3- to 5-celled, septum dividing placenta unequally. Example: *Nicandra.*
- **Tribe** (*iii*) *Solaneae.* Ovary 2-celled. Examples: *Capsicum, Solanum,* etc.

Group B. All stamens or 2–4 stamens fertile. Embryo slightly or almost straight.

- **Tribe** (*i*) *Cestreae.* All 5 stamens fertile. Seeds small, not flat. Example: *Cestrum.*
- **Tribe** (*ii*) *Salpiglossideae.* Flowers zygomorphic. Stamens 2–4, fertile. Examples: *Brunfelsia, Salpiglossis*, etc.

POSITION AND AFFINITY OF THE FAMILY

The Solanaceae was included by Bentham-Hooker as well as Bessey in the Polemoniales. Engler treated the family under the suborder Solanineae to the order Tubiflorae. Hallier regarded the family as a primitive member of the Tubiflorae and together with the Scrophulariaceae to have arisen very likely from the Linaceae. Wettstein put the family in the Tubiflorae along with the Convolvulaceae. Rendle placed the family in the Tubiflorae, assigning a separate position for the Convolvulaceae under the Convolvulales. In Hutchinson's arrangement, the family appeared under the Solanales which also included the Convolvulaceae and Nolanaceae.

The Solanaceae bears a close relationship with the Boraginaceae, the elements of similarity being the alternate leaves, regular flowers and pentamerous stamens. It is related to the Convolvulaceae by the presence of an intraxylary phloem, persistent calyx, twisted corolla and false septum. It is also akin to the Scrophulariaceae, but differs from the latter by the generally actinomorphic corolla, typically pentamerous stamens and obiique position of carpels.

NUMBER IN AND DISTRIBUTION OF THE FAMILY

There are about 90 genera and 2,000–3,000 species in the Solanaceae. Though the members of the family are abundant in the tropical and temperate regions of the world, their principal centres of distribution lie in Australia as well as Central America and South America.

COMMONLY OCCURRING PLANTS OF THE FAMILY

Browallia americana L. is an annual garden herb, having blue or violet flowers.

Queen-of-the-night (*Cestrum nocturnum* L.) is a garden plant with flowers emitting a sweet smell at night.

Tree-tomato [*Cyphomandra betacea* (*Cav.*) *Sendtn.*] is grown at hillstations for its edible fruits.

Thorn-apple (*Datura metel L.*) is a herb with highly poisonous seeds.

Tomato [*Lycopersicon lycopersicum* (*L.*) *Karst.*] is a tall herb, bearing red globose pulpy fruits.

Wild Tobacco (*Nicotiana plumbaginifolia* Viv.), *Physalis minima L.* and *Solanum surattense* Burm. f. are weeds.

Solanum erianthum D. Don is a small tree.

Solanum ferox L., *S. indicum L.* and Black Nightshade (*S. nigrum L.*) are herbs on road sides.

Solanum filicifolium Ort. is a shrub.

Solanum torvum Swartz is a spiny herb or shrub.

Solanum trilobatum L. is a prickly climber, being prevalent in the salt flats near Calcutta.

ECONOMIC ASPECTS OF THE FAMILY

Economically, the members of the Solanaceae are of great importance and yield foods, drugs and ornamentals.

The fruits of *Capsicum annuum*, *C. fastigiatum* and *C. frutescens* arc used as a condiment. The fruits of *Cyphomandra betacea*, *Physalis alkekengi*, *P. heterophylla*, *P. ixocarpa* and *P. peruviana* are edible. A number of plants are used as vegetables, *e.g. Solanum melongena, S.tuberosum* and *Lycopersicon lycopersicum.* The fruits of *S. quitoense* (Andes) form a refreshing beverage.

Some plants are of medicinal value. *Atropa belladonna* (Mediterranean) contains 'atropine', widely used following severe colds. *Hyoscyamus niger* is used as a narcotic and *Withania somnifera* as a nervine tonic. Alkaloids like 'daturin' and 'nicotine' are derived from *Datura stramonium* and *Nicotiana tabacum* respectively; the latter is economically outstanding as the source of tobacco leaves which are turned into cigarettes.

A few plants are cultivated as ornamentals, *e.g. Brunfelsia americana, Cestrum nocturnmum, Petunia hybrida, P. violacea, Nicotiana alata, N. glauca,* etc.

Nolanaceae (Nolana Family)

The Nolanaceae is an anomalous family. On the basis of regular flowers, twisted aestivation and pentacarpellate ovary, Bentham-Hooker placed the family in the Convolvulaceae. Bessey included it, along with the Solanaceae, in his Polemoniales. Wettstein and Rendle recognised the family and set it up in the Tubiflorae. The presence of mericarpous fruits is suggestive of the Boraginaceae. The family is closely allied to the Solanaceae from which it differs by the fruit structure.

This is a family of two genera and 83 species, native to Chile and Peru.

Nolana paradoxa is cultivated as an ornamental plant.

Boraginaceae (Borage Family)

GENERAL FEATURES OF THE FAMILY

Habit	—	Herbs, shrubs or trees, often hispid.
Leaves	—	Alternate, simple, entire, *exstipulate.*
Inflorescence	—	*Scorpioid cymes.*
Flowers	—	Bisexual, regular, hypogynous.
Calyx	—	Sepals 5, united, imbricate.
Corolla	—	Petals 5, united below, limb spreading, imbricate; throat provided with scales.
Androecium	—	Stamens 4–6, epipetalous; filaments equal or unequal; anthers 2-celled, introrse.
Gynoecium	—	Carpels 2, syncarpous; *ovary 4-lobed* with a ring-like nectar secreting disc, 2-*locular with 2 ovules in each loculus or 4-locular with 1 ovule in each loculus*; style gynobasic.
Fruit	—	*Separating into 4 nutlets or drupaceous segments.*

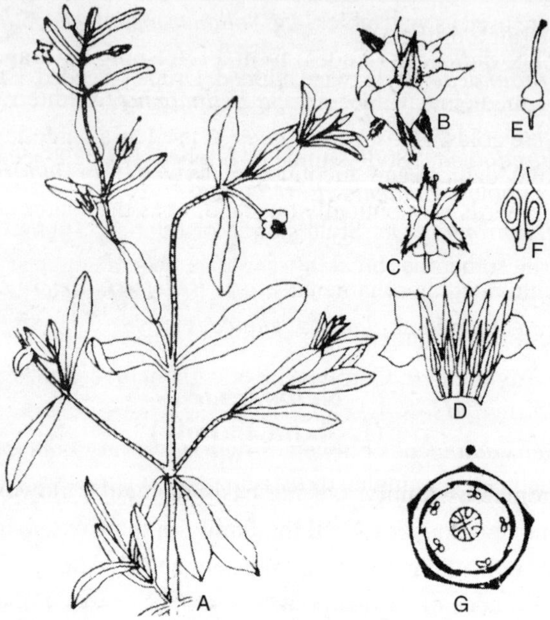

Fig. 73 BORAGINACEAE. *Trichodesma sedgwickianum* Banerjee nom. nov. **A**, portion of plant; **B**, flower; **C**, calyx with carpel; **D**, corolla split open; **E**, carpel; **F**, L.S. of ovary; **G**, floral diagram

FLORAL RANGE IN THE FAMILY

The flowers are usually regular and radially symmetrical. A change in the normal symmetry brings about zygomorphy, as in *Echium*, *Lycopsis* and others.

The sepals are united only at the base (*Lithospermum*) or completely connate (*Myosotis*). The calyx is spreading stellately below the fruit in *Patagonula.* The calyx lobes are deeply divided and hooked in *Harpagonella.*

The corolla may have various forms. It is hypocrateriform in *Anchusa,* infundibuliform in *Cynoglossum* and rotate in *Symphytum.* The corolla limb is bilabiate in *Echiochilon.* The corolla throat is open or closed. Sometimes, there are five pencils of hair, as in *Pulmonaria.*

The anthers are coherent or conivent into a cone in *Trichodesma*, but one anther is large than the other in *Caccinia.*

The carpels are median in position. The stigma is simple in *Myosotis* and bilobed in *Anchusa.* The style is usually terminal in *Cordia*, *Ehretia*, etc. The gynobasic style is met with in *Borago*, *Cynoglossum*, etc. In several cases, the ovule is generally anatropous with an upwardly directed micropyle. In a few instances, the ovule becomes descending with a downwardly directed micropyle.

DIVISIONS OF THE FAMILY

The Boraginaceae is divided into two groups, one with three subfamilies and the other with two subfamilies:

Group A. Style terminal. Fruit a drupe.

Subfamily I. *Cordioideae.* Style twice bilobed. Drupe 4-seeded. Examples: *Auxemma*, *Cordia*, etc.

Subfamily II. *Ehretioideae.* Style simple or bilobed. Drupe 2-seeded or of 1-seeded pyrenes. Examples: *Coldenia*, *Ehretia*, etc.

Subfamily III. *Heliotropioideae.* Style simple or bilobed. Stigma with a ring of hairs below. Fruit sometimes breaking up into nutlets. Examples: *Heliotropium*, *Tournefortia*, etc.

Group B. Style gynobasic. Fruit of nutlets.

Subfamily I. *Boraginoideae.* Corolla with or without inner scales. Nutlets 4 or less. Examples: *Borago*, *Myosotis*, etc.

Subfamily II. *Wellstedioideae.* Flowers 4-merous. Ovary compressed, 2-celled; ovule 1 in each cell. Example: *Wellstedia.*

POSITION AND AFFINITY OF THE FAMILY

Bentham-Hooker treated the Boraginaceae under the Polemoniales which was assigned to a position between the Gentianales and Personel. In Engler's arrangement, the family appeared within the suborder Boragininеae of the order Tubiflorae. Hallier, while discarding Engler's view, circumscribed it to include also the Lennoaceae and Hydrophyllaceae and considered the family as the primitive component of his Campanulales. He was of the opinion that the family arose directly from the Annonaceae or from stocks ancestral to the latter. For the family, Hutchinson created a new order (Boraginales)—an order placed after the Polemoniales and before the Lamiales.

The family bears a close relationship with the Polemoniaceae on the one hand and with the Hydrophyllaceae on the other. It resembles the Solanaceae by the presence of alternate leaves, regular flowers and pentamerous stamens. It forges a link with the Labiatae in having a nectar secreting hypogynous disc and nutlets. It is also related to the Verbenaceae by the possession of general floral organisation and drupaceous fruits.

NUMBER IN AND DISTRIBUTION OF THE FAMILY

The Boraginaceae includes about 100 genera and 2,000 species. The members of this family are widely distributed throughout the tropical and temperate regions of the world. The main centre of the family lies in the Mediterranean area, from where it spreads to Central Europe and Asia. However, it becomes less frequent as one approaches the cold temperate zone.

COMMONLY OCCURRING PLANTS OF THE FAMILY

Arnebia hispidissima (Lehm.) DC. is a xerophytic herb.

Coldenia procumbens L. is a prostrate annual herb with crisped hairy leaves and solitary white flowers.

Clammy Cherry or Indian Cherry [*Cordia dichotoma* var. *wallichii* (Clarke) Maheshwari] is a middle sized low spreading tree with a crooked stem.

Geiger tree (*Cordia sebestena* L.) is a tree which is often planted in gardens for its handsome orange-red flowers.

Cynoglossum lanceolatum Forsk. is an erect annual herb.

Ehretia acuminata R. Br. var. *serrata* (Roxb.) Johnston is a handsome tree.

Heliotropium curassavicum L. is a halophytic fleshy herb.

Heliotrope (*Heliotropium indicum* L.) is a way side weed with small flowers arranged in scorpioid cymes.

Heliotropium ovalifolium Forsk. is a weed prevailing in rice fields.

Forget me not (*Myosotis palustris* Benth.) is found in Darjeeling:

Trichodesma sedgwickianum Banerjee nom. nov. is an annual herb and is seen in waste places.

ECONOMIC ASPECTS OF THE FAMILY

Economically, the members of the Boraginaceae are of some importance. Several are cultivated as ornamentals, *e.g. Anchusa*, *Borago*, *Cordia*, *Ehretia*, *Mertensia*, *Myosotis*, *Pulmonaria*, *Symphytum*, etc. The roots of *Alkanna tinctoria* (Mediterranean) from 'alkanet', a purple dye. *Borago officinalis* (Europe) is raised as a source of nectar for bees. *Cordia alliodora* (tropical America) furnishes valuable timber. The leaves of *C. dichotoma* var. *wallichii* are useful in making the outer covering of Burma cheroots; the dried fruits of this plant produce the 'sebestan' of medicine. The root of *Onosma hispidum* yields a red dye which is used for colouring image, wood and oil. *Symphytum officinale* (Central Europe) is of medicinal value. *S. asperrimum* is cultivated for fodder.

Verbenaceae
(Verbain Family)

GENERAL FEATURES OF THE FAMILY

Habit	—	Herbs, shrubs or trees.
Leaves	—	*Opposite* or whorled, simple or compound, exstipulate.
Inflorescence	—	Cymes or spikes.
Flowers	—	Bisexual or polygamous, irregular, hypogynous.
Calyx	—	Sepals 5, united, campanulate or tubular, *persistent.*
Corolla	—	Petals 5, united, bilabiate, tubular, imbricate.
Androecium	—	Stamens 4, didynamous or 2; staminodes often present; anthers 2-celled, dorsifixed.
Gynoecium	—	Carpels 2, syncarpous; *ovary 4-lobed or entire*, superior, *2- to 4-locular*; *ovules 1 or 2 in each loculus*; *style terminal.*
Fruit	—	Berry or drupe.
Seeds	—	Erect, exalbuminous; testa thin; embryo straight.

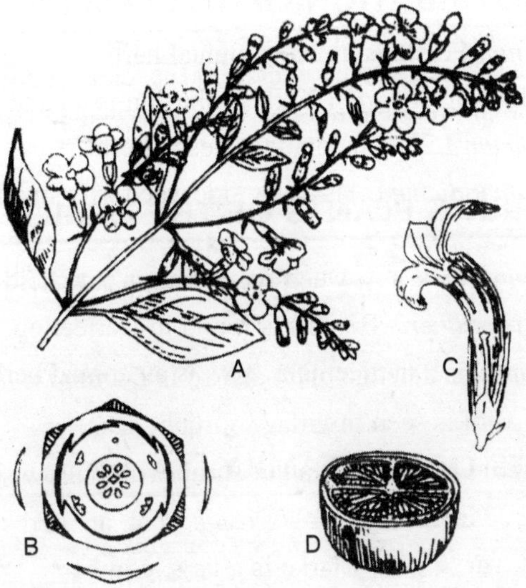

Fig. 74 **VERBENACEAE.** *Daranta repens* L. **A**, flowering shoot; **B**, floral diagram; **C**, V.S. of flower; **D**, T.S. of drupe. (**A–D**, after Rendle)

FLORAL RANGE IN THE FAMILY

In general, the flowers are pentamerous and irregular. The exception is *Physopsis*, where the flowers are tetramerous and regular.

The tendency to zygomorphy is more marked in the corolla than in the calyx. In *Lantana*, *Lippia* and others, the upper lip is formed by the union of the two upper corolla lobes. While valvate aestivation is a characteristic of the calyx, the corolla shows an imbricate aestivation.

All the four stamens may become equal and fertile. In *Geunsia* and *Tectona*, the stamens are 5 in a number. The fifth stamen is usually obsolete and may be represented by a staminode. The stamens may also become diandrous owing to the suppression of staminodes of the two posterior or anterior stamens.

The carpels are frequently 2, but 4 in *Duranta* and 5 in *Geunsia.* The posterior carpel may become abortive, as in *Lantana* and other genera.

The ovary may become round. It is generally somewhat lobed, the lobes being 2 in *Lantana* and 4 in *Clerodendrum* and *Verbena.*

POSITION AND AFFINITY OF THE FAMILY

The Verbenaceae was included in the Lamiales by Bentham-Hooker. In Engler's arrangement, the family appeared under the suborder Verbenineae of the order Tubiflorae. Hallier retained it within the Tubiflorae and sought its origin from the Scrophulariaceae. Hutchinson at first (1926) accepted it as belonging to the Lamiales, but later (1948, 1959, 1969) segregated it as the Verbenales and derived it from rubiaceous stocks.

The Verbenaceae is related to the Labiatae by the presence of the persistent calyx, bilabiate corolla and inferior micropyle of the ovule. It is allied to the Boraginaceae by the nature of inflorescence, calyx and fruit. It bears a relationship with the Acanthaceae, but differs in having a four-celled ovary with one ovule in each cell or two-celled ovary with two ovules in each cell.

NUMBER IN AND DISTRIBUTION OF THE FAMILY

The Verbenaceae embraces about 75 genera and over 3,000 species. Though the members of the family are inhabitants of the tropical or sub-tropical regions, they also extend into the temperate zones.

COMMONLY OCCURRING PLANTS OF THE FAMILY

White Mangrove (*Avicennia alba* Bl.) is a tree with long pneumatophores and viviparous seeds, being noticed in the Sunderbans.

Callicarpa longifolia Lamk. is a hedge plant and so is *Duranta repens* L.

Callicarpa macrophylla Vahl is seen in village shrubberies.

Clerodendrum indicum (L.) O. Kuntze is a tall shrub with hollow stems.

Clerodendrum inerme (L.) Gaertn. and *C. viscosum* Vent. are herbs on waste lands.

Congea tomentosa Roxb. var. *azurea* Clarke is a large climber.

Faradaya splendida F. v. Muell is a tall climber with showy flowers.

Candahar Tree or Comb Tree (*Gmelina arborea* L.) is a moderatesized tree of Orissa jungles.

Holmskioldia sanguinea Retz. and *Petrea volubilis* Jacq. are garden plants, the former with red flowers and the latter with blue flowers.

Lantana camara L. and *L. indica* var. *albiflora* Wt. ex Clarke are weeds.

Phyla nodiflora (L.) Greene is a herb of moist places.

Premna obtusifolia R.Br. is a shrub of the Sunderbans and so is *Stachytarpheta urticaefolia* (Salisb.) Sims.

Indian Oak or Teak (*Tectona grandis* L. f.) is a large deciduous tree.

Chaste Tree of Indian Privet (*Vitex negundo* L.) is a shrub, bearing trifoliate or quadrifoliate leaves.

ECONOMIC ASPECTS OF THE FAMILY

The bark of *Avicennia alba* is utilised for tanning. Some plants are cultivated as ornamentals, *e.g.* *Callicarpa*, *Caroypteris*, *Clerodendrum*, *Petrea*, *Vitex*, etc. The wood of *Gmelina arborea* is used for making drums, sitars, tom-toms and other musical instruments. *Lippia alba* produces a valuable essential oil. *L. citriodora* (tropical America) yields 'verbena oil'. The aromatic leaves of *Premna tomentosa* are employed as dinner plates and the wood for making combs. The tubers of *Priva laevis* (Argentina) are edible. The leaves of *Stachytarpheta dichotoma* are sometimes used as an alternate to tea. *Tectona grandis* supplies a hard and durable timber which is good for ship building and other purposes. *Vitex agnus-castus* produces a valuable oil and *V. celebica* a fine timber. The leaves of *V. negundo* serve as a febrifuge; they are also used for preserving stored grains against insect attacks.

Labiatae (Mint Family)

GENERAL FEATURES OF THE FAMILY

Habit	— Herbs or shrubs, *often aromatic with square stems.*
Leaves	— *Opposite* or whorled, simple, exstipulate, hairy.
Inflorescence	— *Verticillaster* or fascicles of axillary cymes or dense spikes.
Flowers	— Bisexual, irregular, hypogynous, *often bilabiate.*
Calyx	— Sepals 5, united, persistent.
Corolla	— Petals 5, united, 2-lipped, imbricate.
Androecium	— Stamens 4, epipetalous, didynamous or 2.
Gynoecium	— Carpels 2, syncarpous; *ovary 4 lobed, superior,* 4-locular; ovule 1 in each loculus; *micropyle inferior and raphe inward*; *style gynobasic*; stigma bifid.
Fruit	— *Carcerule.*
Seeds	— Erect, exalbuminous; testa thin; embryo with flat cotyledons parallel to fruit-axis.

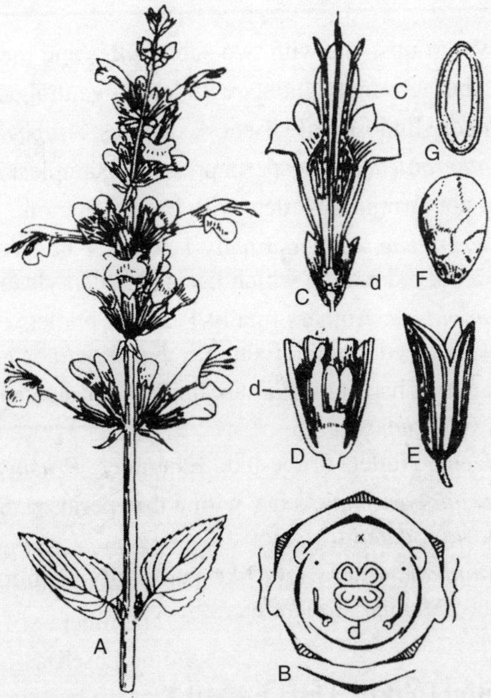

Fig. 75 LABIATAE. *Salvia coccinea* Juss. ex Murr. **A**, flowering shoot; **B**, floral diagram; **C**, flower opened along middle line ofanterior petal (*c,* long connective separating barren and fertile halves of anther); **D**, base of flower; (*d,* large nectary developed from hypogynous disc); **E**, calyx in fruiting stage; **F**, nutlet; **G**, L.S. of nutlet. (**A** and **C–E**, after Rendle; **B**, after Eichler)

FLORAL RANGE IN THE FAMILY

At the initial stage, the verticillaster is a biparous cyme or a dichasium which is succeeded by the uniparous branching of tertiary axis of the next higher order. Here the flowers are sessile and densely crowded. The flowers are also arranged in dichasia, as in some species of *Ocimum* or the flowers may be solitary, as in *Scutellaria.* The inflorescence is a spike in *Teucrium.*

The calyx shows a great variation in its structure. It is campanulate, spherical, bent or tubular. It is regularly 5-toothed (*Orthosiphon*), but may be 2-lipped in *Plectranthus*, *Salvia* and *Thymus.* A fleshy calyx is seen in *Hoslundia.*

The corolla may be tetramerous or pentamerous, as in *Mentha.* It is, in general, bilabiate with a 2/3 arrangement (2 petals forming the upper lip and 3 petals the lower one). It shows a 4/1 arrangement in *Hyptis* and other genera or 0/5 arrangement in *Teucrium.*

The fifth posterior stamen is rarely developed. The two upper stamens are completely reduced (*Lycopus*) or are reduced to staminodes. The anterior pair of stamens may be longer in *Nepeta.* In *Coleus,* the stamens are united into a bundle. The connective is not broad with two parallel theca (Ocimoideae) or transverse with two fertile anther lobes (*Meriandra*). The style is usually gynobasic, but terminal in *Ajuga* and *Teucrium.*

DIVISIONS OF THE FAMILY

The Labiatae is divided into two groups, one with two subfamilies and the other with six:

Group A. Style not gynobasic, but terminal. Nutlets with lateral ventral adnation.

Subfamily I. *Ajugoideae.* Endosperm absent. Examples: *Ajuga*, *Teucrium*, etc.

Subfamily II. *Prostantheroideae.* Endosperm present. Examples: *Prostanthera*, *Wrixonia*, etc.

Group B. Style gynobasic, but not terminal. Nutlets with basal adnation.

Subfamily I. *Catopherioideae.* Anthers finally 1-celled. Nutlets dry. Seeds erect, but embryo having a curved radicle which lies against cotyledons. Example: *Catopheria.*

Subfamily II. *Lavanduloideae.* Anthers finally 1-celled. Nutlets dry. Seeds erect and embryo without a curved radicle. Example: *Lavandula.*

Subfamily III. *Ocimoideae.* Characters like subfamily II, but stamens lie on lower lip. Examples: *Hyptis*, *Ocimum*, etc.

Subfamily IV. *Prasioideae.* Nutlets drupe-like. Examples: *Prasium*, *Stenogyne*, etc.

Subfamily V. *Scutellarioideae.* Nutlets dry with a thin pericarp. Seeds transverse. Examples: *Salazalaria*, *Scutellaria*, etc.

Subfamily VI. *Stachydoideae.* Characters like subfamily II, but stamens lie on upper lip. Examples: *Mentha*, *Salvia*, etc.

POSITION AND AFFINITY OF THE FAMILY

Hallier, Wettstein and Rendle accepted the views of Engler and treated the Labiatae as belonging to the Tubiflorae. On the basis of zygomorphic corolla and gynoecial location, Bessey segregated this family and the Verbenaceae into his Lamiales. Hutchinson did the same thing at first (1926), but later (1948, 1959, 1967) placed the Labiatae under the Lamiales and the Verbenaceae under the Verbenales.

Two subfamilies of the Labiatae resemble the Verbenaceae in having a terminal style. In the same way, several genera of the Verbenaceae have a gynobasic style—an usual character of the other six subfamilies of the Labiatae.

The Labiatae bears a relationship with the Boraginaceae but is set apart by the characters of the ovule. Whereas the micropyle is inferior and raphe inward in the Labiatae, the micropyle is superior and raphe outward in the Boraginacae.

The Labiatae is closely related to the Acanthaceae, but differs from it by the presence of verticillaster inflorescence, gynobasic style and the nature of ovary as well as fruit.

Among the sympetalous families, the Labiatae is reckoned as highly advanced. This is due to the elaboration of floral structure and adaptation for insect pollination.

NUMBER IN AND DISTRIBUTION OF THE FAMILY

The Labiatae comprises about 200 genera and 3,000 species. Though the members of the family are cosmopolitan, the centre of distribution lies in the Mediterranean area.

COMMONLY OCCURRING PLANTS OF THE FAMILY

Calamintha umbrosa (Bieb.) Fisch. & Mey. occurs in Assam, temperate Himalaya and Western Ghats.

Colebrookia oppositifolia Sm. is a shrub with opposite crenulate leaves.

Country Borage (*Coleus amboinicus* Lour.), *Salvia coccinea* Juss. ex Murr. and Sage (*S. plebeia* R. Br.) are garden plants.

Dysophylla verticillata Benth. is a marshy herb with whorled leaves.

Elsholtzia fruticosa (D. Don) Rehder is an undershrub with opposite leaves.

Hyptis suaveolens (L.) Poit. is a rigid annual.

Leonotis neptaefolia (L.) Ait. is an annual herb with flowers borne in axillary whorls.

Motherwort (*Leonurus sibiricus* L.) is a weed with heteromorphic leaves and pink flowers.

Leucas aspera (Willd.) Spreng., *L. mollissima* Wall., *L. procumbens* Desf., *Ocimum basilicum* L., *O. canum* Sims and Sacred Basil (*O. sanctum* L.) are herbs.

Leucosceptrum canum Sm. is a tall tree, bearing long upright spikes of yellow-white flowers.

Ocimum gratissimum L., *Plectranthus ternifolius* D. Don and *Pogostemon bengalense* (Burm. f.) O. Kuntze are bushy plants.

Orthodon diantherus (Buch-Ham. ex Roxb.) Hand. -Mazz. is a fragrant annual herb.

ECONOMIC ASPECTS OF THE FAMILY

Economically, the members of the Labiatae are of importance as the source of volatile aromatic oils and several are cultivated as ornamentals.

An essential oil extracted from the dried leaves and tops of *Hedeoma pulegioides* (USA) is used as an insect repellant. *Lavandula vera* yields 'lavender oil', *Mentha piperita* 'peppermint oil', *Pogostemon heyneanus* 'patchouli oil', *Rosmarinus officinalis* 'rosemary oil' and *Thymus serpyllum* 'thyme oil'. The fruits of *Lycopus europaeus* produce a green dye. *Ocimum kilmandscharicum* is raised for camphor.

Some herbs are in great demand for the flavour or aroma imparted to foods. These include *Majorana hortensis* (Mediterranean), *Marruhbium vulgare* (Canaries to Central Asia), *Melissa officinalis* (southern Europe), *Mentha arvensis*, *M. rotundifolia*, *M. spicata*, *Origanum vulgare* (Europe), *Satureja montana* (Europe), *Thymus vulgaris* (Europe), etc. *Stachys sieboldii* (East Asia) is grown for edible tubers and so is *Coleus parviflorus*. The tubers of *C. rotundifolius* are consumed as a potato substitute.

A few plants are useful in medicine, such as *Hyssopus officinalis* (Europe), *Leonurus cardiaca* (Eurasia), *Ocimum basilicum*, *Stachys officinalis* (western Europe), etc.

Some plants are grown as ornamentals, *e.g. Coleus, Dracocephalum, Lamium, Melissa, Molucella, Monarda, Nepeta, Phlomis, Salvia* and others.

Gesneriaceae
(Gesneria Family)

The family is recognised by the nonparasitic habit, irregular flowers, unilocular ovary and parietal placentation. The family is divided into two subfamilies:

Subfamily I. *Cyrtandroideae.* Ovary superior. Seeds with a copious endosperm. Cotyledons become unequal after germination. Examples: *Cyrtandra*, *Didymocarpus*, etc.

Subfamily II. *Gesnerioideae.* Ovary somewhat inferior. Seeds with little or no endosperm. Cotyledons remain equal after germination. Examples: *Besleria*, *Gesneria*, etc.

As the Gesneriaceae is closely related to the Orobanchaceae, Scrophulariaceae and Bignoniaceae, it is often difficult to separate the four families. The Orobanchaceae is a parasitic and degenerated group of the Gesneriaceae which in itself is nonparasitic. They only difference from the Scrophulariaceae is the presence of unilocular ovary. The Bignoniaceae stands out prominent by the silique-like woody fruit and large generally winged seeds.

The family has about 125 genera and 2,000 species in the tropical and subtropical regions of both hemispheres.

Aeschynanthus bracteata Benth., an epiphyte, grows at an elevation of 3,048 m in the Darjeeling area. *Corallodiscus lanuginosus* (Wall. ex DC.) Burtt is a scapigerous herb. *Platystemma violoides* Wall., a herb, bears the solitary leaf on the plant as a highly developed cotyledon.

A number of the Gesneriaceae are cultivated as ornamentals, *e.g. Achimenes*, *Columnea*, *Episcia*, *Haberlea*, *Jancaea*, *Ramonda*, *Saintpaulia*, etc.

Orobanchaceae
(Broom-rape Family)

The family is known by the root parasitism and lack of green colour. Besides, the flowers are zygomorphic, calyx valvate, stamens didynamous and ovary one-celled with two to six parietal placentae.

Due to the presence of unilocular ovary, the Orobanchaceae has been interpreted as a parasitic offshoot of the Gesneriaceae. It is also regarded as a continuation of parasitism which developed in the Scrophulariaceae. Again, the similarities in the development of gametophytes, endosperm, embryo and seed suggest a link between the Orobanchaceae and Scrophulariaceae; the former can be derived from the latter through a form like *Striga orobanchoides*.

With about 14 genera and 180 species, the members of the Orobanchaceae are inhabitants of the north temperate zone.

Aeginetia pedunculata (Roxb.) Wall. is a parasite on grasses. *Orobanche cernua* Loeffl. var. *nepalensis* DC. and *O. indica* Buch.-Ham. ex Roxb. are parasitic on the roots of Brinjal, Mustard, Poppy, Tobacco and other winter field crops on which they cause havoc if well established.

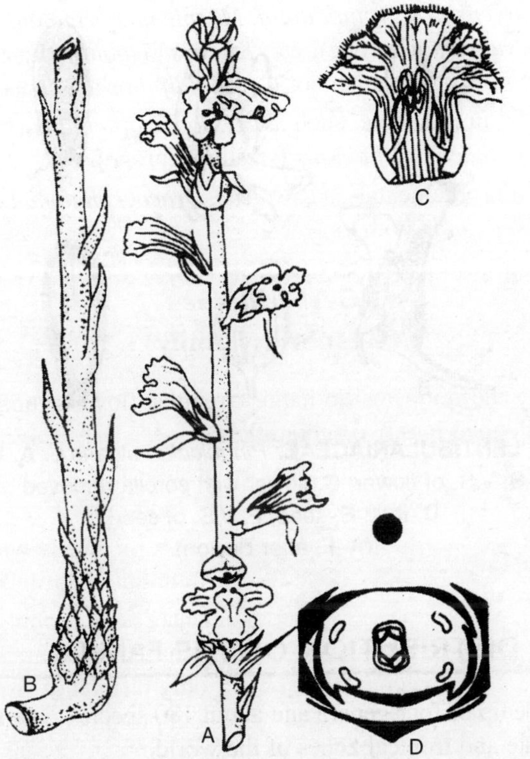

Fig. 76 OROBANCHACEAE. *Orobanche indica* Buch-Ham ex Roxb. **A**, upper part of stem bearing flowers; **B**, lower part of stem bearing scale leaves; **C**, corolla cut open; **D**, floral diagram. (**A–C**, after Reichenbach; **D**, after Eichler)

Lentibulariaceae
(Bladderwort Family)

GENERAL FEATURES OF THE FAMILY

Habit	—	*Insectivorous herbs,* aquatic or marshy.
Leaves	—	When submerged finely dissested, bearing bladders or *utricles*, or a rosette of simple and entire leaves when not submerged.
Flowers	—	Bisexual, *irregular*, hypogynous.
Calyx	—	Sepals 5, united, often 2-lipped.
Corolla	—	Petals 5, united, 2-lipped, lower lip spurred or saccate.
Androecium	—	*Stamens 2, epipetalous; anthers 1-celled.*
Gynoecium	—	*Carpels 2, medianly placed,* syncarpous; *ovary superior*, unilocular; ovules anatropous, many in each loculus; *placentation free central*; stigma sessile, 2-lobed with posterior lobe much reduced.
Fruit	—	Capsule.
Seeds	—	Small, 1 or many, exalbuminous.

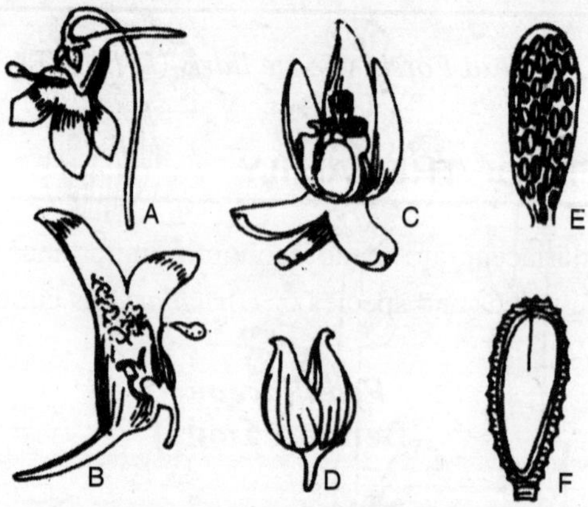

Fig. 77 LENTIBULARIACEAE. *Pinguicula vulgaris* L. **A**, flower; **B**, V.S. of flower; **C**, flower with corolla removed; **D**, fruit; **E**, seed; **F**, V.S. of seed. (**A–F**, after Baillon).

NUMBER IN AND DISTRIBUTION OF THE FAMILY

The Lentibulariaceae is made up of four genera and about 180 species. The members of this family are widely found in the temperate and tropical zones of the world.

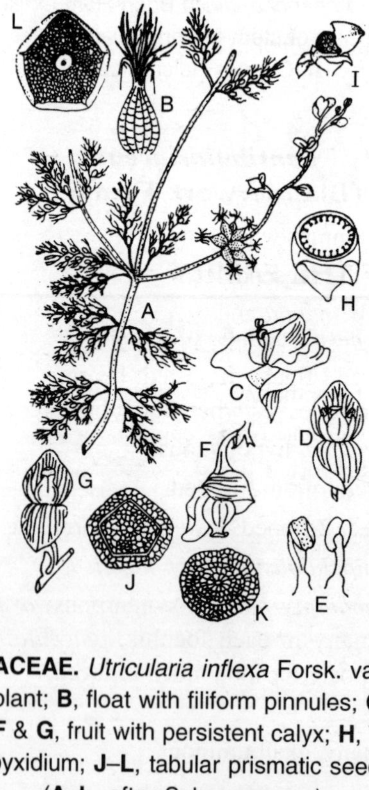

Fig. 78 LENTIBULARIACEAE. *Utricularia inflexa* Forsk. var. *stellaris* (L. f.) Taylor. **A**, portion of plant; **B**, float with filiform pinnules; **C** & **D**, flower; **E**, stamens; **F** & **G**, fruit with persistent calyx; **H**, T.S. of ovary: **I**, pyxidium; **J–L**, tabular prismatic seeds. (**A–L**, after Subramanyam)

COMMONLY OCCURRING PLANTS OF THE FAMILY

Utricularia aurea Lour. and *U. inflexa Forsk. var. ste llaris* (*L. f.*) P. Taylor are floating herbs.

ECONOMIC ASPECTS OF THE FAMILY

The members of the Lentibulariaceae are of no economic importance except for raising species of *Pinguicula* as ornamentals in gardens and species of *Utricularia* as curiosities in aquaria.

Pedaliaceae
(Benne Family)

GENERAL FEATURES OF THE FAMILY

Habit	—	*Herbs or undershrubs, with mucilage containing glandular hairs.*
Leaves	—	Opposite or spirally arranged, simple.
Flowers	—	Bisexual, *irregular*, hypogynous.
Calyx	—	Sepals 5, united.
Corolla	—	Petals 5, united.
Androecium	—	Stamens 4, didynamous; small staminodes present.
Gynoecium	—	Carpels 2, syncarpous; *ovary superior*, *2- to 4-locular*, *loculi somewhat completely divided by false septa*, *1 to many ovules in each loculus*; *placentation axile.*
Fruit	—	Capsule or nut, *commonly spiny.*

POSITION AND AFFINITY OF THE FAMILY

Saponins are lacking in all except *Sesamum indicum* where their doubtful presence is shown. *Pedalium murex* and *S. indicum* are devoid of tannins which doubtfully exist in *Martynia annua* and *S. laciniatum. S. indicum* and *S. laciniatum* show the presence of steroids, where as both steroids and triterpenes are absent in *M. annua* and *P. murex. M. annua* and *S. laciniatum* differ from *P. murex* and *S. indicum* in possessing leucoanthocyanins.

It is evident from the above facts that *M. annua* resembles the other taxa of the Pedaliaceae in a majority of chemical features (Parvati and Narayana, 1978). On the basis of distributional pattern of phenolic acids, Das *et al.* (1985) supported the separation of *Martynia* from the Pedaliaceae into an independent family Martyniaceae.

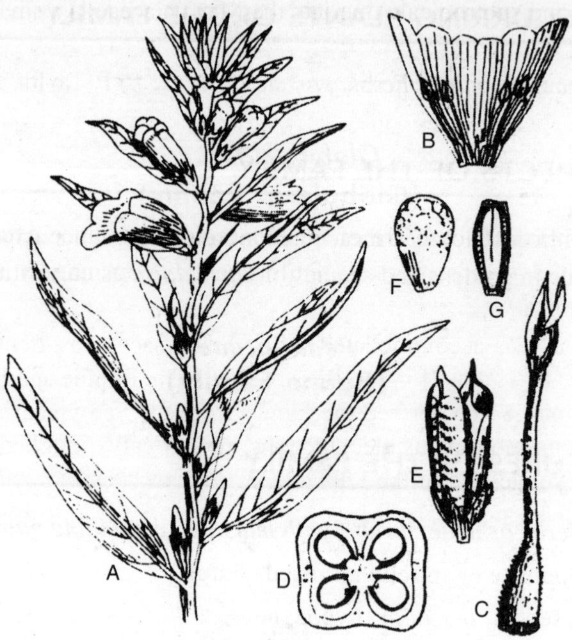

Fig. 79 PEDALIACEAE. *Sesamum indicum* L. **A**, portion of flowering shoot; **B**, corolla spread out; **C**, pistil; **D**, T.S. of ovary; **E**, dehiscing fruit; **F**, seed; **G**, L.S. of seed. (**A** & **C–E**, after Stapf; **B**, **F** & **G**, after Rendle)

NUMBER IN AND DISTRIBUTION OF THE FAMILY

The Pedaliaceae is composed of 12 genera and about 50 species. The members of this family are largely seashore or desert plants occurring in Indo-Malaysia, Malagasy, South Africa and tropical Australia.

COMMONLY OCCURRING PLANT OF THE FAMILY

Gingelly or Sesame (*Sesamum indicum* L.) is an oil seed crop.

ECONOMIC ASPECTS OF THE FAMILY

Economically, the Pedaliaceae is important as the source of 'sesame oil' which is extracted from the seeds of *Sesamum indicum.* Oil is also extracted from *S. angustifolium* seeds.

The leaves of *Ceratotheca sesamoides* and *Pedalium murex* are used as vegetables.

Martyniaceae
(Devil's-claw Family)

Glutinous-villous herbs, exstipulate leaves, zygomorphic flowers, unilocular ovary, parietal placentation and woody capsule with a pair of horn-like prongs are distinctive of the Martyniaceae.

Found in tropical and subtropical America, the Martyniaceae is a small family with five genera and ten species.

Martynia annua L. is an annual herb.

Globulariaceae
(Globularia Family)

The diagnostic characters of the Globulariaceae include the presence of exstipulate leaves, zygomorphic flowers surrounded by an involucre of bracts, reniform anthers opening by a single slit and unilocular ovary with pendulous ovules.

The Globulariaceae was accommodated in the Selaginaceae by Bentham-Hooker and merged with the Scrophulariaceae by Hallier. Engler, Bessey, Wettstein, Rendle and Hutchinson recognised the separate entity of the family. Some derived the family from the Scrophulariaceae and others from the Lamiales; the latter appears more plausible, since the tubular calyx, markedly two-lipped corolla and didynamous stamens are suggestive of the Labiatae.

Two genera and about 30 species make up the family Globulariacdae which abounds in South Europe and the Mediterranean.

The family is of no major economic importance. A few species of *Globularia* are occasionally cultivated as rock garden plants.

Scrophulariaceae
(Snapdragon Family)

GENERAL FEATURES OF THE FAMILY

Habit — Herbs or shrubs, *with collateral vascular bundles.*

Leaves — Opposite or whorled, *exstipulate.*

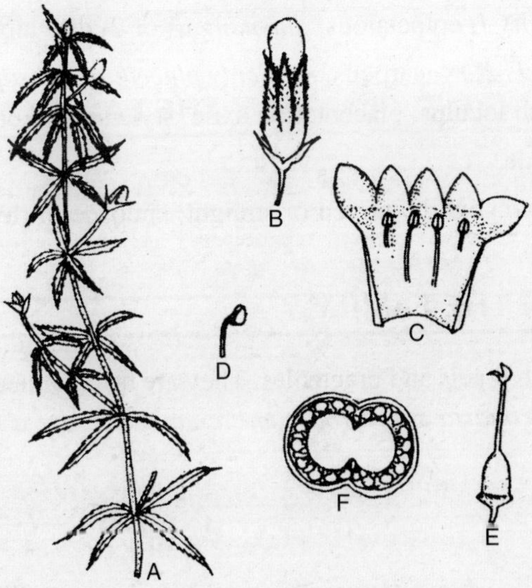

Fig. 80 SCROPHUIARIACEAE. *Limnophila indica* (L.) Druce. **A**, portion of plant; **B**, flower bud; **C**, corolla split open; **D**, stamen; **E**, carpel; **F**, T.S. of ovary

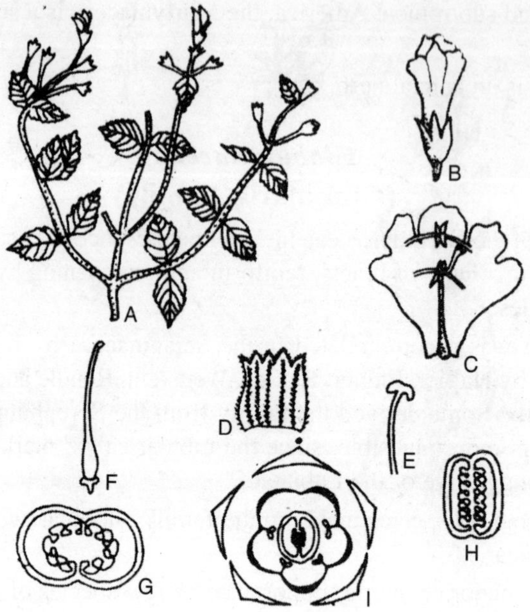

Fig. 81 SCROPHULARIACEAE. *Lindernia Crustacea* (L.) **F**. v. Muell.**A**, portion of plant; **B**, flower; **C**, corolla spread out; **D**, calyx; **E**, stamen; **F**, pistil; **G**, T.S. of ovary; **H**, L.S. of ovary; **I**, floral diagram

Inflorescence	—	Cymose or racemose.
Flowers	—	Bisexual, *irregular*, hypogynous.
Calyx	—	Sepals 5, united, persistent.
Corolla	—	Petals 5, united, sometimes personate, *often spurred.*
Androecium	—	*Stamens 4*, epipetalous, *didynamous* or 2; disc annular or unilateral.
Gynoecium	—	*Carpels 2*, syncarpous, *medianly placed*; *ovary superior* 2-locular; ovules many in each loculus; placentation axile; style simple or bilobed.
Fruit	—	Capsule.
Seeds	—	Embryo slightly curved or straight; endosperm fleshy.

FLORAL RANGE IN THE FAMILY

The flowers are provided with bracts and bracteoles. They are usually medianly zygomorphic, although regular flowers are seen in *Scoparia* and *Verbascum.*

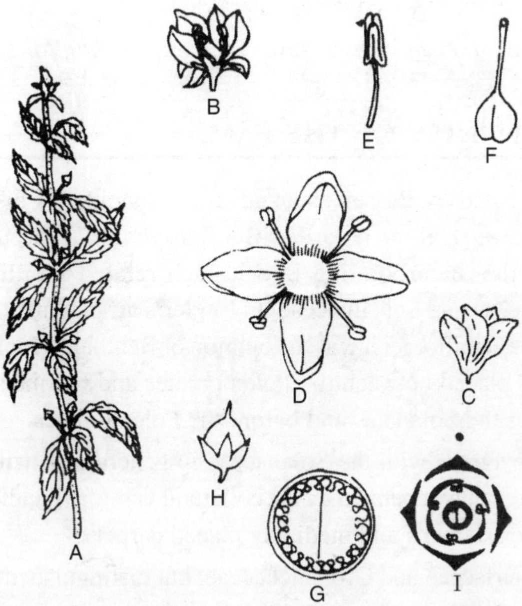

Fig. 82 SCROPHULARIACEAE. *Scoparia dulcis* L. **A**, portion of shoot; **B**, flower; **C**, calyx; **D**, corolla with stamens; **E**, stamen; **F**, carpel; **G**, T.S. of ovary; **H**, fruit; **I**, floral diagram

The calyx is usually pentamerous, but the suppression of the posterior sepal takes place in *Euphrasia* and many species of *Veronica.* In *Calceolaria*, the two anterior sepals are entirely connate.

The corolla shows a considerable variation in its form. It is large and campanulate, as in *Digitalis purpurea.* In *Scrophularia*, it is small and suberect. It is four-lobed in *Gratiola* and other genera. It is furnished with spreading limbs in *Veronica*, the tube being almost absent. The lateral anterior petals may be spurred, as in *Diascia.*

The corolla can also form a pair of closed lips, *e.g. Antirrhinum* and *Linaria.*

All the five stamens are fertile and equal, as in *Verbascum.* In *Russelia* and others, the stamens are 4 in number with the reduction of the posterior stamen. In some cases, the anterior pair of stamens becomes longer than the posterior pair. Only 2 stamens are found in *Veronica* and other genera.

The carpels are generally 2 and equal, but the anterior carpel becomes large in *Antirrhinum.*

DIVISIONS OF THE FAMILY

The Scrophulariaceae is divided into two groups, one with two subfamilies and the other with a single subfamily:

Group A. Posterior lobes of carpels cover two lateral lobes.

Subfamily I. *Antirrhinioideae.* Leaves opposite, at least lower ones. Stamens 4. Examples: *Antirrhinum, Calceolaria, Gratiola, Torenia,* etc.

Subfamily. II. *Pseudosolanioideae.* Leaves alternate. Stamens 5. Example: *Verbascum.*

Group B. Posterior lobes of carpels not covering lateral lobes.

Subfamily I. *Rhinanthoideae.* Stamens 4 or 2. Examples: *Digitalis*, *Rhinanthus*, *Veronica*, etc.

POSITION AND AFFINITY OF THE FAMILY

The Scrophulariaceae was treated by Bessey as an advanced member of his Scrophulariales, probably arising from the Bignoniaceae. Hallier regarded the Scrophulariaceae to be one of the primitive components of the Tubiflorae, incorporating into it such related families as the Globulariaceae, Lentibulariaceae, Plantaginaceae and Selaginaceae. In Engler's arrangement, the family appeared under the suborder Solanineae of the Tubiflorae. It was the opinion of Bentham-Hooker that the family belonged to the Personales which was placed between the Polemoniales and Lamiales. Hutchinson included the family in the Personales after the Solanales and before the Polemoniales.

The Scrophulariaceae agrees with the Solanaceae in general and with the tribe Salpiglossideae in particular in floral structure, but is separated by the collateral vascular bundles, nonplicate zygomorphic corolla, reduction of posterior stamens and medianly placed carpels.

It is allied to the Gesneriaceae and Orobanchaceae, but distinguished from them by the bilocular ovary. It bears an affinity with the Pedaliaceae and Bignoniaceae, but is set apart by the presence of endosperm in seeds. It is related to the Acanthaceae and Labiatae, but differs from them by the alternate leaves, absence of bracts and the characters of the corolla, ovary and fruit.

NUMBER IN AND DISTRIBUTION OF THE FAMILY

The Scrophulariaceae is compounded of about 220 genera and 3,000 species. The members of this family attain the peak of their development in the temperate zones of the world. However, they decrease in number as one approaches the tropical regions.

COMMONLY OCCURRING PLANTS OF THE FAMILY

Snapdragon (*Antirrhinum majus* L.), *Russelia equisetiformis* Schlecht. & Cham., *Torenia fournieri* Linden ex Fourn. and *Verbascum celosioides* Benth. are garden plants.

Bacopa monnieri (L.) Pennel, *Dopatrium junceum* (Roxb.) Buch.-Ham. ex Benth., *Limnophila indica* (L.) Druce, *Lindernia ciliata* (Colsm.) Pennell, *L.crustacea* (L.) F.v. Muell and *Verbascum chinensis* (L.) Santapau are weeds.

Toad-flax [*Kickxia ramosissima* (Wall.) Janch.] is a prostrate herb, having sagittate leaves and yellow flowers.

Limnophila heterophylla Benth. is an aquatic plant, bearing submerged dissected leaves.

Lindenbergia indica (L.) Vatke is a herb with yellow flowers, growing on window sills and old brick walls.

Mazus pumilus (Burm. f.) Steen. is an annual herb, having tufted radical leaves.

Mecardonia dianthera (Swartz) Pennell is an erect herb with square stems, ovate-crenate leaves and white or whitish-yellow flowers.

Scoparia dulcis L. is a rigid perennial herb with nearly regular flowers whose corolla throat is densely bearded.

Striga densiflora Benth., a herb, is parasitic on the roots of Sorghums.

ECONOMIC ASPECTS OF THE FAMILY

Economically, the members of the Scrophulariaceae are of limited value. The plants that are cultivated for showy flowers include *Antirrhinum, Calceolaria, Collinia, Hebe, Kickxia, Mimulus, Pentastemon, Russelia, Veronica, Wulfensia*, etc. *Bacopa monniera* is of medicinal importance for its nervine properties. *Curanga felterrae* is used as a febrifuge. The leaves of *Digitalis purpurea* yield the drug 'digitalin', a powerful stimulant for the heart. The species of *Euphrasia* were formerly utilised as a source of eye wash; the European species of *Scrophularia* were once reputed to cure scrofula.

Bignoniaceae
(Bignonia Family)

The family is characterised by the predominantly lianous habit, compound leaves, zygomorphic flowers, anthers connivent in pairs, numerous ovules, silique-like woody capsule, larged winged seeds and absence of endosperm in seeds.

The Bignoniaceae is separated into four tribes:

Tribe (*i*) *Bignonieae.* Usually tendrillate. Ovary completely bilocular, compressed parallel to septum. Capsule septifragal with winged seeds. Examples: *Glaziovia, Oroxylum*, etc.

Tribe (*ii*) *Crescentieae.* Usually erect, plants. Ovary unilocular or bilocular. Fruit a berry or indehiscent. Seeds not winged. Examples: *Crescentia, Kigelia, Parmentiera*, etc.

Tribe (*ii*) *Eccremocarpeae.* Tendrillate. Ovary bilocular. Capsule splits from below up. Seeds winged. Example; *Eccremocarpus.*

Tribe (*iv*) *Tecomeae.* Rarely tendrillate. Ovary bilocular, compressed at right angles to septum or cylindrical. Capsule loculicidal with winged seeds. Examples: *Catalpa, Jacaranda, Tecoma*, etc.

Having about 120 genera and 650 species, the Bignoniaceae is largely a tropical family and is highly developed in northern South America where climbers constitute an important part of the forest flora.

Indian Cork tree (*Millingtonia hortensis* L. f.), *Oroxylum indicum* (L.) Vent. and *Stereospermum personatum* (Hassk.) Chatt. are road side trees. *Tecoma stans* (L.) HBK. is a scandent garden shrub.

Some plants of the Bignoniaceae produce useful wood, *e.g. Catalpa bignonioides, C. bungei, C. speciosa* and *Cybistax donnellsmithii.* The leaves of *Cybistax antisyphilitica* are used as the source of a blue dye.

Many are ornamentals like the species of *Bignonia. Crescentia, Doxantha, Jacaranda, Kigelia, Paulownia, Tabebuia*, etc. The fruits of *Crescentia cujete* form 'calabash' after the removal of the pulp and are used for holding liquids. The seeds of *Oroxylum indicum* are used as linings for hats and as stuffing material between layers of wicker work to make umbrellas.

Acanthaceae
(Acanthus Family)

GENERAL FEATURES OF THE FAMILY

Habit	— Herbs or shrubs, *often with cystoliths.*
Leaves	— Opposite, simple, entire, exstipulate.
Inflorescence	— Cymes or spikes.
Flowers	— Bisexual, *irregular, hypogynous with prominent bracts and bracteoles.*
Calyx	— Sepals 4 or 5, united, unequal.
Corolla	— Petals 5, *often bilabiate*, imbricate or twisted.
Androecium	— Stamens 4, epipetalous, didynamous or 2; staminodes often present; anthers 2-celled, spurred, hairy.
Gynoecium	— Carpels 2, syncarpous; *ovary unlobed, superior, 2-locular*; ovules 2 to many in each loculus; placentation axile; style terminal; stigmas 2.
Fruit	— *Elastically dehiscent bivalved capsule.*
Seeds	— Many, exalbuminous, *attached to hook-like supports* (*retinacula or jaculators*); embryo curved.

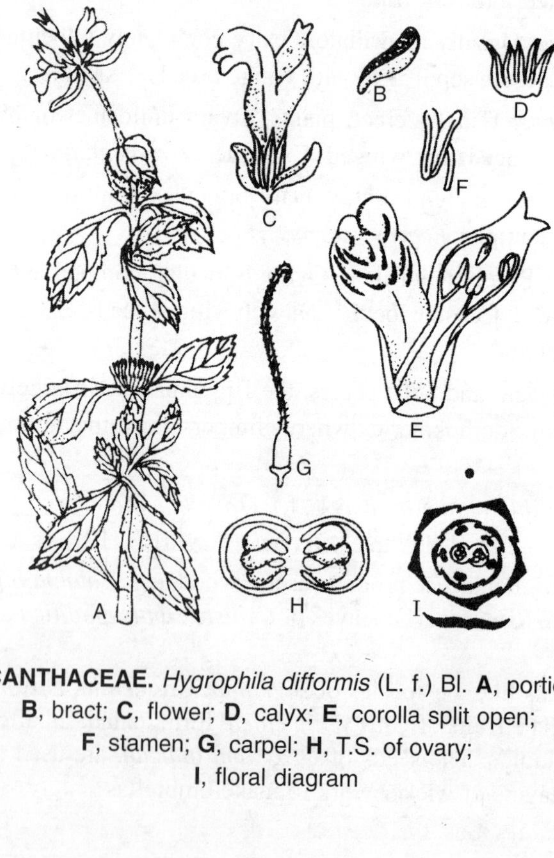

Fig. 83 ACANTHACEAE. *Hygrophila difformis* (L. f.) Bl. **A**, portion of plant; **B**, bract; **C**, flower; **D**, calyx; **E**, corolla split open; **F**, stamen; **G**, carpel; **H**, T.S. of ovary; **I**, floral diagram

FLORAL RANGE IN THE FAMILY

The floral structure is somewhat uniform in the Acanthaceae. Large bracteoles occur in *Boutonia* where they are united into a tubular involucre enclosing a solitary flower. The lower lip of the corolla is rolled in *Symplectochilus*, but the tube is long and slender in *Ecbolium* and other related genera. In *Ruellia*, *Thunbergia* and others, the corolla limb is equally 5-lobed. The stamens are 2 in number in *Adhatoda* and *Justicia.* The anther may be one-celled with the rudiments of a second cell, as in *Chaetothylax tocantinus.* The connective often splits up into two branches and the anther lobes become wide apart, *e.g. Beloperone*, *Salviacanthus*, etc. The pollen grains are of different size and shape with a marked orientation.

DIVISIONS OF THE FAMILY

The Acanthaceae is divided into four subfamilies:

Subfamily I. *Acanthoideae.* Ovules 2 to many. Fruit a capsule. Retinacula present. Examples: *Acanthus*, *Barleria*, etc.

Subfamily II. *Mendoncioideae.* Ovules 4. Fruit a drupe. Seeds not more than 2. Retinacula absent. Example: *Mendoncia.*

Subfamily III. *Nelsonioideae.* Ovules many. Fruit a capsule. Retinacula absent. Example: *Nelsonia.*

Subfamily IV. *Thunbergioideae.* Ovules 4. Fruit a capsule. Retinacula absent. Example: *Thunbergia.*

POSITION AND AFFINITY OF THE FAMILY

There seems to be agreement among botanists to retain the Acanthaceae as a distinct family. It has been derived from the Scrophulariaceae or stocks ancestral to the Scrophulariaceae. Bentham-Hooker as well as Hutchinson treated it as the most advanced member of the Personales. In Engler's arrangement, the family appeared under the suborder Acanthineae of the order Tubiflorae. Bessey included it at the top of his Scrophulariales.

The Acanthaceae is related to the Bignoniaceae and Scrophulariaceae, but differs from them by the possession of prominent bracts and bracteoles, imbricate calyx of unequal sepals, beaked ovary and exalbuminous seeds with a curved embryo. It is akin to the Pedaliaceae in the presence of irregular flowers and the nature of fruits. It is closely allied to the Labiatae, but distinguished by the bracteate inflorescence, bilocular ovary, terminal style and many-seeded capsule.

NUMBER IN AND DISTRIBUTION OF THE FAMILY

The Acanthaceae constitutes about 250 genera and 2,500 species. Though the members of the family are inhabitants of the tropical regions, their centres of distribution lie in Central America, Brazil, Indo-Malaysia and Africa.

COMMONLY OCCURRING PLANTS OF THE FAMILY

Acanthus ilicifolius L. is a mangrove plant which has prickly leaves and handsome flowers.

Adhatoda vasica Nees is a dense shrub, having bracteated spikes.

Andrographis echioides (L.) Nees is a herb seen on old brick walls.

A. paniculata (Burm. f.) Wall. ex Nees is a herb with square stems.

Asystasia gangetica (L.) T. and., *Dicliptera roxburghiana* Nees, *Hemigraphis hirta* T. and., *H. latebrosa* Nees var. *heyneana* Brem., *Hygrophila difformis* (L. f.) Bl., *Justicia diffusa* Willd., *J. simplex* D. Don, *Phaulopsis dorsiflora* (Retz.) Santapau, *Rostellularia procumbens* (L.) Nees and *Rungia pectinata* (L.) Nees are weeds.

Barleria prionitis L. is a shrub, bearing two long thorns at each node.

Crossandra infundibuliformis (L.) Nees is a small shrub.

Dipteracanthus dejectus Nees and Menow-weed (*Ruellia tuberosa* L.) are found in and about gardens.

Ebermaiera glutinosa Wall. is a trailing herb met with in rice fields.

Ecbolium viride (Forsk.) Alston is a small shrub of bushes.

Hygrophila auriculata (Schumach) Heine is an erect highly spinous marshy herb.

Justicia bentonica L., *J. gendarussa* L.f., *Rhinacanthus nasuta* (L.) Kurz and *Thunbergia grandiflora* (Roxb. ex Rottl.) Roxb. are garden plants.

Lepidagathis incurva Buch.-Ham. ex D. Don is a suberect perennial herb.

Mackenziea integrifolia (Dalz.) Brem. is a much branched viscid herb, producing blue flowers.

Nelsonia canescens (Lamk.) Spreng. is a diffuse herb of grassy places.

Peristrophe bicalyculata (Retz.) Nees is a branched herb with six-angled stems.

Pseuderanthemum malabaricum (Clarke) Gamble is found in evergreen forests of Deccan and Western Peninsula.

Pteracanthus alatus (Wall. ex Nees) Brem. is an erect shrub.

Sericocalyx scaber (Nees) Brem. is a herb.

ECONOMIC ASPECTS OF THE FAMILY

Economically, the Acanthaceae is of importance for ornamentals and as a source of medicinal plants. Those grown for showy flowers include *Acanthus montanus*, *Beloperone guttata*, *Jacobinia tinctoria*, *Justicia gendarussa*, *Pachystachys lutea*, *Sanchezia nobilis*, *Thunbergia grandiflora*, etc.

The roots of *Acanthus ebracteatus* (Malaysia) is utilised as a cough medicine. The roots of *A. mollis* (Europe) are used to treat diarrhoea. *Adhatoda vasica* is a remedy for coughs and *Andrographis paniculata* for liver troubles. The leaves of *Blechum pyramidatum* (Central & South America) are used for the treatment of coughs and fevers. *Graptophyllum pictum* is tried in skin diseases and also for cuts. *Hygrophila auriculata* is a specific for diarrhoea and *Justicia pectoralis* for lung troubles. The leaves of *Strobilanthes cusia* constitute the source of a blue dye.

Order 31. *Plantaginales*

The order contains a single family, the Plantaginaceae:

Plantaginaceae (Plantain Family)

GENERAL FEATURES OF THE FAMILY

Habit	— Herbs.
Leaves	— Alternate or opposite, simple, *seemingly parallel-veined, basal, sheathing at base.*
Inflorescence	— *Capitate or spicate on wiry or stout scapes.*
Flowers	— Bisexual, regular, hypogynous, *tetramerous.*
Calyx	— 4-lobed, herbaceous, persistent.
Corolla	— 3- to 4-lobed, scarious or membranous, generally salverform.
Androecium	— *Stamens* 4, rarely 2 or 1, inserted on corolla tube and alternate with lobes, *much exserted*; anthers 2-celled, opening lengthwise; *pollen grains dry, abundant and powdery.*
Gynoecium	— *Carpels* 2; united; ovary superior, 1- to 4-locular; style 1, filiform.
Fruit	— *Circumscissile capsule* or bony nut.
Seeds	— Embryo straight; endosperm fleshy.

POSITION AND AFFINITY OF THE FAMILY

The Plantaginaceae is an anomalous taxon and hence of doubtful relationship. Bessey included it in his Primulales as allied to and originating from the Plumbaginaceae. Hallier as well as Wettstein placed the family in the Tubiflorae as related to the Scrophulariaceae. Rendle, Lawrence and Core recognised the family and retained it in the Plantaginales. Hutchinson also did likewise, but derived the order from primulaceous ancestors. The family, according to Porter, "probably represents an offshoot from the Polemoniales or perhaps from the Primulales". The family is very likely a reduced one, arising from a gamopetalous group with entomophilous flowers.

NUMBER IN AND DISTRIBUTION OF THE FAMILY

The Plantaginaceae comprises three genera and about 223 species. The members of this family are world wide in distribution, but found mainly in the temperate zone and mountains of the tropics.

COMMONLY OCCURRING PLANT OF THE FAMILY

Plantago asiatica L. occurs at wet places and by roadsides in the hilly regions.

ECONOMIC ASPECTS OF THE FAMILY

The family is of medicinal importance for *Plantago indica*, *P. ovata* and *P. psyllium*, producing seeds which contain mucilaginous substance and are used as a mild laxative. Several species of *Plantago* are considered as troublesome weeds.

Order 32. *Rubiales*

The *Rubiales* is a natural taxon. This is reflected in the leaves being opposite, inflorescence cymose, flowers actinomorphic to zygomorphic, ovary inferior and anthers distinct.

As conceived by Engler and Diels, the Rubiales is composed of five families: Rubiaceae, Adoxaceae, Dipsacaceae, Caprifoliaceae, Valerianaceae. This circumscription was also accepted by Rendle, Gundersen, Core and Benson. However, Lawrence did not account for the Adoxaceae. Hutchinson accepted the Rubiales to contain the Dialypetalanthaceae and Rubiaceae. He created the Valerianales in order to accommodate the Valerianaceae, Dispacaceae and Calyceraceae. He removed the Adoxaceae to the Saxifragales and the Caprifoliaceae to the Araliales. Porter included the Rubiaceae and Caprifoliaceae in the Rubiales and placed the Adoxaceae, Dipsacaceae, Calyceraceae and Asteraceae (Compositae) under the Asterales. Cronquist retained the Rubiaceae in the Rubiales, but transferred the Caprifoliaceae, Adoxaceae, Valerianaceae, Dipsacaceae and Calyceraceae to the Dipsacales.

A perusal of the characters of the Umbellales, Rubiales and Campanulales points out that the three taxa are closely related. These characters include (*i*) a shift from the woody arboreal habit to the herbaceous one; (*ii*) a reduction in the number of floral parts associated with an increasing tendency to zygomorphy; (*iii*) an occurrence of epigynous flowers and epigynous disc; (*iv*) a tendency to attain conspicuousness by the aggregation of flowers into umbels, cymose umbels and capitate inflorescences; (*v*) a loss of the protective function of the calyx and its development into pappus and scales as organs for the distribution of fruits, seeds and (*vi*) the presence of a bicarpellary ovary which tends to abort all but one of the loculi, resulting in the development of a few- or one-seeded fruits.

The origin of the Rubiales, according to Rendle, can be traced among the epigynous tetracyclic Dialypetalae. In fact, the order may be considered as a sympetalous development of a stock akin to the Umbelliflorae. Moreover, a comparison of the genera of the Rubiaceae and Caprifoliaceae with the Cornaceae reveals a close connection. Only the polypetalous feature of the Cornaceae seems to make the family distinct; otherwise, the three taxa look exactly alike. Hutchinson (1959), however, derived the Rubiales from loganiaceous ancestors and the Valerianales from the Saxifragales stock.

The usual five families of the *Rubiales*, according to Core (1955), may be recognised by the following characters:

(1) Stamens as many as corolla lobes	(2)
(1) Stamens twice as many as or fewer than corolla lobes	(3)
(2) Leaves opposite, stipulate	*Rubiaceae*
(2) Leaves opposite or perfoliate exstipulate	*Caprifoliaceae*
(3) Stamens twice as many as corolla lobes	*Adoxaceae*
(3) Stamens fewer than corolla lobes	(4)
(4) Ovary trilocular	*Valerianaceae*
(4) Ovary unilocular	*Dipsacaceae*

Rubiaceae
(Madder Family)

GENERAL FEATURES OF THE FAMILY

Habit — Herbs, shrubs or trees.
Leaves — Opposite, simple, entire, *stipulate* (stipules interpetiolar).

Inflorescence	—	Cymes or decussate panicles, sometimes heads.
Flowers	—	Bisexual, regular, *epigynous*, 4- or 5-merous.
Calyx	—	Sepals 4 or 5, united.
Corolla	—	*Petals 4 or 5*, united.
Androecium	—	*Stamens 4 or 5, epipetalous*; *anthers introrse.*
Gynoecium	—	Carpels 2, connate, crowned by a fleshy disc: *ovary inferior*, *2-locular*; *ovules 1 or more in each loculus*; style bifid or multifid.
Fruit	—	Berry, capsule or drupe.
Seeds	—	Embryo straight; endosperm fleshy, horny or cartilaginous.

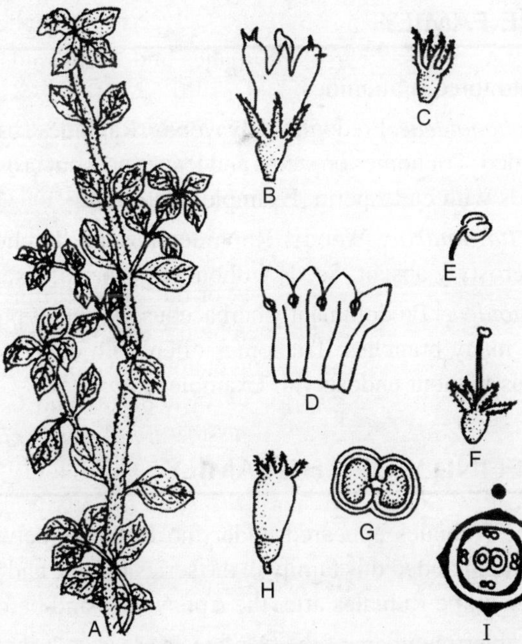

Fig. 84 RUBIACEAE. *Borreria articularis* (L. f.) F.N. Will. **A**, portion of plant; **B**, flower; **C**, calyx; **D**, corolla spread out; **E**, stamen; **F**, carpel with calyx; **G**, T.S. of ovary; **H**, fruit; **I**, floral diagram

FLORAL RANGE IN THE FAMILY

The flowers are solitary in *Gardenia.* They may be aggregated into globose heads, as in *Adina*, *Anthocephalus*, *Morinda,* etc. Although flowers are generally regular, irregular flowers are also found (*Henriquezia*, *Posoqueria*). Unisexual flowers are met with in *Anthospermum* and *Coprosma.*

Variation in the calyx structure has been noted. In *Mussaenda*, *Nematostylis* and *Warszewiczia*, one of the sepals becomes enlarged, showy and leafy. The calyx limb is plumose in *Gaillonia.*

The corolla is somewhat bilabiate in *Capirona*, *Dichilanthe* and *Ferdinandusa* and partite to the base in *Aulacodiscus*. The corolla lobes are appendaged or horned (*Corynanthe*, *Rudgea*) or dextrorsely contorted (*Cosmibuena*, *Deppea*).

The stamens are declinate (*Tammsia*) or monadelphous (*Solenandra*). The filaments are very unequal in *Pallasia* and *Remijia* as well as the anthers are somewhat united in *Acranthera* and *Neurocalyx*.

The ovary is inferior, but semi-inferior in *Synaptanthera* and superior in *Gaertnera* and *Pagamea*. A bilocular ovary is a general rule in the family. *Hamelia*, *Knoxia*, *Meyna*, etc. have an unilocular ovary with a single ovule. *Pavetta* also represents an uniovulate condition with the ovule sunken in the fleshy funiculus. The ovules are paired in the loculi, one pendulous and the other ascending, *e.g. Scyphiphora*. In *Cinchona*, *Gardenia*, *Oldenlandia*, *Rondeletia*, etc., there are many ovules in each loculus. The ovules may be attached to the base of the ovary, as in *Hydnophytum*, *Myrmecodia*, *Paederia*, *Uragoga*, etc. The ovules are also fixed to the septum in *Asperula, Galium, Rubia* and related genera.

DIVISIONS OF THE FAMILY

The Rubiaceae is divided into three subfamilies:

Subfamily I. *Cinchonoideae.* Predominantly woody. Raphides absent in leaves; stipules rarely divided. Trichomes on stems and leaves without cross walls. Heterostyly absent. Seeds with endosperm. Example: *Cinchona.*

Subfamily II. *Guettardoideae.* Woody. Raphides absent. Trichomes without cross walls. Heterostyly absent. Seeds without endosperm. Example: *Guettardia.*

Subfamily III. *Rubioideae.* Predominantly herbaceous. Raphides present; stipules often divided into many branches. Trichomes often with cross walls. Heterostyly common. Seeds without endosperm. Example: *Rubia.*

POSITION AND AFFINITY OF THE FAMILY

In Engler's arrangement, the Rubiales appeared under the Rubiales between the Plantaginales and Cucurbitales. Bentham-Hooker included this family in the series Inferae and just before the Compositae. Hutchinson placed the family in the Rubiales after the Apocynales and before the Bignoniales.

"The Rubiaceae resembles members of the Gentianales (particularly the Loganiaceae) in having well developed stipules which often bear special glands (colleters) and in possessing certam alkaloids; on the other hand, it differs in the absence of internal phloem and in its inferior ovary. The two features are typical of the Dipsacales, but the latter tend to lack stipules and do not synthesize chemicals" (Kupicha, 1978).

The Rubiaceae is closely related to the Cornaceae in such features as the cymose umbel-like inflorescence, epigyny, calyx suppression, etc. It also bears a relationship with the Adoxaceae and Caprifoliaceae in the presence of inferior ovary. It establishes a link with the Compositae, since the head of the Compositae can be derived from the head-like inflorescence of some members of the Rubiaceae and Dipsacaceae.

NUMBER IN AND DISTRIBUTION OF THE FAMILY

There are about 500 genera and 7,000 species in the Rubiaceae. The members of this family dominate the warmer parts of the world with extensions into the temperate and frigid zones.

COMMONLY OCCURRING PLANTS OF THE FAMILY

Adina cordifolia (Roxb.) Hook. f. ex Brandis is a large deciduous tree which is often buttressed.

Anthocephalus cadamba (Roxb.) Miq. is a large tree, having horizontal branches.

Borreria articularis (L. f.) F. N. Will. is a perennial herb.

Canthium angustifolium Roxb. is a spreading shrub of Sunderbans.

Dentella repens (L.) Forst., a herb found in waste places, bears white flowers.

Cape Jasmine (*Gardenia jasminoides* Ellis), *Hamelia patens* Jacq., *Ixora arborea* Roxb. ex Sm. and *I. coccinea* L. are garden plants.

Knoxia sumatrensis (Retz.) DC. is an erect annual.

Meyna laxiflora Robyns is a highly spinous tree.

Mitragyna parviflora (Roxb.) Korth. is a large pubescent or nearly glabrous tree.

Morinda citrifolia L. is a garden plant, producing multiple fruits.

Mussaenda frondosa L. is a small shrub with an interesting flower where one of the sepals develops into a large petaloid yellow leaf.

Neanotis calycina (Wall. ex Hook. f.) Lewis is a slender annual herb.

Oldenlandia biflora L., *O. corymbosa* L. and *O. umbellata* L. occur in rice fields.

Paederia scandens (Lour.) Merr. is a twining slender foetid shrub.

Pavetta indica L. is a small tree.

Petunga racemosa (Roxb.) K. Schum. is an evergreen shrub met with in the Sunderbans.

Randia spinosa Poir. is a large spinous shrub.

Indian Madder (*Rubia tinctorum* L.) is a climbing herb, having big stipules.

Uncaria macrophylla Wall. is a strong climber.

Wendlandia exserta (Roxb.) DC. is a small crooked tree.

Xeromphis spinosa (Thunb.) Keay is a small tree, bearing dimorphic flowers.

ECONOMIC ASPECTS OF THE FAMILY

A number of plants of the Rubiaceae is grown ornamentally, *e.g.*, *Asperula*, *Bouvardia*, *Coprosma*, *Gardenia*, *Ixora*, *Mitchella*, *Nertera*, *Pentas*, etc.

The fruits of *Anthocephalus cadamba* are edible; a spirit is distilled from the flowers of this plant. The seeds of *Coffee arabica*, *C. liberica* (Liberia) and *C. robusta* (Congo) are ground for the beverage, 'coffee'. The ripe fruits of *Randia spinosa* are eaten after roasting.

The juice of *Adina cordifolia* is used to kill maggots in sores. The roots of *Cephaelis ipecacuanha* is the source of 'vinum ipecac'. The bark of *Cinchona calisaya, C. ledgeriana, C. officinalis, C. pitayensis* and *C. succirubra* yields 'quinine'. *Oldenlandia corymbosa* and *O. umbellata* are useful, the farmer as a febrifuge and the latter as an expectorant. The leaves of *Paederia scandens* form a good stomachic. *Uncaria gambier* (Indonesia) is the source of 'gambier', a resinous substance useful in medicine.

Calycophyllum candidissimum and *Ixora ferrea* furnish 'lance wood' and 'iron wood' respectively. *Elaeagia utilis* (Colombia) produces a resin, used for making paint. Some species of *Galium* were once employed for filling mattress due to their pleasant odour. The bark of *Hymenodictyon excelsum* is astringent and used for tanning. From the bark of *Mitragyna parviflora* is obtained a kind of fibres for making cordage. A yellow dye, extracted from the bark of *Morinda angustifolia* and *M. coreia*, is used for dyeing carpets, turbans, etc. The roots of *Rubia cordifolia* serve as the source of a red dye, 'Turkey red', used for colouring medicinal oils.

Caprifoliaceae
(Honeysuckle Family)

The family is characterised by the opposite exstipulate leaves, stamens as many as corolla lobes and trilocular inferior ovary.

The inflorescence is a regular cyme in *Leycesteria formosa*, 2-flowered cyme in the species of *Lonicera* and a spike or raceme in *Diervilla* and *Symphoricarpos.* The flowers are regular in *Sambucus* and *Viburnum*, but slightly zygomorphic in *Linnaea* and *Symphoricarpos* as well as markedly zygomorphic in *Lonicera.* The stamens are didynamous in *Dipelta* and *Linnaea.* The carpels are 5 in *Leycesteria* and some species of *Sambucus*, 4 in *Symphoricarpos*, 3 in most of the remaining genera and 2 in *Diervilla.* The ovules are numerous in *Leycesteria*, but the ovules may be 1 in each loculus (*Sambucus*, *Triosteum*, *Viburnum*).

The Caprifoliaceae comprises about 18 genera and 450 species. The members of this family are highly developed in eastern North America and eastern Asia.

Lonicera ligustrina Wall., a shrub of Khasi Hillis, bears flowers in pairs with connate ovaries. *Sambucus javanica* Reinw. ex Bl. is a large shrub whose leaves are imparipinnate.

Economically, the Caprifoliaceae is important for ornamentals, *e.g.*, *Abelia*, *Diervilla*, *Kolkwitzia*, *Leycesteria*, *Linnaea*, *Lonicera*, *Sambucus*, *Symphoricarpos*, *Viburnum* and *Weigela.* The stem of *Abelia triflora* is used for making walkingsticks. Wine is prepared from the ripe fruits of *Sambucus nigra.* Edible fruits are borne by some species of *Viburnum.*

Adoxaceae
(Moschatel Family)

The family contains a solitary species, *Adoxa moschatellina* L., which is widespread throughout the north temperate zone. It is a perennial herb with a small head of greenish flowers in which the stamens are twice as many as corolla lobes.

The Adoxaceae is an anomalous taxon, being of doubtful affinity. Most systematists have recognised it as a separate family. But the monotypic genus *Adoxa* was retained in the Caprifoliaceae by Bentham-Hooker. The general habit of the plant, coupled with the absence of a calyx and the fission of the stamens, would warrant its removal from the Caprifoliaceae. The half inferior ovary and perhaps superficial resemblance of the flower to such a form as *Chrysosplenium* are suggestive of the Saxifragaceae. The inconspicuous flowers, associated with the half inferior isomerous ovary and the solitary pendulous ovule, call for an alliance with the Araliaceae.

Valerianaceae
(Valerian Family)

The family is recognised by the opposite exstipulate leaves, irregular spurred or saccate flowers, annulate calyx, stamens fewer than corolla lobes and trilocular inferior ovary with a single fertile loculus.

The Valerianaceae is composed of thirteen genera and about 400 species. The members of this family mostly occur in the north temperate region, but are also found in the Andes of South America.

Nardostachys jatamansi (D. Don) DC. occurs in alpine Himalaya, while *Valeriana officinalis* Linn. and *Valerianella dentata* Poll. in Kashmir.

A few plants of the Valerianaceae are cultivated as ornamentals, *e.g., Centranthus ruber* and *Feedia eriocarpa. Nardostachys jatamansi*, *Valeriana excelsa* and *V. officinalis* are used medicinally. *Valerianella chenopodifolia*, *V. locusta* and *V. olitoria* are raised for their edible leaves.

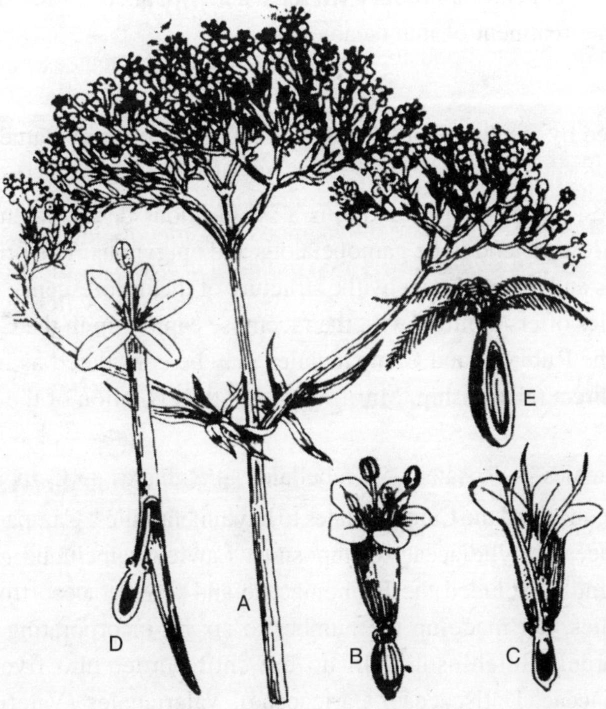

Fig. 85 VALERIANACEAE. *Valeriana officinalis* Linn. **A**, inflorescence; **B**, flower in male stage with stigmas still undeveloped; **C**, V.S. of flower in later stage with unexpanded stigmas and anthers shed; **D**, flower of *Centranthus ruber* with lower portion in section; **E**, V.S. of fruit of *V. officinalis.* (**A–D**, after Rendle)

Dipsacaceae
(Teasel Family)

The family is readily identified by the opposite exstipulate leaves, compact flower heads, calyx divided into pappus-like segments, stamens fewer than corolla lobes and unilocular inferior ovary.

The family is divided into two tribes:

Tribe (*i*) *Dipsaceae.* Flowers borne in capitula, outer often enlarged. Calyx bristly or scaly. Stamens 4, rarely 2, similar. Example: *Dipsacus.*

Tribe (*ii*) *Morineae.* Flowers borne in verticillasters in upper leaf axils. Calyx foliaceous. Stamens 2 or 4, dissimilar. Example: *Morina.*

The Dipsacaceae includes 11 genera and 350 species. The members of this family are inhabitants of the eastern Mediterranean and Balkan areas, but they extend eastwards across the Russian Steppes into India.

Several species of *Cephalaria. Morina*, *Pterocephalus* and *Scabiosa* are cultivated for ornament. The ripened fruiting inflorescences of Fuller's Teasel (*Dipsacus sativus*) are utilised for raising the nap (upon cloth); the dried flower heads are often used for winter bouquets. Some species of *Scabiosa* were formerly employed in the treatment of itch (scabies).

Order 33. *Campanulales*

The order is distinguished by the epigynous flowers, connate or coherent stamens, bithecal anthers and inferior ovary.

The Campanulales, according to Rendle, is a sympetalous development of a stock akin to the Passiflorales where the flowers tend to be gamopetalous and epigynous. The origin of the order may be sought from the Rubiales and this is shown by the structure of the inflorescence. But the cymose feature of the head in the Rubiales offers contrast with the racemose capitulum in the Campanulales. However, resemblances between the Rubiales and Campanulales may be considered as indicative of convergent evolution rather than a direct relationship. Mitra suggested the evolution of the Campanulales from the Ranales thus:

Ranales → Rosales → Umbellales → Rubiales → Campanulales.

Engler and Diels conceived the Campanulales to have six families: Campanulaceae, Brunoniaceae, Calyceraceae, Goodeniaceae, Stylidiaceae, Compositae. Lawrence included all these families in his concept of the order. Rendle excluded the Brunoniaceae and Calyceraceae from the order. Core also ignored these two families, but made up the number to six by incorporating the Cucurbitaceae and Lobeliaceae into the order. Hutchinson split up the entire order into five groups: Cucurbitales (Cucurbitaceae, Begoniaceae, Datiscaceae, Caricaceae), Valerianales (Valerianaceae, Dipsacaceae, Calyceraceae), Campanales (Campanulaceae, Lobeliaceae), Goodeniales (Goodeniaceae, Brunoniaceae, Stylidiaceae), Asterales (Compositae). Cronquist distributed his families to two orders; Campanulales (Pentaphragmataceae, Sphenocleaceae, Campanulaceae, Stylidiaceae, Brunoniaceae, Goodeniaceae) and Asterales (Compositae). The present author, employing largely the contention of Core (1955), takes into account five families with their characters as follows:

(1) Flowers in involucrate heads	*Compositae*
(1) Flowers not in involucrate heads	(2)
(2) Flowers generally imperfect	*Cucurbitaceae*
(2) Flowers generally perfect	(3)
(3) Stamens 5	(4)
(3) Stamens 2–3	*Stylidiaceae*
(4) Carpels 2–5	*Campanulaceae*
(4) Carpels 2	*Goodeniaceae*

Cucurbitaceae (Gourd Family)

GENERAL FEATURES OF THE FAMILY

Habit — Herbs or shrubs, *climbing by solitary extra axillary tendrils.*

Leaves — Alternate, simple, *palminerved, cordate, lobed.*

Flowers — Unisexual*, regular, *epigynous.*

Calyx — Sepals 5, united, campanulate or tubular, imbricate.

Corolla — Petals 5, united, inserted on calyx tube.

Androecium — *Stamens apparently 3 but 5, being 2 paired and 2 free* (in male flowers); anthers free or syngenesious, 1 usually one-celled and other 2 two-celled each, straight or sinuous (twisted like a transverse S).

Gynoecium — Carpels 3 (in female flowers), syncarpous; *ovary inferior, 3-locular with 1 to many ovules in each loculus. placentation parietal*; style 1; stigmas 3, forked.

Fruit — Berry or *pepo.*

Seeds — Exalbuminous; embryo; straight; cotyledons large and oily.

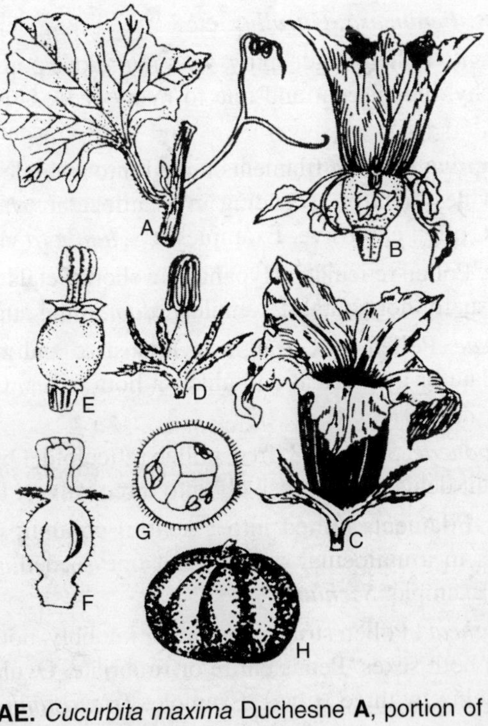

Fig. 86 CUCURBITACEAE. *Cucurbita maxima* Duchesne **A**, portion of shoot; **B**, female flower; **C**, male flower; **D**, stamens fused together; **E**, pistil; **F**, L.S. of pistil after corolla tube removed; **G**, T.S of ovary; **H**, fruit

* Bisexual flowers occur only in *Schizopepon*.

FLORAL RANGE IN THE FAMILY

In the Cucurbitaceae, the flowers are extremely dissimilar in *Dimorphochlamys.* The calyx lobes are serrate in *Benincasa* and pectinate in *Raphidiocystis.* The calyx tube is covered with scales in *Momordica* and *Thladiantha.* The petals are free, as in *Lagenaria* and *Luffa.* The petals may also be united only at the base, as in *Momordica* or united throughout, as in *Coccinia, Cucurbita,* etc. The stamens are free (*Actinostemma, Zanonia*) or united into a synandrium (*Cucurbita*). In *Cyclanthera,* 3 stamens are completely united into a central column or in *Thladiantha* 4 stamens are united and the fifth is free. But filaments are united only at the base in *Gomphogyne.* The styles are remote, *e.g., Fevillea* or radiately multifid, *e.g., Cucurbitella.*

DIVISIONS OF THE FAMILY

The Cucurbitaceae is divided into two subfamilies, one with eight tribes and the other with a single tribe:

Subfamily I. *Cucurbitoideae.* Tendril unbranched or two to seven branched from lower part, spiralling above point of branching. Pollen various. Style 1. Seeds unwinged.

Tribe (*i*) *Benincaseae.* Pollen sacs convoluted and pollen usually reticulate. Hypanthium in female flowers short. Ovules horizontal. Fruit usually smooth and indehiscent. Examples: *Benincasa, Citrullus,* etc.

Tribe (*ii*) *Cucurbiteae.* Pollen large, spiny, with numerous pores. Ovules erect or horizontal. Fruit fleshy, indehiscent and one to many seeded. Examples: *Cucurbita, Sicana,* etc.

Tribe (*iii*) *Cyclanthereae.* Stamen filaments united into a single column; pollen punctate, not spiny. Ovules erect or ascending in a unilocular ovary. Fruit often spiny, usually dehiscent, often explosive. Examples: *Cyclanthera, Marah,* etc.

Tribe (*iv*) *Joliffieae.* Pollen reticulate. Hypanthium short. Petals fimbriate or with basal scales. Ovules usually horizontal. Examples: *Momordica* and *Telfairia.*

Tribe (*v*) *Melothrieae.* Pollen sacs straight or almost so and pollen reticulate. Hypanthium campanulate or cylindrical and alike in both sexes. Ovules horizontal. Examples: *Cucumis, Zehneria,* etc.

Tribe (*vi*) *Schizopeponeae.* Stamens 3, free; pollen reticuloid. Ovules pendulous in a trilocular ovary. Fruit dehiscing explosively into three valves. Example: *Schizopepon.*

Tribe (*vii*) *Sicyoeae.* Filaments united into a central column; pollen spiny. Ovules single, pendulous in a unilocular ovary. Fruit one seeded, indehiscent, usually hard or leathery. Example: *Sechium.*

Tribe (*viii*) *Trichosantheae.* Pollen striate, smooth or knobbly, not spiny. Hypanthium long and tubular in both sexes. Petals entire or fimbriate. Ovules horizontal. Fruit fleshy or dry, dehiscing by three valves. Example: *Trichosanthes.*

Subfamily II. *Zanonioideae.*

Tribe (*i*) *Zanonieae.* Tendrils two branched from near apex, spiralling above or below point of branching. Pollen small, striate, uniform. Styles 3, Seeds often winged. Examples: *Alsomitra, Zanonia,* etc.

POSITION AND AFFINITY OF THE FAMILY

There is a lot of controversy about the systematic position of the Cucurbitaceae. Hitherto, the taxon has been by a universal consent arranged amongst families with trilocular ovary and parietal placentation. It is difficult to agree with this suggestion, since the fruit is quite *sui generis* having the placenta recurved from the axil to the carpellary wall. The stamen morphology as well as the fruit morphology is so unique that the Cucurbitaceae cannot logically be placed in association with the other epigynous groups which bear parietal placentation.

Lindley included the Cucurbitaceae in his epigynous groups amongst a suit of orders having axile placentation, but distinguished it by the possession of unisexual flowers, sinuous stamens and exalbuminous seeds. Endlicher considered it between the Begoniaceae and Passifloraceae, the affinities of which have not yet been clearly understood. The intimate union of the calyx and corolla, which led A. Jussieu to claim the flowers as apetalous and to range the order under his class Apetalae, is a point that cannot be over-ruled in determining relationships. According to Wight, de Jussieu's view is essentially incorrect and indicates an alliance with the Euphorbiaceae in striking respects. As a family, the Cucurbitaceae is demarcated by its stamens, but more specifically by its anthers, the cells of which, in most cases, are very long, winding upwards and downwards on the external surface of the connective. The stamens may be monadelphous, syngenesious and variously disposed upon the filament, usually vertical, but sometimes horizontal or inclined. The anther cells exhibit many fantastic types of configuration, varying from the simple to the curved, variously convoluted and annular patterns. The structure of the anther, coupled with the construction of the ovary, the habit of plants and lateral tendrils, widely separates the family from any other in the Plant Kingdom.

Robert Brown, the de Candolles and Naudin felt that the Cucurbitaceae is related to the Passifloraceae and hence put the family along with the members of the perigynous Polypetalae in the Calyciferae. Wettstein reckoned the family to be very advanced on account of the presence of sympetalous corolla and synandrous androecium and so favoured its position after the Rubiales and before the Campanulales. Eichler treated it as an appendage to his series Campanulinae and next to the Campanulaceae. In support, he marshalled arguments that the family contains typically epigynous pentamerous flowers and that it has the tendency of the connation of stamens and calyx, which though narrow, sometimes display foliaceous characters. The findings of Brown and others were confirmed by Bentham-Hooker. Engler, while corroborating the facts of Eichler, set up the family by itself in the series Cucurbitales and next to the Campanulales. That the Cucurbitaceae is akin to the Campanulaceae was indicated by the serodiagnostic tests of Mez; a positive reaction is shown with the Campanulaceae and a negative one with the Passifloraceae. Fairly sounder reasons may, however be advanced in favour of the older concept. The structure of the ovule with a large persistent nucellus, extensive tapetal tissue and distinct integuments, reason its placement amongst the sympetalous families. But the constantly unisexual flowers, inferior ovary with the tendril, etc. bring the family close to the Passifloraceae rather than to the Metachlamydeae to which it was assigned by Engler. Hutchinson (1959) studied the family from these aspects and believed it to be appearing under the Cucurbitales in his newly created group Lignosae. Rendle accepted its position within the Peponiferae and after the Parietales. It was thought by Vuillemin that the Cucurbitaceae must be relegated to the apetalous group of orders and hence he classed them with the Balanophoraceae, Rafflesiaceae, Datiscaceae, Nepenthaceae and Aristolochiaceae. He interpreted the family as apetalous, pointing out that the corolla is nothing but the inner whorl of calyx.

The present author, in line with Core (1955), is inclined to accommodate the Cucurbitaceae in the Campanulales. In the arrangement of Benson (1957), the family appeared under the Cucurbitales after the Begoniales and before the Rubiales. Porter (1967) followed the practice of Benson, but the former's order was lying in between the Rubiales and Campanulales. Cronquist (1968) imparted the position of the family in the Violales between the Sarraceniales and Salicales.

Chakravarty (1966) expressed the opinion that the Cucurbitaceae is a highly evolutionary group which has progressed tremendously onwards with the consequent metamorphosis of the structures associated with the reduction, amalgamation, connation, adnation and sterilisation of the sex organs. It finds a close connection with the Passifloraceae amongst the flowering plants in having the placenta retained on the wall, single chambered ovary, fleshy fruit, tendril structure and extra floral nectaries. Though the Cucurbitaceae compares very well with the Passifloraceae on the one hand and the Begoniaceae on the other, Chakravarty felt that it deserves a separate rank by itself, as its associates cannot easily be bridged by the morphological features, especially of the male sex organ and the typical gourd type of fruit.

NUMBER IN AND DISTRIBUTION OF THE FAMILY

The Cucurbitaceae contains about 90 genera and 700 species. "It is well represented in the moist and moderately dry tropics of both Old and New Worlds, particularly in rainforest areas of South America and wood, grass and bushland areas of Africa".

COMMONLY OCCURRING PLANTS OF THE FAMILY

Pumpkin or White Gourd [*Benincasa hispida* (Thunb.) Cogn.], Water Melon [*Citrullus lanatus* (Thunb.) Mats. & Nakai], Scarlet Gourd [*Coccinia grandis* (L.) Voigt], Melon (*Cucumis melo* L.), Cucumber (*C. sativus* L.), Sweet Gourd (*Cucurbita maxima* Duchesne), Bottle Gourd [*Lagenaria siceraria* (Mol.) Standl.], Ribbed Gourd [*Luffa acutangula* (L.) Roxb.], Bath Sponge or Loofah [*L. cylindrica* (L.) M.J. Roem], Bitter Gourd (*Momordica charantia* L.), *M. cochinchinensis* (Lour.) Spreng., Squash [*Sechium. edule* (Jacq.) Sw] *Trichosanthes bracteata* (Lamk.) Voigt and Palwal [(*T. dioica* Roxb.) Swartz] grow in different parts of India.

Hodgsonia macrocarpa (Bl.) Cogn. is a gigantic climber whose stem sometimes attain a length of 25 m.

Mukia madraspatana (L.) Roem. and *Solena heterophylla* Lour. are found in hedges.

ECONOMIC ASPECTS OF THE FAMILY

Economically, the Cucurbitaceae is important for edible fruits and vegetables, *e.g., Benincasa hispida, Cucumeriopsis marii, Cucumis anguria, C. melo, C. sativus, Cucurbita filicifolia, C. maxima, C. mixta, C. moschata, Cyclanthera pedata, Hodgsonia macrocarpa, Lagenaria siceraria, Momordica charantia, Praecitrullus fistulosus, Sicana odorifera, Sechium edule* and *Telfairia occidentalis.* The vascular bundle network of the pericarp of *Luffa cylindrica* fruit is utilised as a bath sponge.

A few plants are of medicinal value. 'Bryony' is produced from *Bryonia dioica.* The dried pulp of *Citrullus colocynthis* fruit yields 'colocynth', a powerful laxative. A purgative called 'elaterium' is obtained from the fruits of *Ecballium elaterium. Luffa echinata* produces a bitter alkaloid which finds use as a cure for dropsy.

Some plants of the Cucurbitaceae are grown as ornamentals. These include some species of *Corallocarpus, Ibervillea, Kedrostis, Xerosicyos,* etc.

Campanulaceae
(Bell-flower Family)

GENERAL FEATURES OF THE FAMILY

Habit	—	Herbs, shrubs or trees, with a *watery or milky sap.*
Leaves	—	Alternate, or opposite, simple, *exstipulate.*
Inflorescence	—	Monochasial or dichasial cymes.
Flowers	—	Bisexual, regular or irregular, *epigynous.*
Calyx	—	Sepals 5, united, imbricate or valvate.
Corolla	—	Petals 5, united, *valvate.*
Androecium	—	*Stamens 5,* alternate with petals, free or *variously united;* anthers 2-celled.
Gynoecium	—	Carpels 2–5; *ovary inferior, unilocular; ovules many; placentation axile.*
Fruit	—	Berry or capsule.
Seeds	—	Embryo small; endosperm fleshy.

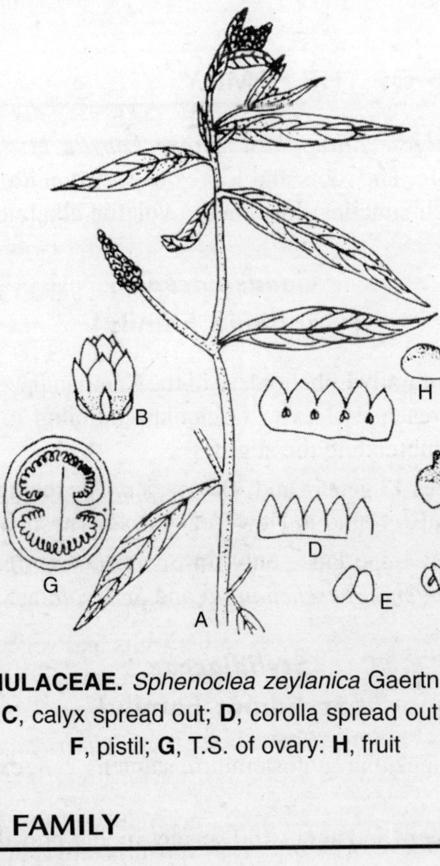

Fig. 87 CAMPANULACEAE. *Sphenoclea zeylanica* Gaertn. **A**, portion of plant; **B**, flower; **C**, calyx spread out; **D**, corolla spread out; **E**, stamens; **F**, pistil; **G**, T.S. of ovary: **H**, fruit

DIVISIONS OF THE FAMILY

The Campanulaceae is divided into three subfamilies:

Subfamily I. *Campanuloideae.* Flowers actinomorphic, rarely slightly zygomorphic. Anthers generally free. Examples: *Campanula, Michauxia, Phyteuma,* etc.

Subfamily II. *Cyphioideae.* Flowers zygomorphic. Stamens with filaments united, but anthers free. Example: *Cyphia.*

Subfamily III. *Lobelioideae.* Flowers markedly zygomorphic, rarely almost actinomorphic. Anthers united. Examples: *Lobelia* and *Siphocampylus.*

NUMBER IN AND DISTRIBUTION OF THE FAMILY

A family of about 35 genera and 600 species, the Campanulaceae is mostly subtropical or temperate and also has representatives in mountains of the tropics.

COMMONLY OCCURRING PLANTS OF THE FAMILY

Lobelia alsinoides Lamk., *Pratia begonifolia* Lindl. and *Wahlenbergia marginata* (Thunb.) A.DC. are herbs.

Sphenoclea zeylanica Gaertn., an aquatic fleshy herb, bears alternate leaves, bracteate spikes and white flowers.

ECONOMIC ASPECTS OF THE FAMILY

Adenophora, Campanula, Codonopsis, Edrianthus, Jasione, Lobelia, Michauxia, Ostrowskia, Platycodon, etc. are cultivated as ornamentals. The roots and leaves of *Campanula rapunculus* are edible. *Lobelia inflata* is a medicinal plant which contains 'lobeline', a volatile alkaloid.

Goodeniaceae
(Goodenia Family)

The family is very close to the subfamily Lobelioideae of the Campanulaceae. However, it is distinguished by the absence of milky sap, presence of extra fascicular cambium in the stem and development of indusiate pollen collecting cup subtending the stigmas.

The Goodeniaceae includes 13 genera and 300 species. The members of this family are primarily distributed in Australia, but are also found in New Zealand, Polynesia and other tropical coasts.

The family is of economic importance only for *Scaevola koenigii* which yields 'Taccada pith'. Many species of *Dampiera, Goodenia, Leschenaultia* and *Scaevola* are cultivated as ornamentals.

Stylidiaceae
(Stylidium Family)

The family is recognised by the peculiar gynostemium, stamens 2–3, extrorse anthers and free central placentation.

The family covers six genera and about 150 species, being prevalent in tropical Asia, Australia, New Zealand and South America.

Stylidium kunthii Wall. is found in Khasi Hills and Sikkim.

A few species of *Stylidium* are cultivated for ornament.

Compositae
(Sunflower Family)

GENERAL FEATURES OF THE FAMILY

Habit	—	Herbs or shrubs.
Leaves	—	Alternate or opposite, simple, exstipulate.
Inflorescence	—	*Head, surrounded by an involucre of bracts.*
Flowers	—	All regular (*tubular*) or all irregular (*ligulate*) or both (forming *disc* and *ray florets*), *epigynous.*
Calyx	—	Absent or reduced to a tuft or hairs (*pappus*) or scales.
Corolla	—	Petals 4 or 5, *united*, tubular.
Androecium	—	Stamens 4 or 5, epipetalous, *syngenesious*; anthers 2-celled; connective prolonged upwards.
Gynoecium	—	Carpels 2, syncarpous; *ovary inferior, unilocular with a single basal ovule*; style bifid at top and recurved.
Fruit	—	*Cypsela, crowned with a persistent pappus.*

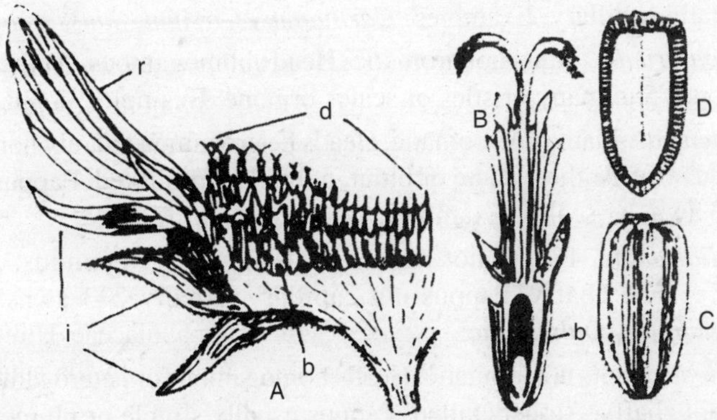

Fig. 88 COMPOSITE. *Helianthus annuus* L. **A**, L.S. of portion of head (*i*, bracts of involucre; *b*, bract subtending a single floret; *d*, disc floret; *r*, ray floret); **B**, V.S. of disc floret; **C**, fruit; **D**, V.S. of fruit. (**A–D**, after Rendle)

FLORAL RANGE IN THE FAMILY

The peduncle is generally convex or flat. It is elongated and cane-like in *Rudbeckia* and *Spilanthes.* It is concave and cup-like in *Epaltes.*

The head is 1-flowered in *Vernonia uniflora* or few flowered in *Mikania* or many flowered in a number of other genera. It may contain one type of tubular and perfect florets. But the central florets are sterile in *Schoenla.* Unisexual florets are present in *Xanthium.* All florets in a head may be ligulate (*Cichorium, Lactuca*) or may have a bilabiate corolla (*Mutisia*).

DIVISIONS OF THE FAMILY

The Compositae* is divided into two subfamilies, one with a single tribe and the other with 12 tribes:

Subfamily I. *Lactucoideae.* Latex present. Corolla of all florets ligulate.

Tribe (*i*) *Cichorieae.* Examples: *Cichorium*, *Crepis*, *Lactuca,* etc.

Subfamily II. *Asteroideae.* Latex rare. Corolla of all disc florets not ligulate.

Tribe (*i*) *Anthemideae.* Plants aromatic. Heads heterogamous, rarely homogamous. Receptacle chaffy or not chaffy. Anthers blunt or acute, not tailed, basifixed. Pappus absent or reduced to a ring or cup. Examples: *Achillea*, *Artemisia*, *Chrysanthemum*, etc.

Tribe (*ii*) *Arctotideae.* Plants aromatic. Heads heterogamous. Anthers blunt pointed at base. Pappus none or not capillary. Examples: *Arctotis* and *Gazania.*

Tribe (*iii*) *Astereae.* Plants aromatic. Heads heterogamous, rarely homogamous. Receptacle not chaffy. Anthers blunt and basifixed. Pappus various. Examples: *Aster*, *Erigeron*, *Solidago*, etc.

Tribe (*iv*) *Calenduleae.* Plants not aromatic. Heads heterogamous. Receptacle not chaffy. Anthers pointed at base. Pappus none. Example: *Calendula.*

Tribe (*v*) *Cynareae.* Plants not aromatic. Heads homogamous, but rays may be female or neuter. Receptacle chaffy, with many bristles. Anthers usually tailed. Pappus usually capillary. Examples: *Carthamus*, *Cirsium*, *Saussurea*, etc.

Tribe (*vi*) *Eupatorieae.* Plants not aromatic. Heads homogamous. Anthers blunt and basifixed. Pappus 5 to many bristles or scales or none. Examples: *Ageratum*, *Eupatorium*, etc.

Tribe (*vii*) *Helenieae.* Plants not aromatic. Heads heterogamous, rarely homogamous. Receptacle not chaffy. Anthers acute or blunt, not tailed, basifixed. Pappus not capillary, usually 2–3 awns or scales. Examples: *Gaillardia* and *Helenia.*

Tribe (*viii*) *Heliantheae.* Plants not aromatic. Heads heterogamous, rarely homogamous. Receptacle chaffy. Pappus not capillary, usually 2–3 awns or scales. Examples: *Helianthus*, *Dahlia*, etc.

Tribe (*ix*) *Inuleae.* Plants not aromatic. Heads homogamous or heterogamous. Receptacle chaffy or not chaffy. Anthers tailed. Pappus usually simple or plumose bristles. Examples: *Blumea, Gnaphalium, Inula,* etc.

Tribe (*x*) *Mutisieae.* Plants not aromatic. Heads homogamous or heterogamous. Anthers long tailed. Pappus capillary. Examples: *Gerbera* and *Mutisia.*

Tribe (*xi*) *Senecioneae.* Plants not aromatic. Heads homogamous or heterogamous. Receptacle usually not chaffy. Anthers acute or blunt, not tailed, basifixed. Pappus capillary. Examples: *Doronicum*, *Senecio*, etc.

Tribe (*xii*) *Vernonieae.* Plants not aromatic. Heads homogamous. Anthers blunt, sagittate or short tailed. Pappus setose and copious. Examples: *Elephantopus* and *Vernonia.*

POSITION AND AFFINITY OF THE FAMILY

The Compositae was the last family of the dicotyledons by Engler and was included in the order Campanulatae under the Metachlamydeae. Bentham-Hooker placed the family in the cohort Asterales

* "The classification of the Compositae is in a state of transition... is now seen to be in need of modification, in the light of recent discoveries in biochemistry, pollen analysis, micromorphology, anatomy and cytology" (Jeffrey, 1978).

under the series Inferae of the subclass Gamopetalae. In Hutchinson's arrangement, the Compositae appeared under the Asterales. All these authors are in agreement to assign the position of the family after the Rubiaceae.

The Compositae has been thought to be at or very near the peak of dicot evolution. It indicates similarity with the Dipsacaceae and Valerianaceae. It is akin to the Rubiaceae in the aggregation of flowers into heads or heads-like structures and in the reduction of calyx lobes. If the polyphyletic origin of the Compositae is taken into account, the origin of the Rubiales and Asterales is to be sought in the Umbellales where the small flowers tend to aggregate into crowded umbels.

The available chemical data suggest that the Umbelliferae and Campanulaceae are the extant families closest to the ancestors of the Compositae (Mabry and Bohlmann, 1977). Though over 90% of more than 600 known sesquiterpene lactones belong to the Compositae, nine taxa of the Umbelliferae contain structurally simple types—guaianolides. A second connecting link between the two families is the presence of many biogenetically related acetylenes. Like the sesquiterpene lactones, more than 90 % of the known acetylenes are from the two families. The Campanulaceae can be connected to the Compositae by its acetylenic chemistry and the Araliaceae and Pittosporaceae to the Umbelliferae. The Calyceraceac, allied by some workers with the Compositae, chemically does not belong to the family. While the Calyceraceae lacks acetylenes, it contains secoiridoids similar to those occurring in the Gentianales; such compounds are completely absent from the Umbelliferae, Campanulaceae and Compositae. In view of the various plausible relationships, Stebbins (1977) concluded that "the Compositae cannot be regarded as descended from or closely related to any other modern family". As a matter of fact, this is implied in the cytochrome *c* data of Boulter *et al.* (1972). Such an aspect is also consistent with the fossil histoiy of the group (Turner, 1977).

The Compositae is considered as the most advanced family of the flowering plants. The reasons are as follows:

1. The preponderance of herbaceous habit, the number of shrub or tree members being exceedingly limited.
2. The presence of about 25,000 species or roughly about 10 percent of the known dicotyledonous plants which are cosmopolitan, being found, often in abundance, all over the world.
3. The admirable adaptation for cross pollination by insects, easy access to nectar and protection of nectar from the elemental forces of Nature; the aggregation of florets into heads with ligulate marginal florets, acting as flags for the attraction of insects; the pollination of all disc florets by one single insect within a very short time.
4. The occurrence of epigynous flowers in conjunction with the syngenesious and protandrous stamens.
5. The ability of florets to shirk self pollination by brushing system or piston mechanism, together with the same procedure in the stigmas.
6. The protection of tender florets, suppression of calyx, reduction in carpel number and presence of an inferior uniovulate ovary.
7. The pappus forming a parachute for the dispersal of fruits and seeds.

NUMBER IN AND DISTRIBUTION OF THE FAMILY

The Compositae is the largest family of flowering plants, comprising about 1,100 genera and 25,000 species. The members of this family are found in all corners of the globe, being poorly represented in the tropical rain forests and absent only from the Antartic main land.

COMMONLY OCCURRING PLANTS OF THE FAMILY

Adenostemma lavenia (L.) O. Kuntze, *Ageratum conyzoides* L., *Blumea mollis* (L.) Druce, *Caesulia axillaris* Roxb., *Cirsium arvense* (L.) Scop., *Eclipta prostrata* (L.) L., Elephant's foot (*Elephantopus scaber* (L.), *Emelia sonchifolia* (L.) DC., *Erigeron asteroides* Roxb., *Eupatorium odoratum* L., *Launaea asplenifolia* Hook, f., *L. sarmentosa* (Willd.) Alston, *Senecio nudicaulis* Buch-Ham. ex D. Don, *Sonchus asper* (L.) Hill., *Sphaeranthus indicus* L., *Spilanthes paniculata* Wall. ex DC., *Synedrella nodiflora* (L.) Gaertn., *Tridax procumbens* L., *Vernonia cinerea* (L.) Less., Cocklebur (*Xanthium strumarium* L.) and *Youngia japonica* (L.) DC. are weeds.

Sage Brush (*Artemisia caruifolia* Roxb.) is a soft stemmed plant, about a metre or so in height.

Centratherum anthelminticum (L.) O. Kuntze is a tall robust annual.

Chrysanthemum (*Chrysanthemum coronarium* L.), Sunflower (*Helianthus annuus* L.), *Silybum marianum* (L.) Gaertn., Marigold (*Tagetes erecta* L.) and Garden Zinnia (*Zinnia elegans* Jacq.) are garden plants.

Echinops echinatus Roxb. is a branched annual.

Enhydra fluctuans Lour. occurs in marshes.

Ethulia conyzoides L. f. is an erect annual herb.

Gnaphalium indicum L., *G. luteo-album* L. and *Grangea maderaspatana* (L.) Poir. are found in dry rice fields.

Climbing Hempwood [*Mikania cordata* (Burm.) B. L. Robinson] is a climber, bearing cordate leaves.

ECONOMIC ASPECTS OF THE FAMILY

The Compositae is of economic importance, "but for its size the importance is not as great as might be expected".

The family contains a number of drug plants. *Artemisia absinthium* is used as an aromatic tonic. *A. cina*, *A. maritima* and *A. nilagarica* yield 'santonine', used for expelling worms. The roots of *Doronicum roylei* are used as an aromatic tonic. *Matricaria chamomilla* (Mediterranean) is the source of 'chamomile', a medicine. Other medicinal plants include *Anthemis nobilis*, *Arnica montana*, *Calendula officinalis*, *Cnicus benedictus*, *Filago germanica*, *Petasites hybridus*, *Tanacetum vulgare*, *Vernonia anthelmintica*, *Wedelia calendulacea*, etc.

There are many food plants in the Compositae. The leaves of *Artemisia dracunculus* are used for flavouring. *Cichorium endivia* (North America), *Cynara cardunculus* and *Lactuca sativa* are grown for lible leaves; *Cynara scolymus* for edible heads; *Helianthus tuberosum* for edible tubers; *Scolymus hispanicus* (Spain), *Scorzonera hispanica* (Europe to Central Asia) and *Tragopogon porrifolius* (South Europe) for edible roots. The roots of *Cichorium intybus* yield 'chicory' which is used to adulterate coffee. *Enhydra fluctuans* is eaten as a pot herb.

The members of the family find uses in a variety of ways. *Adenostemma tinctorium* produces a colouring matter. *Brachylaena huillensis* and *Montanoa quadrangularis* (Colombia & Venezuela) as well as some species of *Vernonia* furnish timber. Camphor is distilled from *B. balsamifera* (China). A type of red dye is extracted from the flowers of *Carthamus tinctorius.* Oil is obtained from the seeds of *C. oxyacantha, Guizotia abyssinica* and *Helianthus annuus.* The flowers of *Chrysanthemum carneum*, *C. cinerariaefolium* and *C. roseum* are used to make insect powder. The leaf juice of *Eclipta prostrata* is used for tattooing the skin bluish-black. Rubber is derived from *Parthenium argentatum*, *Solidago leavenworthii* and *Taraxacum kok-saghyz*. The roots of *Saussurea lappa* yield 'pachuk' of commerce, useful in perfumery, especially for the hair. The wood of *Tarchonanthus camphoratus* (South Africa) is used for musical instruments.

"A paging through a seed catalogue or a visit to a florist shop will show members of this family in their full grandeur" (Robbins *et al.*, 1958). *Aster*, *Chrysanthemum*, *Cosmos*, *Dahlia, Gaillardia*, *Onopordon*, *Tagetes*, *Zinnia* and many others are such representatives of the Compositae. *Helichrysum arenarium* and *H. bracteatum* are raised for their dried flowers.

Several members of the Compositae have become noxious weeds far removed from their original homes, *e.g.*, *Ageratum conyzoides*, *Bidens pilosa*, *Chondrilla juncea*, *Cirsium arvanse*, *Parthenium hysterophorus*, *Sonchus oleraceus*, *Tridax procumbens*, *Xanthium strumarium* etc. The species of *Senecio* are troublesome weeds of pasture and are responsible for deaths of domestic stock. The wind-borne pollen grains of *Ambrosia artemisifolia* and *A. trifida* are one of the chief causes of hay fever in the areas of North America where these species are prevalent.

SELECTED BIBLIOGRAPHY

ALSTON, R.E. and TURNER, B.L. 1966. Biochemical methods in systematics. *In.* W.A. Jensen and L.G. Kavaljian (eds.). *Plant Biology Today*, pp. 92–113. Macmillan, London.

AYENSU, E.S. and STEARN, W.L. 1964. Systematic anatomy and ontogeny of the stem in Passifloraceae. *Contr. U. S. Nat Herb.*, **34:** 45–73.

BAKER, H.G. 1948. The relationship in the Plumbaginaceae. *Nature* (*Lond.*), **161:** 400.

BARLOW, B.A. 1959. Chromosome number in Casuarinaceae. *Aust. J. Bot.*, **7:** 230–237.

BATE-SMITH, E.C. 1962. The phenolic constituents of plants and their taxonomic significance. *J. Linn. Soc.* (*Bot.*), 58: 95–173.

BECHTEL, A.R. 1921. The floral morphology of the Urticales. *Am. J. Bot.,* **8:** 386–410.

BENSON, I. and WEISFORD, E.J. 1909. The morphology of the ovule and female flower of *Juglans regia* and a few allied genera. *Annls. Bot.,* **23:** 632–633.

BERRIDGE, E.M. 1914. The structure of the flower of Fagaceae and its bearing on the affinities of the group. *Annls. Bot*, **28:** 509–526.

BERRY, E.W. 1915. The origin and distribution of Myrtaceae. *Botan. Gaz.,* **59:** 484–490.

BISALPUTRA, T. 1960. Anatomical and morphological studies in the Chenopodiaceae. *Aust. J. Bot.*, 8: 226–242.

BREMEKAMP, C.E.B. 1966. Remarks on the position, the delimitation and the subdivision of the Rubiaceae. *Acta. bot. Neerl.,* **15:** 1–33.

BRETT, D.W. 1964. The inflorescence of *Fagus* and *Castanea* and the evolution of the cupules of the Fagaceae. *New Phytol.,* **63:** 96–117.

BUXBAUM, F. 1944. Untersuchungen zur Morphologie der Kakteen bute. *Bot. Arch.,* **45:** 190–247.

CAROLIN, R.C. 1967. The concept of the inflorescence in the order Campanulales. *Proc. Linn. Soc. N.S. Wales*, **92:** 7–26.

CHAKRAVARTY, H.L. 1959. Monograph on Indian Cucurbitaceae (Taxonomy and Distribution). *Rec. Bot. surv. India,* **17:** 1–234.

CHAKRAVARTY, H.L. 1966. Monograph on the Cucurbitaceae of Iraq. *Tech. Bull.,* no 133. Mins. Agri., Baghdad

CHANDA, S. 1965. The pollen morphology of Droseraceae with special reference to taxonomy. *Pollen et Spores,* **7:** 509–528.

CHANDA, S. 1969. A contribution to the playnotaxonomy of Casuarinaceae. *J. Sen Mem. Vol.*, pp. 191–208. J. Sen Memorial Committee, Calcutta.

CHANNELL, R.B. and WOOD, C.E. Jr. 1962. The Leitneriaceae in the southeastern United States. *J. Arnold Arb.*, **43:** 435–438.

CHODAT, R. 1891-1893. Monographia polygalacearum. *Mem. Soc. Phys. Hist. Nat. Geneve,* vols. 30–31 et Suppl.

CHOPRA, R.N. and KAUR, H. 1965a. Embryology of *Bixa orellana* Linn. *Phytomorphology,* **15:** 211–214.

CHOPRA, R.N. and KAUR, H. 1965b. Some aspects of the embryology of *Cornus. Phytomorphology,* **15:** 353–399.

COPELAND, H.E. 1947. Observations on the structure and classification of the Pyroleae. *Madrono,* **9:** 65–102.

CORNER, E.J.H. 1962. The classification of Moraceae. *Gard. Bull. Singapore,* **19:** 187–252.

CRONQUIST, A. 1955. Phylogeny and taxonomy of the Compositae. *Am. Midl. Nat.,* **53:** 478–511.

DATTA, R.M. 1951. Occurrence of hermaphrodite flower in *Arceuthobium minutissimum* Hook, f., the smallest dicotyledonous plant. *Nature (Lond.),* **167:** 203.

DATTA, R.M. and MITRA, J.N. 1947. The systematic position of the family Moringaceae based on a study of *Moringa pterygosperma* Gaertn. (*M. oleifera* Lam.). *J. Bom. nat. Hist. Soc.,* **47:** 355–357.

DATTA, S.C. 1959. Weeds and weed control. *Indian Agricult.,* **3:** 26–36.

DATTA, S.C. 1964. Carnivorous plants, *Uttar Pradesh,* **21** (Special No.): 55–59.

DATTA, S.C. and BHATTACHARYA, J. 1969. Flowering of Calcutta trees. *J. Bengal nat. Hist. Soc.,* **35:** 64–72.

DATTA, S.C. and MAJUMDER, N.C. 1966. Flora of Calcutta and vicinity. *Bull, botan. Soc. Bengal,* **20:** 16–120.

DATTA, N.M. and MITRA, D. 1961. Three newly recorded plants of W. Bengal. *Indian Forester,* **87:** 304–308.

EAMES, A.J. and WILSON, C.L. 1928. Carpel morphology in Cruciferae. *Am. J. Bot.,* **15:** 251–270.

ENGIER, A. and DIELS, L. 1936. *Syllabus der Pflanzenfamilien.* Aufl. 11 Berlin.

EDLIN, H.L. 1935. A critical revision of certain taxonomic groups of the Malvales. *New Phytol.,* **34:** 122–143.

FAHN, A. and ARZEE, T. 1959. Vascularization of articulated Chenopodiaceae and the nature of the fleshy cortex. *Am. J. Bot.,* **46:** 330–338.

FISHER, M.J. 1928. The morphology and anatomy of the Salicaceae. *Am. J. Bot.,* **15:** 307–326.

FORMAN, L.L. 1964. *Trigonobalanus,* a new genus of Fagaceae, with notes on the classification of the family. *Kew Bull.,* **17:** 381–396.

FRANKS, J.W. and WATSON, K. 1963. The pollen morphology of some critical Ericales. *Pollen et Spores,* **5:** 51–68.

GAGNEPAIN, F. and BOUREAU, E. 1946. Une nouvelle famille de Gymnosperms: Sarcopodacees. *Bull. Soc Bot. Fr.,* **93:** 313–320.

GANAPATHY, P.S. and PALSER, B. 1964. Studies of floral morphology in the Ericales. VII. Embryology in the Phyllodoceae. *Botan. Gaz.,* **125:** 280–297.

GILIY, C.L. 1943. Studies in Sapotaceae. II. The Sapodilla-Nispro complex. *Trop. Woods,* **73:** 1–22.

GOPAL KRISHNA, G. and PURI, V. 1962. Morphology of the flower of some Gentianaceae with special reference to placentation. *Botan.Gaz.,* **124:** 42–57.

GREGORY, W.C. 1941. Phylogenetic and cytological studies in the Ranunculaceae. *Trans. Am. Phil. Soc. N.S.,* **36:** 443–521.

HABER, T.M. 1959. The comparative anatomy and morphology of flowers and inflorescence of Proteaceae. I. Some Australian taxa. *Phytomorphology,* **9:** 325–358.

HEIMSCH, C. 1942. Comparative anatomy of the secondary system in the "Gruinales" and "Terebinthales" of Wettstein, with reference to taxonomic grouping. *Lilloa,* **8:** 83–198.

HENDERSON, M.W. 1920. A comparative study of the structure and saprophytism of the Pyrolaceae and Monotropaceae with reference to the derivation from the Ericaceae. *Contr. Bot. Lab. Univ. Penn.,* **5:** 42–109.

HEYWOOD, V.H. (Ed.) 1978. *Flowering Plants of the World.* Oxford University Press, Oxford.

HICKEY, M. and KING, C. 1981. *100 families of Flowering Plants.* Cambridge University Press, Cambridge.

HILL, A.F. 1952. *Economic Botany.* McGraw-Hill, New York.

HJELMQVIST, H. 1948. Studies on the floral morphology and phylogeny of the Amentiferae. *Bot. Notiser,* Suppl., **2:** 1–171.

JOHNSON, J. A. S. and BRIGGS, B.G. 1963. Evolution in the Proteaceae. *Aust. J. Bot.,* **11:** 21–61.

JOSHI, A. C. 1933. Some abnormal flowers of *Argemone mexicana* L. *J. Indian botan. Soc,* **12:** 258–269.

KAJALE, L.B. 1954. A contribution to the embryology of the Phytolaccaceae. *J. Indian botan. Soc,* **33:** 206–225.

KAUSIK, H.B. 1938. Cytology and floral morphology of *Grevilea robusta* A. Cunn. *Annls. Bot. (N.S.),* **2:** 81–91.

KENG, H. 1962. Comparative morphological studies in Theaceae. *Univ. Calif. Pub. Bot.,* **33:** 269–384.

KHAN, R. 1954. A contribution to the embryology of *Utricularia flexuosa* Vahl. *Phytomorphology,* **4:** 80–117.

LAM, H.J. 1941. Note on the Sapotaceae-Mimusoideae in general and far eastern Manilkara allies in particular. *Blumea,* **4:** 323–358.

LEPPIK, E.E. 1964. Floral evolution in the Ranunculaceae. *Iowa State J. Sci.,* **39:** 1–101.

LIOYD, F.E. 1942. *The Carnivorous Plants.* Chronica Botanica, Waltham, Mass.

MAHESHWARI, P., JOHRI, B.M. and DIXIT, S.N. 1957. The floral morphology and embryology of the Loranthoideae (Loranthaceae). *J. Madras Univ.,* **27B** (Centenary No.): 121–136.

MAJUMDER, G.P. and PAL, P. 1958. The stipules of the Rubiaceae—a review. *Trans. Bose Res. Inst.,* **22:** 57–67.

MANNING, W.E. 1940. The morphology of the flower of the Juglandaceae. The pistillate flowers and fruit. *Am. J. Bot.,* **27:** 839–852.

MARGRAF, F. 1955. Uber Laubblat-Homologien and verwandtchaftliche zusammenhange bei Sarraceniaies. *Planta,* **46:** 414–446.

MITRA, J.N. 1956. On the systematic position of the family Cactaceae as based from the studies of morphological characters of the flowers of *Selenicereus grandiflorus* Brit, et Rose and *Opuntia dillenii* Haw. *Sci. cult.,* **21:** 460–461.

MOHANRAM, H.Y. and WAHDI, M. 1965. Embryology and the delimitation of the Acanthaceae. *Phytomorphology,* **15:** 201–205.

MOORE, N.A. 1936. Floral anatomy and phylogeny in Rutaceae. *New Phytol,* **35:** 318–322.

MOSELEY, M.F. 1948. Comparative anatomy and phylogeny of Casuarinaceae. *Botan. Gaz,* **110:** 237–277.

MURTY, Y.S. 1959. Studies in the order Piperales. IV. A contribution to the study of vegetative anatomy of three species of *Piper. Proc. nat. Inst. Sci. India,* **25B:** 31–38.

NAGARAJ, M. 1952. Floral morphology of *Populus deltoides* and *P. tremuloides. Botan. Gaz.,* **114:** 222–243.

NAIR, N.C. and JOSHI, R.K. 1958. Floral anatomy of some members of the Simaroubaceae. *Botan. Gaz.,* **120:** 88–99.

NAIR, N.C. and MANI, K.V. 1960. Organography and floral anatomy of some species of Vitaceae. *Phytomorphology,* **10:** 138–144.

NAIR, N.C. and NARAYANAN, K.R. 1962. Studies on the Aristolochiaceae. I. Nodal and floral anatomy. *Proc, Nat. Inst. Sci. India,* **28B:** 211–227.

NAIR, N.C. and SARMA, V.S. 1961. Organography and floral anatomy of some members of the Rhamnaceae. *J. Indian Botan, Soc,* **40:** 47–55.

NARAYANA, L.L. 1959. Floral anatomy of Meliaceae. *J. Indian Botan. Soc,* **38:** 288–295.

NAST, C.G. 1935. Morphological development of the fruit of *Juglans regia. Hilgardia,* **9:** 345–381.

NORRIS, T. 1941. Torus anatomy and nectary characteristics as phylogenetic criteria in the Rhoeadales. *Am. J. Bot.,* **28:** 101–113.

ORR, M.Y. 1921. Observations on the structure of the seed in the Capparidaceae and Resedaceae. *Notes Roy. bot. Gdn. Edinb.,* **12:** 259–260.

PALSER, B. and MURTY, Y.S. 1967. Studies of floral morphology in the Ericales. VIII. Organography and vascular anatomy in *Erica. Bull. Torrey Botan. Club,* **94:** 243–320.

PATERSON, B.R. 1961. Studies of floral morphology in the Epacridaceae. *Botan. Gaz.,* **122:** 259–279.

PAX, F. and HOFFMANN, K. 1931. Callitrichaceae. *In:* A. Engler and K. Pranti, *Die naturlichen Pflanzenfamilien.* 19C. Leipzig.

PERRY, F. 1972. *Flowers of the World.* Galahad Books, New York.

PHILIPSON, W.R. 1953. The relationships of the Compositae as illustrated by the morphology of the inflorescence in the Rubiales and Campanulatae. *Phytomorphology,* **3:** 391–404.

PURI, V. 1945. Studies in floral anatomy. II. Floral anatomy of the Moringaceae with special reference to gynoecium constitution. *Proc. nat. Acad. Sci. India,* **15:** 74–91.

RAIMANN, R. 1898. Hydrocaryaceae. *In:* A. Engler and K. Prantl, *Die naturlichen Pflanzenfamilien.* III(7). Leipzig.

RAIZADA, M.B. 1958. Name changes in common Indian plants. *Indian Forester,* **84:** 467–538.

RAIZADA, M.B. 1966. Nomenclatural changes in Indian plants. *Indian Forester,* **92:** 299–339.

RAM, B. 1961. Pollen morphological studies in the Acanthaceae. *Grana Palynologica,* **3:** 3–108.

RAJ, M. 1959. Morphological and embryological studies in the family Santalaceae with discussion on its systematic position. *Phytomorphology,* **9:** 17-19, 31–32.

RAO, D. 1965. Floral anatomy of the Erythroxylaceae. *Proc. Nat. Inst. Sci. India,* **35B:** 156–162.

RAO, V.S. 1952. The floral anatomy of some Verbenaceae with special reference to the gynoecium. *J. Indian Botan. Soc,* **31:** 297–315.

REED, C. 1955. The comparative morphology of the Olacaceae, Opiliaceae and Octoknemaceae. *Mem. Soc. Brot.,* **10:** 29–79.

RICKETT, H.W. 1945. Nyssaceae. *North Amer. Flora,* **28B:** 313–316.

ROY TAPADAR, N.N. 1964. Cytotaxonomic studies in Apocynaceae and delineation of the different evolutionary tendencies operating within the family. *Caryologia,* **17:** 103–138.

ROUSSEAU, D. 1927. Contribution a L' Anatomie comparatif des Piperacees. *Arch. Bot. Univ. Liege,* **7:** 45.

RYDBERG, P.A. 1914. Pyrolaceae. *North Amer. Flora,* **29:** 21–32.

SANTAPAU, H. and WAGH, S.K. 1963. Nomenclatural changes in Indian plants. *Bull, Botan. Surv. India,* **5:** 102–109.

SAXENA, N.P. 1964. Studies in the family Saxifragaceae. I. A contribution to the morphology and embryology of *Saxifraga diversifolia* Wall. *Proc. Indian. Acad. Sci.,* **B. 60:** 38–51.

SCHELLENBERG, G. 1932. Uber Systembildung und uber Reihe der Santalales. *Ber at. Bot. Ges.,* **50a:** 136–145.

SEN, J. and NASKAR, J.N. 1965. Non-herbaceous phanerogams of the Indian Botanic Garden, Calcutta. *Bull, Botan. Surv. India,* **7:** 31–61.

SENN, H.A. 1943. The relation of anatomy and cytology to the classification of Leguminosae. *Chron. Bot.,* **7:** 306–308.

SHARMA, H.P. 1954. Studies in the order Centrospermales. I. Vascular anatomy of the flower of certain species of the Portulacaceae *J. Indian Botan. Soc.* **33:** 98–111.

SINGH, R.P. 1954. Structure and development of seeds in Euphorbiaceae: *Ricinus communis* L. *Phytomorphology,* **4:** 118–123.

SINGH, U., WADHWANI, A.M. and JOHRI, B.M. 1983. *Dictionary of Economic Plants in India.* Indian Council of Agricultural Research, New Delhi.

SMALL, J.K. Monotropaceae. *North Amer. Flora,* **29:** 11–18.

STAUFFER, H.U. 1961. Beiträge zum Blütendiagramm der Santalales. *Verh. Schweiz. Naturf. Ges.,* **141:** 123–125.

STONE, D.E., REICH, J. and WHITFIELD, S. 1964. Fine structure of the walls of *Juglans* and *Carya. Pollen et Spores,* **6:** 379–392.

SUBRAMANYAM, K. 1951. A morphological study *of Stylidium graminifolium. Lloydia.* **14:** 65–81.

SUBRAMANYAM, K. 1962. *Aquatic Angiosperms.* Council of Scientific & Industrial Research, New Delhi.

SWAMY, B.G.L. 1948. A contribution to the life history of *Casuarina. Proc. Am. Acad.* **77:** 1–32.

SWAMY, B.G.L. 1949. The comparative embryology of the Santalaceae: Node, secondary xylem and pollen. *Am. J. Bot.,* **36:** 661–673.

SWEET, H.R. and BARKELEY, F.A. 1936. A most useful plant family, Anacardiaceae. *Bull. Mo. Bot. Gdn.,* **24:** 216–229.

TIPPO, O. 1938. Comparative anatomy of the Moraceae and their presumed allies. *Botan. Gaz.,* **100:** 1–99.

UENO, J. 1963. On the fine structure of the pollenwalls of Angiospermae. III. *Casuarina. Grana Palynologica,* **4:** 189–193.

VAN TIEGHEM, P. 1869. Anatomie de la fleur femelle et du fruit dunoyer. *Bull. Soc. Bot. Fr.,* **16:** 412.

VENKATA RAO, C. 1952. Floral anatomy of some Malvales and its bearing on the affinities of families included in the order. *J. Indian Botan. Soc,* **31:** 171–203.

VENKATA RAO, C. 1960. Studies in the Proteaceae. I. Persoonieae. *Proc. Nat. Inst. Sci. India,* **26B:** 300–337.

VENKATA RAO, C. 1967. Morphology of the nectary in Proteaceae. *New Phytol.:* **66:** 99–107.

VENKATESH, C.S. 1956. The curies anther of *Bixa*—its structure and dehiscence. *Am. Midi. Nat.,* **55:** 473–476.

VESTAL, P.A. 1937. The significance of comparative anatomy in establishing the relationship of Hypericaceae to the Guttiferae and their allies. *Phillip, J. Sci.,* **64:** 199–256.

VISHNU-MITTRE. 1963. Pollen morphology of Indian Amaranthaceae. *J. Indian Botan. Soc,* **42:** 86–101.

VIJAYARAGHAVAN, M.R. 1964. Morphology and embryology of a vesselless dicotyledon—*Sarcandra irvingbaileyi* Swamy, and the systematic position of the Chloranthaceae. *Phytomorphology,* **14:** 429–441.

WARBURG, E.F. 1938. Taxonomy and relationship in the Geraniales in the light of their cytology. *New Phytol,* **37:** 189–210.

WHITEHEAD, D.R. 1963. Pollen morphology in the Juglandaceae. I. Pollen size and pore number variation. *J. Arnold Arb.,* **44:** 101–110.

WHITEHEAD, D.R. 1965. Pollen morphology in the Juglandaceae. II. Survey of the family. *J. Arnold Arb.* **46:** 369–410.

WITHNER, CI. 1941. Stem anatomy and phylogeny of the Rhoipteleaceae. *Am. J. Bot.,* **28:** 872–878.

WOODSON, R.E. 1941. The North American Asclepiadaceae. I. Perspective of the genera. *Ann. Mo. Bot. Gdn.,* **28:** 193–244.

ZOHARY, M. 1950. Evolutionary trends in the fruiting head of Compositae. *Evolution,* **4:** 103–109.

PART FOUR

Systematic Survey of the Monocotyledoneae

Systematic Survey of the Monocotyledoneae

PLANTS WITH a woody simple or branched axis, giving rise to a shrub or tree habit, are widely distributed. Among the orchids, the terrestrial genus *Sobralia* is almost shrubby. The Velloziaceae contains many shrubby forms including *Vellozia* which may be as tall as 6 m and branched. The Bromeliaceae bears largely rosette plants, but massive columnar tall forms are produced in *Puya.* The Bambusoideae is mostly tall and woody, a branched shrubby habit being common among grasses of drier zones. A few shrubby species are present in *Paepalanthus* (Eriocaulaceae), *Prionium* (Juncaceae) and *Microdracoides* (Cyperaceae). Though the woody habit (*Ravenala*) is rare in the Musaceae, the family comprises such tall plants as the herbaceous *Musa.* In the Palmae, there are many species with massive columnar trunks. In contrast, branching of the trunk is common in the Pandanaceae which offers a habit akin to that of *Hyphaene* (Palmae). However, several species of *Freycinetia* are neither shrubs nor trees but embrace mostly epiphytic rhizomatous herbs. While the Araceae are predominantly herbaceous, *Alocasia* and *Philodendron* have "columnar stems upto 2 m tall surmounted by a crown of leaves". There are grass trees (*Xanthorrhoea*), rosette trees (*Agave*, *Furcraea*, *Yucca*, etc.) and tree lilies (*Dracaena*). Scandent shrubs include species of *Petermannia*, *Philesia*, *Ripogonum* and *Ruscus.* Only the Iridaceae has any woody members showing actual secondary growth, *e.g. Klattia*, *Nivenia* and *Witsenia.*

The distribution of bulbs is almost exclusively restricted to the Liliaceae. Bulbs also exist in *Triglochin bulbosum* which is associated with the unusual habit (saline swamps subjected to seasonal drying out). Corms are known mostly from *Colchicum* and are of somewhat general occurrence in the Iridaceae. Hypocotyledonary tubers form a general feature in the Dioscoreaceae and appear to occur in many Araceae.

The genera of the Potamogetonaceae occur in brackish or fresh water, but *Ruppia* sometimes in salt ponds with distinctly high salinity. Species of *Crinum*, *Pancratium* and *Triglochin* grow in saline environments. Several species of *Pandanus* are seashore plants. Among the Palmae, *Nypa* thrives in banks of tidal waters and *Phoenix* is highly tolerant of saline soil.

In the stem, the vascular bundles are scattered throughout and are not arranged in a cylinder. Secondary thickening growth is usually lacking except in the woody Amaryllidaceae, Iridaceae and Lillaceae.

Various methods of seed germination have been described for the monocots. In the *Alisma type*, the radicle is little or not developed at the start of germination; the cotyledon comes out of the seed and becomes straight and elongated. In the *Allium type*, the cotyledon comes out and assumes a green cylindrical structure. In the *grass type*, the cotyledon does not come out of the seed and forms a green structure; it remains within the seed and functions as a sucker. In the *palm type*, the upper half of the cotyledon remains permanently within the seed and the lower half forms a huge tubular sheath covering the axis. In the *orchid type*, certain cellular mass in the embryonic region grows out and forms a green primary tubercle and capillary roots from the lower part.

The Cyperaceae and Juncaceae possess chromosomes with diffuse centromeres. This chromosomal type has developed in common ancestors of these taxa, supporting their affinity.

Species accumulating aluminium (excess of 1,000 ppm dry weight of the element in over dried material) are rare amongst monocotyledons. One species of *Eleocharis*, several species of *Aletris* and most members of the Rapateaceae are considered as accumulators.

Cyanogenic compounds are known from various taxa. The cyanogenesis in nearly all monocotyledons is tyrosine derived, exceptions being *Chlorophytum capense* and *C. comosum*, with holocalin, *i.e.* the phenylalanine pathway. There is a certain tendency for concentration of cyanogenic compounds in the Araceae, Gramineae and Palmae, whereas the Liliaceae seems to be poor in these compounds.

Chelidonic acid occurs mainly in the Dioscoreaceae and Haemodoraceae. However, the Iridaceae is surprisingly poor and the Liliaceae apparently deficient in the compound. A number of records are known in the Gramineae and one in *Canna.*

While in the Gramineae the saponins are chiefly of the triterpene type, they are of the steroid type in the Commelinaceae and *zingiberaceae*. In the palms, they may be both of triterpene and steriod types. Steriod saponins are generally poisonous and serve mainly as a device to protect plants. This may explain their probable absence in the alkaloid rich Amaryllidaceae where they would otherwise be expected from a phylogenetic viewpoint.

The flavonol kaempferol and quercetin are extremely widespread in plants. A third common flavonol, myricetin, are generally less frequent in the monocotyledons. It has a very restricted distribution: Iridaceae, Marantaceae, Restionaceae, Sparganiaceae and Zingiberaceae. While it is a useful marker at tribal level within the last family, its overall distribution is so sporadic that no general significance can be attached to its existence in these plants. Among the various flavones, tricin is almost univeral in the Gramineae. It is also very common in the Cyperaceae and Palmae. The two common flavones of angiosperms include apigenin and luteolin. Their frequency varies widely, with as high as 95 percent in the Juncaceae to as low as six percent in the Araceae and 1 percent in the Orchidaceae. These flavones also occur in plants in *C*-glycosidic combination. Glycoflavones are the most common class of flavonoid encountered in the Araceae, Commelinaceae, Gramineae and Palmae. There is a geographical element in their frequency of occurrence, being more common in tropical and subtropical species of the Orchidaceae and restricted to South African members of the Restionaceae. Finally, *o*-methylation may take place at the 5-position of luteolin and produce luteolin 5-methyl ether; the latter compound is unique to the Cyperaceae and Juncaceae, forging a link between the two families. 6-Hydroxy flavonoids are regular components of several families: Cyperaceae and Commelinaceae (with 6-hydroxyluteolin), Eriocaulaceae (with quercetagetin and its 6-methyl either patuletin), Orchidaceae (with 6-hydroxyapigenin methyl ethers) and Bromeliaceae. In contrast, 8-hydroxy flavonoids have been reported in two families: Restionaceae and Bromeliaceae. While the occurrence of sulphated flavonoids may sometimes be associated with a fresh water or saline habitat, they are of some interest as taxonomic markers. Flavone sulphates occur abundantly in the Gramineae, Juncaceae and Palmae; they are also reported in the Cyperaceae and Restionaceae. Moreover, they are accompanied by flavonol sulphates in the grasses and palms. Interestingly, sulphates found in the Alismataceae and Hydrocharitaceae differ slightly from those of the grasses and palms; these families are distinctive in possessing sulphates of simpler phenolic acids, *e.g.* caffeic acid.

There is a great taxonomic regularity in the distribution of dehydroquinate dehydrolyases. Its occurrence in the Cyperaceae, Gramineae and Juncaceae is of interest on account of the existence of certain flavonoid compounds in these families.

Attempts to correlate the host-preferences of parasitising fungi with monocotyledon taxonomy and phylogeny have been made by Savile (1979). Because of the host distribution of *Uromyces spargani*, the Typhales is thought to be closely allied to the Arales (*Acorus*). The Commeliniflorae is considered a relatively homogeneous group with the exclusion of the Typhales. The Juncales is treated as derived from ancestral Typhales as viewed from their *Puccinia* and *Uromyces* rusts. The Liliiflorae is assumed to be phylogenetically younger than the Commeliniflorean core groups (Cyperales, Juncales, Poales) and unrelated to them.

The distribution of angiosperms has been surveyed by Raven and Axelrod (1974). Many families are restricted in their distribution to the southern hemisphere and thus were possibly differentiated on the Gondwanaland. Others are partly or entirely concentrated on the northern hemisphere and appear to have their past history on the Laurasian block. The fact that the Bromeliaceae and Cyclanthaceae are restricted to America in itself does not necessarily mean that they made their appearance after the period with an Antarctic connection. They might have prevailed in a limited portion of Gondwanaland and hence had a narrow range.

The monocots have been variously treated by taxonomists: 12 orders and 42 families (Lindley, 1853); seven series and 16 natural orders (Bentham and Hooker, 1883); four orders and 27 families (Van Tieghem, 1891); 10 orders and 42 families (Engler, 1892); nine orders and 46 families (Wettstein, 1901); seven orders and 66 families (Lotsy, 1911); seven orders and 36 families (Hallier, 1912); nine orders and 39 families (Wettstein, 1912); eight orders and 45 families (Bessey, 1915); nine orders and 33 families (Rendle, 1930); four orders and 48 families (Skottsberg, 1940); ten orders and 35 families (Lawrence, 1951); 12 orders and 53 families (Novak, 1954); 11 orders and 35 families (Core, 1955); eight orders and 64 families (Deyl, 1955); 28 orders and 64 families (Kimura, 1956); 29 orders and 69 families (Hutchinson, 1959); 18 orders and 58 families (Emberger, 1960); 14 orders and 51 families (Melchior *et al.*, 1964); 18 orders and 61 families (Cronquist, 1968); 20 orders and 67 families (Takhtajan, 1969); 19 orders and 60 families (Stebbins, 1974); 26 orders and 84 families (Dahlgren, 1975); 11 orders and 45 families (Thorne, 1976); 18 orders and 33 families (Ehrendorfer, 1978) and 29 orders and 93 families (Dahlgren and Clifford, 1982). The present author takes into account 11 orders and 33 families.

The various monocot orders may be distinguished by the following characters (Lawrence, 1951; Core, 1955):

(1) Carpels free	(2)
(1) Carpels united	(7)
(2) Inflorescence not a true spadix	(3)
(2) Inflorescence a dense glomerule or fleshy spadix	(5)
(3) Perianth scale-like or bristle-like	*Pandanales*
(3) Perianth sepaloid or petaloid	(4)
(4) Plants bear chlorophyll	*Helobiales*
(4) Plants lack chlorophyll	*Triuridales*
(5) Leaves never plicate	*Arales*
(5) Leaves generally plicate	(6)

(6) Carpel 1-ovuled — *Palmales*
(6) Carpels many-ovuled — *Synanthales*
(7) Flowers borne in axils of dry chaffy scales — *Graminales*
(7) Flowers not borne in axils of dry chaffy scales — (8)
(8) Seeds very small; endosperm absent — *Orchidales*
(8) Seeds not small; endosperm present — (9)
(9) Endosperm mealy — *Farinales*
(9) Endosperm fleshy or horny — (10)
(10) Functional stamens 3–6 — *Liliales*
(10) Functional stamen 1 — *Scitaminales*

Order 1. *Helobiales*

The Helobiales is characterised by the aquatic habit, hemicyclic to cyclic flowers, free carpels, seeds without endosperm and seedlings with a large hypocotyl.

The order is regarded as the most primitive among the monocotyledons points favouring such a concept are six-fold: (*i*) the presence of floral parts which are free and disposed in several whorls; (*ii*) the abortion of perianth and reduction in stamen number in some genera of the Najadaceae; (*iii*) the occurrence of numerous stamens and carpels, showing the transition from the spiral to whorled arrangement; (*iv*) the prevalence of wide lamina-like filaments with basifixed anthers in the members of the Juncaginaceae and Potamogetonaceae; (*v*) the incidence of laminal placentation in some members of the Alismataceae and Hydrocharitaceae and (*vi*) the incomplete closure of carpels in some taxa of the Hydrocharitaceae.

In the monocots, the solitary bracteole is either median posterior or median lateral. This type of bracteole occurs in the Nymphaeaceae and Ranunculaceae, attaching significance to the idea that the Helobiales is the direct descendent of early dicots and direct ancestor of other monocots. In fact, Mitra (1974) conjectured such an origin from the Ranales:

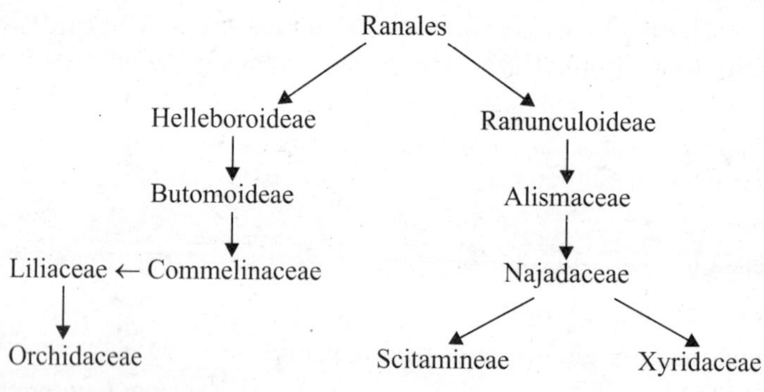

The Helobiales, also called Fluviales and Najadales, represents an unnatural assemblage, contrary to the dictum of Engler. Bessey included most of the families in the Liliales. Rendle placed the order after the Pandanales, but did not consider its derivation from the latter. Hutchinson split up the entire taxon into six orders, such as Butomales, Alismatales, Juncaginales, Aponogetonales, Potamogetonales and Najadales. He opined that the Aponogetonales was ancestral to the Najadales. Moreover, he treated the Potamogetonales as an off-shoot somewhat more primitive than the Juncaginales. The studies of Cheadle (1942) and Uhl (1947) revealed the primitiveness of the Aponogetonaceae, Scheuchzeriaceae and Lilaeaceae as well as the probable highly advanced condition of the Najadaceae. Cronquist (1968) created three orders and 12 families: Alismatales (Butomaceae, Limnocharitaceae, Alismataceae), Hydrocharitales (Hydrocharitaceae), Najadales (Aponogetonaceae, Scheuchzeriaceae, Juncaginaceae, Najadaceae, Potamogetonaceae, Ruppiaceae, Zannichelliaceae, Zosteraceae). Stebbins (1974) followed likewise, but Takhtajan (1969) placed two extra families: Posidoniaceae and Cymodoceae—under the Najadales. But Dahlgren (1975) removed the Butomaceae from the Alismatales and placed it in the Hydrocharitales along with the Hydrocharitaceae and Apcnogetonaceae; here the Najadales was meant to accommodate the Najadaceae and seven families (Scheuchzeriaceae, Juncaginaceae, Potamogetonaceae, Zosteraceae, Posidoniaceae, Zannichelliaceae, Cymodoceaceae) remained in the Zosterales. Thome (1976) accepted three orders and 10 families: Alismatales (Butomaceae, Alismataceae, Hydrocharitaceae), Zosterales (Aponogetonaceae, Scheuchzeriaceae incl. Juncaginaceae, Potamogetonaceae, Posidoniaceae, Zannichelliaceae, Zosteraceae) and Najadales (Najadaceae). Ehrendorfer (1978) favoured the Alismatales for the Butomaceae and Alismataceae, the Hydrocharitales for the Hydrocharitaceae and the Najadales for the Scheuchzeriaceae, Juncaginaceae, Potamogetonaceae, Zosteraceae, Zannichelliaceae and Najadaceae. Dahlgren and Clifford (1982) proposed the Hydrocharitales for the Butomaceae, Aponogetonaceae, Hydrocharitaceae (incl. Thalassiaceae and Halophilaceae), the Alismatales for the Alismataceae (incl. Limnocharitaceae) and the Zosterales for the Scheuchzeriaceae, Juncaginaceae (incl. Lilaeaceae), Najadaceae, Potamogetonaceae (incl. Ruppiaceae), Zosteraceae, Posidoniaceae, Cymodoceaceae and Zannichelliaceae.

The members of the Helobiales show trends in the floral evolution from the simple haplochlamydeous unisexual to the complex diplochlamydeous bisexual types. In the Najadaceae, the flowers are axial structures consisting of a solitary stamen or an uniovuled ovary which is either perianthless or provided with a sac-like perianth. In the Potamogetonaceae, the flowers may be unisexual (*Zannichellia*) or bisexual (*Zostera*). Although the perianth is present in *Althenia* and absent, in other members of the Potamogetonaceae, the connective tends to become petaloid in *Posidonia* or *Potamogeton*. In other families, the floral structure confirms to or is derived from the regular trimerous arrangement. There is a tendency to multiplication of the components of the androecium and gynoecium either by doubling or by increment of the whorl numbers. The Juncaginaceae and Alismataceae contain diplochlamydeous, bisexual, regular and hypogynous flowers. The epigynous flowers of the Hydrocharitaceae are a step towards further advancement, reaching a climax in *Vallisneria* where the flowers are irregular.

The six families under the order, according to Core (1955), may be identified as follows:

(1) Ovary superior	(2)
(1) Ovary inferior	*Hydrocharitaceae*
(2) Perianth absent	(3)
(2) Perianth present	(4)
(3) Perianth absent or sac-like	*Najadaceae*
(3) Perianth absent but with sepaloid appendages on the stamens	*Potamogetonaceae*
(4) Perianth undifferentiated	(5)
(4) Perianth differentiated	(6)
(5) Perianth sepaloid, in two whorls	*Juncaginaceae*
(5) Perianth petaloid, in one whorl	*Aponogetonaceae*
(6) Perianth made up of calyx and corolla	*Alismataceae*

Najadaceae
(Naiad Family)

GENERAL FEATURES OF THE FAMILY

Habit	—	*Annual submerged herbs.*
Stem	—	Nodal, much branched, filiform or muricate.
Leaves	—	Opposite, linear, entire or toothed, sometimes with auricles.
Flowers	—	Unisexual, minute, sessile, *wholly submerged.*
Perianth	—	Absent or present.
Androecium	—	Stamen 1 (in male flowers), enveloped in a bottle shaped bract; anthers 1- to 4-celled.
Gynoecium	—	Carpel 1 (in female flowers), *ovary superior*, *unilocular*; *ovules solitary*, *erect*, *anatropous*; style cylindric; stigma subulate.
Fruit	—	Achene, enclosed by a foliaceous sheath.
Seeds	—	Embryo straight; endosperm absent.

NUMBER IN AND DISTRIBUTION OF THE FAMILY

The Najadaceae comprises one genus (*Najas*) and 50 species which are distributed throughout the northern and southern hemispheres.

COMMONLY OCCURRING PLANTS OF THE FAMILY

Najas gramines Del., *N. indica* (Willd.) Cham., *N. lacerata* Rendle and *N. minor* (Pers.) All. are seen in fresh or brackish water.

ECONOMIC ASPECTS OF THE FAMILY

In warmer regions, the species of *Najas* have been reported to be a nuisance in irrigation ditches and rice fields. However, they are important food plants for fish. They are often used as packing material and as a green fertiliser.

Potamogetonaceae
(Pondweed Family)

GENERAL FEATURES OF THE FAMILY

Habit	—	Fresh water or salt water *perennial submerged herbs.*
Stem	—	Nodose, jointed, bearing roots at lower nodes and leaves at upper nodes.
Leaves	—	Alternate or opposite, distichous, stipulate, basally vaginate.
Inflorescence	—	Axillary spike.

Flowers	—	Bisexual or unisexual, actinomorphic, hypogynous, *floating*.
Perianth*	—	*Variable*, segments 4–6, *sepaloid or lacking*.
Androecium	—	*Stamens 1–4*; anthers 1- to 2-celled, extrorse.
Gynoecium	—	*Carpels 1–4*; *ovary superior*; *ovules solitary*, *apical*, *orthotropous*.
Fruit	—	Indehiscent drupelet or nutlet.
Seeds	—	Embryo axile; endosperm absent.

DIVISIONS OF THE FAMILY

The Potamogetonaceae is divided into two groups, one with two tribes and the other with three tribes:

Group A. Flowers solitary or in cymes.

Tribe (*i*) *Cymodoceae*. Perianth absent. Pollen thread-like. Examples: *Cymodocea* and *Diplanthera*.

Tribe (*ii*) *Zannichellieae*. Perianth absent or present. Pollen globose. Examples: *Althenia* and *Zannichellia*.

Group B. Flowers in spikes.

Tribe (*i*) *Posidonieae*. Spike branched; rachis cylindrical. Example: *Posidonia*.

Tribe (*ii*) *Potamogetoneae*. Spike unbranched; rachis cylindrical. Examples: *Potamogeton* and *Ruppia*.

Tribe (*iii*) *Zostereae*. Rachis compressed. Examples: *Phyllospadix* and *Zostera*.

POSITION AND AFFINITY OF THE FAMILY

Bentham-Hooker did not consider the Potamogetonaceae as a distinct family, but incorporated it into the Najadaceae. Other workers treated it as a separate family under the Helobiae. Hutchinson broke up the taxon into the Potamogetonaceae and Ruppiaceae; in his classification, the Zannichelliaceae was included in the Najadales. On anatomical and morphological grounds, the Potamogetonaceae is related to the Najadaceae and Hydrocharitaceae. Lawrence believed the family to be intermediate in position, being more developed than the Aponogetonaceae or Scheuchzeriaceae and less developed than the Najadaceae or Alismaceae.

NUMBER IN AND DISTRIBUTION OF THE FAMILY

The Potamogetonaceae is composed of nine genera and 125 species, being world wide in distribution.

COMMONLY OCCURRING PLANTS OF THE FAMILY

Pondweed (*Potamogeton nodosus* Poir.) is a rooted aquatic herb with the upper or all leaves floating.

Ruppia maritima L., found in brackish water, forms tangled masses of filiform leafy flaccid stems and branches.

Zannichellia palustris L. ssp. *pedicellata* Wahlen. & Rosen is common in salt marshes and fresh water lagoons.

* It has been explained as a sepaloid connective and as bract-like in origin.

ECONOMIC ASPECTS OF THE FAMILY

Potamogeton species can be problematical weeds in canals and ditches. However, they provide a valuable food for many animals. *P. natans* is edible, the starchy rootstock and turions being consumed by human beings.

Juncaginaceae
(Arrowgrass Family)

GENERAL FEATURES OF THE FAMILY

Habit — *Marshy scapigerous herbs.*
Stem — Grass-like.
Leaves — Radical, linear, sheathing at base.
Inflorescence — Raceme or spike.
Flowers — Bisexual or unisexual, regular, hypogynous, *ebracteate.*
Perianth — *Segments 6, in 2 whorls of 3, sepaloid*, green.
Androecium — Stamens 6; *anthers 2-celled, subsessile.*
Gynoeciura — Carpels 3–6, free or united; ovules 1 or 2, erect, anatropous; *style short, stout* or *absent; stigma often papillose or plumose.*
Fruit — Follicle.
Seeds — Embryo straight; endosperm absent.

NUMBER IN AND DISTRIBUTION OF THE FAMILY

The Juncaginaceae includes three genera and 14 species. The members of this family are distributed throughout the temperate and cold parts of the world.

ECONOMIC ASPECT OF THE FAMILY

The leaves of *Triglochin maritima* and rhizomes of *T. procerum* are edible.

Aponogetonaceae
(Latticeplant Family)

GENERAL FEATURES OF THE FAMILY

Habit — *Laticiferous perennial aquatic herbs.*
Leaves — Radical, oblong to linear, *reticulately veined with a few primary parallel veins and many secondary transverse veins.*
Inflorescence — *Simple or 2- to 8-branched spikes.*
Flowers — Bisexual, actinomorphic, minute.
Perianth — None or *represented by 1–3 petaloid or spathaceous segments.*
Androecium — Stamens 6, in 2 whorls; anthers 2-celled, basifixed.

Gynoecium — Carpels 3–6, free; *ovary superior*; *ovules 1–8*, *basal*, *anatropous*; style slender tapering; stigma papillate.

Fruit — Follicle.

NUMBER IN AND DISTRIBUTION OF THE FAMILY

The Aponogetonaceae contains one genus (*Aponogeton*) and about 45 species. The members of the family spread from tropical and southern Africa to north-eastern Australia and New Guinea.

COMMONLY OCCURRING PLANTS OF THE FAMILY

Aponogeton crispum Thunb. is an aquatic herb whose leaves are all submerged and whose follicles are oblong.

Aponogeton natans (L.) Engl. & Krause is an aquatic herb whose leaves are mostly floating and whose follicles are subglobose.

ECONOMIC ASPECTS OF THE FAMILY

The tubers of many species are consumed by man and animals. Decorative aquarium plants are made by *Aponogeton distachyos.*

Alismataceae
(Water-plantain Family)

GENERAL FEATURES OF THE FAMILY

Habit — *Aquatic or marshy herbs with a stout rhizome.*

Leaves — Radical or clustered at nodes, entire with a broad petiole.

Inflorescence — Panicled or umbellate in scapes.

Flowers — Bisexual or unisexual, regular, hypogynous.

Perianth — *Segments 6, outer 3 sepaloid and inner 3 petaloid, free.*

Androecium — *Stamens 6, 9 or indefinite, free.*

Gynoecium — *Carpels 3, 6 or more, free*; *ovary superior*; *ovules 1 to many*; *placentation marginal* or *superficial*.

Fruit — *Etaerio of achenes or follicles.*

Seeds — Embryo horseshoe shaped with a single bent cotyledon; endosperm absent.

FLORAL RANGE IN THE FAMILY

The flowers are unisexual in *Sagittaria.* The number of stamens may vary from 3 to 9, as in *Butomopsis.* In *Alisma*, the stamens are arranged in three pairs in one whorl with 2 stamens in each pair. All the stamens are not fertile, the outer being sterile, *e.g. Hydrocleys.* The stamens and carpels are indefinite in *Sagittaria. Butomus* is in possession of 6 carpels in two whorls. There are 6–10 carpels in *Damasonium*, where they are united at the base. The ovules are many in each carpel in the Butomoideae or single in each carpel in the Alismatoideae.

DIVISIONS OF THE FAMILY

The Alismataceae is divided into two subfamilies:

Subfamily I. *Alismatoideae.* Ovules solitary in each carpel, rarely 2 or a few; placentation marginal. Examples: *Alisma*, *Sagittaria*, etc.

Subfamily II. *Butomoideae.* Ovules many in each carpel; placentation almost superficial. Examples: *Butomopsis*, *Hydrocleys*, etc.

POSITION AND AFFINITY OF THE FAMILY

The Alismataceae was placed in the sixth series Apocarpeae by Bentham-Hooker. On the other hand, Engler as well as Rendle included the family in the second order Helobiae just after the Pandanales. In Hutchinson's arrangement, the family appeared under his second order Alismatales of the monocots.

The Alismataceae is closely allied to the Hydrocharitaceae, but differs in having flowers arranged in whorls and ovary with one or more basal ovules. It resembles the Commelinaceae, but is set apart by the aquatic habit, panicled or umbellate scapose inflorescence, two-seriate perianth and exalbuminous seeds. It bears a relationship with the Nymphaeaceae through the floral characters of *Hydrocleys nymphaeoides*, particularly due to a superficial placentation; but further affinity is difficult to establish. It is akin to the Ranunculaceae on the basis of such features as hypogynous flowers with free sepals and petals, numerous stamens and carpels as well as etaerio of achenes or follicles. The flowers of the Ranunculaceae are somewhat acyclic which is not the case in the Alismataceae.

The Alismataceae has evoked considerable interest among botanists. It has been thought to be one of the primitive of living monocot families. Although Sharma (1964) debated the homogeneity of the Alismatales, he suggested that *Alisma* contained the characteristic features of ancient germs like low chromosome number, long chromosomes and symmetrical karyotype. The primitiveness of the family is also manifested in the hypogynous flowers, numerous stamens and free carpels. Hutchinson noted that *Ranalisma*, a member of the Alismataceae, differed primarily from *Ranunculus* only in having a single cotyledon and in lacking endosperm. "It is rather a coincidence, a similar type of flower having developed in the two distinct groups" (Rendle, 1930). Cheadle (1942) believed that vessels with simple or porous perforations are important traits of the Alismataceae. This anatomical condition makes the family more advanced than most other families of the Helobiales. The possession of bisexual flowers with many stamens and many carpels in the Alismataceae "mark it as more primitive than its position in the Engler system would indicate". The presence in the Alismataceae of an advanced type of vascular anatomy is in contradiction with any theory that will attempt the derivation of other taxa of the Helobiales from it. Lawrence (1951) stated that the Alismataceae, together with the Butomaceae and Hydrocharitaceae, "terminate a lateral branch from primitive elements" of the Helobiales "that has not given rise to more advanced families of this order".

NUMBER IN AND DISTRIBUTION OF THE FAMILY

The Alismataceae covers 11 genera and about 100 species. The members of this family are widely distributed in the temperate and tropical regions of the globe. They are mostly seen in fresh water swamps and streams.

COMMONLY OCCURRING PLANTS OF THE FAMILY

Water plantain (*Alisma plantago-aquatica* L.) is an aquatic plant with bisexual flowers.

Limnophyton obtusifolium (L.) Miq. is an aquatic plant with unisexual white flowers.

Arrow head (*Sagittaria sagittifolia* L.) is a tank herb with sagittate leaves, globose receptacles and unisexual flowers.

Tenagocharis lanceolata (D. Don) Buchen. is an aquatic herb with oblanceolate leaves and many flowered umbels.

ECONOMIC ASPECTS OF THE FAMILY

Most genera provide food for wildlife. The tubers of *Alisma plantago*, roots of *Sagittaria latifolia* (North America) and corms of *S. sagittifolia* (China & Japan) are edible. *S. latifolia* and *S. subulata* are sometimes used as aquarium plants. Many species of *Alisma*, *Echinodorus* and *Sagittaria* are cultivated as pool side decorative plants.

Hydrocharitaceae
(Frog's-bit Family)

GENERAL FEATURES OF THE FAMILY

Habit — *Aquatic herbs*, partly or completely submerged.
Leaves — Opposite or whorled, basal or cauline, variable.
Flowers — Bisexual or unisexual, regular, *epigynous*, enclosed in a spathe.
Perianth — Segments 6, usually in 2 series, outer 3 green and inner 3 coloured.
Androecium — Stamens 3–15 (in male flowers), in 1–5 whorls, inner being reduced to staminodes.
Gynoecium — Carpels 2–15 (in female flowers); *ovary inferior, unilocular; placentation parietal.*
Fruit — Berry-like.
Seeds — Embryo large; endosperm absent.

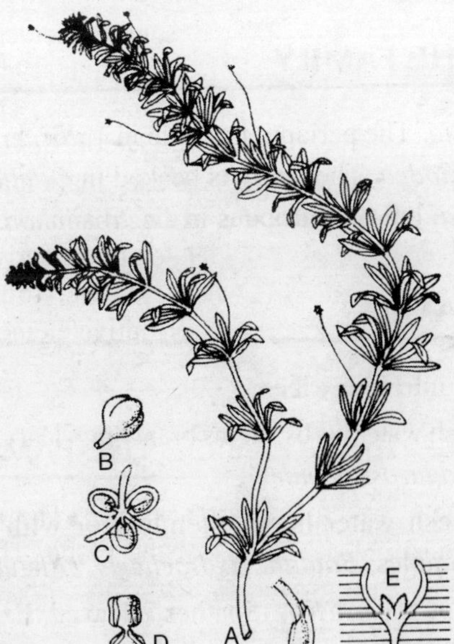

Fig. 89 HYDROCHARITACEAE. *Hydrilla verticillata* (L.f.) Royle **A**, portion of female plant; **B**, male flower bud; **C**, male flower prior to anthesis (as seen from above); **D**, lateral view of male flower on water surface; **E**, sheath of peduncle of female flower; **F**, base of peduncle of female flower. (**A–F**, after Subramanyam)

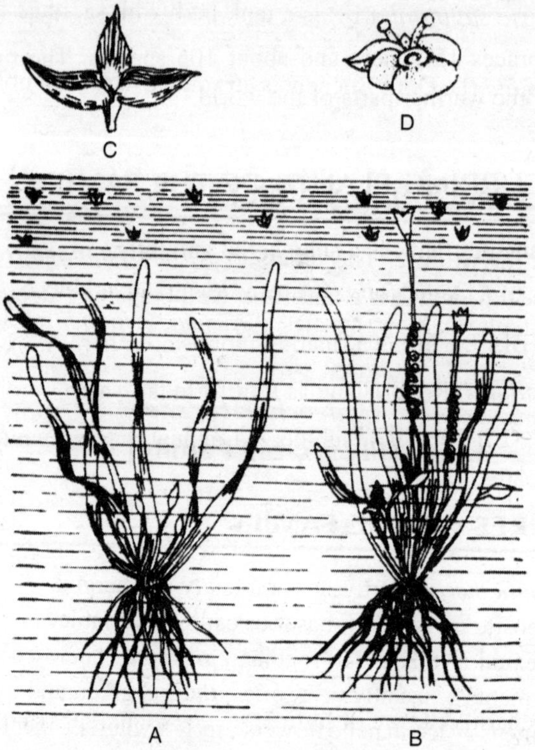

Fig. 90 HYDROCHARITACEAE. *Vallisneria spiralis* L. **A**, male plant; **B**, female plant; **C**, male flower; **D**, female flower

FLORAL RANGE IN THE FAMILY

The flowers are bisexual in *Ottelia.* The perianth is single in *Vallisneria*, but double in *Blyxa.* There are 3 stamens in *Hydrilla*, but 9 in *Elodea.* The ovary is beaked in *Ottelia*, but not so in *Hydrocharis.* The ovules are anatropous in *Hydrilla*, but orthotropous in *Lagarosiphon.*

DIVISIONS OF THE FAMILY

The Hydrocharitaceae is divided into four tribes:

Tribe (*i*) *Hydrilleae.* Fresh water herbs; stems branching, leafy. Spathes small, sessile. Examples: *Hydrilla* and *Lagarosi phon.*

Tribe (*ii*) *Stratioteae.* Fresh water herbs; stemless or with a creeping root stock. Spathes peduncled. Examples: *Boottia*, *Hydrocharis*, *Ottelia* and *Stratiotes.*

Tribe (*iii*) *Thalassieae.* Salt water herbs. Spathes 2-leaved. Examples: *Enhalus* and *Halophila.*

Tribe (*iv*) *Vallisnerieae.* Fresh water herbs, stemless or with stolons only. Spathes on long scapes. Examples: *Blyxa* and *Vallisneria.*

NUMBER IN AND DISTRIBUTION OF THE FAMILY

The Hydrocharitaceae embraces 15 genera and about 106 species. The members of this family are mostly seen in the water of the warmer parts of the world.

COMMONLY OCCURRING PLANTS OF THE FAMILY

Blyxa octandra (Roxb.) Planch ex Thw. is a submerged and tufted annual.

Hydrilla verticillata (L.f.) Royle is a tank herb with branching stems and whorled leaves.

Hydrocharis dubia (Bl.) Backer is a floating monoecious herb.

Nechamandra alternifolia (Roxb.) Thw. is a filiform tank herb.

*Ottelia alsmoideas** (L.) Pers. is a submerged annual, having white flowers and ovoid fruits enclosed in a six-winged spathe.

Eel grass or Tape grass (*Vallisneria spiralis* L.) is a stoloniferous dioecious aquatic weed with ribbon shaped leaves.

ECONOMIC ASPECTS OF THE FAMILY

The members of the Hydrocharitaceae are of little or no economic importance. Yet, *Halophila ovalis* is used as a manure for coconut plantations. The leaves of *Vallisneria spiralis* are ordinarily utilised for refining crude sugar into white sugar.

Order 2. *Triuridales*

The plants are *yellowish or reddish saprophytic herbs*, growing on rotting tree trunks or rich humus of the forest-floor. The *leaves are small, scaly and non-green.* The *flowers are small and drawn out into tail-like appendages; they are perfect or unisexual.* The *perianth segments are valvate in bud*, numerous and free. The ovules are solitary with a single integument. The seed contains a spherical embryo and fatty endosperm.

The order is represented by a single family, the Triuridaceae, the status of which is doubtful. It is a fairly primitive taxon and often stands second and next to the Helobiales. "The apocarpy and sometimes gynobasic stylodia of the carpels have a superficial similarity to those of...*Alisma*" (Dahlgren and Clifford, 1982). It is allied to the genus *Scheuchzeria* of the Juncaginaceae and *Petrosavia* and *Protolirion* of the Liliaceae.

The Triuridaceae includes seven genera and 80 species. The members of this family extend from Guatemala to Brazil.

Andruris khasiana (Benth. & Hook. f.) Schlecht. is found in Khasi Hills, Meghalaya.

Economic uses for plants of the family are unknown.

---

* According to Datta and Biswas (1967), it exhibits a number of form-variations.

Order 3. *Pandanales*

The order is characterised by the plants dioecious or monoecious, leaves stiffly long and sword-shaped, perianth bristle-like or scale-like, pollen in diads, gynoecium monopistillate and seeds with a copious endosperm.

The Pandanales was treated by Engler and Rendle as the most primitive of the angiosperms due to the remarkably simple flowers and wind pollination. It also formed the first order of Lawrence's and Core's monocots. Wettstein included the order in his ninth series after the Spadiciflorae. Hutchinson removed the Typhaceae and Sparganiaceae from the Pandanales and incorporated them into the Typhales near the Arales, but retained the Pandanaceae in the Pandanales next to the Palmales. Mitra placed it in his sixth monocot order between the Spathiflorae and Juncales. Cronquist and Stebbins put the Typhales in his subclass Commelinidae and the Pandanales in his subclass Arecidae. Takhtajan accommodated the Pandanales and Typhales in the superorder Arecanae under the subclass Arecidae. Thorne transferred the Sparganiaceae and Typhaceae to the Arales, keeping the Pandanaceae within the Pandanales. Dahlgren shifted the Typhales to the Typhanae and the Pandanales to the Arecanae. Ehrendorfer considered the Typhales under the Juncanae of the subclass Liliidae and the Pandanales under the subclass Arecidae.

Hutchinson reckoned the Pandanales as a very advanced group, somewhat parallel with the palms. He also interpreted the Typhales as a reduced series which took up the aquatic habit and derived from the Liliaceae. But the Pandanales has a very indistinctive development of tap root in the seedling during germination and a primitive type of cotyledon without the differentiation into a lamina and base (Eames, 1961).

The application of cytological principles in the Pandanales has produced interesting results (Sharma, 1964). In *Sparganium* and *Typha*, the chromosome number is $x = 15$, while in *Pandanus*, it is 30, *i.e.*, a multiple of 15. The similarity in chromosome number of the three genera is striking. Moreover, the homogeneity of chromosomes is evident in very small chromosome size with identical types of constriction. On the basis of their karyotypes, it is very difficult to distinguish the trio as they look very similar to each other. Hence, the evidence from chromosome studies justify their inclusion under the Pandanales as done by Engler. Hutchinson split up Engler's Pandanales into two taxa: the Pandanales comprising *Pandanus* and the Typhales embracing *Sparganium* and *Typha*, the former being nearly terrestrial and the latter aquatic types. Remarkable cytological resemblance and genetical relationship exhibited by these genera reveal the artificiality of Hutchinson's circumscription and the futility of stressing unduly on habitat. But the primitive status of the Pandanales as assigned by Engler cannot be approved. The high chromosome number, coupled with small chromosome size, represents undoubtedly an advanced level of evolution. Cytological data confirm the homogeneity of the Pandanales, but in no way indicate its primitive condition. *Pandanus* differs from *Typha* in the absence of catechol-tannins and leuco-anthocyanins, presence of alkaloids and positive activity of enzyme polyphenolase. Gibbs (1974) reported the absence of such phenolics compounds as kaempferol, cyanidin, sinapic and ferulic acids in *Pandanus* and presence of the same in *Typha*. *Sparganium* shows more similarities in chemical characters with *Typha* than with *Pandanus*. Thus, there appears to be a chemical homogeneity among these three taxa. The totality of evidence and numerical assessment of the characters drawn from diverse disciplines lend support to the close kinship among the three taxa and their retention under one Englerian taxon (Radhakrishnaiah *et al.*, 1984).

Due to their unisexual flowers and spathe formation, the Pandanaceae suggests an affinity with the Cyclanthaceae and Palmae. The coalescence of the fruits of *Pandanus* recalls similar infructescences in *Artocarpus.*

The Pandanales is a natural taxon. The Pandanaceae contains tropical shrubs or trees, whereas the Typhaceae and Sparganiaceae represent marshy herbs. In all the three families, the form and arrangement of flowers are strikingly alike. The flower is often reduced to a sporophyll, consisting of a single stamen or carpel at the axil of a bract. In *Pandanus* and *Sparganium*, the flowers are crowded in heads to form a compound racemose inflorescence. The occasional union of carpels in *Sparganium* speaks for their characteristic cohesion in *Pandanus.* Furthermore, the method of stem branching in *Sparganium* can be compared to the apparent dichotomy in *Pandanus.*

The families under the Pandanales, according to Core (1955), may be characterised as follows:

(1) Woody plants	*Pandanaceae*
(1) Herbs with a rhizome	(2)
(2) Flowers borne in terminal spikes	*Typhaceae*
(2) Flowers borne in axillary heads	*Sparganiaceae*

Pandanaceae
(Screw-pine Family)

GENERAL FEATURES OF THE FAMILY

Habit	—	*Palm-like branched shrubs or trees, supported by stilt roots; dioecious.*
Leaves	—	*Spiral, 4-ranked, long, narrow, spinous* at apex, margin and back of midrib.
Inflorescence	—	Individual flowers often crowded on axis and enclosed initially by foliaceous or spathaceous bracts.
Flowers	—	*Unisexual*, small.
Perianth	—	Absent.
Androecium	—	Stamens many (in male flowers); anthers 2-celled, basifixed; pollen grains ulcerate.
Gynoecium	—	Carpels 1 to many (in female flowers); *ovary superior, unilocular; ovules 1 or many, anatropous.*
Fruits	—	*Syncarp.*
Seeds	—	Embryo minute, linear; endosperm fleshy.

NUMBER IN AND DISTRIBUTION OF THE FAMILY

There are three genera and 700 species in the Pandanaceae which is a palaeotropic family favouring coastal or marshy regions from Africa through Indonesia to Australia.

COMMONLY OCCURRING PLANTS OF THE FAMILY

Screw-pine (*Pandanus odoratissimus* L. f.) and Nicobar Bread fruit or Umbrella Tree (*P. tectorius* Sol. ex Parkin.) are gregarious shrubs.

Pandanus foetidus Roxb., a bush with no proper stem, is found in village thickets and hedges.

ECONOMIC ASPECTS OF THE FAMILY

The leaves of *Pandanus* are important in paper making and thatching; they are also made into hats, mats and umbrellas; fibres extracted from the leaves are valuable for brushes, bristles, cordage, fishing lines, fishing nets and hunting nets. The fibrous aerial roots are utilised for making corks and for tying baskets. The male inflorescences are in demand for the manufacture of 'keora attar'. The fragrant floral leaves and orange-red pulp of the fruits are edible.

The ornamentals include *Freycinetia banksii* and *Pandanus veitchii.*

Typhaceae
(Cat-tail Family)

GENERAL FEATURES OF THE FAMILY

Habit	—	*Tall aquatic or marshy herbs with a perennial creeping rhizome.*
Leaves	—	Radical, *erect*, *long-linear*, *distichous*, *sheathing.*
Inflorescence	—	*Dense terminal superposed spike* in which upper portion is staminate and lower pistillate, *turning fuzzy-brown at maturity.*
Flowers	—	*Unisexual*, surrounded by a number of hairs.
Perianth	—	*Represented by slender jointed threads or elongated spathulate scales.*
Androecium	—	Stamens 2–5 (in male flowers); filaments free or variously united; anthers basifixed.
Gynoecium	—	Carpel 1 (in female flowers); ovary fusiform, stipitate, unilocular; ovule 1, pendulous, anatropous; style fili form, stigma linear to spathulate or rhomboidal, unilateral.
Fruit	—	Minute nutlet with a persistent style.
Seeds	—	Embryo straight; endosperm fleshy or floury.

NUMBER IN AND DISTRIBUTION OF THE FAMILY

The family contains one genus *(Typha)* and 15 species, being distributed in the temperate and tropical regions and occurring in habitats of shallow freshwater.

COMMONLY OCCURRING PLANT OF THE FAMILY

Typha australis Schum. & Thonn. is a robust plant of riparian and estuarian marshes.

ECONOMIC ASPECTS OF THE FAMILY

The leaves of *Typha latifolia* are used for weaving baskets, chair seats, bassocks and mats. The plant is also raised "for the ornamental value of its large brown cylindrical fruiting spears".

Sparganiaceae
(Bur-reed Family)

GENERAL FEATURES OF THE FAMILY

Habit — *Rather low aquatic or marshy herbs with a perennial creeping rhizome.*
Leaves — *Erect or floating, linear, sessile, distichous, sheathing.*
Inflorescence — *Dense axillary globose heads*, upper of which are male and lower female.
Perianth — *Represented by chaffy or glumaceous scales.*
Androecium — Stamens 3–6 (in male flowers); filaments free or partially united; anthers basifixed.
Gynoecium — Carpel 1 (in female flowers); ovary sessile, narrowed at base, unilocular, style simple or forked; stigma unilateral.
Fruit — Drupaceous.
Seeds — Embryo straight; endosperm mealy.

NUMBER IN AND DISTRIBUTION OF THE FAMILY

The Sporganiaceae comprises one genus (*Sparganium*) and about 20 species. The members of this family are confined to the frigid and temperate regions of the northern hemisphere, Australia and New Zealand.

COMMONLY OCCURRING PLANT OF THE FAMILY

Sparganium ramosum Huds. is found in Kashmir.

Order 4. *Palmales*

The order contains a solitary family, the Palmae. The characters of the order are the same as those for the family.

Palmales
(Palm Family)

GENERAL FEATURES OF THE FAMILY

Habit — Shrubs or trees, solitary or gregarious with an unbranched erect, climbing or trailing stem.
Leaves — Large with a well developed petiole (*haft*) and persistent sheathing base, *plaited in bud.*
Inflorescence — *Simple or compound spadix, enclosed in 1 or more spathes.*
Flowers — Unisexual, regular, hypogynous, sessile, arranged in close spirals on inflorescence axis.
Perianth — Small, segments 6, in two whorls, *somewhat leathery or fleshy*, persistent.
Androecium — Stamens 6 (in male flowers), in 2 series; filaments free or united; anthers 2-celled, versatile.

Gynoecium — Carpels 3 (in female flowers), free or united: *ovary superior, trilocular with* 1 *anatropous or 3 orthotropous ovules in each loculus.*

Fruit — Berry or drupe.

Seeds — Embryo minute; *endosperm ruminate.*

FLORAL RANGE IN THE FAMILY

In general, the flowers are unisexual. But bisexual flowers exist in *Livistona*, *Trachycarpus*, etc. The female flowers are much larger than the male flowers, *e.g.*, *Borassus*. The male and female flowers may be present on the same plant (*Cocos*) or on different plants (*Phoenix*). The two types of flowers are intermixed in *Goenoma*, while male and female flowers are arranged in the upper and lower parts respectively.

The perianth segments are usually free, but sometimes united. The inner whorl of perianth segments are larger in size than the outer whorl, *e.g.*, *Areca catechu.*

The stamens are generally 6, but 3 stamens are noted in *Nypa*, where there is a suppression of stamens of the inner whorl. The stamens may also become numerous, as in *Caryota*, *Pinanga*, etc. They may be united at the base into a ring (*Roscheria*) or slightly cohering with the petals (*Oncosperma*).

Generally, there are 3 carpels in a palm flower. A solitary carpel develops in *Cocos nucifera* due to the abortion of the other two carpels. The ovary is usually trilocular, but unilocular in *Areca catechu.*

DIVISIONS OF THE FAMILY

The Palmae is divided into seven tribes:

Tribe (*i*) *Areceae.* Leaves pinnatisect. Flowers bisexual or in branched intrafoliar spadices. Fruit a berry or fibrous drupe. Examples: *Areca*, *Roystonea*, etc.

Tribe (*ii*) *Borasseae.* Leaves palmatisect. Flowers unisexual, males and females vary considerably in size. Examples: *Borassus*, *Hyphaene* and *Lodoicea.*

Tribe (*iii*) *Cocoeae.* Leaves pinnatisect. Flowers bisexual. Fruit a fibrous drupe, not covered with scales. Examples: *Cocos*, *Elaeis*, etc.

Tribe (*iv*) *Lepidocaryeae.* Leaves palmatisect or pinnatisect. Spadices terminal or interfoliar. Fruit covered with imbricate shiny scales. Examples: *Calamus*, *Metroxylon*, etc.

Tribe (*v*) *Phoeniceae.* Leaves pinnatisect. Flowers unisexual. Fruit a 1-seeded berry. Example: *Phoenix.*

Tribe (*vi*) *Phytelephanteae.* Leaves pinnatisect. Male and female inflorescences separate. Flowers naked or perianth small. Seeds with a hard endosperm. Examples: *Nypa*, *Phytelephas*, etc.

Tribe (*vi*) *Sabaleae.* Leaves palmatisect. Flowers unisexual or bisexual, polygamous. Examples: *Corypha*, *Licuala*, etc.

POSITION AND AFFINITY OF THE FAMILY

Based on the presence of sepaloid calycine perianth, the Palmae was considered under the fourth series Calycinae by Bentham-Hooker. Engler included the family in his fifth order Principles due to the possession of homochlamydeous trimerous unisexual flowers. Rendle placed the family, together with the Araceae and Lemnaceae, in the Spadiciflorae on account of the universal occurrence of spadix and usual unisexual flowers. Hutchinson treated it as comprising the Palmales, an order interpreted by him to originate from the liliaceous stock, perhaps through a part of his Agavales. Cronquist, Ehrendorfer and Takhtajan set up the subclass Arecidae to tackle the Arecales containing the Arecaceae (= Palmae). Thorne created the superorder Ariflorae to accommodate the Arecales which lies between the Arales and Cyclanthales. Dahlgren and Clifford constructed the Areciflorae for placing the Arecales, Cyclanthales and Pandanales.

The Palmae is closely related to the Liliaceae. The habit of the palms resemble those of *Cordyline*, *Dracaena* and *Yucca.* The structure of the pollen grains in the palms and *Dracaena* are essentially similar. The characters of the Palmae, such as perianth segments and stamens in two series as well as a tricarpellary superior ovary, recall those of the Liliaceae.

Wettstein suggested that the Palmae is related to the Pandanaceae on the one hand and to the Cyclanthaceae on the other. The family bears a relationship with the Araceae in having spadix, copious endosperm and minute embryo, but is set apart by the habit as well as features of the ovary, seed and fruit.

NUMBER IN AND DISTRIBUTION OF THE FAMILY

The Palmae consists of 212 genera and 2,780 species. The members of this family are distributed throughout the tropics of both hemispheres, particularly in the Indo-Malayan region and in the Guianas and Brazil. Their habitats range "from perhumid lowland rainforest to deserts and from mangrove swamps to high mountain thickets".

COMMONLY OCCURRING PLANTS OF THE FAMILY

Areca-nut or Betel-nut Palm (*Areca catechu* L.) is a graceful single-stemmed palm with a straight trunk and a tuft of feathery leaves at the top.

Fan-Palm or Palmyra-Palm (*Borassus flabellifer* L.) is a tall black palm, swollen above the middle and again contracted upwards.

Rattan-cane or Rotang-cane (*Calamus rotang* L.) and Cane Palm (*C. tenuis* Roxb.) are plants with climbing and slender stems.

Fish-tail Palm or Kittul Palm (*Caryota urens* L.) is a handsome palm with a cylindric and annulated trunk.

Coconut-palm or Porcupine-wood (*Cocos nucifera* L.) is a tall stately palm, marked with ring-like leaf scars.

Talipot-palm (*Corypha elata* Roxb., *C. umbraculifera* L.) and Royal Palm [*Roystonea regia* (HBK) O.F. Cook] are planted in gardens.

Licuala peltata Roxb. is a gregarious shrub.

Livistona chinensis R. Br. ex Mart. is a tall tree.

Nannorhops ritchieana H. Wendl. is a low gregarious palm.

Nypa-Palm or Water-coconut (*Nypa fruticans* Wurmb.) is a stemless palm of the Sunderbans.

Date-palm or Edible Date (*Phoenix dactylifera* L.) and Date Sugar or Wild Date-palm [*P. sylvestris* (L.) Roxb.] are tall palms with trunks rough from the persistent bases of leaf-stalks.

Phoenix paludosa Roxb. is a gregarious palm of the Sunderbans.

ECONOMIC ASPECTS OF THE FAMILY

"To residents of the tropics, this family is second only to the grasses in economic importance, providing food, shelter, clothing and other necessities" (Core, 1955).

The seeds of *Areca catechu* are used as a masticatory together with Betel leaf. *Arenga saccharifera* (Indomalaysia) yields jaggery and fibres. *Astrocaryum murumuru*, *A. tucuma*, *A. vulgare* and *Syagrus coronata* produce 'palm kernel oil'. *Attalea cohune* (Central & South America) forms Cohune nut which is a source of oil. *A. funifera* and *Leopoldinia piassaba* provide brush fibres like 'Bahia piassaba' and 'Para piassaba' respectively. *Borassus flabellifer* is tapped for the sap which is consumed as a beverage or fermented into a palm wine or made into malasses; the young seedlings and ripe fruits are edible, while the terminal bud of the tree is a culinary delicacy; the timber of the plant is valuable for making bows, joints, pillars, posts, rafters, etc. The stems of *Calamus rotang* furnish canes which are utilised for making baskets and chairs; the tender shoots and seeds of this plant are eaten. *C. scipionum* and *Eugeissona minor* furnish quality canes. *Caryota urens* gives us a kind of toddy; it provides commercial 'kittul' and 'salopa' fibres; the woolly substance present on the leaf-stalk is employed as a tinder and for caulking boats, the seeds are made into beads. *Ceroxylon andicola* (northern Andes) forms wax on the trunk which is scrapped off and used commercially. 'Crin végétal', a type of twisted and shredded fibre, is derived from the leaves of *Chamaerops humilis. Cocos nucifera* is "man's most useful tree"; food is secured from the 'meat' (endosperm) of the fruit and a cooling drink from the 'milk', while the juice tapped from the inflorescence spathe, forms toddy and produce a saccharine matter; the dried kernel is the 'copra' of commerce, yielding oil and the resulting cake 'poonac' is a cattle-feed; the fallen immature nuts are convetted into glazed boards and marketed under the trade name of 'coconite'; the husk of the fruit furnishes 'coir' which finds use in upholstery, saddles and brushes; the shell serves as a resonant backing for musical instruments and is also used as a hubble-bubble (hookah); the leaves are turned into ornamental combs, knitting needles, skewers and tooth-picks; the wood is suitable for ridgepoles, spear-handles, shingles, etc. *Copernicia cerifera* supplies the hardest highest melting natural wax which is good for making candles, gramophone records and so forth, The leaves of *Corypha umbraculifera* (Sri Lanka) are used as umbrellas. A red resin is extracted from the unripe fruits of *Daemonorops draco* (Indo-Malaysia). *Elaeis guineensis* (West Africa) produces a lubricating oil from the fruit. The species of *Mauritia* provide fruits, wines, wood, etc. *Metroxylon laeve* and *M. rumphii* (Thailand to New Guinea) form 'sago starch' which is obtained from the pith by crushing and washing. The leaves of *Nannorhops ritchieana* (Pakistan) are used for making baskets, brooms, fans, mats, sandals, etc. the leaves form fibres which are valuable for making cordage, ropes, paper and paste-board; the woolly tomentum, covering the petiole, is used as a tinder when mixed with salt petre or with the juice of Mulberry leaves; the young leaves, inflorescences and fruits are edible. The leaves of *Nypa fruticans* are used for umbrella-covering and as a thatching material. 'Babassu oil', a kind of palm oil, is derived from *Orbignya martiana* and *O. oleifera.* The fruits of *Phoenix dactylifera* form a staple diet in the countries of the Middle East; 'tara water' for making sherbets, is distilled from the fresh flower-heads, while a medicinal gum, 'kukm chil', is derived from the palm stem; a treacle, made out of the ripe fruits, is used for coating leather; the pinnate leaves are made into bags, baskets, fans, mats, thatch, etc. *P. sylvestris* is noted for its vitamin-rich juice which is consumed either as a beverage or converted into molasses. The hard endosperm of *Phytelephas* is used as a substitute for ivory in chessman, inlays, toilet articles, etc. and for making billiard balls. Strong fibres are extracted from *Raphia ruffia* (Malagasy) and *R. vinifera* (West Africa). *Roystonea oleracea* (Guyana) contains a young head of edible leaves. The leaves of *Sabal causiarum* is used for making the so called Puerto Rican hats. *S. palmetto* yields a valuable fibre termed 'palmetto'. The leaf-sheaths of *Trachycarpus excelsa* (China) are used for cordage, mats and the like.

Some members of the Palmae are planted along avenues or cultivated as ornamentals, *e.g.*, *Chamaedora elegans*, *Cocos nucifera*, *Howea belmoreana*, *Jubaea spectabilis*, *Livistona chinensis*, *Microcoelum weddellianum*, *Roystonea regia*, *Trachycarpus fortunei* and *Washingtonia filifera.* "The graceful trunks ending in a crown of large leaves are among the most characteristic features of tropical scenery" (Core 1955).

Order 5. *Synanthales*

This is a monotypic order with the characters of its one family, the Cyclanthaceae. The order is equivalent to the Cyclanthales of modern taxonomists.

The Cyclanthaceae (Panama hat-palm family) is distinguished by the herbs of palm-like habit, leaves deeply bilobed, flowers minute, unisexual, apetalous and densely crowded in a spadix, carpels united and many-ovuled and fruit a fleshy syncarp.

The family is divided into two subfamilies:

Subfamily I. *Carludovicoideae.* Leaves bifid, flabelliform or entire. Male and female flowers in spirally arranged groups. Fruiting spadix not screw-like. Example: *Carludovica*.

Subfamily II. *Cyclanthoideae.* Leaves deeply 2-partite, with forked costa. Male and female flowers in separate alternating whorls or sometimes part spirals. Fruiting spadix screw-like. Example: *Cyclanthus.*

The Cyclanthaceae indicates great advancement and specialisation in the floral structure; this is evident in the axillary spadix and much-reduced unisexual flowers. This family seems to stand as an intermediate between the palms and aroids in evolutionary development—a view advocated by Engler and Diels (1936) and supported by Harling (1946). Hutchinson (1959) regarded the taxon as a very advanced climax group, equivalent to the Araceae in so far as the line of development is concerned. "It seems likely that the similarity of the Cyclanthales to the palms in leaf ontogeny and structure reflects an independent realization of similar potentialities rather than direct inheritance from a common ancestor. Neither group could have given rise to the other" (Cronquist, 1968).

A family of 11 genera and 180 species, the Cyclanthaceae abounds in tropical America and West Indies.

The leaves of *Carludovica angustifolia* (Peru) are used for thatching native huts and those of *C. sarmentosa* (Guyana) for making brooms. The leaves of *C. insignis* (Eucador) are made into Panama hats.

Order 6. *Arales*

The order is characterised by the presence of a thickened spadix subtended by a single large herbaceous spathe.

The Arales includes two families which, according to Core (1955), can be identified as follows:

(I) Rhizomatous or tuberous herbs, sometimes woody — *Araceae*

(1) Small free-floating thalloid aquatic plants — *Lemnaceae*

Araceae
(Arum Family)

GENERAL FEATURES OF THE FAMILY

Habit	—	*Herbs with an acrid watery latex*, sometimes epiphytic with aerial roots or shrubby with corms, rhizomes and tubers.
Leaves	—	Radical or cauline, simple or compound, often *broad, long-petioled, net-veined.*
Inflorescence	—	*Spadix, subtended by a spathe.*
Flowers	—	Bisexual or unisexual, small, often with a foetid smell.
Perianth	—	Absent or present.
Androecium	—	Stamens 2, 4 or 8, opposite to perianth segment, free or united; anthers mostly 2-celled.
Gynoecium	—	Carpels 1–3; ovary sunk on peduncle of spadix, ovary *superior*, 1- to 3-locular; *ovules few to many in each loculus*; placentation parietal.
Fruit	—	Berry.
Seeds	—	*Albuminous*, embedded in a mucilaginous pulp or exalbuminous; embryo minute, straight.

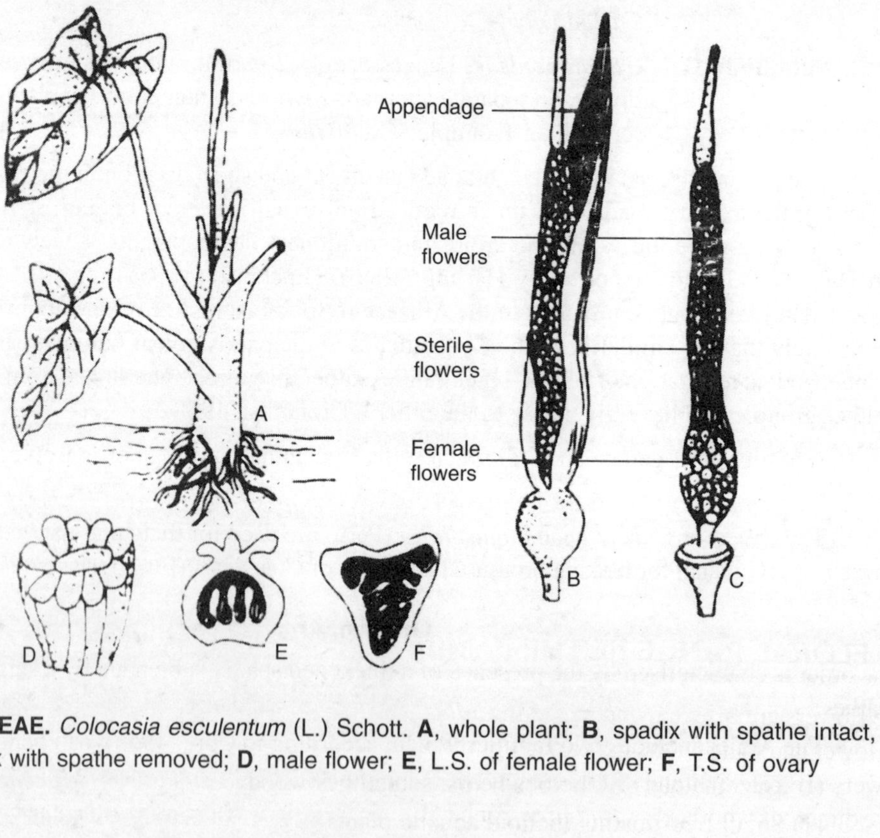

Fig. 91 ARACEAE. *Colocasia esculentum* (L.) Schott. **A**, whole plant; **B**, spadix with spathe intact, **C**, spadix with spathe removed; **D**, male flower; **E**, L.S. of female flower; **F**, T.S. of ovary

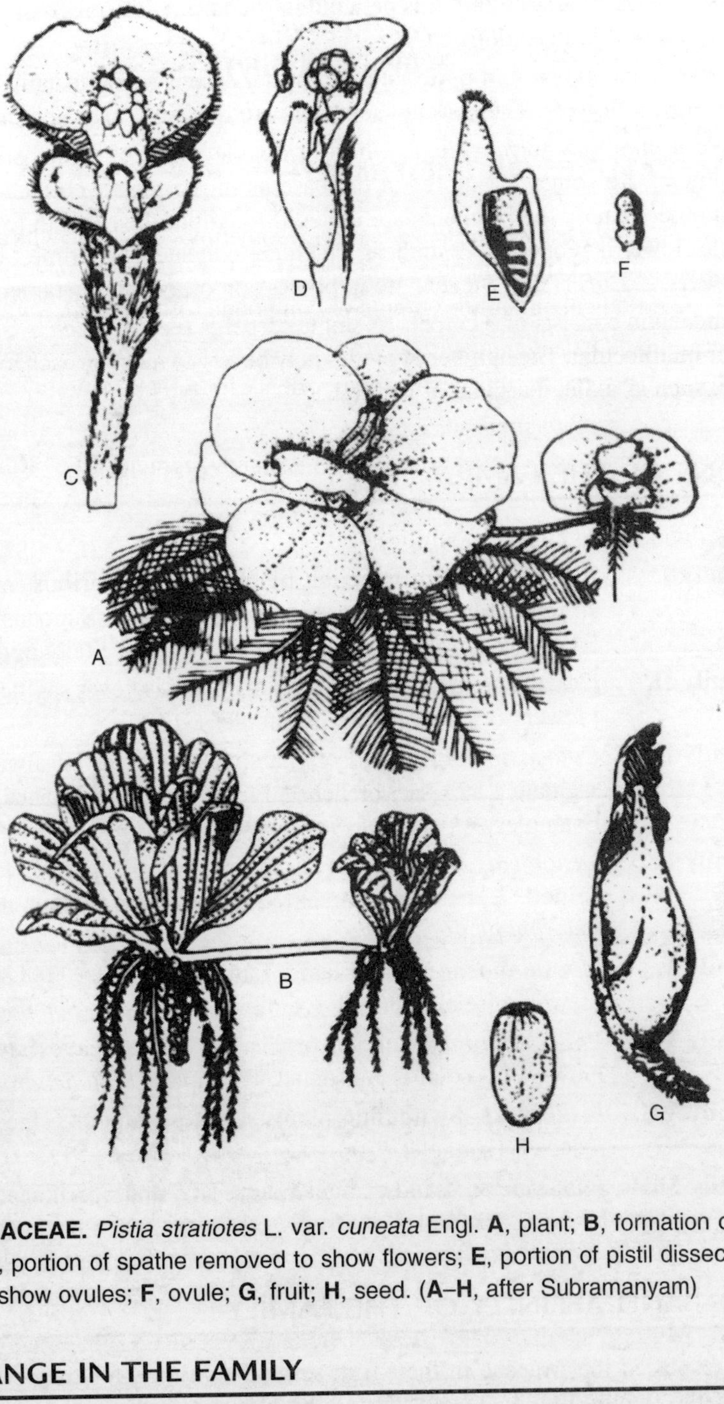

Fig. 92 ARACEAE. *Pistia stratiotes* L. var. *cuneata* Engl. **A**, plant; **B**, formation of offset; **C**, spadix; **D**, portion of spathe removed to show flowers; **E**, portion of pistil dissected out to show ovules; **F**, ovule; **G**, fruit; **H**, seed. (**A–H**, after Subramanyam)

FLORAL RANGE IN THE FAMILY

The flowers are bisexual and naked in *Calla.* They may be unisexual and naked with female flowers on the lower and male on the upper parts of the spadix, as in *Arum*, *Alocasia*, *Colocasia*, etc. In *Acorus*, the flowers are bisexual and characterised by a trimerous arrangement. On the contrary, the flowers may have dimerous whorls, *e.g.*, *Anthurium* or have dimerous whorls with a solitary carpel, *e.g.*, *Gymnostachys.*

The spathe is of various colour. It is petaloid in *Richardia.* It is green in *Alocasia*, *Colocasia* and related genera. It is scarlet in *Anthurium.*

The perianth may be present or absent; the former case is found usually in bisexual flowers and the latter in unisexual flowers. The perianth segments are free in *Acorus* and united in *Spathiphyllum.*

The androecium may form a single whorl or double whorl of stamens. When there is a single whorl, the filaments are somewhat united at the base as in *Arisaema*, *Dracunculus*, etc. In *Alocasia*, *Colocasia* and other genera, the filaments are united throughout their length to form a synandrium. In *Pistia*, the male flower is reduced to 2 anthers which are connate. Sometimes, there are staminodes in the female flowers. *Arisarum* and *Biarum* are in possession of a solitary stamen.

The gynoecium consists of 3 carpels. A single carpel is seen in *Arum*, *Pistia*, etc. The ovary may be unilocular or multilocular. The number of ovules may be few to many in each loculus. The placentation is also diverse, such as axile, marginal or parietal.

DIVISIONS OF THE FAMILY

The Araceae is divided into eight subfamilies:

Subfamily I. *Aroideae.* Land or marsh plants. Leaves various, net-veined. Latex-sacs straight. Flowers with or without perianth. Stamens free or in synandria. Examples: *Arum*, *Typhonium*, etc.

Subfamily II. *Colloideae.* Land or marsh plants. Leaves never sagittate. Latex-sacs present. Flowers bisexual, naked. Example: *Calla.*

Subfamily III. *Colocasioideae.* Land or marsh plants. Leaves always net-veined, never sagittate. Latex-sacs branched. Flowers unisexual, naked. Stamens in synandria. Examples: *Alocasia*, *Colocasia*, etc.

Subfamily IV. *Lasioideae.* Land or marsh plants. Leaves sagittate, often much-lobed, net-veined. Latex-sacs present. Flowers bisexual or unisexual. Examples: *Amorphophallus*, *Lasia*, etc.

Subfamily V. *Monsteroideae.* Land plants. Latex-sacs absent, but spicular cells present. Flowers bisexual, naked. Examples: *Monstera*, *Scindapsus*, etc.

Subfamily VI. *Philodendroideae.* Land or marsh plants. Leaves always parallel-veined. Flowers bisexual or unisexual. Examples: *Philodendron*, *Richardia*, etc.

Subfamily VII. *Pistioideae.* Swimming plants. Leaves parallel-veined. Flowers extremely reduced. Example: *Pistia.*

Subfamily VIII. *Pothoicieae.* Land plants. Latex-sacs and spicular cells absent. Flowers bisexual. Examples: *Acorus*, *Pothos*, etc.

POSITION AND AFFINITY OF THE FAMILY

Bentham-Hooker placed the Araceae in their fifth series Nudiflorae for the perianth being absent in many genera of the family. Engler treated the family, along with the Lemnaceae, as comprising his seventh order Spathiflorae on account of the universal presence of a spathe. Rendle included the family in his fifth order Spadiciflorae on the basis of the spadix inflorescence and unisexual flowers. In Hutchinson's arrangement, the family appeared under his 97th order Arales; the criteria for assigning such a position were the same as those used by Rendle. Both the Araceae and Lemnaceae have been considered under the superorder Arecanae and the order Arales by Takhtajan and Dahlgren.

The Araceae is closely related to the Palmae on such grounds as small flowers arranged in a spadix and subtended by a spathe as well as the relative size of the embryo and endosperm. It is also akin to the Lemnaceae which is a replica of the aroids, though an extremely reduced one.

The origin of the Araceae has been subject of much dispute. Lotsy suggested that the family, together with the Palmae and Pandanaceae, arose from the Piperales. Engler felt that the family was a derivative of the Palmae via he Cyclanthaceae. Wettstein regarded the family to be more advanced than the Orchidaceae and to be originating from the Helobiae-Liliiflorae stocks. Bessey opined that the Palmae and Araceae represented parallel or diverge development from liliaceous ancestors. Hutchinson expressed the idea that the family developed directly from the Liliaceae through the tribe Aspidistreae and parallel with the Palmae but along different lines.

NUMBER IN AND DISTRIBUTION OF THE FAMILY

The Araceae is composed of about 110 genera and 2,000 species. The members of this family are mostly abundant in tropical regions, but also found in temperate zones.

COMMONLY OCCURRING PLANTS OF THE FAMILY

Sweet Flag (*Acorus calamus* L.) is an erect aromatic marshy herb, having a thick creeping rootstock.

Alocasia indica (Roxb.) Schott. is a terrestrial herb with a suberect thick rhizome.

Teliga Potato [*Amorphophallus campanulatus* (Roxb.) Bl. ex Decne] is a terrestrial herb with a roundish warty thick corm.

Snake Plant [*Arisaema tortuosum* (Wall.) Schott.], seen in Darjeeling, is characterised by a greenish purple spathe which expands over the spadix like the hood of a snake; it is an instance of mimicry.

Caladium bicolor (Ait.) Vent is often planted in gardens.

Common Arum or Taro [*Colocasia esculentum* (L.) Schott.] is a terrestrial herb with a tuberous rhizome.

Cryptocoryne spiralis (Retz.) Fisch. ex Wyndler, an aquatic herb on the muddy banks of the Ganges, exhibits vivipary.

Gonatanthus pumilus (D. Don) Engl. & Krause is found in eastern Himalaya.

Lasia spinosa (L.) Thw. is a stout spinous herb in swampy areas.

Water-cabbage or Water-soldier (*Pistia stratiotes* L. var. *cuneata* Engl.) is a floating stoloniferous herb, bearing a rosette of sessile obcordate cuneate leaves and emitting a muriatic odour.

Pothos scandens L. is a gigantic root-climber.

Scindapsus officinalis Schott. is a garden climber.

Typhonium trilobatum (L.) Schott. is a tuberous herb of waste places, having a bright red appendage.

ECONOMIC ASPECTS OF THE FAMILY

The Araceae is of some importance in the economic sense. The leaves and rootstocks of *Acorus calamus* are utilised for making perfume and hair powder as well as for giving flavour to beer, gin, vinegar, snuff and other preparations; the rhizomes yield a drug. The underground stems of *Alocasia indica*, *A. macrorrhiza*, *Amorphophallus campanulatus*, *Colocasia esculentum*, *C. nymphaeifolia*, *Cyrtosperma chamissonis*, *Xanthosoma atrovirens*, *X. sagittifolium* and *X. violaceum* are edible. The tender leaves of *Lasia spinosa* are consumed as a food. *Monstera deliciosa* yields edible fruits. Some plants are cultivated as ornamentals, *e.g.*, *Aglaonema*, *Anthurium*, *Caladium*, *Dracunculus*, *Philodendron*, *Pothos*, *Zantedeschia*, etc.

Lemnaceae (Duckweed Family)

GENERAL FEATURES OF THE FAMILY

Habit	—	*Minute, free-floating fresh water plants without roots or with thread-like roots.*
Leaves	—	*Plant body reduced to a small oval or oblong, flat or globose thallus which acts as a leaf.*
Flowers	—	*Unisexual, naked or initially enclosed in a membranous sheathing spathe.*
Perianth	—	Absent.
Androecium	—	*Stamens 1–2* (in male flowers); anthers 1- to 2-celled.
Gynoecium	—	*Carpel 1* (in female flowers); ovary sessile, unilocular; ovules 1–7, basal, erect, anatropous or orthotropous; style short; stigma funnel-shaped.
Fruit	—	*Utricle.*

POSITION AND AFFINITY OF THE FAMILY

Enlger included the Lemnaceae in his seventh order Spathiflorae under the monocots. Bessey treated the family, along with the Araceae and Cyclanthaceae, within the Arales. Bentham-Hooker considered the genera of the Lemnaceae as members of the Aroideae in the series Nudiflorae. According to Cronquist (1968), the Lemnaceae "occupy a distinctive ecological niche which has scarcely been entered by other angiosperms. Their whole structure reflects their habitat, and they are so highly reduced and specialized, that they could easily be treated as a distinct order Lemnales".

The Lemnaceae is regarded by most botanists as a degenerate offshoot of the Araceae, the genus *Pistia* providing a link between the two families. In both *Pistia* and *Lemna*, the shoots undergo metamorphosis and reduction and the floral structures are similar (Arber, 1920). On the basis of anatomical, cytological and morphological studies, Brooks (1940) concluded that the Lemnaceae arose as a result of a repressive evolution in such sequence as *Pistia* → *Spirodela* → *Lemna* → *Wolffiela* → *Wolffia*. Lawalree (1945) stated that the Lemnaceae is not akin to the Araceae but is of helobial origin. His idea was based on the presence in *Spirodela* of *squamulae intravaginales*—an object of common occurrence in most members of the Helobiales. The embryological studies of Maheshwari (1956) indicate a close alliance between the Lemnaceae and Araceae. The common features include the successive division of microspore-mother cells, presence of a true plasmodium, three-celled pollen-grains, periclinal divisions in nucellar epidermis, nucellar cap, endosperm caecum at the chalazal end and operculum. However, the two families differ as regards the development of embryosac; in, the Lemnaceae, the embryosac is generally bisporic and in the Araceae it is monosporic. The recent work by McClure and Alston (1966) on the Lemnaceae represents an effort to supplement available taxonomic criteria by the use of comparative chemical data concerning the flavonoids. If *Spirodela* can be assumed as primitive in the family and. *Wolffia* the most advanced, then there is a general tendency for the flavonoid chemistry to be more simplified in the supposedly derived genera.

NUMBER IN AND DISTRIBUTION OF THE FAMILY

The Lemnaceae represents six genera and about 43 species of cosmopolitan distribution except the Arctic region.

COMMONLY OCCURRING PLANTS OF THE FAMILY

Lemna perpusilla Torr, *Spirodela polyrhiza* (L.) Schleid. and *Wolffia arrhiza* (L.) Wimm. are found in tanks, ponds, old wells and still waters as floating aquatic plants. The last-named species in the smallest known flowering plant.

ECONOMIC ASPECTS OF THE FAMILY

Although direct uses for the Lemnaceae are unknown, they constitute serious weeds of still water and are of importance as a source of food for fish and water fowl.

Order 7. *Graminales*

The order is distinguished by the prominent nodes, narrow leaves with a sheathing base and small flowers arranged in spikelets and borne in the axils of dry chaffy bracts.

The Gramineae is a very advanced taxon in which the apparently simple floral structure represents a drastic reduction from an unknown ancestor, possibly from a primitive liliaceous stock.

The usual two families of the Graminales may be characterised as follows:

(1) Culm mostly solid, triangular; leaf-sheath closed; fruit an achene — *Cyperaceae*

(1) Culm mostly hollow, cylindrical; leaf-sheath open; fruit a caryopsis — *Gramineae*

Cyperaceae
(Sedge Family)

GENERAL FEATURES OF THE FAMILY

Habit — Annual or perennial grass-like herbs with a sympodial rhizome and a *solid triangular culm.*

Leaves — Linear, *tristichous without a ligule*; *leaf-sheath closed.*

Inflorescence — Spike or head of spikelets or panicle.

Flowers — Bisexual or unisexual, small, *subtended by imbricating bracts.*

Perianth — *Absent or represented by 3 or 6 scales or bristles.*

Androecium — Stamens 3, in 1 whorl; *anthers basifixed.*

Gynoecium — Carpels 3, united; ovary superior, unilocular; ovules single, erect, anatropous; *stigma tripartite, papillose.*

Fruit — *Achene or nut.*

Seeds — *Albuminous*; cotyledon comes out of sheath during germination; embryo embedded in endosperm.

FLORAL RANGE IN THE FAMILY

The spikelets are 1-flowered in *Diplacrum* or many-flowered in *Schoenus.*

In general, the flowers are bisexual. But unisexual flowers exist *in Carex.* In *Elyna*, the secondary axis bears male flowers above the females. The secondary axis may be elongated and forms a bristle-like projection, as in *Uncinia.*

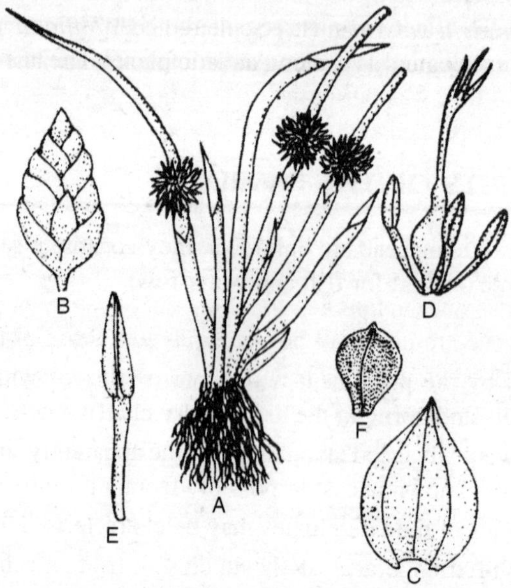

Fig. 93 CYPERACEAE. *Scirpus articulatus* L. **A**, portion of plant; **B**, spike; **C**, bract; **D**, androccium and gynoecium; **E**, stamen; **F**, fruit

The flowers are naked or the perianth is reduced to bristles in *Scirpus* or hairs in *Eriophorum. Oreobolus* has 6 perianth segments arranged in two whorls of 3 each. In *Rhynchospora* and other genera, the perianth segments are numerous.

The androecium generally contains 6 stamens, but sometimes 3 stamens are present. A single stamen is found in *Bisboeckelera* and *Hemicarpha.*

The gynoecium consists of 2 or 3 carpels, being noted in the species of *Carex.*

DIVISIONS OF THE FAMILY

The Cyperaceae is divided into three subfamilies:

Subfamily I. *Caricoideae.* Spikes many-flowered. Flowers unisexual, naked. Carpels borne in a sac-like structure called *perigynium.* Examples: *Carex*, *Kobresia*, etc.

Subfamily II. *Rhynchosporoideae.* Spikelets 1- to few-flowered, cymosely arranged. Flowers bisexual or unisexual. Examples: *Bisboeckelera*, *Rhynchospora*, etc.

Subfamily III. *Scirpoideae.* Spikelets many-flowered. Flowers bisexual. Examples: *Cyperus*, *Scirpus*, etc.

POSITION AND AFFINITY OF THE FAMILY

Bentham-Hooker placed the Cyperaceae (along with the Gramineae) in their seventh and last series of monocots, the Glumaceae, on account of flowers being subtended by glumaceous bracts. Using the same basis, Engler included the family in the Glumiflorae which also contained the Gramineae. Hutchinson segregated the families into two distinct orders, the Cyperales and Graminales of the division Glumiflorae and interpreted them as being "derived from liliaceous ancestors via the Juncaceae complex". Hutchinson's circumscription of the two orders was followed by Stebbins as well as Dahlgren and Clifford. Takhtajan and Dahlgren treated the family under the last order Poales of the Commelinidae and Commelinanae respectively. But Cronquist considered the cyperaceae and Gramineae under the Cyperales lying between the Juncales and Typhales. According to Thorne, the two families are components of the Commelinales which in turn was put in the Commeliniflorae. Ehrendorfer created the Juncanae in order to accommodate the Cyperales and the Commelinanae for the Poales.

The Cyperaceae resembles the Gramineae in so far as the habit is concerned. But the features of flower, fruit and seed as well as the method of germination sharply delimit the two families. That is why there is a growing tendency not to regard the Gramineae as a close ally of the Cyperaceae. Recent studies have indicated that phylogenetic significance should not be attached to the superficial grass-like habit, since it has made its appearance in families which are not related to the Cyperaceae and Gramineae. The spikelets of the Cyperaceae are not at all homologous with those of the Gramineae, as they vary widely in arrangement in the former and have remained reasonably constant in the latter. The florets of the Cyperaceae are axillary in position, while in the Gramineae (excluding the bracts) they are borne terminally. A definite perianth can be somewhat demonstrated in the Cyperaceae, but the fruits with two or three carpels are less specialised than the caryopsis of the Gramineae. The basic placental condition of the Cyperaceae is a derivative of the ancestral free-central type, whereas it has arisen from a parietal type in the Gramineae. The ovules are chiefly anatropous in the Cyperaceae and chiefly orthotropous in the Gramineae. The embryo of the Cyperaceae is embedded at the base of the endosperm and not outside of it, as in the Gramineae. In the Cyperaceae, the terminal portion of the cotyledon comes out of the seed and the other portion remains in the seed to function as a sucker; in the Gramineae, neither the cotyledon nor any part comes out of the seed.

In view of the above considerations, it has been thought that the two families belong to two separate orders and that they do not indicate bond of phyletic relationship (Hutchinson, 1959). However, Cronquist (1968) believed that "it is most useful to group the Gramineae and Cyperaceae in a single order Cyperales". It is quite possible, according to Blaser (1940), that the Cyperaceae arose from "small-flowered, few-seeded, hypogynous Liliales with axile placentation ... the transition having occurred through loss of septa and placental axis and may have occurred through some line not now recognisable".

NUMBER IN AND DISTRIBUTION OF THE FAMILY

The Cyperaceae is a very large family which is made up of approximately 90 genera and 4,000 species. The members of this family are chiefly subarctic and temperate in distribution, although some are found in the tropics. They are seen in relatively wet situations, such as marshes, bogs and swamps.

COMMONLY OCCURRING PLANTS OF THE FAMILY

Bulbostylis barbata (Rottb.) Clarke is an annual herb.

Slough Grass (*Carex indica* L.) is seen in sandy areas.

Cyperus brevifolius (Rottb.) Hassk. is a weed, bearing white globose heads.

Sedge (*Cyperus rotundus* L.) is a weed of cultivated lands.

Fimbristylis complanata Link. is a weed, having a glabrous stem flattened under the umbel.

Fimbristylis junciformis Kunth. inhibits wet places.

Scirpus articulatus L. is a glabrous herb with flowering glumes spirally arranged.

ECONOMIC ASPECTS OF THE FAMILY

Economically, the Cyperaceae is of considerable importance. *Carex atheroides* (USA) is utilised as a haygrass. The leaves and stems of *C. brizoides* (Central Europe) are used as a packing material. The leaves of *C. dispalatha* (Japan) are used to make hats. *C. paniculata* and *C. riparia* are employed in stables instead of straw. The stems of *Cladium effusum* (tropical America) constitute a source of cheap paper. *C. mariscus* (North Africa) are used for thatching houses. The stems of *Cyperus articulatus* and *C. longus* bear sweet-scented rhizomes and roots which can be use in perfumery. The tubers of *C. esculentus* (Europe), *Eleocharis aulcis*, *E. tuberosa* (Japan) and *Scirpus grossus* are edible. The pith of Papyrus Plant (*Cyperus papyrus*) is the classical material used by the Egyptians for making paper. The split stems of *C. malacopsis*, *C. tegetifornis*, *Lepironia articulata* and *Juncellus inundatus* are woven into mats. *C. rotundus* is a remedy for dysentery, whereas *Mariscus sieberianus* (Sumatra) is used as a vermifuge and *Scirpus articulatus* as a purgative. The stems of *Eleocharis austro-caledonica* are used for basket work, those of *Lepironia muoronata* for packing, *Scirpus lacustris* for chairseats and *S. totara* for canoes and rafts.

Gramineae
(Grass Family)

GENERAL FEATURES OF THE FAMILY

Habit — Annual or perennial herbs, rarely shrubs or trees, often with a *hollow cylindrical culm.*

Leaves — Alternate, *distichous with a ligule*; *leaf-sheath split open.*

Inflorescence — Spike or panicle of spikelets.

Flowers — Bisexual or unisexual, grouped in small clusters called *spikelets* which may consist of 1 to a dozen tiny florets; each spikelet enclosed at base 2 bracts called *empty glumes*, one placed a little above another; these glumes are followed by 1 or more (*flowering glumes*) which enclose small flowers or florets arising from a short stem-axis or *rachilla*; flowering glume may bear an awn which may be attached to back (*dorsal awn*) or apex (*terminal awn*); each floret has at its base 2 bracts named *lemma* (outer) and *palea* (inner).

Perianth — *Represented by 2 or 3 minute scales* (*lodicules*), placed within lemma and palea.

Androecium — Stamens 3 or 6; *anthers versatile.*

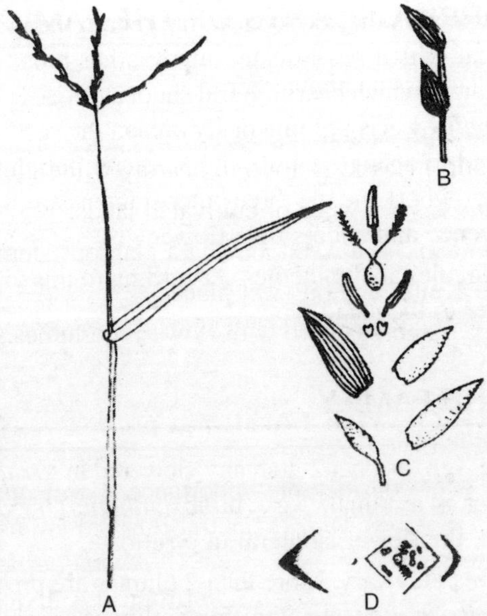

Fig. 94 GRAMINEAE. *Digitaria sanguinalis* (L.) Scop. **A**, portion of plant; **B**, spikelets; **C**, dissected spikelet; **D**, floral diagram..

Gynoecium	—	Carpel 1; ovary superior, unilocular; ovules single, erect, anatropous; *stigmas* 2, *lateral and feathery.*
Fruit	—	*Caryopsis.*
Seeds	—	Albuminous; cotyledon does not come out of sheath during germination; embryo outside endosperm.

MORPHOLOGY OF THE GRASS SPIKELET

The morphology of the grass spikelet is a controversial topic. The spikelet is considered as a basic floral unit, *i.e.*, a single flower and not an inflorescence. Such a flower agrees closely with the liliaceous type, showing a modification in the scarious perianth and suppression in the number of floral whorls. The resemblance can be traced to the flower of the Juncaceae in which the liliaceous type is apparent in the membranous perianth. The outer perianth or sepaloid whorl is represented by the lower and upper palea, the latter of which signify two sepals that have become connate. This is supported by the fact that the upper palea contains two main midveins (vascular traces) and is often incised to various degrees. In a species of *Bromus*, this incision extends to the base with the result that there is a whorl of three perianth segments. The midveins in the glumes and palea (perianth segments) situated as they are at an angular distance of 120° lend support to this view. In proper alteration with these midribs occur the corolla, represented by the lodicules. If one is suppressed, the two lodicules occupy a space vacant between the two midribs of the upper palea. The androecium generally shows the suppression of the inner whorl, leaving three stamens which alternate with the lodicules. Suppression is evident in the pistil, but the original three carpels are often suggested by three stigmas. This means that the grass flower is trimerous. It is built on the same plan as a flower of the Liliaceae, proving the origin of the Gramineae from the liliiflorean stock.

Other botanists do not agree with the above view, but regard the grass flower to have undergone reduction in a way similar to that which has affected many other monocots, especially the Araceae. Accordingly, the outer palea is a bract which bears a floral shoot at its axil. The upper palea is a bracteole which is inserted a little above the flowering glume or lemma on the rachilla. The lodicules represent a rudimentary perianth now reduced to a single whorl, the posterior perianth member being suppressed. The perianth members of the outer whorl and the three outer stamens are wholly suppressed. Hence, the grass spikelet is a short inflorescence and not a single flower.

A third view is that the lodicules are bracteoles. According to this view, the flower is perianthless and its relationship is more nearer to the naiad type.

FLORAL RANGE IN THE FAMILY

The spikelet may be 1-flowered in *Agrostis*, but 2- to many-flowered in *Arrhenatherum*, *Bromus*, *Festuca*, etc. In many spikelets, the flower is terminal, *e.g.*, *Anthoxanthum*, *Oryza*, etc. In many others, as in *Apera*, *Deyeuxia* and *Gastridium*, the flower is lateral in position.

The number of basal glumes may vary. More than 2 glumes are present in *Anthoxanthum.* There is 1 barren glume in *Reimaria.* In *Briza*, *Poa*, etc., the empty glumes are similar to the flowering glumes or are different in form in *Avena* where they are much larger enclosing the rest of the spikelets. The empty glumes are less than half the length of the spikelets in *Oryza.* In *Stipa* and other genera, there is a third posterior lodicule. A solitary lodicule is found in *Melica* and several lodicules in *Ochlandra.*

Variations occur in the number of stamens. In *Uniola* and *Vulpia*, the stamen is 1. There are 2 stamens in some species of *Sporobolus*, 3 in most grasses, 6 in *Leersia*, *Leptaspis*, etc. and 10–40 in *Pariana.* In *Microlaena* and *Tetrarrhena*, there are two alternating dimerous whorls of stamens. The stamens are generally free, but monadelphous in *Neohouzeaua.* Each stamens consists of a long slender filament which carries a 2-celled anther. The anther cells are minutely bearded at the apex in *Duthiea* and *Pennisetum*, open by pores in *Bambusa* or open by longitudinal slits in most grasses.

The number of styles is generally 2, but, 1 in *Duthiea*, *Leptaspis* and *Nardus* or 3 in *Pseudodanthonia.* In *Urochondra*, the style-bases are united into a beak which is persistent and fixed to the pericarp.

DIVISIONS OF THE FAMILY

The Gramineae is divided into two subfamilies, one with ten tribes and the other with three tribes:

Subfamily I. *Festucoideae.* Spikelets 1- to several-flowered; sterile florets, if any, usually above fertile florets in spikelet. Spikelets laterally compressed; glumes and lemmas generally folded lengthwise or sometimes spikelets torete. Articulation usually above glumes and between florets; glumes remain on plant when seeds drop.

Tribe (*i*) *Agrostideae.* Plants herbaceous. Inflorescence an open or narrow panicle or sometimes a spike-like raceme. Spikelets strictly 1-flowered; lemmas awnless or awned from tip or back. Articulation usually above glumes which are not reduced in size. Examples: *Agrostis*, *Stipa*, etc.

Tribe (*ii*) *Aveneae.* Plants herbaceous. Inflorescence a panicle, sometimes narrow and spike-like. Spikelets 2-to several-flowered, with glumes longer than first lemma. Example: *Avena.*

Tribe (*iii*) *Bambuseae.* Plants woody, mostly large trees or shrubs. Examples: *Arundinaria*, *Bambusa*, etc.

Tribe (*iv*) *Chlorideae.* Plants herbaceous. Inflorescence of 1 or more spikes which are one-sided. Lowest floret in spikelet fertile, upper ones being sterile and much reduced. Example: *Bouteloua.*

Tribe (*v*) *Festuceae.* Plants herbaceous. Spikelets 2- to several-flowered, in open, narrow or spike-like panicles. Lemmas awnless or awned; glumes shorter than first lemma Examples: *Festuca*, *Poa*, etc.

Tribe (*vi*) *Hordeae.* Plants herbaceous. Inflorescence of 1 or more spikes which are symmetrical. Spikelets 1- to several-flowered. Articulation above glumes or rachis sometimes disarticulating. Examples: *Hordeum*, *Secale*, *Triticum*, etc.

Tribe (*vii*) *Oryzeae.* Plants herbaceous. Inflorescence a panicle Spikelets 1-flowered. Articulation below glumes which are reduced or lacking. Example: *Oryza.*

Tribe (*viii*) *Phalarideae.* Plants herbaceous. Inflorescence an open or narrow panicle, sometimes spike-like, but never a true spike. Spikelets with 1 perfect floret and 2 sterile florets below it. Example: *Phalaris.*

Tribe (*ix*) *Zizanieae.* Plants aquatic, mostly rather tall. Male and female spikelets on same plant, but in different parts of panicle. Spikelets 1-flowered, falling entire, terete. Example: *Zizania.*

Tribe (*x*) *Zoysieae.* Plants herbaceous. Inflorescence a symmetrical spike. Spikelets sessile or sub-sessile. Articulation below glumes, rachis not disarticulating. Example: *Hilaria.*

Subfamily II. *Panicoideae.* Spikelets with 1 perfect terminal floret and a sterile floret below it; sterile floret usually represented by empty glume. Spikelets dorsally compressed; glumes and lemmas flat or sometimes spikelets terete. Articulation below glumes; whole spikelet falls when seeds drop.

Tribe (*i*) *Andropogoneae.* Spikelets in pairs, 1 sessile and fertile, other pedicellate and sterile. Glumes indurate. Examples: *Andropogon*, *Saccharum*, etc.

Tribe (*ii*) *Paniceae.* Spikelets not paired, with 1 perfect terminal floret and 1 sterile floret below it. Glumes herbaceous or membranaceous, not indurate. Examples: *Cenchru*, *Digitaria*, *Panicum*, etc.

Tribe (*iii*) *Tripsaceae.* Spikelets unisexual, male and female spikelets in separate inflorescences or in different parts of same inflorescence. Examples: *Coix*, *Zea*, etc.

POSITION AND AFFINITY OF THE FAMILY

Bentham-Hooker included the Gramineae in the seventh series Glumaceae (which also contained the Eriocaulaceae, Centrolepideae, Restionaceae and Cyperaceae). Rendle placed the family after the Triuridales due to the wind-pollinated perianthless flowers. Hutchinson went to the extent of creating a new order (Graminales) and treating the family under it.

Cronquist (1968) wrote: "Most recent authors have considered the Gramineae to be related to the Restionales or to a broadly defined order Commelinales in which the Restionales are submerged. Takhtajan has emphasized the significance of *Streptochaeta* as a primitive grass which suggests a close relationship to the Flagellariaceae, and this view has also been adopted by Butzin".

"The Cyperaceae and Gramineae are both large and successful families with reduced flowers adapted to wind pollination. The Gramineae are notably the larger and more successful of the two, in terms of both the number of species and the number of individuals. If the formal morphological characters usually cited as distinguishing the two families have any great adaptive significance, the fact remains to be established. One may reasonably speculate instead that it is the intercalary meristem of the grass leaf which plays the major role in the greater success of the grasses. The leaves of monocots in general usually mature from the tip downwards; in the grasses the base of the leaf blade remains more or less permanently immature and meristematic. Some of the Cyperaceae, notably species of *Carex*, share this same feature (teste Koyama), but in general it is much less developed in the Cyperaceae than in the Gramineae" (Cronquist, 1968).

Of late, there has been a trend to seek the origin of the Gramineae from other families of the Glumiflorae where the floral envelope is scale-like. Hutchinson suggested that the grasses developed parallel with the sedges from liliaceous stocks through the Juncaceae complex. Ziegenspeck (1938) also derived the Graminales from the Liliales, but the common stock underwent division into two branches, one of which gave rise to the Commelinaceae and the order to the Juncaceae and Cyperaceae. "After Commelinaceae has branched off, the ancestral stock of the Gramineae, the Urgramineae is left. Reduction and specialisation are then supposed to go to work to produce the family as we know it" (Bor, 1960).

NUMBER IN AND DISTRIBUTION OF THE FAMILY

The Gramineae contains about 650 genera and 9,000 species. The members of this family are seen in every vegetation type, dominating areas where the rainfall is neither high to support trees nor so low that only arid plants may grow. These areas include the Pampas, Prairie, Steppe, Veldt or other types of grassland communities in every continent.

COMMONLY OCCURRING PLANTS OF THE FAMILY

Love Grass [*Chrysopogon aciculatus* (Retz.) Trin.], Job's-tear (*Coix lacryma-jobi* L.), Dub Grass [*Cynodon dactylon* (L.) Pers.], *Dactyloctenium aegyptium* (L.) Beauv., *Desmostachya bipinnata* (L.) Stapf *Eleusine indica* (L.) Gaertn., *Eriochloa nubica* (Steud.) Hack. & Stapf, ex Thell., *Eulaliopsis binata* (Retz.) C.E. Hubb., *Imperata cylindrica* (L.) Beauv. var. *major* (Nees) C.E. Hubb. ex Hubb. & Vaughan, *Panicum frumentaceum* Roxb., *Pennisetum typhoides* (Burm. f.) Stapf & C.E. Hubb., *Phragmites karka* Trin. ex Steud., *Sclerostachya fusca* (Roxb.) A. Camus and *Vetiveria zizanioides* (L.) Nash are different types of grass.

Dendrocalamus strictus (Roxb.) Nees is a deciduous arborescent bamboo.

Digitaria sanguinalis (L.) Scop and *D. tenuiflora* Beauv. are found everywhere.

Duthiea bromoides Hack. occurs in rock clefts at arid regions.

Hygroryza aristata (Retz.) Nees ex Wt. & Arn. is seen floating on ponds.

Leptochloa panicea (Retz.) Ohwi, *Panicum flavescens* Swartz and *Setaria glauca* (L.) Beauv. are characterised by the hygroscopic movement of their awns.

Spinifex littoreus (Burm. f.) Merr. is a sand-binder.

Thysanolaena maxima (Roxb.) O. Kuntze is a tall handsome grass.

ECONOMIC ASPECTS OF THE FAMILY

"Not only man, but mammals in general, have been associated with the grass family for far more years than those of recorded history. It is probably not an overstatement to say that without the grass family and man's ability to learn to exploit it, civilization as we know it, could not have developed. The three great civilizations domesticated different grasses, the wheats, rye, barley and oats by the Mediterranean people; maize by the Aztecs, Mayans, and Incas and other natives of the Western Hemisphere; and rice and millet by the Chinese. Grasses are grown for human and animal consumption, for ornament and for use in the arts and industry" (Rabbins *et al.*, 1957).

In an economic sense, the Gramineae is the most important family of the monocots and perhaps of all the angiosperms. According to their utility, grasses may be conveniently divided into six groups:

1. *Arundo donax* is used for making walking sticks, musical pipes, fishing rods, etc. Many species of *Bambusa* (Orient) as well as *Dendrocalamus strictus*, *Guadua angustifolia* (tropical America) and *Gynerium sagittatum* (South America) are utilised for construction purposes. *Imperata cylindrica*, *I. exaltata* and *Saccharum spontaneum* are used for thatching and *Lygeum spartum* (Mediterranean) for basketary.

   The stony fruits of *Chionachne koenigii* are employed as rosary beads. *Ampelodesma tenax*, *Eulaliopsis binata* and *Stipa tenacissima* find use in the manufacture of paper. The fibres obtained from *Saccharum munja*, *S. spontaneum* and *Themeda arundinacea* are good for cordage and ropes. The split stems of *Phragmites karka* are woven into mats.
2. The cereals include Oat (*Avena sativa*), Japanese Millet (*Echinocloa frumentacea*), Ragi (*Eleusine coracana*), Finger Millet (*E. crusgalli*), Taff (*Eragrostis abyssinica*), Barley (*Hordeum vulgare*), Rice (*Oryza sativa*), True Millet (*Panicum miliaceum*), Little Millet (*P. miliare*), Kodo Millet (*Paspalum scorbiculatum*), Pearl Millet (*Pennisetum glaucum*), Rye (*Secale cereale*), Italian Millet (*Setaria italica*), Sorghum (*Sorghum vulgare*), Wheat (*Triticum aestivum*), *T. durum*, *T. monococcum*, *T. polonicum*, Maize (*Zea mays*), etc.
3. The sugar-yielding grass is Sugarcane (*Saccharum officinarum*). The canes are cut and crushed in order to extract the juice. Then the juice is boiled down to yield molasses and sugar; the distilled juice is rum.
4. A number of aromatic grasses finds application in perfumery. Lemon-grass (*Cymbopogon citratus*) yields 'lemon-grass oil', Ginger-grass (*C. martini*) 'ginger-grass oil', *C. nardus* 'oil of citronella' and Khuskhus (*Vetiveria zizanioides*) 'oil of vetiver'.
5. The sod-forming grasses are used in lawns and meadows, *e.g.* *Anthoxanthum odoratum*, *Cortaderia argentea*, *Hierochloe odorata*, *Lagurus ovatus*, etc.

   Ammophila arenaria is used for binding and consolidating drifting sand.
6. Grasses supply forage for animals, such as *Agrostis alba*, *Bromus inermis*, *Cynodon dactylon*, *Danthonia racemosa*, *Panicum maximum*, *Pennisetum purpureum*, *Trisetum flavescens*, etc.

Order 8. *Farinales*

The order is known by the flowers regular or nearly so, ovary superior and endosperm mealy.

The Farinales is not a homogeneous taxon. Engler considered thirteen families, belonging to six suborders. His basis in creating the order solely on the endosperm character was rejected by Bessey, Wettstein, Hutchinson and others. The families under the order have been distributed by them elsewhere among the monocots and there is no uniformity in their treatments. Within the superorder Commeliniflorae, Dahlgren and Clifford (1982) recognised eight orders: Commelinales, Eriocaulales, Typhales, Juncales, Cyperales, Hydatellales, Restionales and Poales.

The present author follows the plan of Core (1955) and considers five families in the Farinales:

(1) Perianth various, generally sepaloid	(2)
(1) Perianth not various, generally petaloid	*Pontederiaceae*
(2) Plant usually with a grass-like habit	(3)
(2) Plants not usually grass-like	(4)
(3) Flowers inconspicuous	*Eriocaulaceae*
(3) Flowers conspicuous	*Xyridaceae*
(4) Plants not epiphytic	*Commelinaceae*
(4) Plants usually epiphytic	*Bromeliaceae*

Eriocaulaceae
(Pipewort Family)

In the Eriocaulaceae, the plants are small acaulescent herbs with a dense tuft of basal, linear and grass-like leaves and 1 or more scapes, bearing involucrate heads of inconspicuous flowers.

The family is divided into two subfamilies:

Subfamily I. *Eriocauloideae.* Stamens twice as many as petals. Example: *Eriocaulon.*

Subfamily II. *Paepalanthoideae.* Stamens as many as petals. Example: *Paepalanthus.*

Hutchinson derived the Eriocaulaceae from the Commelinaceae by way of reduction in the floral structure. He included the family in the Eriocaulales—an order related to but advanced over his Xyridales. Bentham-Hooker placed it under the seventh series Glumaceae. The Eriocaulaceae is close to the Xyridaceae in having a cluster of basal leaves and a terminal racemose head on a long scape or peduncle. This has led Cronquist to question "why the Eriocaulaceae might not have been derived directly from the Xyridaceae or from similar common ancestors". The Eriocaulaceae is called the Compositae of the monocot group on account of the characteristic inflorescence.

The Eriocaulaceae comprises 13 genera and about 1,200 species. The members of this family occur chiefly in bogs and along wet shores of the tropics, but a few species extend into the temperature regions.

Eriocaulon capillusnaiadis Hook. f., *E. cristatum* Mart., *E. luzulaefolium* Mart., *E. quinquangulare* L. and *E. trilobum* Buch.-Ham-ex Wall. are marshy scapigerous herbs.

Economically, the Eriocaulaceae is of slight importance *Eriocaulon setaceum* is used in the treatment of skin diseases. Some species of *Eriocaulon* prevail as weeds in rice fields. The inflorescences of *Syngonanthus* are marketed as 'everlasting flowers'.

Xyridaceae
(Yellow-eyed Grass Family)

The family is recognised by the tufted perennial marsh plants, showy irregular flowers in bracteate heads, biseriate perianth, hairy staminodes and three-valved loculicidal capsule.

The Xyridaceae covers two genera (*Abolboda*, *Xyris*) and about 200 species of the warmer regions and saline marshes.

Xyris indica L. and *X. pauciflora* Willd. are tufted aquatic herbs and are seen in Bangladesh, West Bengal and other parts of India.

Fig. 95 XYRIDACEAE. *Xyris indica* L. **A**, V.S. of flower; **B**, corolla spread open; **C**, pistil; **D**, T.S. of ovary; **E**, fruit; **F**, seed. (**A–F**, after Hutchinson)

A few species of *Xyris* are utilised as aquarium plants. The roots and leaves of *X. ambigua* and *X. caroliniana*, both of North America, constitute domestic remedies in treating colds and skin diseases respectively.

Commelinaceae (Spiderwort Family)

GENERAL FEATURES OF THE FAMILY

Habit	— *Terrestrial herbs with a jointed stem.*
Leaves	— *Alternate*, entire, *sheathing.*
Inflorescence	— Monochasial cyme, generally a cincinnus.
Flowers	— Bisexual or polygamous, somewhat irregular, hypogynous, often enclosed in a spsthe.
Perianth	— *Segments 6, in 2 series, outer sepaloid and inner petaloid.*
Androecium	— *Stamens 6–8, all perfect or some reduced to staminodes*; filaments often bearded with moniliform hairs.
Gynoecium	— Carpels 3, united; *ovary superior, trilocular*; ovules solitary or few; *placentation axile.*
Fruit	— Capsule or nut.
Seeds	— Angled, albuminous; embryo small, located beneath a disc-like callosity (*embryostega*) on seedcoat.

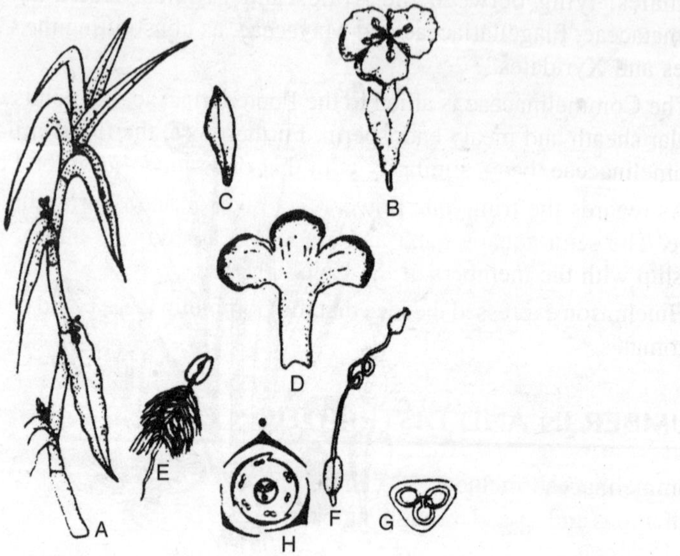

Fig. 96 COMMELINACEAE. *Amischophacelus axillaris* (L.) Rao et Kamathy.
A, portion of plant; **B**, flower; **C**, sepaloid perianth segment; **D**, petaloid perianth-segment; **E**, stamen; **F**, carpel; **G**, T.S. of ovary; **H**, floral diagram

FLORAL RANGE IN THE FAMILY

In many genera, 1 of the three petals is reduced to a scaly structure. All the six stamens are fertile in *Cyanotis*, *Pollia*, *Rhoeo*, etc. or 3–4 may be sterile and the rest fertile or 2–3 may be sterile and the rest fertile. *Callisia* has 1 functional stamen and *Aneilema lineolatum* has 2 such stamens. Out of the three loculi, 1 may be abortive (*Floscopa*).

DIVISIONS OF THE FAMILY

The Commelinaceae is divided into two subfamilies, each with two tribes:

Subfamily I. *Commelinoideae.* Guard cells mostly 4. Flowers zygomorphic.

Tribe (*i*) *Declinateae.* Flowers buds bent downwards and away from inflorescence axis. Examples: *Commelina*, *Floscopa*, etc.

Tribe (*ii*) *Inclinateae.* Flowers buds bent upwards and towards inflorescence axis. Examples: *Cochliostema*, *Geogenanthus*, etc.

Subfamily II. *Tradescantioideae.* Guard cells mostly 2. Flowers actinomorphic.

Tribe (*i*) *Hexandreae.* All 6 stamens fertile. Examples: *Pollia*, *Zebrina*, etc.

Tribe (*ii*) *Triandreae*. Only 3 stamens fertile, the rest being sterile or absent. Examples: *Murdannia*, *Palisota*, etc.

POSITION AND AFFINITY OF THE FAMILY

In the arrangement of Bentham-Hooker, the Commelinaceae appeared under the series Coronarieae. Engler, Rendle and Lawrence included it in the Farinosae just after the Spathiflorae. Core placed it in the Farinales, lying between the Arales and Liliales. Hutchinson treated it, together with the Cartonemataceae, Flagellariaceae and Mayaceae, as constituting the Commelinales and connecting the Najadales and Xyridales.

The Commelinaceae is allied to the Pontederiaceae, the points of resemblance being the ochrea-like ligular sheath and mealy endosperm. Furthermore, the floral structure of *Spironema*, a member of the Commelinaceae, bears similarity with that of *Pontederia* of the Pontederiaceae.

As regards the trimerous flowers and nuclear endosperm, the Commelinaceae is related to the Liliaceae. The semi-aquatic habit and amoeboid tapetum of anthers of the Commelinaceae suggest a relationship with the members of the Helobiales.

Hutchinson expressed the idea that the Commelinaceae could be derived from both the Alismatales and Butomales.

NUMBER IN AND DISTRIBUTION OF THE FAMILY

The Commelinaceae includes 38 genera and about 600 species. The members of this family prefer damp situations and spread throughout the tropical and subtropical regions, but are absent from temperate Asia and Europe.

COMMONLY OCCURRING PLANTS OF THE FAMILY

Amischophacelus axillaris (L.) Rao et Kamathy is a semi-aquatic herb with linear leaves.

Commelina appendiculata Clarke, *C. diffusa* Burm. f. and *C. obliqua* Buch.-Ham. ex D. Don are weeds of ditches and other moist places.

Floscopa scandens Lour. bears lilac or pink flowers.

Murdannia nudiflora (L.) Brenan and *M. vaginatum* (L.) Bruckner are field-weeds.

Rhoeo spathacea (Sw.) Stearn, characterised by a slow circulation of protoplasm in the cells of the staminal hairs, is a garden herb.

ECONOMIC ASPECTS OF THE FAMILY

An extract of leaves and stems of *Aneilema beninense* (tropical Africa) finds application as a laxative. The leafsap of *Floscopa scandens* (tropical Asia) is used to treat inflammation of eyes. The species of *Commelina*, *Cyanotis* and *Floscopa* are cultivated for the flowers and *Zebrina pendula* for the striped leaves. *Commelina bengalensis* yields edible leaves, the young shoots and leaves of *Tradescantia virginiana* can be used in salads.

Bromeliaceae
(Pineapple Family)

GENERAL FEATURES OF THE FAMILY

Habit — *Acaulescent epiphytes* or eminently xerophytes.

Leaves — Radical, rosette-forming, linear, spinulose, fleshy, *coloured towards base, covered with peltate or stellate scales or hairs.*

Inflorescence — Panicle, raceme, spike or head, *often with brightly coloured bracts.*

Flowers — Bisexual, regular, *trimerous.*

Perianth — *Biseriate, outer series calyx-like, and inner corolla-like,* usually with appendages within, forming a corona.

Androecium — Stamens 6, free or adnate to perianth segments; *anthers 2-celled, versatile*; pollen grains furrowed, pitted, punctate or reticulate.

Gynoecium — Carpel 1; *ovary with septal glands, superior to inferior*, trilocular; ovules many on axile placentae; style slender; stigma spirally twisted.

Fruit — Fleshy indehiscent or a multiple of berries or capsules.

Seeds — Embryo small, sometimes caudate or winged; endosperm mealy.

FLORAL RANGE IN THE FAMILY

Unisexual flowers are rare, as in *Catopsis*, *Hechtia* and *Prionophyllum.* The flowers are usually trimerous, but tetramerous in *Billbergia nutans.* Sometimes, 2 posterior perianth segments are united (*Tillandsia*). Zygomorphy occurs in *Pitcairnia* where the inner perianth segments form a hooded structure over the stamens.

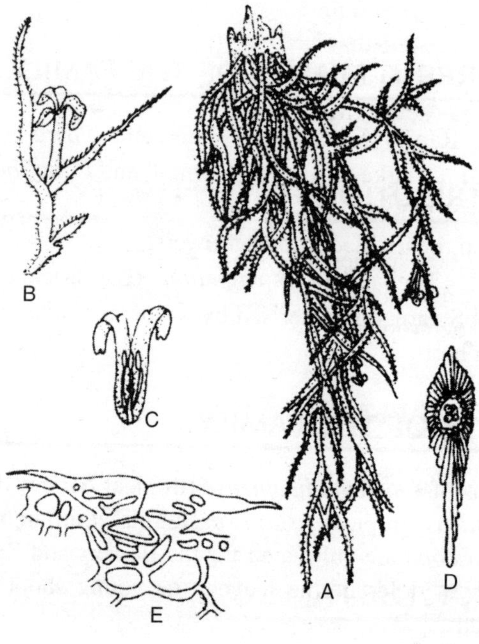

Fig. 97 BROMELIACEAE. *Tillandsia usneoides* L. **A.** whole plant hanging from tree-trunk; **B**, flower; **C**, L.S. of flower; **D**, peltate scale; **E**, T.S. of scale. (**A–E**, after Harms)

DIVISIONS OF THE FAMILY

The Bromeliaceae is divided into four subfamilies:

Subfamily I. *Bromelioideae.* Leaves toothed or spinose. Ovary inferior. Fruit a berry. Seeds neither tailed nor winged. Examples: *Ananas*, *Bromelia*, etc.

Subfamily II. *Navioideae.* Leaves spiny to dentate. Ovary superior. Fruit a capsule. Seeds neither appendaged nor winged. Example: *Navia.*

Subfamily III. *Pitcairnioideae.* Leaves entire or spinose. Ovary superior to half-inferior. Fruit a capsule. Seeds nonplumosely tailed or winged. Examples: *Dyckia*, *Pitcairnia*, etc.

Subfamily IV. *Tillandsioideae.* Leaves entire. Ovary superior. Fruit a capsule. Seeds with long plumose pappus-like appendages. Examples: *Tillandsia*, *Vriesia*, etc.

POSITION AND AFFINITY OF THE FAMILY

The Bromeliaceae was placed in the Farinosae by Engler, Rendle and Lawrence; the Farinales by Core; the Commelinales by Mitra and the Bromeliales by Cronquist, Takhtajan, Stebbins, Hamann, Huber, Dahlgren and Ehrendorfer.

Hutchinson contended that the Bromeliaceae "represent the climax of a line of descent wherein the calyx and corolla have remained distinct or fairly distinct from each other, a feature retained from the Dicotyledonous stock". In order to accommodate the family, he created an order, the Bromeliales, which he thought as related to but more advanced than the Commelinales. The advanced features include epiphytic habit, coloured bracts, coloured petals, ovary with septal glands and appendaged or winged seeds. Moreover, the scaly appendages within the base of petals are homologous with the corona of the Amaryllidaceae. Smith (1934) suggested that the strongest ties of the Bromeliaceae are with the Rapateaceae, the two families probably arising from a common stock. The genus *Navia* serves a link between the Bromeliaceae and Rapateaceae (Mitra, 1974).

NUMBER IN AND DISTRIBUTION OF THE FAMILY

A family of about 50 genera and 2,000 species, the Bromeliaceae is almost entirely indigenous to the New World* from Virginia to Chile. The plants form the prominent epiphytes of tropical rainforests and the family "as a whole is adapted to xerophytic conditions".

COMMONLY OCCURRING PLANT OF THE FAMILY

Ananas comosus (L.) Merr., a cultivated herb, bears spiny-serrate leaves and blue flowers arranged in a dense capitate spiral spike.

ECONOMIC ASPECTS OF THE FAMILY

The Bromeliaceae is of supreme importance for Pineapple (*Ananas comosus*), a delicious fruit; the leaves of the plant is used for making 'Pina cloth'. A number of plants are grown as ornamentals, *e.g.*, *Billbergia*, *Guzmania*, *Nidularium*, *Vriesia*, etc. Fibers like 'Pita floja' are obtained from the leaves of *Bromelia magdalenae*, 'Caraguata' from *B. serra* and 'Caroa' from *Neoglaziovia variegata.* The leaf-spines of *Puya chilensis* are used as hooks for catching fish. The dried stems and leaves of *Tillandsia usneoides* are used for car-cushions and upholstery.

* *Pitcairnia feliciana* occurs in West Africa.

Pontederiaceae
(Pickerel-weed Family)

GENERAL FEATURES OF THE FAMILY

Habit	—	*Erect or floating fresh water herbs.*
Stem	—	Very short, unbranched, mostly enveloped by sheathing leaf-bases.
Leaves	—	Opposite or whorled, simple, *lanceolate to broadly ovate or elliptic*, sometimes with a spongy inflated petiole.
Inflorescence	—	*Panicle or raceme, subtended by a spathe-like or tubular leaf-sheath.*
Flowers	—	Bisexual, somewhat zygomorphic, hypogynous.
Perianth	—	Obscurely 2-seriate, petaloid, free or united.
Androecium	—	Stamens 6, 3 or 1, *often unequal*; anthers 2-celled, introrse.
Gynoecium	—	Carpels 3, united; *ovary superior, trilocular with axile placentae or unilocular with parietal placentae*; ovules 1 to many, pendulous, anatropous.
Fruit	—	Many-seeded capsule or 1-seeded utricle.
Seeds	—	Ellipsoid or ovoid, longitudinally ribbed; embryo cylindrical, straight; endosperm copious, mealy.

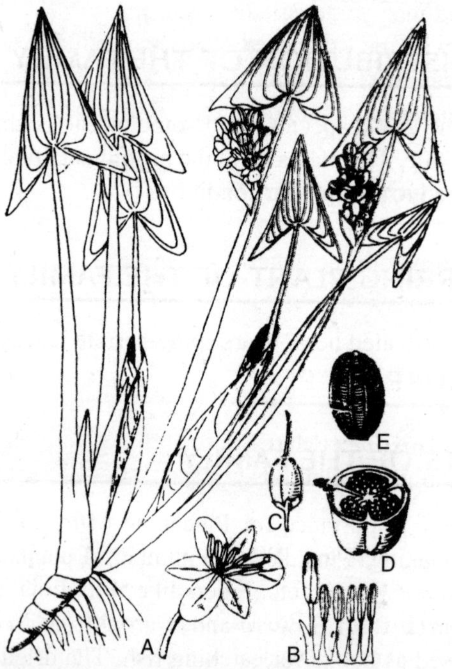

Fig. 98 PONTEDERIACEAE. *Monochoria hastata* (L.) Solms. **A**, flower; **B**, stamens; **C**, fruit; **D**, T. S. of ovary; **E**, Seed (**A–E**, after Hutchinson)

FLORAL RANGE IN THE FAMILY

The flowers are dimorphic in *Eichhornia crassipes*, but trimorphic in *E. azurea* and *Pontederia*. *Heteranthera* bears cleistogamous flowers where the number of stamens may be reduced to 1. In *Hydrothrix*, the stamen number is 1. The ovary is 1-ovuled in *Pontederia*.

DIVISIONS OF THE FAMILY

The family is divided into three tribes:

Tribe (*i*) *Eichhornieae*. Stamens 6. Ovary trilocular. Fruit many-seeded. Example: *Eichhornia*.

Tribe (*ii*) *Heterantherеae*. Stamens 3 or 1. Ovary trilocular. Examples: *Heteranthera*, *Hydrothrix* and *Monochoria*.

Tribe (*iii*) *Pontederieae*. Ovary unilocular. Fruit 1-seeded. Examples: *Pontederia* and *Reussia*.

POSITION AND AFFINITY OF THE FAMILY

Bentham-Hooker included the Pontederiaceae in the third series Coronarieae. Engler, Rendle and Lawrence placed it in the Farinosae. Core assigned, its position to Farinales. Hutchinson accommodated it within the Liliales.

The small embryo and mealy endosperm of the Pontederiaceae suggest a relationship with the Commelinaceae. Its affinity with the Liliaceae is established due to the petaloid six-partite perianth. Its resemblance to the Araceae can be seen from the structure of the inflorescence with a spathe-like bract. The mode of branching of *Eichhornia* similar to that of *Zostera*, a member of the Potamogetonaceae.

NUMBER IN AND DISTRIBUTION OF THE FAMILY

The Pontederiaceae is composed of nine genera and 34 species, being found in the warmer parts of the world.

COMMONLY OCCURRING PLANTS OF THE FAMILY

Water-hyacinth [*Eichhornia crassipes* (Mart.) Solms], a native of Brazil, is a floating herb with large broadly ovate or cuneate deep green glabrous coriaceous leaves and with a spiciform scape of lilac or pale violet flowers.

Monochoria hastata (L.) Solms, a robust aquatic herb, bears tringular-ovate hastate leaves and pale blue flowers.

Monochoria vaginalis (Burm. f.) Presl ex Kunth is an aquatic herb with leaves of variable shape and dark blue flowers.

ECONOMIC ASPECTS OF THE FAMILY

Eichhornia crassipes is a troublesome aquatic weed, since it spreads over water-courses and greatly impedes navigation and fishing. *Monochoria hastata* and *M. vaginalis* are used for medicinal purposes. *Pontederia cordata* is cultivated as an aquatic ornamental.

Order 9. *Liliales*

The Liliales is characterised by the bisexual pentacyclic trimerous flowers and fleshy endosperm.

There is a number of views concerning the origin of the Liliales. Some have regarded it as originating from the Ranales. Others have derived it from the Commelinales (Farinales). Another idea accepted its derivation from the Triuridales. Still a few have supported its helobial origin.

The Liliales is of phyletic significance, as it is the general basal stock from which prolific evolution of other monocots like the Arales, Palmales, Amaryllidales, Iridales, Haemodorales, Agavales, Orchidales and much-reduced Glumiflorae (Juncales, Cyperales, Graminales) have taken place (Hutchinson, 1959).

The order is considered as a primitive group among the monocots, since it has not advanced far in the floral make-up except for the entomophilous habit and the tendency towards zygomorphy as well as epigyny. The primitiveness of the order is also supported by anatomical, morphological and palynological features.

In Engler's arrangement, the order comprised three suborders and nine families: Juncineae (Juncaceae), Liliineae (Liliaceae, Stemonaceae, Haemodoraceae, Amaryllidaceae, Velloziaceae, Taccaceae, Dioscoreaceae), Iridineae (Iridaceae). Under this order, Rendle considered six of these families by ignoring the Stemonaceae, Haemodoraceae and Velloziaceae. Core treated seven families in the order, but omitted the Stemonaceae and Taccaceae. Hutchinson constructed several separate orders (Liliales, Amaryllidales, Iridales, Dioscoreales, etc.) and placed them apart from each other. Ehrendorfer considered six families in the order, Huber seven, Thorne as well as Dahlgren and Clifford eight each, Hamann ten, Cronquist 13, Stebbins 14 and Takhtajan 20.

The present author takes into account six families with their principal characters given below:

(1) Perianth sepaloid	(2)
(2) Perianth petaloid (at least the inner one)	(3)
(2) Ovary superior	*Juncaceae*
(2) Ovary inferior	*Dioscoreaceae*
(3) Ovary superior	*Liliaceae*
(3) Ovary inferior	(4)
(3) Stamens 6	*Amaryllidaceae*
(4) Stamens 3	(5)
(4) Stamens opposite to petals	*Haemodoraceae*
(5) Stamens alternate with petals	*Iridaceae*

Juncaceae
(Rush Family)

The family is characterised by the hygrophilous plants with a grass-like habit, small anemophilous flowers associated with bracts and bracteoles in a many-flowered inflorescence, glumaceous perianth and superior ovary.

In *Rostkovia* and *Juncus trifidus*, the flowers are solitary. The inflorescence may be a small head-like cluster (*J. articulatus*) or a cincinnus (*J. bufonius*) or a monochasial cyme of the fan type

(*J. inflexus*). The perianth is woody in *Marsippospermum*, but petaloid in some species of *Juncus*, *Luzula lactea*, *L. nivea* and *L. purpurea*. The ovary may be unilocular, as in *Luzula* and some species of *Juncus*.

Rendle regarded the Juncaceae as "a less elaborated group of Liliiflorae adapted to a simpler, more uniform environment than the versatile Liliaceae". They resemble each other in their typical trimerous hypogynous flower and cotyledon behaviour (the tip of which acts as a sucker and is retained in the seed) during germination. The relation between these taxa is often so close that it is difficult to draw the line of demarcation. In fact, several genera (*Calectasia*, *Xanthorrhoea*, etc.) with membranous perianth are incorporated indifferently in the Juncaceae or Liliaceae. Though some have accepted the Juncaceae as a degenerate group reduced from liliaceous stocks, others considered it as ancestral to the Liliaceae. In general habit, glumaceous perianth, feathery stigmas and wind-pollination, the Juncaceae is related to the Cyperaceae and Gramineae, but differ from them in having a six-parted perianth. Hutchinson placed the Juncaceae in the Juncales. a component of his Glumiflorae. He also treated it as advanced over his Liliales and assumed it as a progenitor of his Cyperales and Graminales.

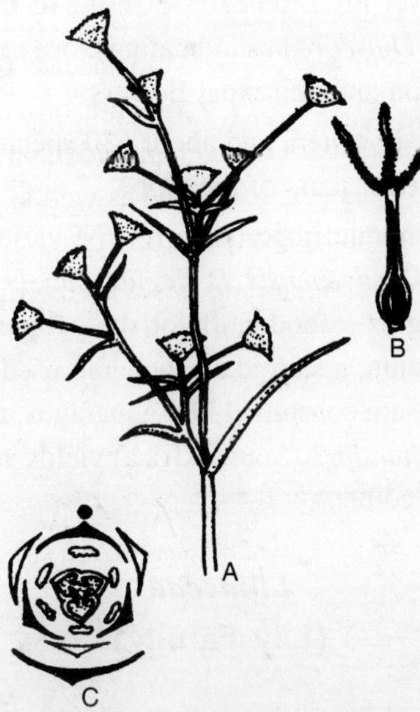

Fig. 99 JUNCACEAE. *Juncus articulatus* L. **A**, portion of plant with head-like inflorescence; **B**, pistil; **C**, floral diagram. (**A–C**, after Vierhapper)

The Juncaceae consists of nine genera and about 400 species, being largely distributed in cold temperate or montane regions of the southern hemisphere and found in damp or wet habitats.

Economically, the Juncaceae is of some importance. The stems of *Juncus effusus* and *J. squarrosus* are used for making hats, mats, cushions, etc. Several species of *Luzula* serve as a cute for kidney diseases. The leaf-fibres of *Prionium serratum* are of value in the textile industry, being the source of 'palmite'.

Dioscoreaceae
(Yam Family)

In this family, we find climbing plants with a fleshy tuberous root-stock, reticulate leaves, generally unisexual flowers and inferior ovary.

Bisexual flowers are seen in *Stenomeris*, *Trichopus*, etc. The stamens may be monadelphous, as in *Dioscorea monadelpha.* In *Rajania*, out of 3 carpels, 2 are abortive. The placentation is parietal in *Petermannia.*

The family is divided into two tribes:

Tribe (*i*) *Dioscoreae.* Plants dioecious. Flowers unisexual. Examples: *Dioscorea*, *Epipetrum*, *Tamus, etc.*

Tribe (*ii*) *Stenomerideae.* Plants monoecious. Flowers bisexual. Examples: *Petermannia*, *Stenomeris*, etc.

The Dioscoreaceae is generally placed in the Liliiflorae (Liliales). Hutchinson put it in an order by itself, treating it as advanced over his Liliales as a result of the reduction in the components of essential whorls. Rendle considered *Dioscorea* as indicating close relationship with *Smilax* in respect of the climbing habit, reticulate venation and unisexual flowers.

The Dioscoreaceae includes six genera and about 630 species. The members of this family are widely dispersed throughout the warmer parts of the globe.

The Dioscoreaceae is of economic importance for the various types of Yam, tubers produced from *Dioscorea alata*, *D. bulbifera*, *D. cayenensis*, *D. dumetorum*, *D. esculenta*, *D. hispida*, *D. opposita*, *D. rotundata* and *D. trifida*, serving as a food-stuff for the people in the Far East. Some species of *Dioscorea* form the source of diosgenin, a steroidal sapogenin used in oral contraceptives. The shoots of *Tamus communis* (Mediterranean) are consumed like Asparagus; the plant is also known medicinally as 'black bryony'. *Testudinaria elephantipes* (South Africa) yields an enormous tuber called 'Hottentot bread' which is rich in starch and is edible.

Liliaceae
(Lily Family)

GENERAL FEATURES OF THE FAMILY

Habit — Perennial herbs with a bulbous or rhizomatous stem, sometimes climbing or arborescent.

Leaves — Cauline or radical, alternate or whorled, mostly parallel veined.

Inflorescence — *Racemose, sometimes cymose.*

Flowers — Bisexual, *regular*, *hypogynous*; bracts small, scarious or spathaoeous.

Perianth — Segments 6, in 2 whorls, sepaloid or petaloid, free or united.

Androecium — Stamens 6, in 2 series, free; anthers introrse or extrorse.

Gynoecium — Carpels 3, united; *ovary superior*, *trilocular*; ovules 2 or more in each loculus; *placentation axile.*

Fruit — Berry or capsule.

Seeds — Embryo small; *endosperm fleshy, horny or cartilaginous.*

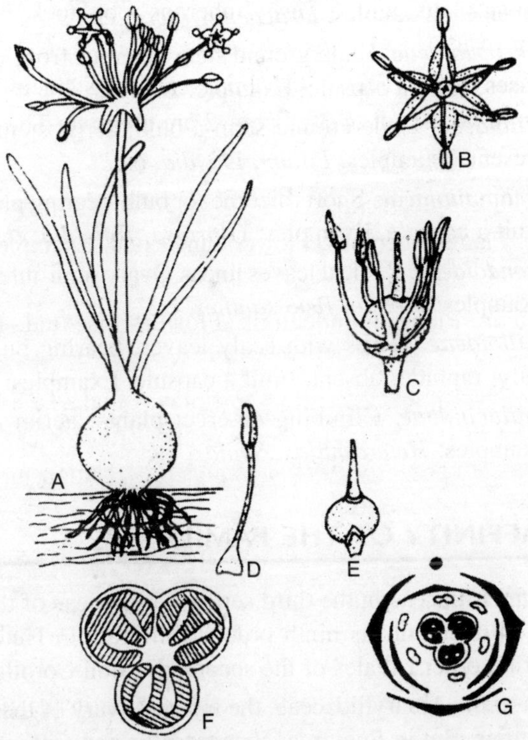

Fig. 100 LILIACEAE. *Allium cepa* L. **A**, whole plant; **B**, flower; **C**, flower after removing perianth; **D**, stamen; **E**, carpel; **F**, T.S. of ovary; **G**, floral diagram

FLORAL RANGE IN THE FAMILY

In the Liliaceae, departure from the normal type of floral structure is rare. The flowers may be unisexual, as in *Lomandra*, *Ruscus*, *Smilax* and other genera. The flowers are also zygomorphic, *e.g.*, *Gilliesia*, *Haworthia* and *Hemerocallis.* The flowers may be tetramerous or pblymerous in *Paris*, The ovary is semi-inferior in *Mondo*. There are 3 stamens in the male flower of *Ruscus*, one whorl being suppressed.

DIVISIONS OF THE FAMILY

The Liliaceae is divided into 11 subfamilies:

Subfamily I. *Aletridoideae.* Radical leaves lanceolate and small. Ovary semi-inferior. Fruit a capsule. Example: *Aletris.*

Subfamily II. *Allioideae.* Underground stem mostly a bulb. Inflorescence an umbellate cyme, always enclosed by a membranous spathe. Example: *Allium*, *Gilliesia*, etc.

Subfamily III. *Asparagoideae.* Underground stem a rhizome and aerial stem with leaves. Fruit a berry. Examples: *Asparagus*, *Paris*, etc.

Subfamily IV. *Asphodeloideae.* Underground stem mostly a rhizome from which a leafless scape arises. Examples: *Asphodelus*, *Hemerocallis*, etc.

Subfamily V. *Dracaenoideae.* Mostly arboreal; stem leafy; bulbs absent. Fruit a berry or capsule. Examples: *Dracaena*, *Yucca*, etc.

Subfamily VI. *Herrerioideae.* Underground stem tuberous from which a climbing aerial stem arises. Fruit a capsule. Example: *Herreria.*

Subfamily VII. *Lilioideae.* Underground stem a bulb; aerial stem scape-like, leafy; raphides present. Examples: *Lilium*, *Lloydia*, etc.

Subfamily VIII. *Melanthioideae.* Short rhizome- or bulb-bearing plants. Inflorescence terminal. Fruit a capsule. Examples: *Gloriosa*, *Tofieldia*, etc.

Subfamily IX. *Mondoideae.* Radical leaves linear. Ovary semi-inferior. Fruit a berry or capsule. Examples: *Mondo*, *Peliosanthes*, etc.

Subfamily X. *Scilloideae.* Herbs with scaly leaves, bearing bulbs: aerial stem scape-like, leafy; raphides absent. Fruit a capsule. Examples: *Scilla*, *Urginea*, etc.

Subfamily XI. *Smilacoideae.* Climbing or erect plants; aerial stems leafy. Fruit a berry. Examples: *Heterosmilax*, *Smilax*, etc.

POSITION AND AFFINITY OF THE FAMILY

Bentham-Hooker included the Liliaceae in the third series Coronarieae of the monocots. Engler placed the family in the suborder Liliineae of his ninth order Liliiflorae. In Hutchinson's arrangement, the family appeared under the first order Liliales of the second division Corolliferae of the monocots.

The Liliaceae is akin to the Amaryllidaceae, the inferior ovary of the latter usually standing as a barrier between the two families. But in *Bomarea*, *Hemerocallis* and *Ophiopogon* of the Liliaceae both superior- and inferior-ovaried flowers are found. Hutchinson's radically changed the concept of the differentiation of the two families based on the position of the ovary. He recognised the importance of umbellate inflorescence as the unifying principle in his classification. For this reason, he included the tribes Agapantheae, Allieae and Gillesieae in the Amaryllidaceae, which were formerly placed in the Liliaceae. Moreover, *Agave* and *Yucca* are quite similar cytologically in having 5 pairs of large and 25 pairs of small chromosomes. Yet, the former was placed in the Liliaceae and the latter in the Amaryllidaceae by earlier taxonomists (Mckelvey and Sax, 1933; Whitaker, 1934; Granick, 1944). On the basis of his studies of the floral anatomy of the Liliales, Anderson (1940) approved of Hutchinson's transfer of the Allieae and Agapantheae to the Amaryllidaceae. Maia (1941) and Cheadle (1942) also came to the same conclusion through their pollen and anatomical studies respectively.

The Liliaceae originated from the Helobiales or its ancestor, since the flowers of some genera of the Liliaceae, such as *Petrosavia* possess semi-free carpels. Moreover, the floral structure of *Helonias*, a member of the Liliaceae, resembles that of the Juncaginaceae, a component of the Helobiales.

The Liliaceae has been considered by botanists to represent the basic monocot stock—a stock which has given rise to many families, formerly thought to be more primitive, by way of reduction. It is usually regarded as the most typical family of the monocotyledonous group.

NUMBER IN AND DISTRIBUTION OF THE FAMILY

The Liliaceae embraces about 250 genera and 3,000 species. The plants belonging to this family are widely distributed over the earth, but abundant in the warm temperate and tropical zones.

COMMONLY OCCURRING PLANTS OF THE FAMILY

Indian Aloe (*Aloe indica* L., *A. perfoliata* Willd.) is a herb with sword-shaped fleshy leaves.

Asparagus racemosus Willd. is a thorny climber, having fasciculated roots and cladodes.

Asphodel (*Asphodelus tenuifolius* Cav.) and Day Lily (*Hemerocallis fulva* L.) are garden plants.

Meadow Saffron (*Colchicum luteum* Baker) is found in the hills.

Dragon Tree (*Dracaena spicata* Roxb.) and Adam's-needle or Dagger Plant (*Yucca gloriosa* L.) are arborescent garden shrubs, characterised by the presence of concentric bundles and secondary thickening of wood.

Glory Lily (*Gloriosa superba* L.), seen in dry places, is a climber with the tips of leaves turned into tendrils.

Bow-string Hemp (*Sansevieria roxburghiana* Schult. f.) is a perennial herb with a rosette of fleshy concave sharp-pointed leaves.

Sarsaparilla (*Smilax zeylanica* L.) is a prickly climber with stipular tendrils and reticulate leaf venation.

ECONOMIC ASPECTS OF THE FAMILY

"Economically, the Liliaceae ranks very high in the number of important ornamentals" (Core, 1955). The species of *Agapanthus*, *Brodiaea*, *Convallaria*, *Dracaena*, *Hemerocallis*, *Kniphofia*, *Muscari*, *Yucca*, etc. are cultivated as ornamentals. The growing of the species of *Hyacinthus*, *Lilium*, *Scilla*, *Tulipa* and other genera is an important industry in Holland.

Several plants of the Liliaceae yield food-stuff. Shallot is derived from *Allium ascalonicum*, Onion from *A. cepa*, Leek from *A. porrum*, Garlic from *A. sativum* and Chives from *A. schoenosprasum. Asparagus officinalis* is raised for the edible young shoots, *Chlorophytum arundinaceum* for the edible roots, *Lapageria rosea* (Chile) for the edible fruits and *Ophiopogon japonicus* for the edible tubers.

Some plants of the Liliaceae are of medicinal value. 'Aloin', important in the drug trade, is secured from *Aloe africana*, *A. barbadensis*, etc. *Colchicum autumnale* yields 'colchicine', an alkaloid used as a drug. 'Sarsaparilla', another medicine, is obtained from the roots of *Smilax medica* (Mexico), *S. officinalis* (Hondurus) and *S. ornata* (Jamaica). Another medicinal plant is *Veratrum album* whose rhizomes yield 'veratrin'.

A few plants of the Liliaceae furnish fibres. 'New Zealand flax' is derived from *Phormium tenax* and 'bow-string hemp' from *Sensevieria roxburghiana. Yucca filamentosa* also yields a kind of tenacious fibres from the leaves.

The plants of the Liliaceae are put to a variety of uses. *Hyacinthus orientalis* (France) is cultivated for 'hyacinth', a perfume. The bulbs of *Scilla* give us a glucoside which is in demand as a rat poison. The bulbs of *Urginea* form 'red squill', used in rodent control. The leaves of *Xanthorrhoea hastilis* (Australia) yield an acaroid resin which is used for making sealing wax, gold size, etc.

Amaryllidaceae
(Amaryllis Family)

GENERAL FEATURES OF THE FAMILY

Habit — Perennial herbs with a bulbous or rhizomatous stem.

Leaves — Radical, linear or strap-shaped.

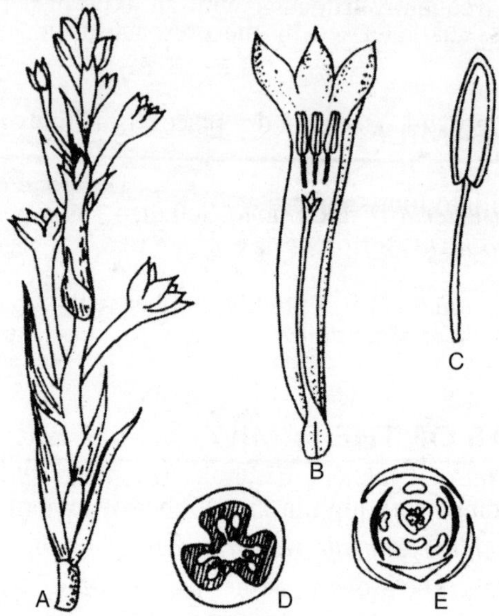

Fig. 101 AMARYLLIDACEAE. *Polianthes tuberosa* Willd. **A**, portion of plant; **B**, L.S. of flower; **C**, stamen; **D**, T.S. of ovary; **E**, floral diagram

Inflorescence — *Umbellate cyme, subtended by a spathe and borne upon a leafless scape.*

Flowers — Bisexual, *regular or irregular, epigynous.*

Perianth — Segments 6, in 2 whorls, similar and petaloid; *sometimes with a corona*, united below into a tube, imbricate or valvate.

Androecium — Stamens 6, in 2 whorls, epiphyllous; anthers introrse.

Gynoecium — Carpels 3, united; *ovary inferior, trilocular with many ovules arranged in 2 series on axile placentae.*

Fruit — Berry or capsule.

Seeds — Embryo small; *endosperm fleshy, never cartilaginous.*

FLORAL RANGE IN THE FAMILY

The flowers are solitary in *Galanthus* and *Zephyranthes.* The inflorescence of *Agave* is large and consists of cymes arranged in a racemose manner.

The perianth segments are usually connate, but free in *Leucoium* and other genera.

The presence of a corona or paracorolla is an important morphological feature of the flowers of the Amaryllidaceae. It is regarded as the basal outgrowth of filaments which are united to form the funnel-shaped structure or it is supposed to be the inner outgrowth of the perianth to constitute the funnel-shaped structure. Hence, according to its origin, the corona is either staminal or perigonal. The staminal nature of the corona is manifested in *Pancratium*, while the perigonal nature in *Cryptostephanus.*

The number of stamens is a multiple of 6 in *Vellozia.* In *Gethyllis*, the stamens (12–18) are united into several bundles.

Although the ovary is frequently trilocular with an axile placentation, it is unilocular with a parietal placentation in *Leontochir.*

DIVISIONS OF THE FAMILY

The Amaryllidaceae is divided into four subfamilies:

Subfamily I. *Agavoideae.* Inflorescence racemose or paniculate. Stamens with introrse anthers. Examples: *Agave*, *Polianthes*, etc.

Subfamily II. *Amaryllidoideae.* Scape leafless bearing an umbellate inflorescence enclosed by a spathe or scape bearing a single flower. Stamens with introrse anthers. Examples: *Amaryllis*, *Leucoium*, etc.

Subfamily III. *Campynematoideae.* Inflorescence various. Outer whorl of stamens at least with introrse anthers. Examples: *Campynema* and *Campynemanthe.*

Subfamily IV. *Hypoxidoideae.* Inflorescence various. Stamens with introrse anthers. Examples: *Curculigo*, *Hypoxis*, etc.

POSITION AND AFFINITY OF THE FAMILY

Bentham-Hooker included the Amaryllidaceae in their second series Epigynae. Engler placed the family in the suborder Liliineae of the order Liliiflorae. Hutchinson split up the entire family retaining the Amaryllidaceae proper under the Amaryllidales and creating orders like the Agavales for the Agavaceae, Haemodorales for the Hypoxidaceae, etc.

The Amaryllidaceae is allied to the Iridaceae by the nature of ovary, but is distinguished by the presence of six stamens. It is related to the Dioscoreaceae, but the climbing habit of the latter stands out prominently. It bears a relationship with the Liliaceae, the genus *Agapanthus* forming a link between the two families. While the ovary is inferior in the Amaryllidaceae, it is superior in the Liliaceae.

Bose (1962) stated that gene mutation, chromosome repatterning, polyploidy, hybridisation and apomixis played prominent roles in evolution and speciation in the Amaryllidaceae.

NUMBER IN AND DISTRIBUTION OF THE FAMILY

The Amaryllidaceae covers about 75 genera and 1,100 species. The members of this family are "widely distributed throughout the world, with the majority of their members occurring on the more level plains, plateaus and steppe areas of the tropics and subtropics".

COMMONLY OCCURRING PLANTS OF THE FAMILY

American Aloe or Century Plant (*Agave americana* L., *A. cantala* Roxb.) is found growing along railway-lines and in gardens.

Crinum asiaticum L., *C. latifolium* L., *Haemanthus coccineus* L., *Pancratium verecundum* Ait. and *Polianthes tuberosa* Willd. are garden plants.

Curculigo capitulata O. Kuntze, a plant of tropical Himalaya, is often mistaken for a dwarf palm due to its palm like corrugated leaves.

Curculigo orchioides Gaertn. is a small herb, having yellow flowers.

Furcraea gigantea Vent. is a short-stemmed large shrub with radical tufted fleshy leaves.

Star-grass (*Hypoxis aurea* Lour.) is a small weed with tubercled seeds.

Zephyr Lily (*Zephyranthes carinata* Herb.) is a bulbous herb.

ECONOMIC ASPECTS OF THE FAMILY

Many members of the Amaryllidaceae yield fibres. 'Henequen' is obtained from *Agave fourcroydes*, 'Jaumave istle' from *A. funkiana*, 'Tula istle' from *A. lophantha*, 'Salvador sisal' from *A. letonae*, 'Sisal hemp' from *A. sisalana*, 'Cabuya' from *Furcraea cabuya*, 'Mauritius hemp' from *F. gigantea* and 'Fique' from *F. macrophylla.*

Flour is derived from the roots of some species of *Alstroemeria* (Chile). In Latin America the species of *Agave* are cultivated as an source of the sugary exudate which is employed as the basis for such distilled liquors as 'mezcal', 'pulque' and 'tequila'. The bulbs of *Hippeastrum equestre* (Java) are used for poulticing the swelling of the neck. The bulbs of *Leucoium aestivatum* are eaten after they are cooked.

The family is of horticultural importance, *e.g. Amaryllis, Clivia, Crinum, Eucharis, Galantlhus, Habranthus, Narcissus, Sprekelia* and *Vallota.*

Haemodoraceae
(Bloodwort Family)

The family is known by the perennial herbs with an orange to red sap. Here the flowers bear generally three stamens which are opposite to the petals and the ovary is inferior.

The Haemodoraceae contains ten genera and 40 species, being limited to South America, South Africa and Australia.

The family is not of any economic importance.

Iridaceae
(Iris Family)

GENERAL FEATURES OF THE FAMILY

Habit — *Perennial herbs with a bulb, corm or rhizome.*

Stem — Solitary, several or none.

Leaves — *Radical, equitant*, linear or ensiform.

Inflorescence	—	Panicle or raceme or solitary.
Flowers	—	*Showy, bisexual, regular or irregular, epigynous.*
Perianth	—	Segments 6, in 2 whorls, petaloid.
Androecium	—	*Stamens 3, alternate with petals*; anther 2-celled, extrorse.
Gynoecium	—	Carpel 1; *ovary inferior, trilocular with axile placentation or unilocular with parietal placentation.*
Fruit	—	*Loculicidal capsule.*
Seeds	—	Angular or roundish by compression; embryo small; endosperm fleshy or hard.

FLORAL RANGE IN THE FAMILY

The flowers are solitary in *Crocus* and *Romulea.* Both regular and irregular flowers are found in *Gladiolus.* The inner series of perianth is usually smaller than the outer. The stamens are united into a tube in *Sisyrinchium grandiflorum.* The ovary is unilocular with a parietal placentation in *Hermodactylon.* The style-arms are short and bifid in *Freesia*, *Watsonia* and related genera; simple (undivided) in *Gladiolus*, *Sisyrinchium*, etc. simple and flattened or very variously divided in *Crocus* and large, broad and petaloid in *Iris* and allied genera.

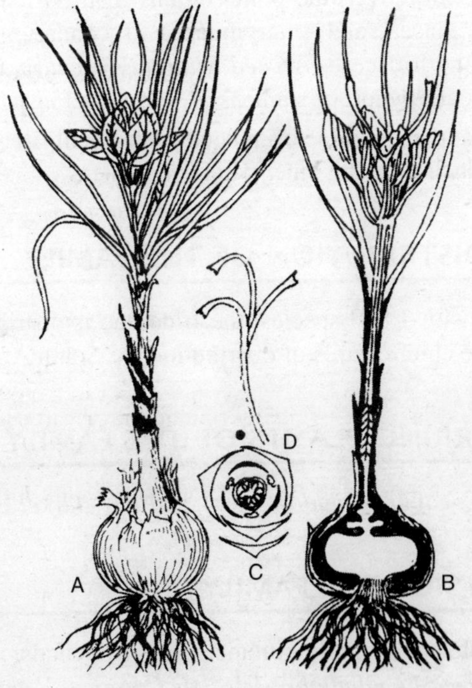

Fig. 102 IRIDACEAE. *Crocus sativus* L. **A**, whole plant with corm, scales and foliage leaves; **B**, V.S. of plant to show short scape terminating into a solitary flower; **C**, floral diagram; **D**, stigma. (**A–D**, after Diels)

DIVISIONS OF THE FAMILY

The Iridaceae is divided into three subfamilies:

Subfamily I. *Crocoideae.* Leaves linear or filiform in several rows. Spathe 1-flowered. Flowers regular. Perianth whorls subequal. Examples: *Crocus* and *Romulea.*

Subfamily II. *Iridoideae.* Leaves equitant, in 2 rows. Spathe 2 to several-flowered. Flowers regular. Perianth whorls unequal. Examples: *Belamcanda, Iris, Sisyrinchium,* etc.

Subfamily III. *Ixioideae.* Leaves equitant, in 2 rows. Spathe 1-flowered. Flowers medianly zygomorphic. Examples: *Gladiolus, Ixia,* etc.

POSITION AND AFFINITY OF THE FAMILY

The Iridaceae is generally included in the Liliiflorae (Liliales). However, Hutchinson placed the family in an order by itself (Iridales). Takhtajan's Iridales included not only the Iridaceae but also the Geosiridaceae, Burmanniaceae and Corsiaceae.

The Iridaceae, while resembling the Amaryllidaceae in the inferior ovary and that family and the Liliaceae in the germination of seeds, is distinguished from other families of the Liliales by the presence of a single whorl of stamens. Moreover, the pollen grains of the Iridaceae and Amaryllidaceae are similar. Like the arborescent Liliaceae and Amaryllidaceae, secondary growth takes place by cambial activity in some members of the Iridaceae (*Aristea Klattia, Witsenia*). The Iridaceae is related to the Haemodoraceae in which the outer whorl of stamens is suppressed instead of the inner whorl.

It has been assumed that the Iridaceae originated from the liliiflorean stock or directly from the Amaryllidaceae by way of reduction in the inner whorl of stamens.

NUMBER IN AND DISTRIBUTION OF THE FAMILY

A family of about 70 genera with 1,800 species, the Iridaceae is distributed over much of the earth except the coldest regions. The chief centres of distribution are South Africa and tropical America.

COMMONLY OCCURRING PLANTS OF THE FAMILY

Belamcanda chinensis (L.) DC, *Cipura paludosa* Aubl. and *Iris clarkei* Bak. are garden plants.

ECONOMIC ASPECTS OF THE FAMILY

The Iridaceae is of considerable economic importance for ornamentals, *e.g.*, *Antholyza, Belamcanda, Crocus, Crocosmia, Eustylis, Freesia, Gladiolus, Iris, Ixia, Moraea, Nemostylis, Neomarica, Romulea, Sparaxis, Sisyrinchium* and *Tigridia.* 'Saffron', a yellow dye, is obtained from the dried stigmas of *Crocus sativus.* The dried rhizomes of *Iris florentina, I. germanica* var. *germanica* and *I. pallida* constitute 'orris root', a violet perfume. The rhizome of *I. foetidissima* serves as a specific for hysteria.

Order 10. *Scitaminales*

From the stand-point of floral structure and adaptation for cross pollination, the Scitaminales seems to be quite advanced in the scale of monocot evolution. It has been considered as ancestral to or parallel with the Orchidales. It has probably sprung from the liliiflorean stock, since the floral construction of *Ravenala* and some wild species of *Musa* resemble that of the Liliiflorae (Liliales).

Bentham-Hooker considered the Scitamineae as a single natural order and put it in their second series Epigynae. Engler, Rendle, Lawrence, Core and Mitra treated the order just before the Microspermae (Orchidales). Bessey placed the families of the Scitaminales under the Iridales. Most taxonomists have divided the order into four distinct families: Musaceae, Zingiberaceae, Cannaceae, Marantaceae. Hutchinson went to the extent of creating a new order, the Zingiberales, and raising the number of families to six by including the Strelitziaceae and Lowiaceae besides the usual four. Cronquist, Takhtajan, Stebbins and Dahlgren used the ordinal name Zingiberales for the Scitamineae and accommodated eight families (Strelitziaceae, Lowiaceae, Heliconiaceae, Musaceae, Zingiberaceae, Costaceae, Cannaceae, Marantaceae) within the order. Dahlgren and Clifford set up the superorder Zingiberiflorae to place the Zingiberales.

The Scitaminales is a natural taxon. This is reflected in the habit, leaf structure, epigyny, seeds with perisperm and pollen grains. Furthermore, one can trace an increasing floral complexity in conjunction with the entomophilous or orninthophilous habit. "This development finds expression in the reduction of stamens to petaloid structures to which the attractive function becomes more or less relegated" (Rendle, 1904). As a matter of fact, only 5 stamens are generally fertile in the Musaceae, the sixth being absent or represented by a structure which is not petaloid. Only a single entire stamen is fertile in the Zingiberaceae. But in the Cannaceae and Marantaceae, only half an anther is fertile, the rest of the stamen being somewhat petaloid in form.

Cronquist (1968) pointed out that "... some of the differences between individual families, appear to be intimately related to the mechanism of pollination. A detailed correlation of pollinators with the families and genera of the order remains to the undertaken, however. The adaptive significance of many of the characters of presently unknown or doubtful significance are the arrangement of leaves, and the presence or absence of raphide sacs, laticifers, mucilage ducts and oil cells".

The usual four families of the Scitaminales may be characterised below (Core, 1955):

(1) Staminode 1, not petaloid; fertile stamens 5	*Musaceae*
(1) Staminode 1 or more, some or all petaloid	(2)
(2) Fertile stamen 1; anthers 2-celled	*Zingiberaceae*
(2) Fertile stamen 1; anthers 1-celled	(3)
(3) Ovules several in each loculus	*Cannaceae*
(3) Ovule solitary in each loculus	*Marantaceae*

Musaceae
(Banana Family)

GENERAL FEATURES OF THE FAMILY

Habit	—	*Perennial herbs, often tree-like in appearance.*
Stem	—	Stout, unbranched, ensheathed by petioles.
Leaves	—	*Large, alternate*, entire or lacerated (due to mechanical injury by wind), *distichous or spirally arranged.*
Inflorescence	—	*Spike or panicle, subtended by spathaceous bracts which is often brilliantly coloured.*

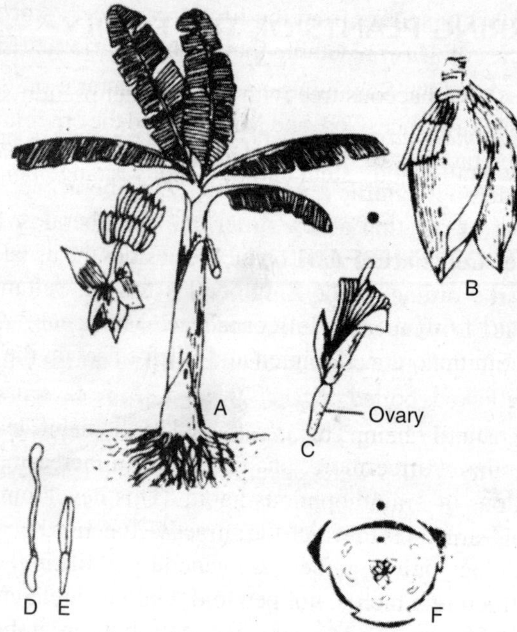

Fig. 103 MUSACEAE. *Musa paradisiaca* L. **A**, whole plant; **B**, inflorescence; **C**, flower; **D**, stamen; **E**, carpel; **F**, floral diagram

Flowers	—	*Bisexual or unisexual, irregular.*
Perianth	—	Segments 6, similar or distinguished into 2 series, free or variously connate.
Androecium	—	*Fertile stamens 5* (*6 in Ravenala*); *staminode 1, not petaloid.*
Gynoeciura	—	Carpels 3, united; *ovary inferior, trilocular with many ovules on axile placentae*; style filiform; stigma capitate or lobed.
Fruit	—	Berry or capsule.
Seeds	—	*Arillate, albuminous with an abundant perisperm*; embryo straight.

DIVISIONS OF THE FAMILY

The Musaceae is divided into three subfamilies:

Subfamily I. *Lowioideae.* Leaves half-alternate. Flowers bisexual. Largest median petal forms a labellum. Example: *Orchidantha.*

Subfamily II. *Musoideae.* Leaves spiral. Flowers bisexual or unisexual. Fruit a berry or leathery, indehiscent. Aril absent. Example: *Musa.*

Subfamily III. *Strelitzioideae.* Leaves half-alternate. Flowers bisexual. Fruit a woody capsule or schizocarpic capsule. Examples: *Heliconia*, *Ravenala* and *Strelitzia.*

NUMBER IN AND DISTRIBUTION OF THE FAMILY

The Musaceae is composed of five genera and about 150 species of wide distribution in the tropics.

COMMONLY OCCURRING PLANTS OF THE FAMILY

Banana (*Musa paradisiaca* L.) is a herbaceous tree with a long erect pseudo-stem.

Traveller's Tree [*Ravenala madagascariensis* (J. F. Gmel) Sonn.], a native of Malagasy and a garden plant, is characterised by a tall woody stem and distichous arrangement of leaves.

ECONOMIC ASPECTS OF THE FAMILY

Ensete superbum, *Heliconia metallica*, *Musa nepalensis* and *Strelitzia reginae* are plantea as ornamentals. The leaf-stalks of *Musa ensete* (Ethiopia) are consumed as food. The unripe fruits of *M. paradisiaca* are commonly devoured after being baked, boiled or fried; the ripe fruits are eaten fresh as a dessert. The leaf-stalks of *M. textilis* yield 'Manila hemp' of commerce. Another fibre yielding plant is *Ensete ventricosa. Ravenala madagascariensis* has water accumulated in the leaf-bases which can be used for drinking.

Zingiberaceae
(Ginger Family)

GENERAL FEATURES OF THE FAMILY

Habit — *Perennial herbs with a fleshy rhizome.*

Stem — Bracted, scapose or leafy, short or elongated.

Leaves — Radical or cauline, alternate, *distichous, with a ligule at junction of blade and sheath.*

Inflorescence — *Open raceme or compact spike or solitary, subtended by prominent bracts.*

Flowers — *Bisexual, irregular, epigynous.*

Perianth — Segments 6, in 2 series, outer sepaloid and inner petaloid.

Androecium — *2 lateral stamens of outer whorl modified into staminodes*; *1 median stamen of inner whorl fertile and 2 lateral stamens of same whorl united to form a labellum; anthers 2-celled.*

Gynoecium — Carpels 3, united; ovary inferior, unilocular with parietal placentation or trilocular with axile placentation.

Fruit — Capsule.

Seeds — *Arillate with a smaller endosperm and larger perisperm*; embryo straight.

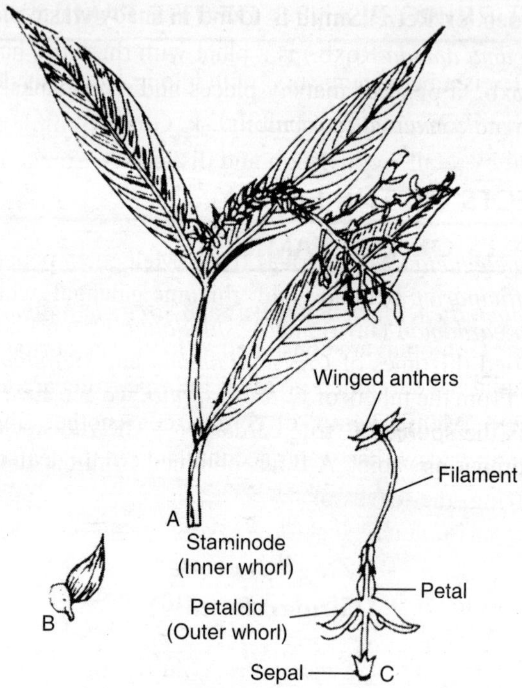

Fig. 104 ZINGIBERACEAE. *Globba bulbifera* Roxb. **A**, portion of plant; **B**, bulbil with bract; **C**, flower with different parts

DIVISIONS OF THE FAMILY

The Zingiberaceae is divided into two subfamilies, one with no tribe and the other with three tribes:

Subfamily I. *Costoideae.* Upper parts of the plant not aromatic. Leaves spiral. Lateral staminodes small or absent. Nectaries replaced by septal glands. Example: *Costus.*

Subfamily II. *Zingiberoideae.* Plants aromatic. Leaves distichous. Lateral staminodes large, small or absent.

Tribe (*i*) *Globbeae.* Lateral staminodes of outer whorl petaloid. Ovary unilocular; placentation parietal. Example: *Globba.*

Tribe (*ii*) *Hedychieae.* Lateral staminodes of outer whorl petaloid and conspicuous. Ovary trilocular. Examples: *Curcuma, Hedychium, Kaempferia,* etc.

Tribe (*iii*) *Zingibereae.* Lateral staminodes of outer whorl linear, teeth-like or absent. Ovary trilocular. Examples: *Ammomum, Elettaria, Zingiber,* etc.

NUMBER IN AND DISTRIBUTION OF THE FAMILY

The Zingiberaceae consists of 49 genera and about 1,300 species. The members of this family are distributed throughout the eastern hemisphere, particularly in the Indo-Malayan area.

COMMONLY OCCURRING PLANTS OF THE FAMILY

Alpinia galanga (L.) Sw., Butterfly Lily (*Hedychium coronarium* Koen. ex Retz.) and *Kaempferia rotunda* L. are garden plants with scented flowers.

Costus speciosus (Koen. ex Retz.) Smith is found in shady waste places.

Mango Ginger (*Curcuma amada* Roxb.) is a plant with rhizomes having the smell of a Mango.

Globba bulbifera Roxb., a plant of marshy places and river-banks, has a terminal panicle, the lower flower-buds of which ate converted into bulbils.

ECONOMIC ASPECTS OF THE FAMILY

The seeds of *Aframomum melegueta* (West Africa) from 'Melegueta pepper', a type of spice. *Alpinia galanga* (Malaya) and *A. officinarum* (China) yield 'rhizome galangal' which is used in medicine. The aromatic seeds of *Amomum cardamon* (Java) and *A. subulatum* (Nepal) are utilised as a substitute for true cardamon. From the dried rhizomes of *Curcuma amada* and *C. longa*, we derive 'mango ginger' and 'turmeric' respectively. From the tubers of *C. angustifolia*, we get 'East Indian arrowroot'. *Elettaria cardamomum* (Indonesia) is the source of 'true cardamon'. The rootstock of *Hedychium spicatum* is used as a perfume and in preparing 'abir'. A flour, obtained from the tubers of *Hitchenia caulina*, is suitable for glues and in sizing; the leaves of this plant are useful in making paper. The rhizomes of *Zingiber cassumunar* give us 'cassumunar ginger', those of *Z. officinale* 'common ginger' and *Z. zerumbet* 'zerumbet'.

Some species of *Alpinia*, *Brachychilum*, *Costus*, *Globba*, *Hedychium* and *Kaempferia* are cultivated as ornamentals.

Cannaceae
(Canna Family)

GENERAL FEATURES OF THE FAMILY

Habit — *Perennial herbs with a tuberous rhizome.*

Leaves — *Large*, cauline, *parallel-veined, without a ligule or pulvinus.*

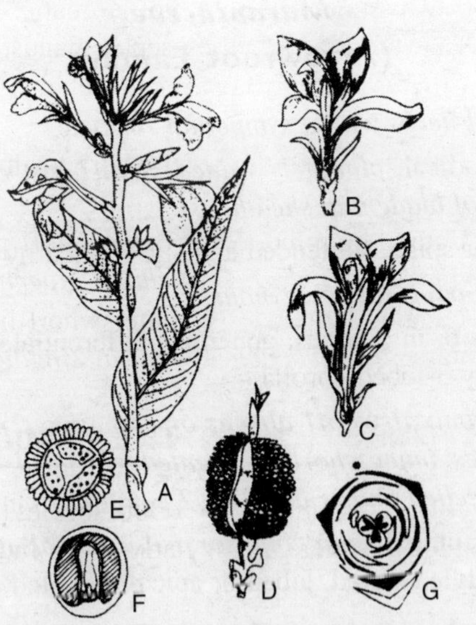

Fig. 105 CANNACEAE. *Canna orientalis* Rosc. **A**, portion of plant; **B**, flower; **C**, L.S. of flower; **D**, fruit; **E**, T.S. of ovary; **F**, L.S. of seed; **G**; floral diagram

Inflorescence — Raceme or panicle.
Flowers — *Showy, bisexual, irregular, epigynous.*
Perianth — Biseriate, of 3 segments in each series, outer sepaloid and inner petaloid.
Androecium — *1-5 petaloid staminodes, 1 of which forms a labellum*; 1 fertile stamen bears a half-anthers.
Gynoecium — Carpels 3, united; *ovary inferior, trilocular, with a series of anatropous ovules at inner angle of each loculus.*
Fruit — Warty capsule.
Seeds — *Subglobose, with a hard endosperm and copious perisperm*; embryo straight.

NUMBER IN AND DISTRIBUTION OF THE FAMILY

The Cannaceae comprises one genus (*Canna*) and 30–35 species. The members of this family are native to Central America and West Indies, but have been widely introduced elsewhere.

COMMONLY OCCURRING PLANT OF THE FAMILY

Indian Shot (*Canna orientalis* Rosc.) is a garden plant with showy flowers.

ECONOMIC ASPECTS OF THE FAMILY

The rhizomes *Canna bidentata* are sometimes considered as emergency foods, while those of *C. gigantea* and *C. speciosa* yield medicinal extracts. The rhizomes of *C. edulis*, called 'Queensland' arrowroot', form an article of food. *C. orientalis* is of ornamental value.

Marantaceae
(Arrowroot Family)

Habit — *Perennial herbs with a sympodial rhizome.*
Leaves — Mostly radical, *pinnately veined, distichous, often with a swollen pulvinus at junction of blade and sheath.*
Inflorescence — Panicle or spike, subtended and surrounded by spathaceous bracts.
Flowers — *Bisexual, irregular, epigynous.*
Perianth — Segments 6, in 2 series, generally differentiated into a 3-sepaled calyx and an irregularly 3-lobed corolla.
Androecium — *Outer staminal whorl absent or sometimes represented by 1 or 2 petaloid staminodes; inner whorl represented by 1 fertile stamen (which is often petaloid) with a 1-celled anther and 1 or 2 petaloid staminodes.*
Gynoecium — Carpels 3, united; *ovary inferior, trilocular or unilocular; ovules solitary in each loculus*, style twisted, lobed or apically dilated.
Fruit — Capsule.
Seeds — *Arillate; embryo curved or folded; endosperm copious.*

DIVISIONS OF THE FAMILY

The Marantaceae is divided into two tribes:

Tribe (*i*) *Maranteae*. Ovary unilocular. Example; *Maranta*.

Tribe (*ii*) *Phrynieae*. Ovary trilocular. Examples: *Clinogyne*, *Phrynium*, etc.

NUMBER IN AND DISTRIBUTION OF THE FAMILY

The Marantaceae includes 30 genera and about 350 species, being distributed in the tropics of the New and Old World.

COMMONLY OCCURRING PLANT OF THE FAMILY

Mat Grass [*Schumannianthus dichotomus* (Roxb.) Gagnep.] is a leafy cane-like shrub of Bangladesh.

ECONOMIC ASPECTS OF THE FAMILY

The tubers of *Calathea allouia* (West Indies) are edible. The bulbs of *Maranta arundinacea* yield 'commercial arrowroot', a diet for children and invalids. The split stems of *Schumannianthus dichotomus* are woven into mats. *Thalia dealbata* is a favourite garden aquatic.

Order 11. *Orchidales*

In the scale of monocot evolution, the Orchidales is the most highly developed group; this is evident in the elaboration, complexity and reduction of floral members. Engler, Rendle, Lawrence, Core, Mitra, Cronquist and Stebbins considered the order as the last evolved and most advanced among the monocots. However, Wettstein did not concur with this view and regarded his Spadiciflorae and Pandanales as more advanced. Hutchinson, Takhtajan, Ehrendorfer as well as Dahlgren and Clifford treated the orchids as phylogenetically more primitive than the grasses.

The order is characterised by the very numerous minute seeds with an undifferentiated embryo and little or no endosperm.

"The lack of differentiation of the embryo is doubtless at least in part a consequence of mycotrophy, but the number and size of seeds and the loss of endosperm are presumably due to other factors.... The plants are physiologically dependent on their fungal symbionts, sometimes even for food, sometimes only for other factors as yet not fully understood, but in any case they can grow only where their symbiont finds the conditions suitable" (Cronquist, 1968).

The order contains two families which may be distinguished as follows:

(1) Flowers generally actinomorphic; seeds with endosperm	*Burmanniaceae*
(1) Flowers zygomorphic; seeds without endosperm	*Orchidaceae*

Burmanniaceae
(Burmannia Family)

Besides the characters already mentioned, the Burmanniaceae has usually 3 stamens.

The Burmanniaceae is divided into three tribes:

Tribe (*i*) *Corsieae.* Perianth zygomorphic. Stamens 6. Example: *Corsia.*

Tribe (*ii*) *Euburmannieae:* Perianth actinomorphic and tube cylindrical. Stamens 3. Example: *Burmannia.*

Tribe (*iii*) *Thismieae.* Perianth actinomorphic and tube oblong to obovoid. Stamens 6. Example: *Thismia.*

The Burmanniaceae resembles the Amaryllidaceae by reason of the regular trimerous flowers. It also represents a line of development parallel to the Orchidaceae, since the plants tend towards the saprophytic habit and produce numerous tiny seeds. But the extent of specialisation has not approached the level of the Orchidaceae. Cronquist (1968) admitted: "The ovary in the Orchidaceae is always inferior and apparently does not have typical septal nectaries, although it is possible that some of the ovarian nectaries in both the Burmanniaceae and the Orchidaceae are derived from septal nectaries". Furthermore, the extended thin membranous testa enclosing a small undifferentiated embryo strongly suggests that of the orchid seeds.

The Burmanniaceae embraces 5–16 genera and 125 species, being most prevalent in damp tropical woods or savannahs as colourless saprophytes.

Burmannia coelestis D. Don is a slender weed of waste places. *B. pusilla* (Wall. ex Miers) Thw. bears bluish violet flowers.

The Burmanniaceae is of no economic significance except that some members of the family are used as adulterants in the composition of green tea.

Orchidaceae
(Orchid Family)

GENERAL FEATURES OF THE FAMILY

Habit	—	Perennial herbs, terrestrial or epiphytic or saprophytic.
Roots	—	Main roots always absent; instead, three types of adventitious roots arise from the nodes: aerial roots, tuberous and normal cylindrical earth roots.
Leaves	—	Thick half-alternate with a closed sheath; ligules and stipules lacking.
Flowers	—	*Showy, bisexual, irregular, epigynous.*
Perianth	—	Segments 6, in 2 whorls, free; outer segments consist of nearly equal 3 sepals and inner of somewhat dissimilar 3 petals; 2 lateral petals alike and simulate sepals, but median petal larger and constitutes a *labellum*; labellum of diverse shape, size and colour, normally posterior but rendered anterior by resupination, *i.e.*, twisting of pedicel or ovary through 180°.
Androecium	—	Stamens 1 or 2; filament adheres to style (*gynandrous*) and forms a *column* or *gymnostemium* which springs from top of ovary and ends in a beak known as *rostellum*; pollen grains united into masses termed *pollinia*; each pollinium bears a cord (*caudicle* or *translator*) which terminates in a sticky disc or gland called *retinaculum* or *viscidium.*
Gynoecium	—	Carpels 3, united; *ovary inferior, twisted, unilocular with many ovules*; placentation parietal; stigmas 3, discoid, glutinous, situated below rostellum and facing labellum.

Fruit — Capsule.
Seeds — Very minute*, exalbuminous; *embryo undifferentiated.*

FLORAL RANGE IN THE FAMILY

The Orchidaceae is characterised by a wide range of variations in the floral structure.

The sepals and petals may be nearly alike or somewhat different in form. In general, the sepals are smaller and inconspicuous; but they may become larger and prominent, *e.g.*, *Masdevallia.* The sepals are free or somewhat united. The odd sepal is spurred in *Disa*, but joins with the lateral petals to form a hood-like structure in *Haemaria.*

In *Coryanthes*, the lateral petals are lost. The petals are usually filiform in *Epicranthes* and smaller than the labellum, but sometimes become larger in *Oncidium.* The petals may assume the shape of a ribbon, as in *Cypripedium.* In some species of *Paphiopedilum*, the petals may attain a length of 30 cm. The petals may or may not be spurred. The labellum is sometimes minute and narrowed or curved in *Cattleyd* or large with spreading limbs in *Odontoglossum* or may look like a slipper in *Cypripedium* or may form a cage in *Stanhopea* or may assume the form of a bucket in *Coryanthes.* The labellum is simple in *Thelymitra* or trifid or tripartite or variously incised.

The outer whorl of stamens generally becomes fertile, as in *Arudina pentandra*; but they may form a fleshy staminode in *Diuris.* The staminodes are represented by small auricles in *Cypripedium* and related genera and form a leaf-like staminode in *Orchis.* The lateral pair of the inner whorl of stamens is fertile in *Cypripedium*, but foliaceous in *Thelymitra* or form staminodes of diverse shape in *Epipactis.* The anther is generally bent over the inner free of the column, but the anther may become erect and free on the top of the column (*Ophrys*) or the anther may hang downwards vertically from the top (*Coelogyne*). The anthers are 2-, 4- or 6-chambered with the corresponding number of pollen masses. The pollen masses are free and granular in *Cephalanthera* or united into packets by elastic nets in *Orchis* or waxy pollinia cohere in a chamber in *Cattleya.* The pollinia are provided with a caudicle, but sessile in *Malaxis.*

All the three stigmas are fertile in *Cypripedium*, *Paphiopedilum* and other genera. The ovary is generally spindle-shaped or cylindrical, but furnished with lines, ridges or wings.

DIVISIONS OF THE FAMILY

The Orchidaceae is divided into two subfamilies, one with two tribes and the other with four tribes.

Subfamily I. *Diandroideae.* Stamens 2 or rarely 3. Functional stigma 3.

Tribe (*i*) *Apostasieae.* Flowers with an almost radical symmetry. Stamens 2 or 3, fertile. Ovary trilocular, Examples: *Apostasia*, *Neuwiedia*, etc.

Tribe (*ii*) *Cypripedieae.* Flowers with a well-marked median symmetry. Stamens 2, fertile; odd stamen of outer whorl forms a large staminode Ovary unilocular or incompletely or completely trilocular. Examples: *Cypripedium*, *Paphiopedilum*, etc.

Subfamily II. *Monandroideae.* Stamen 1, fertile. Median stigma rudimentry or forms a rostellum.

Tribe (*i*) *Epidendreae.* Stamen lifting up like a little cap. Pollinia soft and waxy in texture, ends sometimes drawn out into sticky portions but with no distinct sticky discs. Examples: *Coelogyne*, *Dendrobium*, etc.

* Each weighing as low as 0.004 mg.

Tribe (*ii*) *Neottieae.* Stamen attached to back of a column, but easily lifted. Pollen soft and easily breaking up. Examples: *Neottia, Spiranthes,* etc.

Tribe (*ii*) *Ophrydeae.* Stamen firmly attached to column and immovable. Pollinia consist of small masses attached to a central stalk with one or two sticky discs. Examples: *Habenaria, Orchis,* etc.

Tribe (*iii*) *Vandeae.* Stamen lifting up like a little cap. Pollinia attached to one or two distinct sticky discs and removed with them, texture more frequently hard and bony but sometimes softer and waxy. Examples: *Cymbidium, Vanda,* etc.

POSITION AND AFFINITY OF THE FAMILY

The Orchidaceae was included by Bentham-Hookcr as the third family of the Microspermae at the beginning of the monocotyledons. Engler placed the family in the suborder Gynandrae of the order Microspermae which was at the end of the monocot class. In Hutchinson's arrangement, the family appeared under the Orchidales which marked the end of his second division Corolliferae of the monocots.

The Orchidaceae is related to the epigynous components of the Liliiflorae (Liliales). Hutchinson thought that the family originated either from the Haemodoraceae or Amaryllidaceae as a result of complex development of floral parts and reduction of stamens.

Furthermore, the Burmanniaceae and Apostasiaceae of his Orchidales form a connecting link with the families of the epigynous Liliales. The Orchidaceae has also been believed to arise from the Musaceae or its ancestor; this suggestion is based on the similar structure and construction of flowers of most genera of the Orchidaceae and *Orchidantha*, a typical Musaceae.

The Orchidaceae is agreed upon by almost all botanists to represent the most advanced family in the monocots for the following reasons:

1. The orchids comprise an enormously richly differentiated group of plants in terms of habit, size, vegetative features and floral peculiarities.
2. All the orchids are herbaceous plants, there being rarely shrubs or no tree forms. They exist ordinary rooted herbs, climbing vines, epiphytes or saprophytes.
3. "The Orchidaceae, … together with the Asteraceae or Compositae comprise the largest family of flowering plants as regards number of species. They are chiefly tropical but numerous are subtropical, temperate, and boreal, and some species even arctic …" (Dahlgren and Clifford, 1982).
4. "In the reproductive region, the specialization is very diverse. The inflorescence is racemose and the flowers thus lateral which condition is often connected with zygomorphy. The tepals are petaline and the labellum may or may not be supplied with a spur".
5. "The flowers are always epigynous. Epigyny may have evolved as a means of protection of the gynoecium and possibly also greater stability of the perigone. Nectar production or probably more frequently the production of various volatile substances lead to the attraction of particular groups of insects … . Specialization in the shape, size, colour and pubescence of the perigone indicates a long period of co-evolution with the pollen vector".
6. "Further adaptations of style, stigma and androecial parts have secured a position of the functioning anther and the stigmatic areas to facilitate pollination".
7. "The pollen grains are numerous and usually where in massulae or pollinia, which are further adapted for dispersal by the occurrence of stipes, caudicle and a retinaculum which serves to

keep the pollen aggregated. As the pollen is often sticky there is no need for a thick, sculptured exine or any distinct apertures".

8. In the Orchidaceae, "... the above character states are operating in interaction and are logical consequences of the specialized mode of pollination".

NUMBER IN AND DISTRIBUTION OF THE FAMILY

The Orchidaceae comprises about 750 genera and 18,000 species. The members of this family are of wide distribution, but most abundant in tropical forests. They are concentrated in three areas, notably, tropical America, Indo-Malaysia and eastern Himalaya.

About the orchids, Cronquist (1968) commented thus: "... they are not among the most abundant plants in terms of number of individuals, and not many of the species have a wide geographical range of any ecological amplitude". "Orchid species with surprisingly similar flowers can be found from the damp mountain sides of Norway and New Zealand to the dry savannahs of tropical Africa and South America, and from the coastal strands of the South Pacific to rocky ledges in the Himalayas".

COMMONLY OCCURRING PLANTS OF THE FAMILY

Acampe praemorsa (Roxb.) Blatt. & McCann and *Vanda tessellata* (Roxb.) Hook. ex G. Don are epiphytic on trees.

Ania viridifusca (Hook.) Tang & Wang occurs in Assam hills and *Anoectochilus rotundifolia* (Blatt.) Balakr. in South India.

Calanthe veratrifolia (Willd.) R. Br. occurs in the hilly tracts of South India.

Dendrobium nobile Lindl. is seen in the hills.

Didymoplexis pallens Griff., *Pogonia carinata* Lindl. and *P. plicata* Lindl. are found in shady thickets.

Geodorum dilatatum R. Br. is a terrestrial herb of grassy lawns.

Zeuxine strateumatica (L.) Schlecht. is a grass-like herb which is ubiquitous in the plains.

ECONOMIC ASPECTS OF THE FAMILY

"For use in corsages and as an ornamental the orchids are more sought after than any other type of plant, and not only is an industry built upon orchid culture but growing orchids is also a widespread hobby" (Russell, 1958). Genera commonly grown in glasshouses include *Cymbidium*, *Dendrobium*, *Epidendrum*, *Miltonia*, *Odontoglossum*, *Oncidium*, *Paphiopedilum* and *Phalaenopsis*.

The leaves of *Calanthe veratrifolia* contain a glycoside, 'indican', which on hydrolysis, yields 'indigo blue'. The fibres from the stemcortex of *Dendrobium crumenatum* are used as a braiding material for hats. The tubers of *Eulophia epidendraea* are used as vermifuge. The rootstocks of *Geodorum densiflora* are crushed and rubbed on cattle to kill flies. The dried leaves of *Jumellea fragrans* constitute 'Faham tea'. The roots of *Vanda tessellata* are employed as a remedy for scorpionbites and for rheumatism. The capsules of *Vanilla planifolia* yield commercial 'vanilla', a flavouring agent.

SELECTED BIBLIOGRAPHY

AMES, O. 1946. The evolution of the orchid flower. *Am. Orchid. Soc. Bull.*, **14:** 355–360.

ARBER, A. 1920. *Water Plants. A Study of Aquatic Angiosperms.* University Press, Cambridge.

AREKAL, G.D. and RAMASWAMY, S.N. 1973. Embryology of *Burmannia pusilla* (Wall, ex Miers) Thw. and the systematic status. *Beitr. Biol. Pflaz.* **49;** 35–46.

AYENSU, E.S. 1964. Notes on the anatomy of the Dioscoreaceae. *Ghana J. Sci.* **5:** 19–3.

BATE-SMITH, E.C. 1968. The phenolic constitution of plants and their taxonomic significance II. Monocotyledons. *J. Linn. Soc. (Bot.),* **60:** 325–356.

BEHNKE, H.D. 1972. Sieve-tube plastids in relation to angiosperm systematics. An attempt towards a classification by ultrastructural analysis. *Botan. Rev.*, **38:** 155–197.

BELVAL, H. 1938. A propos des idées de Hutchinson sur les Amaryllicacées. *Bull. Soc. Botan. Fr.*, **85:** 486–489.

BESSEY, C.A. 1897. Phylogeny and taxonomy of the angiosperms. *Botan. Gaz.*, **24:** 145–178.

BOR, N.L. I960. *The Grasses of Burma, Ceylon, India and Pakistan.* Pergamon Press, London.

BOSE, S. 1962. Cytotaxonomy of Amaryllidaceae. *Bull. bot. Surv. India*, **4:** 27–38.

BOYD, L. 1932. Monocotyledonous seedlings. *Trans. Proc. bot. Soc. Edin.*, **31:** 1–224.

BROOKS, J.S. 1940. The cytology and morphology of the Lemnaceae, Ph. D. Thesis, Cornell University.

BURKILL, I. 1960. The organography and the evolution of the Dioscoreaceae, the family of the yams. *J. Linn. Soc. (Bot.)*, **56:** 319–412.

BUTZIN, F. 1965. Neue Untersuchungen über die Blute der Gramineae: *Iraug. Diss. Fr. Univ. Berlin*, 1–183.

CAMPBELL, D.H. 1930. Phylogeny of monocotyledons. *Annls. Bot.*, **44:** 311–330.

CAVE, M.S. 1964. Cytological observations of some genera of the Agavaceae. *Madroño*, **17:** 163–170.

CHAKRAVARTY, A.K. 1948. On the morphology of flower of *Musa. Bull. Botan. Soc. Bengal*, **2:** 36–40.

CHAKRAVARTY, H.L. and MITRA, G.C. 1948. Development studies in the tendrils of *Smilax macrophylla L. Bull. Botan. Soc. Bengal*, **2:** 22–27.

CHEADLE, V.I. 1942. The occurrence and types of vessels in the various organs of the plant in the monocotyledoneae. *Am. J. Bot.*, **29:** 441–450.

CHEADLE, V.I. 1955. The taxonomic use of specialization of vessels in the metaxylem of Gramineae, Cyperaceae, Juncaceae and Restionaceae. J. *Arnold Arb.*, **36:** 141–157.

CLIFFORD, H.T. 1961. Floral evolution in the family Gramineae. *Evolution*, **15:** 455–460.

CULLEN, J. 1978, A preliminary survey of ptyxis (vernation) in the angiosperms. *Notes Roy. Bot. Gdn.*, **37:** 161–214.

CZAJA, A.T. 1978. Structure of starch grains and the classification of vascular plant families. *Taxon*, **27:** 463–470.

DAHLGREEN, R.M.T. and CLIFFORD, H.T. 1982. *The Monocotyledons. A Comparative Survey.* Academic Press, London.

DATTA, S.C and BISWAS, K.K. 1567. Form-variation in *Ottelia alismoides* (L.) Pers. *Brotéria*, **36:** 63–69.

DATTA, S.C. and BISWAS, K.K. 1969. Physiology of germination in *Pistia stratiotes* seeds. *Biologia*, **24:** 70–79.

DATTA, S.C. and BISWAS, K.K. 1970a. Germination pattern and seedling morphology *of Pistia stratiotes* L. *Phyton*, **27:** 157–161.

DATTA, S.C. and BISWAS, K.K. 1970b. Aquatic angiosperms. *J. Sci. Club (Calcutta)*, **23:** 68–75.

DAUMAN, E. 1964. Zur Morpholegie der Blüte von *Alisma plantagoaquatica* L. *Preslia*, **36:** 226–239.

DEYL, M. 1955. The evolution of plants and the taxonomy of monocotyledons. *Acta Mus. Nat. Prague*, *XI* B, No. 6 Bot. No. **3:** 1–143.

DRESSLER, R.L. and DODSON, C.H. 1960. Classification and phylogeny in the Orchidaceae. *Annls. Missouri Bot. Gdn.*, **47:** 25–68.

DUNN, D.B., SHARMA, G.P. and CAMPBELL, C.C. 1965. Stomatal patterns of dicotyledons and monocotyiedons. *Am. Midi. Nat.*, **74:** 185–195.

EAMES, A. 1953. Neglected morphology of the palm leaf. *Phytomorphologyy* **3:** 172–189.

GARAY, L.A. 1960. On the origin of the Orchidaceae. *Botan. Mus. Leafl. Hlary.*, **19:** 57–96.

GRANICK, E.B. 1944. A karyosystematic study of the genus *Agave. Am. J. Bot.*, **31:** 283–298.

GREEN, P.S. and SOLBRIG, O. 1966. *Sciaphila dolichostyla* (Triuridaceae). *J. Arnold Arb.*, **47:** 266–269.

HAMANN, U. 1961. Merkmalsbestand und Verwandtschaftsbeziehungen der Farinosac. *Willdenowia*, **2:** 643–768.

HARE, C.L. 1950. The structure and development of *Eriocaulon septangulare* With. *J. Linn. Soc.* (*Bot.*), **53:** 422–448.

HARLING, G. 1946. Studien über den Blutenbau und die Embryologic der Families Cyclanthaceae. *Svensk. Bot. Tidskr.*, **40:** 257–272.

HILLMAN, W.S. 1961. The Lemnaceae, or duckweeds. *Botan. Rev.*, **27:** 221–272.

HOLTTUM, R E. 1955. Growth-habits of monocotyledons—variations on a theme. *Phytomorphology*, **5:** 399–411.

ISLAM, A.S. 1950. A contribution to the life history of *Ottelia alismoides* Pers. *J. Indian Botan. Soc*, **29:** 79–91.

JONES, S.B. and LUCHSINGER, A.E. 1979. *Plant Systematics.* McGraw-Hill, New York.

JONKER, F.P. 1938. A monograph of the Burmanniaceae. *Med. Bot. Mus. Rijksuniv. Utrecht.*, **51:** 1–279

JOSHI, A.C. and PANTULU. J.V. 1941. Cytology and embryology of *Polyanthes tuberosa* L. *J. Indian Botan. Soc.*, **20:** 37–66.

KOYAMA, T. 1961. Classification of the family Cyperaceae (I). *J. Fac. Sci. Univ. Tokyo*, Sect. III. **8:** 37–148.

KUDRYASHOL, L.V. 1964. The origin of monocotyledony (as illustrated by the example of Helobiae). *Botan. Zhurnal SSSR*, **49:** 473–486.

LAKSHMANAN, K.K. 1963. Embryological studies, in the Hydrocharitaceae. III *Nechamandra alternifolia. Phyton*, **20:** 49–58.

LAWALREE, A. 1945. La position systematique des Lemnaceae et leur classification. *Bull. Soc.* [*Roy.*] *Bot. Belgique*, **77:** 27–38.

LOTSY, P. 1910. Phylogeny of plants. *Botan. Gaz.*, **37:** 460–461.

LOWE, J. 1961. The phylogeny of monocotyledons. *New Phytol.*, **60:** 355–387.

MAHESHWARI, P. and JOHRI, B.M. 1950. The occurrence of persistent pollen tube in *Hydrilla*, *Ottelia* and *Boerhaavia* together with discussion of possible significance of this phenomenon in the life history of angiosperms. *J. Indian Botan. Soc*, **29:** 47–57.

MAIA, L. D'O. 1941. Le grain de pollen dans 1 identification et la classification des plantes. I. Sur la position systematique du genre *Allium. Bull. Bot. Soc. Portugaise Sci. Nat.*, **13:** 135–147.

McCLURE, J.W. and ALSTON, R.L. 1966. A chemotaxonomic study of Lemnaceae. *Am. J. Bot.* **53:** 849–860.

McKELVEY, S.D. and SAX, K. 1933. Taxonomic and cytological relationships of *Yucca* and *Agave. J. Arnold. Arb.*, **14:** 76–81.

MEEUSE, A.D. J. 1974. Aspects on the evolution of the monocotyledons. *Acta Bot. Neerl.*, **24:** 421–436.

MITRA, J.N. 1952. Review and revision of Commelinaceae of eastern India. *Bull. Botan. Soc. Bengal*, **6:** 53–57.

MITRA, J.N. 1955. A critical review on the importance of the order Liliiflorae from phylogenetic point of view. *Bull. Botan. Soc. Bengal*, **9:** 60–61.

MITRA, J.N. 1958. *Flowering Plants of Eastern India.* Vol. I. *Monocotyledons* World Press, Calcutta.

MOORE, H.E. 1973. The major groups of palms and their distributions. *Genetics Herb.*, **11:** 27–141.

PAI, R.M. 1966. Studies in the floral morphology of the Burmanniaceae. I. Vascular anatomy of the flower of *Burmannia pusilla* (Wall, ex Miers) Thw. *Proc. Indian Acad. Sci.*, B. **63:** 301–308.

RANGASAMY, K. 1941. A morphological study of the flower of *Blyxa echinosperma* Hook. f. *J. Indian Botan. Soc*, **20:** 123–133.

RAO, Y.S. 1953. Karyosystematic studies in Helobiales. I. Butomaceae. *Proc. Not. Inst. Sci. India*, **19B:** 563–581.

RAIZADA, M.B. 1958. Name change in common Indian grasses. *Indian Forester*, **85:** 473–509.

ROBBINS, W.W., WEIER, T.E. and STOCKING, C.R. 1957. *An Introduction to Plant Science*. John Wiley, New York.

ROWLEF, W.W. 1898. The morphological significance of the lodicules of grasses. *Botan. Gaz.*, **15:** 199–203.

SACHS, T. 1978. Phyletic diversity in higher plants. *Plant. Syst. Evol.*, **130:** 1–11.

SAHA, B. 1952. Morphology of the ligule and sickle of *Oryza saliva* L. *Bull, Botan. Soc. Bengal*, **6:** 49–52.

SANE, Y.K. 1939. A contribution to the embryology of Aponogetonaceae. *J. Indian Botan. Soc*, **18:** 79–91.

SATO, D. 1942. Karyotype alteration and phylogeny in Liliaceae and allied families. *Jap. J. Bot.*, **12:** 57–161.

SATTLER, R. 1965. Perianth development of *Potamogeton richardsonii. Am. J. Bot.*, **52:** 35–41.

SAVILLE, D.B.O. 1979. Fungi as aids in higher plant classification. *Botan. Rev.*, **45:** 377–503.

SHARMA, A.K. and BAL, A.K. 1956. A cytological investigation of some members of the Cyperaceae. *Phyton*, **6:** 7–22.

SMITH, L.B. 1934. Geographical evidence on the lines of evolution in the Bromeliaceae. *Bot. Jb.*, **66:** 446–468.

STANT, M.Y. 1964. Anatomy of the Alismataceae. *J. Linn. Soc.* (*Bot.*) **59:** 1–42.

STELBINS, G.L. 1974. *Flowering Plants Evolution Above the Species Level.* Edward Arnold, London.

SWAMY, B.G.L. and LAKSHMANN, K.K. 1962. Contributions to the embryology of the Najadaceae. *J. Indian Botan. Soc*, **41:** 247–267.

TILAK, V.D. and PAI, R.M. 1966. Studies in the floral morphology of the Marantaceae. I. Vascular anatomy of the flower of *Schumannianthus virgatus* Rolfe, with special reference to the labellum. *Can. J. Bot.*, **44:** 1365–1370.

TOMLINSON, P.B. 1960. Seedling leaves in palms and their morphological significance. *J. Arnold Arb.*, **41:** 414–428.

TOMLINSON, P.B. 1962. Phylogeny of the Scitamineae—morphological and anatomical considerations. *Evolution*, **16:** 192–213.

UHL, N.W. 1947. Studies in the floral morphology and anatomy of certain members of the Helobiae. Ph. D. Thesis, Cornell University.

VENKATANARAYANA, G. 1957. On certain aspects of the development of leaf of *Cocos nucifera L. Phytomorphology*, **7:** 297–305.

VERMEULEN, P. 1966. The system of the Orchidales. *Acta Bot. Neerl.*, **15:** 224–253.

WAGNER, P. 1977. Vessel types of monocotyledons: a survey. *Bot, Notiser*, **130:** 383–402.

WALKER, J.W. 1974. Aperture evolution in the pollen of primitive angiosperms *Am. J. Bot.*, **61:** 1112–1137.

WHITAKER, T.W. 1934. Chromosome constitution in certain mono-cotyledons. *J. Arnold. Arb.*, **15:** 135–143.

WITHNER, C.L. (ed.). 1959. *The Orchids: A Scientific Survey.* Ronald, Press, New York.

YAKOVLEV, M.S. and ZHUKOVA, G.Y. 1980. Chlorophyll in embryos of angiosperm seeds, a review. *Bot. Notiser*, **133:** 323–336.

YOUNG, D.J. and WATSON, L. 1970. The classification of dicotyledons: a study of the upper levels of the hierarchy. *Austr. J. Bot.*, **18:** 387–433.

ZIEGENSPECK, H. 1938. Die phylogenie der Glumiflorae. *Bot. Arch.* **39:** 177–207.

Index

A

B

D

E

F

H

I

J

K

L

M

N

O

P

Q

R

S

T

U

V

W

X

Y

Z